1998

Plant Ecophysiology

Plant Ecophysiology

Edited by
M.N.V. Prasad
Department of Plant Sciences
University of Hyderabad, India

John Wiley & Sons, Inc.
NEW YORK / CHICHESTER / BRISBANE / TORONTO / SINGAPORE / WEINHEIM

Library of Congress Cataloging in Publication Data:
Plant ecophysiology / edited by M.N.V. Prasad.
    p.   cm.
   Includes index.
   ISBN 0-471-13157-1 (cloth : alk. paper)
   1. Plant ecophysiology.   I. Prasad, Majeti Narasimha Vara, 1953–
QK905.P55   1996
581'.5—dc20   96-19477

Printed in the United States of America

10 9 8 7 6 5 4 3 2 1

Contents

v

Preface

Plant ecophysiology is the science of plant interaction with the environment, and the vital underlying acclimation and adaptation processes. Various environmental stresses are the limiting factors for plant productivity, and the prominent environmental stresses of the contemporary world are dealt with in this book's 15 chapters.

Environmental stresses are dealt with under two headings: biotic and abiotic; however, there is always some multifacet interaction between these two. These factors both play crucial roles in the productivity, survival, and reproductive biology of plants. Eventually, of course, plants would be expected to evolve mechanisms of endurance for all manner of environmental stresses—mechanisms that would permit them to adapt and withstand the phenomena. Studies on plant environmental stress are not only important fundamentally to probe the underlying mechanisms for adaptation of plants to surrounding environment but also are immediately relevant to agriculture, silviculture, and horticulture. Thus, investigation of plant-environment interactions is emerging as a major area of scientific investigation, one whose importance is magnified by the contemporary dwindling of natural resources. The knowledge of stress phenomena that has thus far been gained helps plant biologists/biotechnologists to maximize plant productivity by developing plant genotypes better adapted to tolerate today's range of environmental stresses.

At the end of this century and the beginning of the next, plant ecophysiologists face an exciting era of countless opportunities to be of service in maximizing plant productivity for plants subjected to less than optimum environmental circumstances. To achieve success, it is necessary to understand the basic ecophysiological and molecular principles governing the functioning of plant systems in relation to their environment.

Therefore, this book is intended to stimulate researchers in the broad field of plant ecophysiology by providing a state-of-the-art report. I hope that this book will also be useful to environmental biologists and environmental consultants, and to the students of plant sciences, environmental botany, agriculture, silviculture, and horticulture.

I accept the responsibility for factual errors, omissions, and distortions of this book, if any, while full credit goes to authors of various chapters who have contributed cogent reviews. I extend my sincere thanks to my students and colleagues, particularly Professor A. R. Reddy, Dean, School of Life Sciences, for encouraging me to take up this task. I am extremely thankful to Professor G. Mehta, Vice-Chancellor, University of Hyderabad, for his keen interest in this work. I am most grateful to my wife, Savithri, for her unstinting support.

Excellent technical help was received from Dr. Philip Manor, Donna Conte, and the team at John Wiley & Sons, New York; it resulted in timely production of this work, and is gratefully acknowledged.

M.N.V. Prasad

Hyderabad, India

Contributors

RICHARD N. ARTECA, Horticulture Department, The Penn State University, 101 Tyson Building, University Park, PA 16802-001

A. K. DAS, School of Environmental Sciences, Jawaharlal Nehru University, New Delhi 110067, India

S. RAMA DEVI, Department of Plant Sciences, University of Hyderabad, Hyderabad, 500046 India

JEAN-MARC FERULLO, Agriculture and Agri-Food Canada, Soils and Crops Research and Development Centre, 2560 Hochelaga Boulevard, Sainte-Foy, Québec, Canada, G1V 2J3

HELENA M. O. FREITAS, Departamento de Botânica, Universidade de Coimbra, 3000 Coimbra, Portugal

ANIL GROVER, Department of Plant Molecular Biology, University of Delhi, South Campus, Benito Juarez Road, New Delhi 110021, India

JÜRGEN HAGEMEYER, Department of Ecology, Faculty of Biology, Bielefeld University, D-33615, Bielefeld, Germany

PIETER KETNER, Department of Terrestrial Ecology and Nature Conservation, Wageningen Agricultural University, Borbsesteeg 69, 6708 PD, Wageningen, The Netherlands

G. KULANDAIVELU, Department of Plant Sciences, School of Biological Sciences, Madurai Kamaraj University, Madurai 625021, Tamilnadu, India

SERGE LABERGE, Agriculture and Agri-Food Canada, Soils and Crops Research and Development Centre, 2560 Hochelaga Boulevard, Sainte-Foy, Québec, Canada G1V 2J3

GILLES LALIBERTÉ, Agriculture and Agri-Food Canada, Soils and Crops Research and Development Centre, 2560 Hochelaga Boulevard, Sainte-Foy, Québec, Canada G1V 2J3

K. LINGAKUMAR, Department of Plant Sciences, School of Biological Sciences, Madurai Kamaraj University, Madurai 625021, Tamilnadu, India

C. MEL LYTLE, Department of Botany and Range Science, Brigham Young University, 401 WIDB, Box 25181, Provo, Utah 84602-5181

GÉRARD MERLIN, Université de Savoie, ESIGEC, Laboratoire de Biologie et Biochimie Appliquées, Le Bourget-du-Lac F73376, France

NORIYUKI MOMOSHIMA, Department of Chemistry, Faculty of Science, Kyushu University 33, Hakozaka 812, Japan

ASHWANI PAREEK, Department of Plant Molecular Biology, University of Delhi, South Campus, Benito Juarez Road, New Delhi 110021, India

F. PELLISSIER, Université de Savoie, Altitude Ecosystems Dynamics, Scientific Campus, Le Bourget-du-Lac, F73376, France

M.N.V. PRASAD, Department of Plant Sciences, School of Life Sciences, University of Hyderabad, Hyderabad 500046, India

A. PREMKUMAR, Department of Plant Sciences, School of Biological Sciences, Madurai Kamaraj University, Madurai 625 021, Tamilnadu, India

MANCHIKATLA V. RAJAM, Department of Genetics, University of Delhi, South Campus, Benito Juarez Road, New Delhi 110021, India

P. S. RAMAKRISHNAN, School of Environmental Sciences, Jawaharlal Nehru University, New Delhi 110067, India

K. S. RAO, G. B. Pant Institute of Himalayan Environment and Development, Kosi-Katarmal, Almora 263643, India

K. G. SAXENA, School of Environmental Sciences, Jawaharlal Nehru University, New Delhi, 110067, India

SNEH LATA SINGLA, Department of Plant Molecular Biology, University of Delhi, South Campus, Benito Juarez Road, New Delhi 110021, India

BRUCE N. SMITH, Department of Botany and Range Science, Brigham Young University, 401 WIDB, Box 25181, Provo, Utah 84602-5181

KAZIMIERZ STRZAŁKA, Department of Plant Physiology and Biochemistry, The Jan Zurzycki Institute of Molecular Biology, Jagiellonian University, Al. Mickiewicza 3, 31-120 Kraków, Poland

YANHONG TANG, Global Environmental Studies, National Institute for Environmental Studies, 16-2 Onogawa, Tsukuba 305, Japan

LOUIS-P. VÉZINA, Agriculture and Agri-Food Canada, Soils and Crops Research and Development Centre, 2560 Hochelaga Boulevard, Sainte-Foy, Québec, Canada G1V 2J3

CLAUDE WILLEMOT, Agriculture and Agri-Food Canada, Food Research and Development Centre, Sainte-Hyacinthe, Québec, Canada J2S 8E3

Part I

Natural, Abiotic Factors

1

Light

Yanhong Tang

INTRODUCTION

Light is of interest to plant ecologists not only because of its great importance for plants, but also because of its own considerable complexity in nature. Unlike other environmental factors such as temperature, moisture, and wind, light varies in at least four dimensions: quality, quantity, direction, and duration. The great variation of light environments imposes a great demand on plants' responsiveness. Basically, light influences plants in two ways: providing energy input and acting as a medium to transfer information from the environment to the plants.

As an energy source, light is captured and converted to chemical energy through photosynthesis, by which nearly all energy enters our biosphere. Light also plays an important role in energy exchange between plants and their environments, an often neglected feature in ecophysiological studies. If light energy is high enough to induce a change of structure in the genetic material, mutation may result. When light energy is extremely high, under some conditions, light may cause damage or even have a lethal effect on cells or plants.

As a medium of information, light is involved in regulating various processes of growth and developments of plants, such as photomorphogenesis, phototropism, and photoperiodism. Light is an incomparable medium of information, and it transmits a considerable volume of data to plants because of its great variations.

In this chapter, we first introduce the fundamentals of light pertinent to the ecophysiology of plants, and then discuss the light environment of plants. We also describe major physiological processes of plants in response to light.

FUNDAMENTALS OF LIGHT

Light often denotes the range of electromagnetic radiation perceivable by the human eye, but the word "light" can be also used to refer to a wider range of electromagnetic radiation (Fig. 1-1). The wavelength of visible light is from approximately 380 to 800 nm. Visible light is fundamental to plants because it corresponds roughly to photosynthetic photon flux (400–700 nm). Light of wavelengths other than photosynthetic photon flux is also important to plants; for example, far-red light (700–800 nm) influences morphogenesis, while ultraviolet light can have damaging effects on plants. In general, it is considered that electromagnetic radiation with a wavelength between 300 and 1000 nm has some biological effects. This range of wavelengths is called the "biological window" of light.

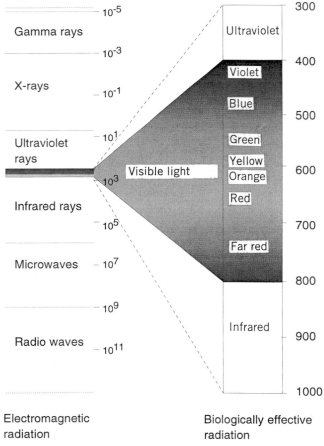

Figure 1-1. *The electromagnetic radiation spectrum (unit of wavelength: nm).*

The Nature of Light

Light has both wavelike and particlelike properties. Light is propagated like waves, but it interacts with matter more like particles.

As early as the seventeenth century Christian Huygen proposed the first scientific theory on the nature of light. He suggested that light could be explained as a wave phenomenon. However, it was not until 1894, when James Clerk Maxwell proved that electric and magnetic fields were propagated together and that their speeds were identical with the speed of light, that the wave theory of light was widely accepted.

Light as a form of electromagnetic wave can be described as oscillations of the local electric and magnetic fields. A wave can be characterized by its wavelength (λ), which is the distance between successive points of the same phase, such as between two successive peaks of a wave. There are three major wavelength regions that are important in the ecophysiology of plants: the ultraviolet (UV), the visible, and the infrared (IR). UV wavelengths are from the low limit of visible light to approximately 10 nm, while IR wavelengths are from the red end of visible region, extending up to about 100 μm. A light wave has also a frequency (ν). The product of the wavelength times the frequency is the velocity (V):

$$V = \lambda \times \nu \tag{1-1}$$

Light can travel in a solid (e.g., certain plastics, glass), in a liquid, in a gas, and even in a vacuum (e.g., the space between the sun and the earth's atmosphere). In a vacuum, the velocity of light is constant, approximately 3.0×10^8 m s^{-1}, independent of wavelength. The speed at which light propagates will be lower when it passes through mediums other than a vacuum.

In addition to wavelike properties, light also has particlelike properties. In 1704 Newton first proposed that light consists of tiny particles. In 1901 Planck suggested that light can be described as discrete packets or quanta. Einstein developed this theory and proposed that light, as well as other forms of electromagnetic radiation, travels as tiny bundles of energy called light quanta or photons. The energy (E) of a photon is inversely proportional to its wavelength (λ) and is a function of its frequency (ν):

$$E = h\nu = hc/\lambda \tag{1-2}$$

where h is Planck's constant (6.6×10^{-34} J$\cdot$s). In ecophysiological studies, the amount of photons is often described in moles. The energy per mole photons is $1.2 \times 10^{-1}/\lambda_{nm}$ (J mol^{-1}). Light energy varies with the wavelength of photons. For example, the energy of light at 660 nm will be 1.8×10^5 J mol^{-1} photons, while the energy of ultraviolet (below 400 nm) with a representative wavelength of 254 nm is about 4.7×10^5 J mol^{-1}.

Emission

To understand light energy available on the earth, it is necessary to consider some properties of the light source. Essentially, all light energy used by plants, and further by all life on earth, is emitted from the sun. According to Planck, any object will emit electromagnetic radiation if its surface temperature is above absolute 0 K. The total radiation of a body is proportional to the fourth power of its absolute temperature; this relationship is expressed by the Stefan–Bolzmann law:

$$\Phi = \varepsilon \sigma T^4 \tag{1-3}$$

where Φ is the total radiation emitted per unit area per unit time; ε is a constant (for a blackbody, which is a perfect absorber or emitter of light at all wavelengths, ε is 1); σ is the Stefan–Bolzmann constant (5.7×10^{-8} W m^{-2} K^{-4}); and T is the absolute temperature (temperature in °C + 273.15) of the surface. The surface temperature of the sun is about 5800 K. Since the sun can be considered as a blackbody, the total radiant energy emitted by the sun is 6.4×10^{-7} W m^{-2}.

Surface temperature not only affects the total emitted energy of radiation, but also results in a difference in energy distribution for a wavelength. The energy distribution at different wavelengths can be calculated by Planck's radiation distribution formula. The derived expression is also known as Wien's displacement law:

$$\lambda_{max} = (3.6 \times 10^6)/K \tag{1-4}$$

where λ_{max} is wavelength for maximum photon flux density (PFD, nm). Therefore, formula (1-4) is a calculation based on photons, and its use is recommended when we discuss the photon-dependent processes, for example, photosynthesis. However, we may be more interested in the spectral distribution of energy per unit wavelength interval when the energy balance of plants is considered. In that case, equation (1-4) becomes:

$$\lambda_{max} = 2897/K \tag{1-5}$$

The surface temperature of the sun gives a wavelength of 620 nm for the maximum PFD. The surface temperature of a plant is about 300 K, which emits infrared radiation with a wavelength around 12 μm for the maximum PFD.

Absorption

Lambert–Beer's Law   The absorption of light at a particular wavelength by a certain species can be quantitatively described using an absorption coefficient k, which is also referred to as an extinction coefficient. When we consider a monochromatic light beam (light of a single wavelength) passing through a solution with a substance dissolved in a nonabsorbing solvent, the quantity of the light beam absorbed (A_λ) by the substance is proportional to the substance concentration (c) and the light path-length through the solution (b):

$$A_\lambda = \log\left(\frac{I_0}{I}\right) = kcb \tag{1-6}$$

where I_0 is the PFD of the incident light beam and I is the PFD after transmission through the solution. This is known as Beer's law, also referred to as Lambert–Beer's law.

When a monochromatic light beam of PFD (I_0) passes through a homogenous medium, we can obtain I according to equation (1-6):

$$I = I_0 \, e^{-kb} \tag{1-7}$$

Equation (1-7) can also be used to describe the attenuation of light in the atmosphere or in water. We further discuss its application for describing the attenuation of light within the plant tissues and plant canopy.

Photochemical Effects   According to Grotthus and Draper's photochemical law, only light that is absorbed can cause a photochemical reaction. In plant ecophysiology, this is true for the light energy converted into other forms of energy, such as the energy fixed in photosynthetic products or the energy dissipated by consumption in metabolic processes, as well as for the light energy used as information stimulus.

The extent of a photochemical reaction induced by light depends only on photon fluency, which has a dimension of mol photons m^{-2}, and not on the photon flux (mol photons $m^{-2} s^{-1}$). If the arithmetical product of flux $\times$ duration is held constant, then short-time light at a high flux should yield the same level of effect as long-time light at a low flux. This is known as the law of reciprocity, proposed by Bunsen and Roscoe in 1850. If the law holds true in a response, it indicates that only one type of photoreceptor is involved.

To understand some important processes in photosynthesis, another photochemical law, Stark and Einstein's photochemical equivalence law proposed in 1905, must be mentioned here. This law states that the absorption of one quantum gives rise to a photochemical change in one molecule, or one atom, or one electron. It indicates that in a strictly photochemical reaction the number of molecules affected should be equal to the number of photons absorbed. The ratio (R_a) of the amount of molecules activated (Q_{ac}) to the amount of photons absorbed (Q_{ab}) is used to evaluate the quantified efficiency of a photochemical reaction:

$$R_a = \frac{Q_{ac}}{Q_{ab}} \tag{1-8}$$

However, since the number of photons actually absorbed is usually difficult to determine experimentally, instead of using the number of photons absorbed, we use the ratio of relative quantum efficiency (R_r):

$$R_r = \frac{Q_{ac}}{Q_{in}} \tag{1-9}$$

where Q_{in} is the number of incident photons.

Photoelectric Effects   Understanding the photoelectric effect is helpful for us in considering the distinction between the light energy and the energy of photons, which is necessary if we are to understand some photobiological processes in photosynthesis and photomorphogenesis. In 1887, H. R. Hertz found that a substance, especially a metal, can emit electrons when photons strike its surfaces. Each electron emitted by the surface is a result of the absorption of one photon.

However, not every photon has sufficient energy to cause a certain photoelectric reaction, that is, a photon of a specific minimum energy is needed to cause a certain photoelectric reactions. For example, in order to lead to a photoelectric effect in sodium, photons with at least 175 kJ mol^{-1} are needed. The wavelengths of these photons must be 683 nm or less to cause the photoelectric effect. If the photons have a wavelength above 683 nm, then no electrons will be ejected from the surface regardless of the total photon energy absorbed or how many photons are involved. Such a photoelectric effect clearly demonstrates that the total light energy is not the energy of its photons.

Therefore, knowing the total light energy does not necessarily mean knowing how many photons are involved or what the individual energies of these photons are, unless we know the wavelength distribution of the light. There are a number of practical applications of the photoelectric effect in ecophysiology; for example, a measurement of total solar energy is not adequate information to estimate the details of photosynthesis.

Reflection and Transmission

In addition to being emitted or absorbed, light can also be reflected or transmitted by a body. Light reflection occurs when light returns from a surface that it strikes back into the medium through which it has traveled. The following laws governing light reflection are important in ecopysiology of plants:

1. The angle of reflection is equal to the angle of incidence.
2. A smooth surface gives regular reflection, in which incident parallel rays remain parallel after reflection. A rough or an uneven surface gives diffuse reflection since reflected rays become scattered and nonparallel.
3. Another important property of light reflection is internal reflection, which occurs when light passes from one medium (e.g., water or leaf tissue) to a less dense medium (e.g., air); the rays reach the boundary of the two mediums and then return to the denser medium instead of going away from it. Such internal reflection may be important in increasing the utilization of light within a leaf.

If light beams impinge upon a surface at angles of incidence other than perpendicular to the surface, the quantitative effects of light distribution arise. The Lambert cosine law formalizes this case, stating that PFD incident on a surface depends on the orientation of the photon flux beam according to:

$$I_c = I_{c0} \cos \beta \qquad (1\text{-}10)$$

where I_c is the PFD at the surface, I_{c0} is the PFD perpendicular to the incident beam, and β is the angle between the beam and a line perpendicular to the surface. The PFD decreases with the increase of the incident angle β, that is, with the increase of incident beam area. It is important to consider the cosine effect of light in light measurement, as well as in studying the light environment at slopes and aspects.

Light passing through a medium will either be reflected, absorbed, or transmitted. At any particular spectrum, the following relation holds:

$$\alpha(\lambda) + \gamma(\lambda) + \tau(\lambda) = 1 \qquad (1\text{-}11)$$

where $\alpha(\lambda)$, $\gamma(\lambda)$, and $\tau(\lambda)$ are, respectively, the absorbance, reflectance, and transmittance at spectrum λ. The partitioning of light absorbance, reflectance, and transmittance depends on a number of factors. Some important factors in plant ecophysiology are discussed in terms of light propagation in the atmosphere and within the plant in the following section.

LIGHT ENVIRONMENT

The energy of sun light perpendicular to the solar beam at the top of earth's atmosphere at the average distance between the earth and the sun (which is also known as the solar constant) is about 1370 W m^{-2}. With the change in the distance between the sun and the earth between July and January, the actual value of the solar constant may vary around $\pm$ 4%. Based on the constant, we can obtain the mean daily energy of sunlight on the top of atmosphere, which is about 30 MJ m^{-2} d^{-1}. Although the electromagnetic radiation emitted from the sun has a continuous wavelength of a wide range, about 98% of the total energy of sunlight at the top of the atmosphere falls within the wavelengths between 300 and 3000 nm. The light within these wavelengths is usually called shortwave radiation, while the light with a wavelength longer than 3000 nm, up to 400 μm, is called longwave radiation.

In this section, we first review briefly the light environment at the top of plant canopies and then summarize the studies on light environments within plant canopies. Finally the optical properties of the plant tissue are briefly reviewed.

Light Environment Above Plant Canopies

When passing through the earth's atmosphere, sun light is modified by the reflection, absorption, and scattering of its various components in the atmosphere. The modification changes the quantity, quality, and direction of light.

Modification of Light Quantity   The atmosphere reduces more than half of solar energy, and only about 47% of the total energy of sun light at the top of the atmosphere reaches the earth's surface. The reduction of light energy varies greatly at different sites on the surface of the earth. Latitude is thus the first important factor that determines the incident light energy at the top of plant canopies at a particular site. Therefore the tropical region, which lies in the low latitudes, less than 23°27′, has a light energy much higher than the average of the earth's surface. The annual total of global light energy is 700–800 GJ km^{-2}. Photosynthetic PFD at midday often reaches 1800–2200 μmol m^{-2} s^{-1}, with a highest value up to 2600 μmol m^{-2} s^{-1} (see Chazdon et al., 1996). Total daily photons range from 40 to 50 mol m^{-2} d^{-1} in the absence of cloud cover. Cloud formation reduces PFD by 75% or more, while local haziness can reduce PFD by 20–55% in Malaysia (Tang et al., 1996). Toward the poles, light availability decreases gradually. The polar regions receive no light in winter.

Annual photoperiod and day-length vary with latitudes. For example, on the winter solstice the day-length is 12 hours at latitude of 23.5°S, but 24 hours at latitude of 66.5°S. With the increase of latitude, the seasonal difference in day-length markedly increases. The annual or diurnal trend in light availability can be approximated by a sine curve:

$$I_t = I_{max} \sin \left(\frac{\pi t}{N} \right) \tag{1-12}$$

where I_t is the irradiance t hours after sunrise and N is the day-length in hours (see Jones, 1992).

In addition, variations in topography, that is, in slope angle and aspect angle, also have considerable influence on the natural light environment. For example, in the northern hemisphere, a south-facing slope always receives greater light energy than a north-facing slope. The difference changes with the degree of the slope. The influence of topography also changes with latitudes and solar elevation.

Modification of Light Quality and Direction   Atmosphere also modifies light quality (Table 1-1). The atmosphere permits the passage of light mainly at the wavelengths from 300 to 1500 nm and is relatively transparent to visible light. There is a strong absorption of light at shorter wavelengths (ozone absorbs most of the UV radiation). A large amount of IR light is also absorbed by H_2O and CO_2.

TABLE 1-1   Modification of Atmosphere on the Wavelength of Light

Type of Light	On Top of Atmosphere (%)	On the Earth Surface (%)
Ultraviolet light	5	2
Visible light	28	45
Infrared light	67	53

Source: Derived from Nobel (1991).

The light energy in the photosynthetic photon flux takes about 47% of direct light, but about 50% of total light energy, because the increase of diffuse light tends to be higher within the wavelength of photosynthetic photon flux.

The light's long path also changes its quality. Daily variation of light wavelength is larger at lower latitudes than at higher latitudes. However, the effect of the atmosphere on the transmittance of light is slightly reduced at higher latitudes since the atmosphere toward the poles becomes thinner than at lower latitudes.

Having passed through the atmosphere, light incident at the top of plant canopies can be distinguished into two parts: direct light and diffuse light (Fig. 1-2). Direct light or incident light is largely unmodified by the particles in the atmosphere and travels directly to the top of plant canopies. The wavelength is also similar to the light incident on the atmosphere, with only some large absorption by H_2O, CO_2, and O_2. On the other hand, diffuse light or skylight consists of light components scatted, or reflected, by various particles.

There are two different forms of scattering by molecules and particles in the atmosphere: Rayleigh scattering and Mie scattering. Rayleigh scattering is due to particles with diameter smaller than the wavelength of light, while Mie scattering occurs owing to the larger particles such as dust and water droplets. After scattering or refection in the atmosphere, sky light is relatively enhanced in blue light. When the sun is below the horizon, almost all light can be sky light. Both direct light and diffuse light are shortwave radiation.

Plants on the earth's surface use this shortwave radiation as light energy for

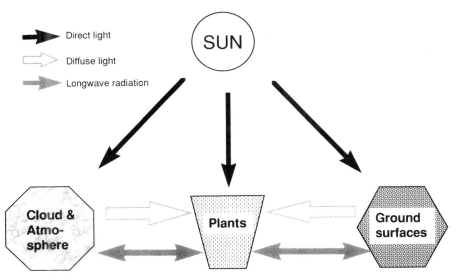

Figure 1-2. Radiation exchange between a plant and its environment. Radiation received by the plant comes from: (1) direct shortwave radiation from the sun, (2) diffuse shortwave radiation reflected from the plant's surroundings, and (3) longwave radiation emitted by surrounding objects with normal terrestrial temperatures. The plant also emits longwave radiation to its environment.

photosynthesis and as information for photomorphogensis. In addition, another important form of radiation, longwave radiation, is also very important for plant energy balance. Instead of being emitted from the sun, longwave radiation is emitted by the objects at normal terrestrial temperatures. Various gases in the atmosphere, especially H_2O and CO_2, are the main source of longwave radiation from the sky. This part of longwave radiation is also called downward longwave radiation. Upward longwave radiation is emitted from the surfaces such as the ground surface (Fig. 1-2).

Light Environment Within Plant Canopies

Light environment within a plant canopy varies considerably both in time and in space. Characterizing the variable light environment has long attracted great interest in plant ecology, as well as in agronomy. The variation of light environment is influenced by many factors, such as solar elevation, sky and wind conditions, and canopy structure, including leaf area and the spatial arrangement of canopy foliage, branches, and stems. Because of these factors, the measurement of spatial and temporal variations of light environments within plant canopies previously faced many difficulties, though laboratory measurement of light has been an important subject and has been relatively established in physics. Only in the past decades has there been a rapid development in the study of the light environment within plant canopies because of the application of new technology such as computers and new types of sensors. In this section, we summarize the main methods of light measurement and then discuss the recent studies on light environments within plant canopies.

Characterizing Light Environment There are several methods used to measure the quality and quantity of light in the ecophysiological studies of plants. Choosing the proper method is very important for the description of light environments, and the choice will depend on the purpose of the study. Details available for light measurement are well described elsewhere (e.g., Jones, 1993; Pearcy, 1989).

Basis of Light Measurement To describe light fully we need to at least consider the following properties of light in ecophysiological studies of plants:

1. Spectrum of the light, which is necessary if we are to describe the "amount" of light at the different wavelengths.
2. Direction of the light, including the two extreme cases: direct light and diffuse light.
3. Time factor, which is often neglected or confused; for example, light can be expressed by an instantaneous or a time integrated quantity.

In addition to the properties of light itself, various biotic and environmental components, such as the structure of plant canopy, and wind and cloud conditions, make light measurements more complicated in the field, though it may not always be practical to give a full description when considering all these factors.

Light can be expressed either by the amount of energy (J) or by the number of photons (mol). Whether an energy or a photon unit should be used depends largely on the study subject. In general, a photon unit is recommended in the study of photosynthesis, while an energy unit should be used when the energy balance of plants is considered. The energy amount or the number of photons can further have additional spatial and temporal dimensions. The term *flux* is the light amount per unit time, while *fluency* is used to describe the light amount per unit (cross-sectional) area. When both spatial and temporal dimensions are considered, we have terms like *flux density* (light amount per unit time per unit area of a plane surface) and *fluence rate*. The term *flux density* is sometimes used as a synonym for *fluency rate* because both mean the light amount per unit time per unit area. Strictly speaking, however, the two terms are different, as fluency rate measures the flux across unit area incident from all directions on a spherical surface. The measurement of fluency rate therefore requires the use of a spherical detector.

Methods for direct measurements of light environment in plant canopies include photophysical methods using various sensors and photochemical methods such as diazo paper. Light environments can also be estimated from photographs or from parameters of canopy structure.

Sensor Measurements   Light environments in the plant canopy can be measured by using various types of sensors. Each type of sensor has its characteristic specifications in spectrum, output, response time, and so on. A careful selection of the sensor type is important to the study. A typical setup for sensor measurements uses one or more sensors connected to a datalog, which can be controlled by a computer. One common method of conducting sensor measurements is instantaneous sampling using a number of sensors that characterize the spatial and temporal distribution of light in various plant canopies.

There are three main types of sensors: (1) quantum sensors, (2) radiometric sensors, and (3) photometric sensors. Some details about the first two groups, which are frequently used in ecophysiological studies of plants, are shown in Table 1-2.

A quantum sensor consists of a photovoltaic cell that transforms light energy into an electric current. The output unit of a quantum sensor is usually mole photons per unit surface area per unit time, for example, μmol photon m^{-2} s^{-1}. This type of sensor is used when knowing the number of photons becomes necessary. Photosynthetic PFD can be measured using a quantum sensor incorporating a set of filters that correct the sensitivity of the sensor between 400 and 700 nm. Quantum sensors are therefore most suitable for studying photosynthesis and other plant photobiology. The model LI-190SA quantum sensor from LI-COR Instruments is commonly used in physiological ecology. Other quantum sensors (Delta-T Instruments Quantum Sensor QS; Skye Instruments model SKP 215; Koito Quantum Sensor KT 25) with similar response characteristics are also commercially available.

Radiometric sensors include mainly thermocouples, thermopiles, or thermistors. These sensors are basically able to absorb all wavelengths and thus are sensitive to radiation from ultraviolet to infrared light. Light energy is expressed in energy per unit area per unit time (J m^{-2} s^{-1}, or W m^{-2}). Filters can also be used in these

TABLE 1-2 Radiometric Sensor and Quantum Sensor Measurements

	Radiometric Sensor Measurement	Quantum Sensor Measurement
Detectors	Thermoelectric sensors such as radiometer, net radiometer	Photoelectric sensors such as silicon cells, photodiodes (GaAsp), quantum sensors
Function basis of detectors	A voltage is generated between two thermocouples that have a temperature difference caused by the absorption of radiation	Generation of an electric current in direct proportion to the photon flux is caused by absorbing photons of detectors
Basic terminology	Joule (J): a force of 1 newton acting through a distance of 1 meter	Mole (mol): 6.02×10^{23} particles (photons)
Instantaneous measurement	Radiation: Radiant energy incident per unit time on a unit surface ($W\ m^{-2}$)	Photon flux density: number of photons incident per unit time on a unit surface ($mol\ m^{-2}\ s^{-1}$)
Important conversion factors	$1\ W\ m^{-2} = 1\ J\ m^{-2}\ s^{-1}$ $1\ W\ m^{-2} = 0.1196/\lambda\ (nm)\ mol\ m^{-2}\ s^{-1}$	$1\ mol\ m^{-2}\ s^{-1} = 6.02 \times 10^{23}$ photons $m^{-2}\ s^{-1}$ $1\ mol\ m^{-2}\ s^{-1} = \lambda\ (nm)/0.1196\ W\ m^{-2}$
Full sun plus sky	$1000\ W\ m^{-2}$	$2000\ \mu mol\ m^{-2}\ s^{-1}$

sensors to measure the energy of photosynthetic PFD. Some important radiation instruments include pyranometers, pyrradiometers, and pyrheliometers. Pyranometers measure both direct and diffuse sunlight, but pyrheliometers measure only direct beams. Pyrradiometers are used to measure both sunlight at wavelengths between 0.3 and 3 μm and longwave thermal radiation.

A photometric sensor is really a "light" sensor in the sense that it mimics human vision, with maximum sensitivity in the green region of wavelength at 500 nm. A photometric sensor, therefore, measures brightness or luminosity, not energy or photons of light. The basic unit for photometric sensors is lux or foot-candles. Measurements based on photometric sensors and the units of lux or foot-candles are currently unacceptable in plant physiological ecology, since the measurements differ for each source of radiant energy.

Diazo Paper Method In this method, a stack of photosensitive diazo paper is used for a period and the integrated light energy can be estimated from the number of paper layers that are bleached. The assumption is that the number of diazo sheets exposed is proportional to the total quantity of light received. Although a linear relationship does exist between the exposed diazo sheets and integrated light energy, a recent study showed that there is also a stronger linear relationship between the number of exposed diazo sheets and maximum instantaneous photosynthetic PFD (Bardon et al., 1995). Therefore, this method is not proper for measurements of light under conditions with a significant temporal variation such as sunflecks. The advantage of this simple technique is that it is inexpensive and easy to use for multiple sampling in the field.

Hemispherical Photographs   Hemispherical photographs are taken by a superwide-angle lenses attached to a camera. Photographs of this type can be used to estimate indirectly the light received at the point where the photograph was taken. Evans and Coombe (1959) first used hemispherical photographs to see whether direct sunlight could reach the forest floor at any particular time. Later, Anderson (1964, 1974) developed this method and first used such photographs to estimate light environments qualitatively in forest canopies. From the analysis of hemispherical photographs, she proposed a method for computing the diffuse site factor, which is the percentage of diffuse light under the canopy to that in the open, and the direct site factor, which is the percentage of direct light that could potentially be received under the canopy. With the recent development of computer techniques, the analysis of hemispherical photographs has become much more convenient and more efficient than Anderson could achieve manually. Pearcy (1989) described this method in detail.

One of the advantages of using hemispherical photographs to characterize light environments in the plant canopy is that the technique can provide relative comparison between a large number of microsites. The method can also estimate diffuse and direct components of the light environment separately for the same site. In fact, a wide range of ecological studies have used this technique in evaluating the light environment in different plant canopies. However, precise temporal and spatial variations of PFD cannot be obtained by this method because of the complicated effects of canopy structure, weather conditions, and solar movement.

Measurements of Canopy Structure   Canopy structure is the spatial arrangement of the aboveground organs of plants. Information from measurements of canopy structure can be used to describe indirectly and understand light environments within the canopy. There are various methods for obtaining canopy structure information (Campbell and Norman, 1990). Direct methods involve the measurements of the size, shape, orientation, and positional distribution of leaves, branches, and stems. Canopy structure can also be measured indirectly from the measurement of light within and above the canopy, which is not discussed here. We now briefly consider several methods that have been widely used for study of light environments in ecology and agronomy.

To study the light environment and productivity in grass canopies, Monsi and Saeki (1953) first introduced the stratified-clip method, which has been widely used for studies of plant canopy structure. In this method, a representative area, usually rectangular or circular, is identified. The sampling area should be large enough to include a relatively large number of plants. In grasslands, the area is usually from 0.25 to 4 m^2, depending on the density of grass canopy. The canopy in the area is stratified vertically into several layers and then clipped. Various parameters such as leaf area, leaf orientation, or the dry weight of leaves or stems can be measured for each layer from the clipped plant materials. The method is very useful for studying the relationship between light environment and matter production in canopies such as grasses or small legumes, but is not suitable for those canopies consisting of high, large, or widely sparse plants.

The dispersed individual plant method was proposed by Ross (1981) to overcome some of the difficulties in the stratified-clip method. This method is basically similar to the stratified-clip approach, but focuses on individual plants and is therefore useful when a canopy is dispersed and plants are large. Light environment and stand productivity can be estimated by measuring and analyzing various characteristics of individual plants, usually from 10 to 30 trees, such as leaf area, branch angle, stem diameter, and plant height. Spatial distribution of plants in the sampling area is also a very important parameter for the estimation of light environment by this method. Samples are stratified according to plant height, leaf angles, or the distance between the plant axis.

Statistical Tools In characterizing light environments under canopies, the sampling requirement is always a severe problem because of the large and complex variations of the light environment. It is suggested that obtaining reliable estimations of direct radiation in deciduous and conifer forests will require at least 18 or 412 sensors, respectively (Reifsnyder et al., 1971). A smaller number of sensors is necessary for estimating a diffuse light regime because of the small spatial variation under diffuse light conditions. On the basis of a liberal assumption that the probability of spatial distribution of light is Gaussian, Baldocchi and Collineau (1994) showed that to measure the light regime with a spatial coefficient variation of 100 for a representative area, which is a conservative estimation for the spatial light distribution in the understory of many forest canopies, 270 and 382 samples are needed to give 10% and 5% of the population mean. However, it seems to be difficult to have a general sampling strategy in light measurement.

The mean value of light flux density is not a good statistical parameter for describing light environment in plant canopies. This is partially because the frequency distribution of light within a canopy is nonnormal in most cases. Probability density functions seem to be a useful tool for studying light transfer through heterogeneous plant canopies. In addition, variance, coefficient of variation, skewness, and kurtosis, which describe how data are dispersed, skewed, and peaked in a population, are very useful parameters to characterize the spatial heterogeneity of light regime.

Statistic information from transect data can provide an insight into the characteristics of spatial variation in light regime within plant canopies. If there is periodicity within a set of data, the information on the periodicity can be provided by autocorrelation and spectral analysis (Tang et al., 1989; Tang and Washitani, 1995). Autocorrelation analysis describes the time or space associated with the persistence of the repetition of a given event. For example, the spatial autocorrelation coefficient (r_k) is a measure of the dependence of the value at one point on the values at neighboring points, which is given by

$$r_k = \frac{\sum_{s=1}^{N-k}(X_s - M)(X_{s-k} - M)}{\sum_{s=1}^{N}(X_s - M)^2,} \qquad k = 0, 1, \ldots, N-1 \qquad (1\text{-}13)$$

where r_k is called the spatial autocorrelation coefficient at lag k for the spatial series X_s with the mean M.

To discover the hidden periodicity in the data of a transect, we can examine the estimation of the spectral density function $I(v)$, which is the Fourier transform of the sample autocorrelation coefficients and is calculated as

$$I(v) = \frac{\left[\left(\sum_{s=1}^{N} x_s \cos 2\pi rs/N\right)^2 + \left(\sum_{s=1}^{N} x_s \sin 2\pi rs/N\right)^2\right]}{N\pi} \tag{1-14}$$

where $v = 2\pi r/N$ is the frequency at period of r for a spatial series (x_s) with total observations of N. The plot of $I(v)$ against v is called the periodogram. Spectral analysis is mainly used to find the contribution made by different spatial properties to the total variance of a data series.

Sunflecks   It has long been recognized that sunflecks might play a very important role in leaf photosynthesis and the growth of forest understory plants. During the past decade, field measurements showed that sunflecks compose a large proportion of total daily PFD and contribute greatly to the heterogeneity of the spatial and temporal variation of light environments in plant canopies. Laboratory experiments also showed that the utilization of sunflecks can account for a large fraction of the photosynthesis in understory leaves (Chazdon, 1988; Pearcy, 1990).

It is difficult to give a strict definition for the term *sunfleck*, because of the great variation in PFD, size, spectrum properties, and duration of sunflecks. Sunflecks have been defined by the PFD exceeding some threshold values just above the background diffuse light, or based on the photosynthetic response of the plants studied (Pearcy, 1983; 1990; Tang et al., 1988). Once a sunfleck threshold is defined, further analysis can be made to obtain various sunfleck parameters such as sunfleck duration, maximum sunfleck PFD, and total sunfleck PFD. Such a definition for sunflecks, however, is both arbitrary and author dependent because of the choice of detection threshold value. A sunfleck threshold value may differ in different plant canopies, or differ at different vertical layers of the same canopy. Therefore, the same threshold of sunflecks cannot be used for various plant canopies, which makes it difficult to compare sunfleck environments between different plant canopies. To overcome this difficulty, one approach, wavelet analysis, has been used recently by Baldocchi and Collineau (1994). This method allows a user to identify and classify patterns or singularities in a data set according to position, size, and shape. No threshold value is needed, but special events, sunfleck position and durations, and so on can be detected in the data. The basic wavelet transform $(T(a, b))$ is defined as a convolution between the original function $(h(t))$ and a wavelet function $(g(t - b)/a)$:

$$T(a,\ b) = \frac{1}{a} \int_{-\infty}^{+\infty} h(t)g\left(\frac{t-b}{a}\right) dt \tag{1-15}$$

where *a* is window width and *b* is the transitional position of wavelet window. In a deciduous forest, wavelet analysis showed that sunfleck events exhibit various thresholds above the background light regimes, which seems due to penumbra effects and the change of diffuse PFD.

Penumbra effects play an important role in sunfleck activity in plant canopies. The sun is not a perfect point source of light, but subtends at an angle of about $0.5°$ on the surface of the earth. The sun will therefore project an image with a numbra and a penumbra when sunlight passes through an opening larger than a pinhole (Fig. 1-3). Such a projected image can be considered as a sunfleck. The numbra, resulting from the overlap of light beams from the opposite edges of the sun, is the center part of the sunfleck with direct light, while the penumbra part is the diffuse shadow-edge surrounding the numbra center. The relative sizes of the numbra and penumbra are determined by the diameter of the opening and the distance of the opening to the incident surface of the projected image. A higher plant canopy or a smaller opening will result in larger penumbra effect, that is, a larger proportion of penumbra. If there are no effects of wind, sunflecks due to canopy openings and the solar movement will be relatively constant during a given period. However, the variation of wind speed and direct light cause a highly complicated spatial and temporal variation of sunflecks in most plant canopies.

With the increase of the size of canopy openings, the penumbra effects on light environment will generally decrease. To characterize the penumbra effects on sunlight penetration in plant communities, Smith et al. (1989) proposed to use a gap-diameter and distance ratio (GDR) in classifying different canopy openings. GDR

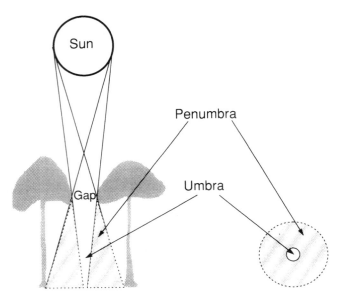

Figure 1-3. *An illustration of the penumbra effect. The relative areas of umbra and penumbra are determined by gap size (diameter) and gap height.*

is determined by the ratio of a hypothetical opening diameter *(G)* of plant canopy and distance to the incident surface*(D)*. The projected sun spot from an opening of canopy will be almost entirely penumbra if GDR < 0.01, while about 80% of the sun spot is numbra when GDR = 0.04. For example, for a 40 m tall tropical forest, the projected image of the sun will be entirely penumbra if a canopy opening is less than about 0.35 m in diameter (40 m × tan 0.5 = 0.349 m). Smith et al. (1989) then define light regimes as sunflecks(GDR < 0.01), sunpatches(0.01 ≤ GDR ≤ 0.05), and gaps(GDR > 0.05), according to GDR. The penumbra effects may have potential ecophysiological influence on understory plants, but little knowledge has been accumulated about this aspect.

Sunflecks usually prevail less than 10% of the daytime, but they can contribute more than 60% of daily total PFD in most plant canopies (see Chazdon, 1988). In general, the contribution of sunflecks to total daily PFD is relatively high in tropical forests or in some dense grass canopies, where diffuse background light is usually very low. Sunflecks can contribute as much as 90% of total daily PFD in some dense canopies. This high contribution of sunfleck PFD is usually due to a few long sunflecks (Pearcy, 1990; Pearcy and Sim, 1994; Tang et al., 1988). Many brief sunflecks, on the other hand, are caused by a larger number of small canopy openings. Pearcy (1990) found that more than 1800 sunflecks per day were received at a given location within a soybean canopy. Most of these sunflecks lasted less than 6 seconds. The short sunflecks are low in PFD because of the effect of penumbra. Canopy movement caused by windy conditions is also an important reason for short and highly frequent sunflecks (Roden and Pearcy, 1993; Tang et al., 1988); for example, leaf flutter may cause high frequency (3–5 Hz) variations of PFD in poplar canopies.

Attenuation of Light   As we move downward into a plant canopy, light decreases more or less exponentially, depending on the amount of total leaves. According to Monsi and Saeki's (1953) modification of the Lambert–Beer extinction law, the average flux density of photons *(I)* on a horizontal surface below certain layers of leaves can be obtained by:

$$I = I_0 \, e^{(-k \, \text{LAI})} \tag{1-16}$$

where I_0 is the flux density of light at the top of the layer. k is a dimensionless parameter that describes the absorption properties of a particular type of foliage. LAI (leaf area index) is the ratio of the accumulative area of total leaves at a given height of canopy to the ground area.

The model requires an estimation of a canopy light extinction coefficient k, which is generally assumed to be a species constant. k is mainly affected by the variation of leaf angle. A larger erectness of canopy leaves will result in a lower k, and therefore a higher PFD transmitted into the lower layers of the canopy. In general, grass canopies, such as *Miscanthus sinensis,* or crop canopies, like rice, have a smaller k than some herb canopies like potato and white clover. Leaf angle may change vertically within a plant canopy. In fact, leaves are more vertical toward to the top of canopy and more horizontal toward the ground in many canopies. Such the canopy structure can increase the amount of light for the lower leaves in a canopy. When LAI or the vertical distribution of canopy foliage varies, the model (1-16) may no longer be appropriate for the whole canopy. Decreased light extinction with increased canopy depth was observed recently (e.g., Gholz et al., 1991). It has also been found that there is a predictable decrease in extinction coefficients with increasing LAI for a range of stand ages and structure in lodgepole pine canopies (Gholz et al., 1991).

There is a great variation in light attenuation in different plant canopies (Fig. 1-4). Tropical forests have a very low transmittance to the forest floor; it can be as low as 0.2% of the mean daily PFD received at the top of the forest (see, e.g., Yoda, 1978). In broadleaf forests or in conifer forests of the temperate region, the average light transmittance to the forest floor is around 3–10% in the growing season, but increases to as high as 50–70% during winter (see Larcher, 1995). Transmittance of light in most grass canopies in temperate regions can often be lower than 4%, and even as low as about 1% during the growing season (Tang et al., 1992).

In addition to the differences in light transmittance, the vertical gradient of light also varies considerably in different plant canopies (Fig. 1-4). In tropical broadleaf forests, a large proportion of light is absorbed in upper canopy layers. The uppermost layer usually comprises less than 20% of the upper canopy, but intercepts more than 70% of incident PFD. However, most PFD is absorbed at the middle layers of a grass canopy (see Iwaki, 1990).

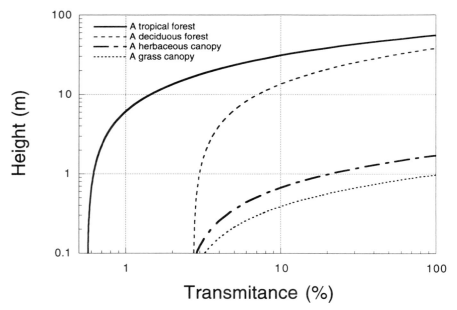

Figure 1-4. *Vertical changes of light transmittance in four different canopies. Data for the tropical forest are derived from Yoda (1978), for the deciduous forest from Parker (1995), and for the herbaceous canopy and the grass canopy from Iwaki (1990).*

Light Microenvironment Within the Plant

Light environments of plants in nature are highly heterogeneous and very complicated, while light microenvironments within the plant are much more complicated and still less understood. The topic of light within the plant has long been neglected in plant ecophysiological studies. However, understanding how light propagates and distributes within the plant, especially within the leaf, is essential to understanding how plants detect the information and utilize the energy from their environmental light.

In this section, we first introduce some important optical properties of plant tissue, and then briefly review the studies on light propagation and distribution within the leaf. For details on the topic, the reader is referred to Vogelmann (1994), Vogelmann et al. (1991), and Terashima (1989).

Optical Effects of Plant Tissue   Leaves in higher plants are the most essential organs that respond to light, but some other plant tissues and organs also play important roles in receiving and utilizing light. Studies on optical properties of various plant tissues and organs, in particular on those of the leaf, have achieved great progress during the past decade. The following major optical effects concerned with the microenvironment of the plant may be important in plant ecophysiology: (1) lens effect; (2) sieve effect; (3) optical waveguide, and (4) light trap effect.

Lens Effect   Some plant cells can act as lenses, a phenomenon found in a various species from algae to higher plants. A lens cell looks like a planoconvex or a cylindrical lens. A lens effect can be caused by liquids such as oil and water within the lens cell.

In higher plants, lens cells have been found on the surfaces of many leaves. When light enters the lens cell from the air, the light can be focused on a microscopic area within the cell. For example, the highly convex epidermal cell in some tropical understory species seems to able to focus light on some of the chloroplast (Bone et al., 1985; Lee, 1986). The effect of epidermal cells on focusing light depends on the curvature of the outer cell wall, cell size, and the structure within the cell. Focusing effects will be larger under the direct light than under diffuse light. Epidermis can also increase the heterogeneity of light distribution by creating the epidermal focal spots within a leaf because of the lens effect. In some species, chloroplast seems to be able to move in or out of high light spots. Under environments of low light availability, the lens effect may increase light utilization efficiency.

In addition, lens effects are also found near the surface of phototropic organs. Calculations of the light transmission through a maize mesocotyl suggests that light can be focused on the distal surface of the mesocotyl (see Vogelmann, 1994).

Sieve Effect   Pigments are important components influencing the propagation of light within the plant. If these pigments are homogeneously distributed within plant tissue, the transmission of light can be predicted to be directly proportional to pig-

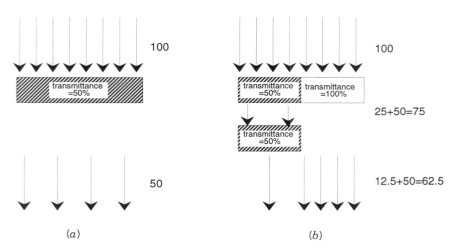

Figure 1-5. *An illustration of the sieve effect. (a) Pigments are homogeneously distributed in a layer that has a transmittance of 50%. If an incident light is 100 units, then a total light of 50 units can be received on the other side. (b) The same amount of pigments as in case a are confined within half of their original area. The transmittance of the pigment layer becomes 50% × 50% = 25%, while that of the other open part is 100%. When 100 units of incident light passes through the heterogenous pigment layer, a total of 62.5 light units (25% × 50 + 100% × 50 = 62.5) should be received on the other side. Heterogeneous distribution of pigments increases the penetration of light. With the decrease of pigment transmittance, the sieve effect will increase; for example, if the transmittance is 5% in case a, then the received light in case b should be 2.5 % × 50 + 100% × 50 = 51.3 units (51.3/5 > 62.5/50).*

ment concentration. However, in most plant tissue, pigments are not uniformly distributed. The heterogenous distribution of pigments tends to increase the light penetration, and thus decrease the light absorption within a cell or tissue. This is called the sieve effect (Fig. 1-5). A higher concentration of pigments will increase the sieve effect, which can be easily understood in Fig. 1-5. Similarly, the sieve effect will increase at the wavelength with a higher absorbance.

Optical Waveguide Effect   Light can be reflected back and forth between two parallel surfaces so that it propagates in a direction parallel to the surfaces. A similar way to direct light can be found in some plant organs, such as stems and roots, or some long cells. Light may be directed between the cell wall and cytoplasm in these organ tissues, or between cytoplasm and vacuoles within the cells.

Light guiding may be of ecological importance in seedling establishment. For example, in the shoot of an oat seedling, early growth and emergence is controlled by photoreceptors that are located at the top of the mesocotyl immediately beneath the node. Light has been found to be directed through the shoot of an oat seedling from the soil surface to a photoreceptive site within the soil; similar light guiding occurs under conditions of low light availability (Mandoli and Briggs 1982). The light guiding may be advantageous for early elongation of the shoot and the seedling emergence. Similar axis light guiding is also involved in other physiological processes such as root formation.

Light Trap Effect   The fluence rate of light within plant cells or tissue can be higher than that of the incident light, due to the light trap effect of plant cells and tissue. There are two optical properties of plant that are important for the light trap:

1. The refractive index of the plant cuticle (1.45) is much higher than that of the air (1.00). Hence the cuticle potentially reflects escaping light back into the interior of the plant, preventing it from escaping.

2. The light scattering in intercellular air spaces and organelles increases the light pathlength and extends the interval of time that photons dwell within the plant. Because of these two properties, light within plants can commonly increase the incident light by two to three times, as found for many species (Vogelmann, 1994).

Light Propagation and Distribution in the Leaf   Light incident on a leaf will be absorbed, transmitted, or reflected (Fig. 1-6). The pattern of light absorbance, transmittance, and reflectance for most green leaves is distinct in three wavelength

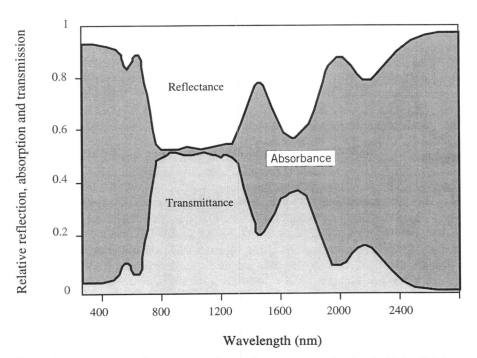

Figure 1-6. *Reflectance, absorbance, and transmittance spectra of a plant leaf (after Knipling, E. D.,* Remote Sensory Environment, *1970, pp. 155–159, with kind permission from Elsevier Science—NL, Sara Burgerhartstraat 25, 1055 KV Amsterdam, The Netherlands).*

regions: visible (300–800 nm), near-infrared (800–1350 nm) and mid-infrared (> 1350 nm). The partitioning of light into absorbed, transmitted, or reflected energy depends on:

1. The interior structure of the leaf
2. Leaf surface features, including leaf surface roughness and leaf pubescence, that is, air-filled hairs, or trichome, and in some species a deposit of salt crystals on the leaf surface
3. Leaf morphology and physiology

A detailed knowledge of optical properties of a leaf will be useful not only for studying leaf ecophysiology, but also for interpreting and processing remotely sensed data, which has become of increasing importance (Nilson, 1991; Vogelmann, 1994).

Reflection and Transmission   Visible light generally has a low reflectance by leaves, about 6–10% of incident light (Walter-Shea and Norman, 1991). Some leaves in warm temperate and tropical rainforests can reflect as much as 15% of the visible light. There is a high reflectance (10–20%) at the wavelength of green light. Little UV light (3%) is reflected by the leaf. Reflection of red and far-red light from leaves is of great importance for the morphological adjustment of individual plants within communities. The reflection of infrared light (ca. 70%) provides important information about the "health state" of leaves or the canopy.

Transmittance by an individual leaf is highly variable, with the range from less than 3% to as high as 40% of incident light. Soft, flexible, thin leaves have a higher transmittance than hard, coarse, thick leaves. Relatively little visible light is transmitted by green leaves. Transmittance of light by a leaf has the lowest value at the wavelength of green light.

Various factors affect the reflection of light by the leaf. Biotically, light reflected by a leaf can be either from leaf surfaces or from the interior of a leaf. Reflection of light from the leaf surface usually does not change in wavelength, but is often polarized. However, the wavelength of light will change when light is reflected from the interior of the leaf. Surface reflectance is large when there is a wax covering, trichome, or other specialized structure such as a salt bladder. Water content of the leaf and environmental moisture influence both surface and interior reflection. Reflectance from the surface of lichens is higher when the leaf is dry than when it is water saturated. In some leaves, there is a wavelength-specific reflection. For example, leaves of some tropical plants show a striking blue iridescence. The wavelength-specific reflection is thought to result from some modification in the cell walls or in the epicuticular wax layer.

Absorption   Absorbance of light by a leaf varies considerably with various characteristics of the leaf. In general, a leaf absorbs 60–80% of incident light. Almost all UV is absorbed by the cuticular and the outer layer of leaf epidermis, as well as by phenolic compounds in the cell sap of the uppermost cell layers, which prevents

UV (usually less than 1% of incident on the leaf surface) from reaching the interior of the leaf. The leaf absorbance increases, while leaf transmittance decreases, with the increase of specific leaf weight. However, the absorbance may decrease even if a leaf thickness increases when the reflective pubescent layer on the leaf surface increases. Absorbance of light by a leaf tends to be higher than a solution having similar pigment content because of the light trap effect.

In higher plants, chlorophylls are the major pigments responsible for absorption of visible light, though other pigments such as carotenoids, xanthophylls, and anthocyanins also contribute to the absorption. Different tissues may play different roles in forming the light gradient within a leaf. Spongy tissue can facilitate light absorption (Terashima and Saeki, 1983), while palisade cells may increase the penetration of light, particularly direct light (Vogelman and Martin, 1993). PFD can decline as much as 80% from the adaxial to the abaxial surface of a leaf. This fact indicates that there is a large gradient of light from one side to the other side of a leaf, especially at wavelengths with a higher absorbance. The light gradient is brought about both by the screening effect of absorbing pigments, and by the scattering of light within the leaf. A recent study suggests that the intercellular reflectance caused about a twofold increase in photosynthesis in shade leaves under low PFD (DeLucia et al., 1996).

Light in Water and Soil

The penetration of light in water bodies depends initially on the solar elevation because of the reflection of water surface. A higher elevation causes a higher penetration of light into water. Light is also absorbed and scattered by various components in an aquatic system: the water itself; dissolved yellow pigments (gilvin); and the photosynthetic biota and inanimate particulate matter (tripton). The degrees of scattering and absorption by these components vary considerably in different bodies of water (see Kirk, 1994). Light of 1% can reach down to about 150 m in a clear water body, but it can only reach about 50 m near the coast, depending on the turbidity of water.

Wavelength of light changes with the depth of water. Longwave radiation is absorbed in the upper few millimeters, IR light in the uppermost centimeters, while UV can reach a meter. In a clear water body, the ratio of red to far-red light (R/FR) will increase linearly with depth because of the relatively higher attenuation of far red by water (see Kirk, 1994).

Only 1–2% of light can transmit through a 2–5 mm layer of sands or soil. Most of the transmitted light is between 700 and 800 nm, but red light can penetrate longer distance in the water. The water in the soil increases the transmission of light. R/FR penetration is also higher in a moist than in a dry sandy soil.

PLANT RESPONSES TO LIGHT

Plants have evolved remarkably sophisticated mechanisms to cope with the great variation in their light environment. By adapting to the quality, quantity, direction, and duration of light, plants are able to optimize their growth and to control various complex processes such as photosynthesis, germination, and flowering. For example, the variation of the light environment causes physiological acclimation of chloroplast, and this acclimation in turn modifies the impact of variation in the light environment on photosynthesis. Plants have not only the capacity of passively receiving the energy and information from environmental light, but also show the ability to respond actively to their light environments (see Aphalo and Ballare, 1995). By detecting the information from the current light environments, plants are able to "know" about their current resource availability, and also to "predict" the potential future changes in their environment.

In this section, we first introduce the pigment system of photoreceptors in the plant and then summarize how light is involved in the major responses of the plant to light: photosynthesis and photomorphogensis.

Pigment Systems of Photoreceptors

To capture light energy and to detect the information in light environment around them, green plants have evolved a number of pigment systems as photoreceptors. These pigments can be functionally classified into two main groups: photosynthetic and photomorphogenic pigment systems.

Photosynthetic Pigments There are three major photosynthetic pigments: the chlorophylls, the carotenoids, and the phycobilins. The most important chlorophyll is chlorophyll *a,* which is found in all photosynthetic organisms except the green and the purple bacteria. The other chlorophylls vary with different organisms and light environments. These chlorophylls include photosynthetic pigments that pass their light excited electrons on to chlorophyll *a.* In higher plants, photosynthetic pigments include chlorophyll *a, b,* and several carotenoids. The chlorophylls are embedded in three chlorophyll-protein complexes: the light harvesting complex, the photosystem I antenna complex (PSI), and the photosystem II antenna complex (PSII).

A chlorophyll molecule has two basic energetic states: the ground state and the excitation state. When a pigment molecule absorbs a photon, its ground state (S_0) can be converted into the excitation state. Since the energy of a photon is different at different wavelengths, an excited state varies also in its energy level. Chlorophyll has two distinct energy states in excitation. The first state (S_1) with low energy is excited by red light with a wavelength of 680 nm, while the second state (S_2) with more energy requires excitation by blue light (430 nm), which has higher energy. The energetic state S_2 is highly unstable and will return to S_1. The transition from S_2 to S_1 is extremely fast and the energy is thus lost in the form of heat. From S_1 to S_0, however, the transition is slower and the energy is lost in the form of fluorescence.

Phytochrome Phytochrome, as one of photomorphogenic pigments exists widely in all higher plants evolved from green algae, where it plays a fundamental role in the life of these plants similar to the role of rhodopsin, a pigment sensitive to red light in the retinal rods of the eyes, in the life of animals. The highest concentrations of phytochrome are found in young, rapidly expanding cells, including those in root caps, epicotyl and hypocotyl hooks, and coleoptilar nodes (Pratt, 1994). Within cells that have never been exposed to light, phytochrome seems to be distributed uniformly throughout the cytosol. However, the details about the location and distribution of phytochrome are still not clear.

There are two photo-interconvertible forms of phytochrome: a red-light absorbing form (Pr) that has an absorption maximum in the red (660 nm), and a far-red absorbing form (Pfr) that has an absorption maximum in the far-red (730 nm). The two forms can be converted into each other upon the absorption of light. Therefore red light or sunlight tends to convert Pr into Pfr, while the far-red light or canopy filtered light has the effect in converting Pfr to Pr form.

The transformations are driven by the energy of photons absorbed with quantum yields from about 0.07 to 0.17. It has been found that there are some intermediates between Pfr and Pr. An irradiation that lasts longer than the formation of the first intermediate will establish a photoequilibrium between Pr and intermediates or Pfr. The photoequilibrium is characterized by the proportion of Pfr among the total phytochrome molecules, which is assumed to be important to determine some photomorphogenic response. The Pfr/Pr ratio ranges from 0.02 under only far-red light to 0.86 when there is monochromatic red light.

In a natural light environment, the Pfr/Pr ratio depends on: (1) the amount of Pfr and Pr, which is determined by the R/FR of incident light; and (2) the amount of total phytochrome, which is determined by the synthesis and destruction process. The physiological processes seem to be regulated by the phytochrome of Pfr form, which is likely to influence gene expression, though the detailed mechanism is unclear.

In addition, there are some other pigment systems of photoreceptors for light information. The B receptor is a little characterized pigment system for UV-B radiation (see Horwitz, 1994; Senger and Schmidt, 1994). The action spectra indicated that the chromophoric group of B receptors could be either flavins or carotenoids. B receptors seem evolutionarily older than phytochrome. In higher plants they seem to have been specialized for sensing light direction and for regulating certain processes related to photosynthesis, such as control of stomatal movement and adaptation of the photosynthetic apparatus to shade conditions.

Photosynthesis

Photosynthesis is a process of converting light energy of the sun into chemical energy of plant tissues. The major reaction of photosynthesis can be represented by the equation:

$$CO_2 + H_2O \xrightarrow[\text{plants}]{\text{light}} (CH_2O) + O_2 \qquad (1\text{-}17)$$

The carbohydrates (CH_2O) formed possess more energy than the starting materials CO_2 and H_2O. The whole photosynthetic process is shown in Fig. 1-7. It has been estimated that photosynthesis fixes approximately 3×10^{18} kJ solar energy annually, which is about 2×10^{11} ton carbon. In this section, only the effects of light on the photosynthetic processes will be discussed.

Light Regulation in Photosynthesis   The primary process of photosynthesis is the absorption of light by the pigments. The light energy absorbed is first transferred by the exited electrons to reaction centers (P700 and P680). Through a series of electron carriers, part of the light energy is stored in ATP (adenosine triphosphate) and NADPH (reduced nicotinamide adenine dinucleotide phosphate). ATP and NADPH are the energy currency, which are further used for CO_2 fixation and photorespiration (Fig. 1-7).

Photosynthetic rate depends on PFD. At low light, photosynthesis increase depends linearly on an increase of PFD. The efficiency of light utilization, that is, the photosynthesis per unit PFD, is constant and maximal under low PFD. Increase of photosynthesis becomes gradually less than proportional to the increase of PFD, and ultimately fails to increase with the increase of PFD. The dependence of photosynthesis on PFD differs in plants or leaves grown under different PFD conditions. In general, shade plants have a lower PFD than sun plants to balance respiration (light composition). The PFD for saturated photosynthesis is lower in shade than in sun plants.

There are two major ways that light controls photosynthesis: light-controlled stomatal movement and light-controlled photosynthetic enzyme activation.

Light controls photosynthesis by regulating the movement of stomata. Pigments within the guard cells absorb light, which directly controls stomatal movements. Evidence shows that red light directly stimulates the opening of stomata through guard cell chlorophyll (see, e.g., Sharkey and Ogawa, 1987). This response starts under a red light of 20 μmol m^{-2} s^{-1}, and will be saturated at a relatively high PFD. Indirectly, PFD can increase photosynthesis and thus decrease the intercellular CO_2, which results in stomatal opening. Transmission of some agent from the mesophyll cells to the guard cells such that photosynthetic CO_2 assimilation in the mesophyll controls the degree of stomatal opening may occur, but there is no direct evidence for the rapid response of stomata to a messenger from the mesophyll.

At least six enzymes in the CO_2 fixation pathway are modulated by light (see Hall and Rao, 1994). Several mechanisms have been proposed to explain the light regulation of these enzymes. For example, illumination of chloroplast increases the stromatal pH from 7 to 8 or higher, and increases Mg^{2+} concentration. Both the increases enhance the catalytic activities of the light-regulated enzymes of the Calvin pathway. In addition, light can also regulate photosynthesis through the control ferredoxin-thioredoxin system.

Not all light energy absorbed by photosynthetic pigments is used for photosynthetic carbon dioxide fixation. Light energy absorbed can be dissipated by consumption in the metabolic process, by remission as fluorescence, or by conversion into heat in the pigment bed (Fig. 1-8). The balance of these processes plays an im-

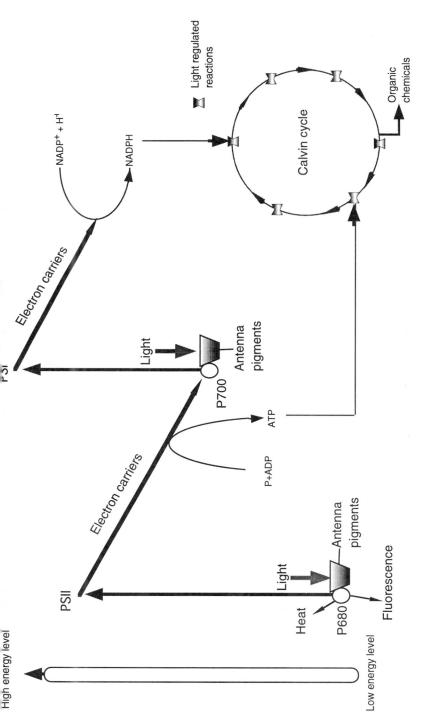

Figure 1-7. A diagram of photosynthesis. In photosystem II (PSII), light energy is absorbed by the antenna pigment complex and then transferred to the reactive chlorophyll a (P680). This energy boosts electrons from P680 to a primary electron acceptor at a higher energy level. Through a series of electron carriers, the energy is then transferred down to P700. Energy released during the transport of electrons is used to power the synthesis of ATP. In PSI, a similar transport of electrons results in the synthesis of NADPH. ATP and NADPH formed in the light-dependant reaction of photosynthesis act as energy currency and are used to drive the Calvin cycle. The immediate product of the Calvin cycle is glyceraldehyde phosphate, which is further used to synthesize a variety of organic chemicals. In addition to supplying energy for the light-dependant reaction, light also regulates and controls several biochemical reactions in the Calvin cycle.

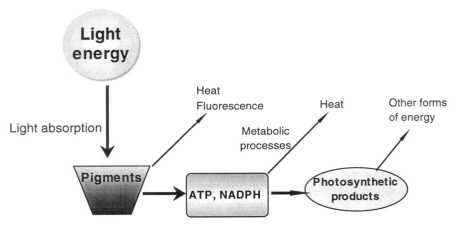

Figure 1-8. *A diagram showing the conversion and utilization of light energy that is absorbed by photosynthetic pigments in the plant. Absorbed light energy can be converted into the following four processes: CO_2 fixation, consumption in metabolism, conversion to heat, and fluorescence.*

portant role in the regulation of light energy in various environments. For example, the ratio of the energy used for CO_2 fixation and photorespiration to the total excitation energy from photosynthetic pigments is found to decrease with increase of water stress in cotton crop (Björkman and Schafer, 1989). The ratio is about 46% under irrigated conditions, and only about 20% under water stress conditions. The amount of excitation energy dissipated by fluorescence is small, at most 3–4% of the total excitation energy. However, the chlorophyll fluorescence emitted from a leaf can be used as a very powerful intrinsic probe to assess the efficiency of energy conversion or the degree of reaction center closure in PSII, as well as the extent of nonradiative energy dissipation (see Björkman and Demming-Adams, 1995).

Photosynthesis Under Transient Lights Light environment in leaf canopies is highly variable under conditions of direct sunlight. Photosynthetic response to the rapid variation of PFD is thus of great importance for leaf carbon gain in a natural light environment. Fig. 1-9 shows a typical photosynthetic response to a simulated sunfleck (lightfleck). There are two components that determine how efficiently a leaf can utilize transient PFD for CO_2 fixation: photosynthetic induction response and postillumination CO_2 fixation (Pearcy, 1990). The induction response indicates the capacity of photosynthesis after a sudden increase of PFD, while the postillumination CO_2 fixation determines how much of the light energy can be used immediately after PFD decreases. The induction response of photosynthesis can be divided into two phases that increase and decay with different time constants (Fig. 1-9). The first phase of induction is very fast and completes within 1–2 min after PFD increase. If light returns to a low PFD, this fast phase also declines rapidly, with a half-time of 2–5 minutes. The second phase is much slower, needing 10–30 minutes or more for completion. The decay of this phase is also very slow when

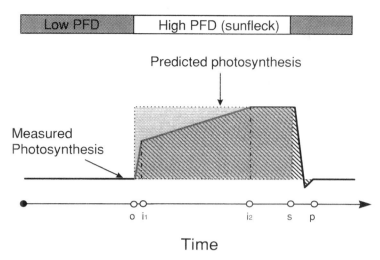

Figure 1-9. *A diagram to illustrate photosynthetic response to a simulated sunfleck (lightfleck). o and s indicate, respectively, the start and end of the lightfleck. i_1 and p are, respectively, the end time of photosynthetic induction response and of postillumination effect. i_1 and i_2 indicate the end time of the first and the second phases of induction response, respectively. The assimilation rate would reach a steady state after i_2 or p. (Derived from Tang et al., 1994 with major modification.)*

PFD is decreased (Kirschbaum and Pearcy, 1988a,b). It has been found that the capacity of RuBP (ribulose 1,5-bisphosphate) regeneration is the major limiting factor for the first fast induction phase. The second phase of induction, however, appears to be limited by stomatal conductance and the light modulation of Rubisco activity (Kirschbaum and Pearcy, 1988b; Sassenrath-Cole and Pearcy, 1992; Woodrow and Mott, 1989).

The limitation of photosynthetic induction can be reduced if the previous PFD is high, or if there is a series of previous sunflecks (Chazdon and Pearcy, 1986; Pearcy et al., 1985). Growth light environments also seem to influence the photosynthetic induction response. Pearcy and Pfitsch (1995) suggested that there may be a higher contribution of sunflecks to carbon gain for plants growing under the light environment of a lower diffuse light. In some species, leaves grown under a low light condition tend to have a rapid response to an increase of PFD, which indicates an efficient utilization of sunflecks (Pearcy and Pfitsch, 1995; Tang et al., 1994).

Postillumination photosynthesis depends on the development of an assimilatory charge acquired during high PFD. This charge consists of the pools of high-energy, Calvin-cycle intermediates such as triose phosphates and RuBP (Laisk et al., 1984; Sharkey et al., 1986). After a sudden decrease of PFD, the pools are able to support a period of continuous photosynthesis. Pons and Pearcy (1992) showed that after a short lightfleck, postillumination CO_2 fixation can continue for 3 seconds at the same photosynthetic rate as under the lightfleck. Postillumination CO_2 fixation has been shown to contribute significantly to the total carbon gain during light flecks.

The contribution is higher in a short than in a long lightfleck (Pons and Pearcy, 1992). There is also evidence that postillumination photosynthesis may be greater in shade leaves than in sun leaves (see, e.g., Chazdon and Pearcy, 1986; Sharkey et al., 1986). This high postillumination photosynthesis would be beneficial for carbon gain of shade leaves under low light availability within plant canopies.

Photosynthetic Acclimation to PFD   Variations of light environments experienced by plants in nature may range from sunflecks changing in seconds or minutes to canopy development or gap formation on a much longer time scale. In response to the short-term variation of light such as sunflecks, plants tend to adjust the metabolic balance in the existing system. These adjustments are very fast and include biochemical and physiological responses such as activation of enzymes or changes of stomatal conductance. However, response to a long-term variation of light regime also appears to take a long time. Photosynthetic acclimation is one of the most important responses of plants to long-term changes in light environment, and it has been studied and reviewed extensively (e.g., Anderson and Osmond, 1987; Pearcy and Sim, 1994).

One important assumption on photosynthetic response to sunny and shady environment is that if all other environmental and biological factors are similar, natural selection may favor plants whose physiological characteristics and morphological features tend to maximize their net rate of photosynthetic energy capture (see, e.g., Givnish, 1988). Various acclimatory traits of plants have been used as proof of this assumption (Table 1-3). It has been found that in general a shade-intolerant species shows a higher morphological and physiological acclimation capacity than a shade-tolerant species, though both shade-intolerant and -tolerant species are able to acclimate to optimize the whole-plant carbon balance in shade.

However, survival of a plant under a particular light regime depends not only on its ability to capture energy, but also on other factors such as the effects of herbivores and pathogens. Plants appear to be able to set the balance between the cost of energy acquisition and that of survival. Pearcy and Sim (1994) suggested that a redistribution of internal resources may increase either the ability to acquire light energy or the capacity to resist light stress. It has also been found that only a weak trade-off exists between ability to achieve high photosynthetic capacity in sun and ability to survive in shade (Kitajima 1994).

Photomorphogenesis

Besides using light energy in photosynthesis, plants are also able to detect the quality, quantity, direction, and duration of light to adjust their growth, and control various biological processes. Photomorphogenesis, in a broad sense, refers to all the nonphotosynthetic responses of plants to light, including photocontrol of seed germination, photomodulation of growth, phototropism, and photomovement.

Seed Germination   Seeds with plentiful reserves can germinate without light, but a remarkably large proportion of higher plant seeds requires light for their germination. For these plants, light is the dormancy-breaking environmental factor,

TABLE 1-3   Characteristics of Plants Acclimating to Shade and Sun Environments

Trait	Shade	Sun
Physiology		
PFD-saturate photosynthetic rate/leaf area	Low	Height
Compensation PFD	Low	Height
Saturation PFD	Low	Height
Dark respiration rate	Low	Height
Quantum yield	Similar	Similar
Electron transport capacity (leaf area basis)	Low	Height
Carboxylation capacity (leaf area basis)	Low	Height
Sensitivity to photoinhibition	High	Low
Mesophyll resistance	Low	Height
Stomatal conductance	High	Low
Leaf longevity	Long	Short
Biochemistry		
Chlorophyll per unit leaf area	Low or similar	Height
Chlorophyll a/b ratio	Low	High
N, rubisco, and soluble protein content/mass	Low	High
Electron transport carrier	Low	High
Rubisco activity	Low	High
Morphology and Anatomy		
Chloroplast size	Large	Small
Stomatal size	Large	Small
Stomatal density	Low	High
Leaf thickness	Low	High
Allocation and Allometry		
Palisade/spongy mesophyll ratio	Low	High
Internal surface area/leaf area	Low	High
Thylakoid/grana ratio	High	Low
Petiole investment/leaf weight	Low	High
Leaf mass per unit area	Low	High
Allocation to leaves	High	Low
Reproductive effort	Low	High
Allocation to roots	Low	High
Plant Architecture		
Leaf area index	Low	High
Leaf orientation	Horizontal	Erect
Twig orientation	Horizontal	Erect
Branching number	Low	High

Source: Derived from Givnish (1988) and Pearcy and Sim (1994).

while for others, light is a major factor inhibiting their germination. In general, the germination of light-sensitive seeds will be stimulated or enhanced under red light, but be inhibited by subsequent treatment with far-red, or a low R/FR ratio. Borthwick et al. (1952) first showed that the phytochrome system is involved in the control of germination by light. They found that a few minutes exposure to dim red light (660 nm) is enough to stimulate imbibed lettuce seed to germinate, while a far-red light (730 nm) reverses the effect of a previously applied red pulse. However, the mechanism involved in the whole process of light control in seed germination is not yet fully understood.

Light control of germination is particularly important in two environments: disturbed fields and under-canopy regimes. Most weed seeds in crop fields will quickly germinate after a soil disturbance such as plowing. Exclusion of light, or cultivation during the night reduced the germination of light-requiring weed seeds in the field. Recently, Scopel et al. (1991) demonstrated that lack of light is only an environmental factor that limits the germination of buried seeds of *Datura ferox,* an annual weed, in the field. This experiment also showed that light needed for stimulating germination is very low: around 70% of seeds germinated after 0.1 second illumination with an equivalent fluence of 50 μmol m^{-2} PFD. Such rapid germination after a weak stimulation of light may be an efficient way for the weeds to use their limited seed reserve. Similarly, germination of light-requiring seeds under canopy is most likely to be successful when a canopy is temporarily open or removed. Increased PFD and decreased R/FR ratio are the major environmental factors in these cases. Seeds of most shade-avoiding plants will germinate if they are exposed to high PFD and/or a low R/FR ratio. Such a mechanism could also be used to adjust the timing of germination to take advantage of seasonal information of light environment, such as under a deciduous forest canopy.

On the other hand, light-inhibited seeds will not germinate under light, which may be important for the seedlings of those seeds to survive under dry conditions since an open and higher light environment is usually associated with a dry soil or air condition. In addition, the fact that some species require only a brief illumination, while others may need repeated exposure for several hours per day, may suggest the importance of sunfleck environment in seed germination. For example, the percentage of germination of *Piper auritum* and *P. umbellatum,* which are tropical understory plants, was significantly higher in microsites with longer sunflecks (Orozco-Segovia and Vazquez-Yanes, 1989).

Morphological Response   After germination, the emerged seedlings will face a great spatial heterogeneity of light environment. Plant seedlings adjust in order to intercept available light under spatially heterogeneous light regimes in different ways: changing the height growth, altering the direction of stem growth, adjusting their branching pattern, or modifying the leaf morphology and leaf position.

Stem Growth   Light may stimulate stem elongation by promoting total growth, but it can also inhibit height growth if PFD is high. The change of height growth has been viewed as one of the most important strategies of plants in response to the variation of light regime by Grime (1979). The response of seedling growth to shade can be classified into two extreme strategies: shade avoidance and shade tolerance. Shade-tolerant plants show relatively slow height growth and conservative utilization of energy and resources, while plants that are shade intolerant have a fast height growth to avoid low light environments. In some light environments, such as in grass canopies, a small increase in plant height would mean a large benefit of light capture due to the large vertical gradient of light.

Under plant canopies, the light environment is characterized by low levels of UV, low visible PFD, and reduced R/FR ratio. All these factors influence the height growth of plants. UV-B and white light inhibit stem elongation in many species; therefore, the reduction of UV-B and visible PFD by a leaf canopy would lead to increased elongation in sensitive species. The reduction of R/FR ratio under canopies is a function of LAI and is therefore correlated with the reduction of PFD. This reduction of R/FR increases stem elongation of shade-avoiding species. Recent studies suggest that the activity of phytochrome B plays a major role in determining the short-term elongation response to shading (Ballare at al., 1991; Lopez-Juez et al., 1992).

The change of height-growth rate in plants happens not only after a shading but also before shading actually occurs. This strategy would be most effective for a plant to avoid shade. R/FR ratio seems to be the important information for such the early detection of light environment. Evidence showed that plants could react to low-R/FR light delivered through fiber-optic probes from the side even when the rest of the plant was exposed to white light of high R/FR (Child and Smith, 1987; Morgan et al., 1980). Field experiments also demonstrated that increased extension rate was due to low-R/FR light reflected from the neighboring grass (Ballare et al., 1988, 1990). A quantitative study of neighbor detection showed that plants not only can detect their neighbors, they can effectively perceive how far away they are and therefore are able to gauge the competitive threat posed (Smith et al., 1990).

In a directionally different light environment, the elongation of stem would be not exactly upright, but bent to the direction of high PFD. The directional response of growth in plants in response to directional variation of light is called phototropism. The directional growth is due to the redistribution of growth in different part of a phototropic organ (e.g., coleoptile tip or stem). When a stem is illuminated from one side, a bending growth to the illumination source is caused by the inhibition of elongation on the illuminated side and stimulation on the shade side.

The detection of the direction of light by plants has been found to be based on fluence rate gradients that occur because of the internal scattering and absorption in the phototropic organ. The absorption of chlorophyll will result in an abrupt decrease in the fluence rate of red light, but a gradual decline of FR. This forms a gradient of R/FR ratio, and thus a photoequilibrium gradient of phytochrome between the illuminated side and shaded side of the organ. Since a low R/FR or a low photoequilibrium will stimulate elongation, the shade side will elongate faster than the illuminated one. Ballare et al. (1992) found that a low R/FR ratio on the side that faces the neighboring canopy may signal elongation organs to bend toward gaps.

Growth and Orientation of Branches A plant produces new branches to use spatial resources within the plant canopy. Reduced total PFD would result in increased apical dominance and reduced branching. Recent studies also showed that the differential reflection of R and FR light by nearby leaves might stimulate grass plants to reduce tillering rate early in canopy development (see, e.g., Casal and Smith, 1989).

Orientation of branches is an important parameter determining plant architecture under a heterogenous light regime. Branches or shoots would grow more erectly in a dense than in a sparse plant canopy. Shoot inclination also seems to be controlled by R/FR ratio.

Leaf Morphological Plasticity and Movement In addition to regulating stem extension and branching pattern, PFD and R/FR ratio can also influence the morphological development of leaves. For example, FR-treated tobacco plants have longer, narrower, lighter leaves with fewer stomata and less chlorophyll per unit leaf area, and a higher photosynthesis rate per unit leaf dry weight than plants grown under a lower FR light condition.

Apart from being able to change the morphological properties of leaves, many plants can also change the spatial orientation of their leaves in response to the change of direction of incident light. Solar tracking is a diaphototropic leaf movement, in which a leaf reorients its lamina throughout the day so as to maintain it nearly normal to the direct light beams. In young leaves, solar tracking has a mechanism similar to phototropic stem bending. In both cases, reorientation depends on redistribution of growth between opposite sides of the petiole or stem. However, in mature leaves, reorientation is due to turgor variation and reversible structure deformations of the pulvinus, located in the basal part of the leaf lamina. Light information for solar tracking seems to be received by the B-light receptor. Solar tracking of leaves would maintains a high and relatively constant PFD interception during the day, by increasing interception of light, or by increasing leaf temperature in the morning and evening.

Photoperiodism Photoperiodism is the phenomenon shown by many plants in which the day-length or photoperiod determines whether or not vegetative apices switch to flower formation. The day-length is constant throughout the year at the equator, but it changes with season in all other parts of the world. Many plants are able to use the changes of day-length as the signals for adjusting their physiological activities.

On the basis of the flowing response to day-length, plants can be classified into three major groups:

1. *Long-Day (LD) Plants:* The flowering of this group of plants, for example, oats *(Avena sativa),* only occurs when the day-length exceeds a certain number of long days. Most LD plants originate in the temperate clime, and they include many temperate crop species.

2. *Short-Day (SD) Plants:* Plants such as *Chenopodium rubrum*, and coffee *(Coffea arabica)* are found to require a certain minimum number of days where the continuous dark period exceeds a certain minimum duration for lower induction to occur. Many SD plants originated in low latitude regions, including important crop species such as maize, millet, rice, and sugarcane.

3. *Day-Neutral Plants:* The flowering of these plants does not respond to day-length, and many of them are tropical in origin.

The plants' response to day-length may change with age; in other words, plants may require a particular sequence of day-lengths. For example, some plants need a period of long days followed by a period of short days (LSD plants), or of short days followed by long days (SLD plants).

Photoperiodism can be interpreted as an adaptation to the sequences of seasons. It is advantageous for a SD plant to flower only if short-day conditions are prevalent at its natural site. By using the day-length information, plants are able to avoid disadvantageous environments and prepare for the coming of advantageous environments.

The information needed for photoperiodism is detected in leaves. Plants differ not only in the ability of leaves to produce the photoperiodic signal, but also in the number of photoperiodic cycles required for flower induction. The qualitative difference in the reaction to the same Pfr signal is genetically determined, but how Pfr stimulates or suppresses the formation of florigen in leaves is not yet known.

SUMMARY

In the past decade, interest has rapidly increased in light environments and their role in plant ecophysiology. Considerable developments have been achieved in our understanding, not only about the great complexity of light environment in plant canopies, but also about the ecological and physiological effects of light on various plant processes. We now know that there is a great heterogeneity of light environments for temporal scales ranging from seconds to years or longer, and from a spatial scale ranging from cellular to global. It is the heterogeneity that plays an important role in the ecophysiological response of plants, which may be linked to ecological organization of plant community and biodiversity. Recent studies on sunflecks and their importance can be used as a good illustration of our understanding of the temporal heterogeneity of light environments and its effects on photosynthesis and plant growth. Additionally, one should mention here that there has been a great technical improvement in light measurement in plant canopies because of the development of pertinent fields, providing an important basis for studying light environments.

However, the quantification of light environment heterogeneity pattern and scale is still an evolving field. Integrated regulation and control of light on various biological and ecological processes of plants at different temporal and spatial scales has received much less attention. Even some basic problems, such as definition of gaps and sunflecks, transient photosynthesis, and light signal sensing, are even more complicated. Further studies on all of these aspects remains a pertinent challenge in plant ecophysiology.

REFERENCES

Anderson, M. C. (1964), *J. Ecol., 52*, 643–663.

Anderson, M. C. (1974), *J. Appl. Ecol., 11*, 691–697.

Anderson, J. M. and C. B. Osmond (1987), in *Photoinhibition* (D. J. Kyle, C. B. Osmond, and C. J. Arntze, Eds), Elsevier, Amsterdam, pp. 1–38.

Aphalo, P. J. and C. L. Ballare (1995), *Func. Ecol., 9*, 5–14.

Baldocchi, D. and S. Collineau (1994), in *Plant physiological ecology: Field methods and instrumentation* (E. W. Pearcy, J. R. Ehleringer, H. A. Mooney, and P. W. Rundel, Eds.), Chapman and Hall, London, pp. 21–72.

Ballare, C. L., R. A. Sanchez, A. L. Scopel, and C. M. Ghersa (1988), *Oecologia, 76*, 288–293.

Ballare, C. L., A. L. Scopel, and R. A. Sanchez (1990), *Science, 247*, 329–332.

Ballare, C. L., A. L. Scopel, and R. A. Sanchez (1991), *Oecologia, 86*, 561–567.

Ballare, C. L., A. L. Scopel, R. R. Radosevich, and R. E. Kendrick (1992), *Plant Physiol., 100*, 170–177.

Bardon, R. E., D. W. Countryman, and R. B. Hall (1995), *Ecology, 76*, 1013–1016.

Björkman, O. and C. Schafer (1989), *Philos. Trans. Roy. Soc. Lond., B323*, 309–311.

Björkman, O. and B. Demming-Adams (1995), in *Ecophysiology of photosynthesis*, Springer, Berlin, pp. 17–48.

Bone, R. A., D. W. Lee, and J. M. Norman (1985), *Appl. Opt., 24*, 1408–1412.

Borthwick, H. A., S. B. Hendricks, M. W. Parker, E. M. Toole, and V. K. Toole (1952), *Proc. Natl. Acad. Sci. USA, 38*, 662–666.

Campbell, G. S. and J. M. Norman (1990), in *Plant canopies: Their growth, form and function* (G. Russell, B. Marshall, and P. G. Jarvis, Eds.), Cambridge University Press, New York, pp. 1–20.

Casal, J. J. and H. Smith (1989), *Plant Cell Environ., 12*, 855–862.

Chazdon, R. L. (1988), *Adv. Ecol. Res., 18*, 1–63.

Chazdon, R. L. and R. W. Pearcy (1986), *Oecologia, 69*, 517–523.

Chazdon, R., R. W. Pearcy, D. Lee, and N. Fetcher (1996), in *Tropical forest plant ecophysiology* (S. S. Mulkey, R. Chazdon, and A. P. Smith, Eds.), Chapman and Hall, London, in press.

Child, R. J. and H. Smith (1987), *Planta, 172*, 219–229.

DeLucia, E. H., K. Nelson, T. C. Vogelmann, and W. K. Smith (1996), *Plant, cell and Environ.* 19: 159–170.

Evans, G. C. and D. E. Coombe (1959), *J. Ecol., 47*, 103–113.

Gholz, H. L., S. A. Vogel, W. P. Cropper, K. McKelvey, Jr., K. C. Ewel, R. O. Teskey, and P. J. Curran (1991), *Ecol. Monogr., 61*, 33–51.

Givnish, T. J. (1988), *Aust. J. Plant Physiol., 15*, 63–92.

Grime, J. P. (1979), *Plant strategies and vegetation processes*, Wiley, New York, pp. 19–79.

Hall, D. O. and K. K. Rao (1994), *Photosynthesis*, 5th ed., Cambridge University Press, New York, pp. 175–189.

Horwitz, B. A. (1994), in *Photomorphogenesis in plants* (R. E. Kendrick and G. H. M. Kronenberg, Eds.), Kluwer Academic, Dordrecht, Netherlands, pp. 327–350.

Iwaki, H. (1990), *Introduction of ecology*, 2nd ed. (in Japanese), Hosoudaigaku, Tokyo, p. 96.

Jones, H. G. (1992), *Plant and microclimate*, 2nd ed., Cambridge University Press, Australia, Victoria, pp. 9–45.

Jones, M. B. (1993), in *Photosynthesis and production in a changing environment: A field and laboratory manual* (D. O. Hall, J. M. O. Scurlock, H. R. Bolhar-Nordenkampf, R. C. Leegood, and S. P. Loog, Eds.), Chapman and Hall, London, pp. 47–64.

Kirk, J. T. O. (1994), *Light and photosynthesis in aquatic ecosystems*, 2nd ed., Cambridge University Press, Cambridge, UK, pp. 46–168.

Kirschbaum, M. U. F. and R. W. Pearcy (1988a), *Plant Physiol.*, *86*, 782–785.

Kirschbaum, M. U. F. and R. W. Pearcy (1988b), *Plant Physiol.*, *87*, 818–821.

Kitajima, K. (1994), *Oecologia*, *98*, 419–428.

Knipling, E. D. (1970), *Remote Sens. Environ.*, *1*, 155–159.

Laisk, A., O. Kiirats, and V. Oja (1984), *Plant Physiol.*, *76*, 723–729.

Larcher, W. (1995), *Physiological plant ecology*, 3rd ed., Springer, Berlin, pp. 31–56.

Lee, D. W. (1987a), *Biotropica*, *19*, 161–166.

Lee, D. W. (1987b), In *On the economy of plant form and function* (T. J. Givinish, Ed.), Cambridge University Press, Cambridge, UK, pp. 105–126.

Lopez-Juez, E., A. Nagatani, K.-I. Tomizawa, M. Deak, R. Kern, R. E. Kendrick, and M. Furuya (1992), *Plant Cell*, *4*, 241–251.

Mandoli, D. F. and W. R. Briggs (1982). *Plant, cell and environ.* 5: 137–145.

Monsi, M. and T. Saeki (1953), *Jap. J. Bot.*, *14*, 22–52.

Morgan, D. C., T. O'Brien, and H. Smith (1980), *Planta*, *150*, 95–101.

Nilson, T. (1991), in *Photon-vegetation interactions: Applications in optical remote sensing and plant ecology* (R. B. Myneni and J. Ross, Eds.), Springer-Verlag, Berlin, pp. 161–190.

Nobel, P. S. (1991), *Physicochemical and environmental plant physiology*, Academic Press, New York, pp. 245–340.

Orozco-Segovia, A. and C. Vazquez-Yanes (1989), *Acta Oecol.*, *10*, 123–146.

Parker (1995), in *Forest canopies* (M. D. Lowman and N. M. Nadkarni, Eds.), Academic Press, New York, pp. 73–108.

Pearcy, R. W. (1983), *Oecologia*, *58*, 19–25.

Pearcy, R. W. (1989), in *Plant physiological ecology: Field methods and instrumentation* (E. W. Pearcy, J. R. Ehleringer, H. A. Mooney, and P. W. Rundel, Eds.), Chapman and Hall, London, pp. 97–116.

Pearcy, R. W. (1990), *Annu. Rev. Plant Physiol. Plant Mol. Biol.*, *41*, 421–453.

Pearcy, R. W. and W. A. Pfitsch (1995), in *Ecophysiology of photosynthesis*, Springer, Berlin, pp. 343–36/

Pearcy, R. W. and D. A. Sim (1994), in *Exploitation of environmental heterogeneity by plants: Ecophysiological processes above- and belowground* (M. C. Martyn and R. W. Pearcy, Eds.), Academic Press, New York, pp. 145–174.

Pearcy, R. W., K. Osteryoung, and H. W. Calkin (1985), *Plant Physiol.*, *79*, 896–902.

Pons, T. L. and R. W. Pearcy (1992), *Plant, Cell Environ.*, *15*, 577–584.

Pratt, L. H. (1994), in *Photomorphogenesis in plants* (R. E. Kendrick and G. H. M. Kronenberg, Eds.), Kluwer Academic, Dordrecht, Netherlands, pp. 163–185.

Reifsnyder, W. E., G. M. Furnival, and J. L. Horowitz (1971), *Agric. Meterorl.*, *9*, 1–37.

Roden, J. S. and R. W. Pearcy (1993), *Oecologia*, *93*, 201–207.

Ross, J. (1981), *The radiation regime and architecture of plant stands*, W. Junk, The Hague.

Sassenrath-Cole, G. F. and R. W. Pearcy (1992), *Plant Physiol.*, *99*, 227–234.

Scopel, A. L., C. L. Ballare, and R. A. Sanchez (1991), *Plant Cell Environ.*, *14*, 501–508.

Senger, H. and W. Schmidt (1994), in *Photomorphogenesis in plants* (R. E. Kendrick and G. H. M. Kronenberg, Eds.), Kluwer Academic, Dordrecht, Netherlands, pp. 301–325.

Sharkey, T. D. and T. Ogawa (1987), in *Stomatal function* (E. Zeiger, G. D. Farquhar, and I. R. Cowan, Eds.), Stanford University Press, Stanford, CA, pp. 195–208.

Sharkey, T. D., J. R. Seemann, and R. W. Pearcy (1986), *Plant Physiol.*, *82*, 1063–1068.

Sims, D. A. and R. W. Pearcy (1993), *Funct. Ecol.*, *7*, 683–689.

Smith, W. K., A. K. Knapp, and W. A. Reiners (1989), *Ecology*, *70*, 1603–1609.

Smith, H., J. J. Casal, and G. M. Jackson (1990), *Plant Cell Environ.*, *13*, 73–78.

Tang, Y. and I. Washitani (1995), *Ecol. Res.*, *10*, 189–197.

Tang, Y., I. Washitani, T. Tsuchiya, and H. Iwaki (1988), *Ecol. Res.*, *3*, 253–266.

Tang, Y., T. Tsuchiya, I. Washitani, and H. Iwaki (1989), *Ecol. Res.*, *4*, 339–350.

Tang, Y., I. Washitani, and H. Iwaki (1992), *Ecol. Res.*, *7*, 97–106.

Tang, Y., H. Koizumi, M. Satoh, and I. Washitani (1994), *Oecologia*, *100*, 463–469.

Tang, Y., N. Kachi, I. Furukawa, and M. Awang (1996), *Forest ecology and management* (in press).

Terashima, I. (1989), in *Photosynthesis* (W. R. Briggs, Ed.), Alan R. Liss, New York, pp. 207–226.

Terashima, I. and T. Saeki (1983), *Plant cell physiol.*, 24: 1493–1501.

Vogelmann T. C. and G. Martin (1993), *Plant, cell and Environ.* 16: 65–72.

Vogelmann, T. C. (1994), in *Photomorphogenesis in plants* (R. E. Kendrick and G. H. M. Kronenberg, Eds.), Kluwer Academic, Dordrecht, Netherlands, pp. 491–535.

Vogelmann, T. C., G. Martin, G. Chen, and D. Buttry (1991), *Adv. Bot. Res.*, *18*, 256–296.

Walter-Shea, E. A. and J. M. Norman (1991), in *Photon-vegetation interactions: Applications in optical remote sensing and plant ecology* (R. B. Myneni and J. Ross, Eds.), Springer-Verlag, Berlin, pp. 230–251.

Woodrow, I. E. and K. A. Mott (1989), *Aust. J. Plant Physiol.*, *16*, 487–500.

Yoda, K. (1978), *Malay. Nat. J.*, *30*, 161–177.

2

UV-B Radiation

G. Kulandaivelu, K. Lingakumar, and A. Premkumar

INTRODUCTION

In the last two decades, there has been a great concern about the depletion of strato-spheric ozone (O_3), caused primarily by the extensive release of some O_3-depleting susbtances. Stratospheric O_3, a thin gaseous layer, strongly attenuates ultraviolet (UV) radiation shorter than 310 nm. Reductions in stratospheric O_3 allow more solar UV radiation, particularly in the UV-B band (280–320 nm), to reach the earth's surface. Although the level of UV-B in solar radiation accounts for less than 0.5% of total radiation, even a small quantitative and qualitative increase in this level has been shown to cause significant reduction in growth and other physiologi-cal responses in many sensitive crops. These conclusions were drawn from both laboratory and field experiments. This chapter summarizes the major physiological and biochemical changes caused by enhanced UV-B radiation.

ABBREVIATIONS

ATP	Adenosine-5′-triphosphate
cab mRNA	Chlorophyll a/b binding-protein mRNA
Chl	Chlorophyll
D_1,D_2	Photosystem II core complex proteins
DCMU	3,-(3,4-dichlorophenyl)-1,1-dimethylurea
DCPIP	2,6-dichlorophenolindophenol
HSPs	Heat shock proteins

IAA	Indole-3-acetic acid
LHCP	Light-harvesting chlorophyll proteins
MV	Methyl viologen
OEC	Oxygen evolving complex
PAR	Photosynthetically active radiation
PEPcase	Phosphoenol pyruvate carboxylase
PPFD	Photosynthetic photon flux density
PS	Photosystem
RC	Reaction center
RuBP	Ribulose-1,5-bisphosphate
RuBPcase	Ribulose 1,5-bisphosphate carboxylase
SDS-PAGE	Sodium dodecyl sulphate polyacrylamide gel electrophoresis
UV-B	Ultraviolet-B

GENERAL PLANT RESPONSES TO UV-B RADIATION

Increased solar UV-B radiation has been reported to cause detrimental effects on different plant species; inhibitions of growth, photosynthesis, and phytomass accumulation are some of the UV-sensitive basic responses. The effects of increased solar UV-B radiation on terrestrial plants have been extensively reviewed by Tevini and Teramura (1989) and Caldwell et al. (1995). As leaves receive the major proportion of UV-B radiation, much attention was paid to the foliar characteristics. Leaf area, which determines the light-capturing efficiency of a plant, was found to reduce significantly in 60% of the species tested for UV-B sensitivity (Biggs and Kossuth, 1978). Among the crop species, soybean, bean, pea, cowpea, cucumber, watermelon, rhubarb, rutabaga, kohlrabi, and brussel sprouts were found to be very sensitive; in these crops the leaf expansion was reduced by 60–70% (Biggs et al., 1981; Tevini et al., 1981). Species like rice, wheat, barley, millet, oats, peanuts, cotton, and sunflower were shown to be UV-B resistant. In a study involving 22 cultivars of rice tested for UV-B sensitivity, nearly 50% of the cultivars showed high inhibitions in leaf area and growth (Barnes et al., 1993; Teramura et al., 1991a). However, sensitivity to UV-B depends on plant species, cultivars, developmental stage of the plant, and UV irradiation procedures.

In general, plants grown in growth chambers and greenhouses are more susceptible to UV-B than those grown under natural field conditions. The visible radiation component (350–700 nm) as well as the UV-B/UV-A/PAR ratios in the field differ from those of growth chambers. In most of the plant species almost 70% of the UV-B radiation reaching the leaf surface is attenuated before reaching the mesophyll tissues. Leaf symptoms like bronzing, scorching, glazing, or chlorosis have been shown to occur upon UV-B radiation (Biggs et al., 1981; Teramura et al., 1983;

Tevini et al., 1981; Vu et al., 1978, 1984). Plant stunting is yet another common UV-B induced phenonmenon found visible under high UV-B and low PAR levels; it is primarily due to short internodes and reductions in node number. UV-B induced stunting could be specifically due to the accumulation of flavins, which act as UV-B photoreceptors and which are also responsible for inhibition of internode elongation (Ballaré et al., 1995; Ensminger and Schäfer, 1992). Furthermore, destruction of endogenous levels of IAA (Kulandaivelu et al., 1989; Tevini and Iwanzik, 1986) and induction of oxidative enzymes associated with growth responses by UV-B irradiation (Ros, 1990) were also reported. High levels of UV-B were shown to induce wilting, which is primarily due to the lack of turgidity, low water content, and reduced xylem vessel number and size, resulting in less water uptake (Lingakumar and Kulandaivelu, 1993b).

In a UV-B enhanced solar irradiation experiment, four cultivars of rice, namely ADT 36, IR 20, IR 50, and MDU 4, collected from the several rice cultivating regions of Tamilnadu, India, were treated with different levels of UV-B corresponding to 20 and 40% O_3 depletion at Madurai (10°N). Figs. 2-1 and 2-2 show the changes in seedling height and shoot fresh weight of these cultivars in response to different levels of UV-B enhanced solar radiation. All the cultivars except MDU 4 exhibited reductions in seedling height and shoot fresh weight, with severe damage noticed in ADT 36 and IR 20 cultivars. MDU 4 showed resistance to these parameters even at highest level of UV-B (40% O_3 depletion). As reported for other species, UV-B sensitivity was found to vary among rice cultivars. Such cultivar differences could be due to the species composition itself.

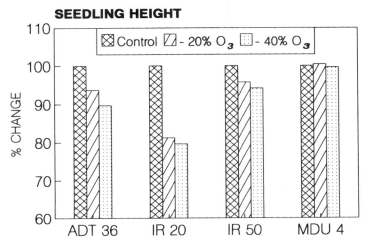

Figure 2-1. Effect of UV-B enhanced solar radiation (simulating a 20 and 40% O_3 depletion) on seedling growth of several different tropical rice cultivars.

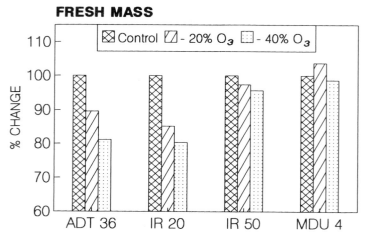

Figure 2-2. *Effect of UV-B enhanced solar radiation (simulating 20 and 40% O₃ depletion levels) on fresh mass of several different tropical rice cultivars.*

EFFECT OF UV-B RADIATION ON BIOCHEMICAL COMPOSITION

High UV-B irradiance combined with low PAR growth levels produced significant reductions in chlorophyll (Chl) concentrations (Garrard et al., 1976; Tevini et al., 1981; Vu et al., 1982b). Increase in the level of chlorophylls under UV-B radiation has also been reported, and it differs among species and cultivars (Mirecki and Teramura, 1984). Such an increase could be considered as a result of decreased leaf growth and in such situations leaf fresh weight basis would be a better parameter for expressing Chl content than leaf area. A similar conclusion was drawn by Bornman (1989). Decrease in Chl content could be due to photooxidation of Chl by UV-B or to inhibition of Chl biosynthesis. Strid and Porra (1992) have reported that UV-B does not affect the early steps in Chl biosynthesis. The Chl *a/b* ratio also varies with species and growth conditions. Tevini et al. (1981) have reported an increase in Chl *a/b* that was due to the preferential damage of Chl *b* by enhanced UV-B radiation. Damage to Chl *b* levels rather than to Chl *a* was supported by the observed low levels of cab mRNA transcripts under UV-B treatment (Jordan et al., 1991). Compared to Chl, carotenoids are less affected by UV-B radiation. Less inhibition of carotenoids and accumulation of carotenoids under certain treatment conditions are suggested to be due to their photoprotective role in the photosynthetic apparatus (Middleton and Teramura, 1993). Both carotenoids and flavonol glycosides are known to have protective roles in safeguarding the photosynthetic apparatus and subcellular organelles from UV damage.

Unlike photosynthetic pigments, UV-B radiation induces a class of nonphotosynthetic pigments, namely phenyl propanoids (Beggs et al., 1986; Tevini et al., 1991; Wellmann, 1983). It is very obvious that epidermal membranes represent the first barrier between the plant and the environment, and hence they become the first target of all incident radiation on the plant surface. Leaf epidermal transmittance of UV radiation was examined in several species, and the adaxial epidermis was found to have the potential to alter the quality and quantity of the radiation penetrating to the mesophyll region (Robberecht and Caldwell, 1978). The transmittance in the UV-B region was found to vary from 0.5 to 69% (Fig. 2-3). The xerophytic species showed a high degree of UV attenuation by their epidermal layer.

The major factor in determining the filtration capacity of leaf epidermis is the presence of UV-B absorbing flavonoids (Caldwell, 1971; Gausman et al., 1975). Flavonoid accumulation in tissues is a protective strategy against UV radiation. The flavonoids are synthesized in vacuoles and are often found in epidermal layers. The content and composition of flavonoids can be altered by various environmental factors. Biosynthesis of flavonoids was shown to be induced through the phyto-chrome system and in response to high intensity of visible radiation or blue and red lights (McClure, 1975). Phytochrome regulation of flavonoid accumulation is dependent upon the sequence of the irradiation program. Simultaneous treatment with far-red and UV-B radiations produced enhanced level of flavonoids in *Vigna sinensis* seedlings (Lingakumar and Kulandaivelu, 1993a). Under low PAR growth

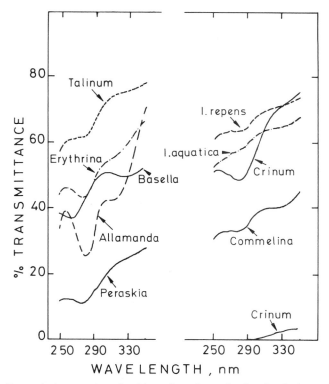

Figure 2-3. *Transmission spectra of epidermal peelings of a few tropical species in the UV region. Epidermal peeling without any mesophyll cell contamination were removed from the leaves and mounted on a plexiglass frame for spectral analysis.*

regimes and high UV-B irradiance, Tevini et al. (1981) observed high concentration of flavonoids in moderately UV-B resistant plants (barley and radish). Flavonoids protect organelle DNA against UV-B damage by forming thymine dimers (Beggs and Wellmann, 1994).

Similar to flavonoids, anthocyanins also act as in vivo UV-B attenuators in protecting the cellular components against radiation damage. Anthocyanin-deficient mutants of maize were shown to be more sensitive to UV-B than wild plants (Stapleton and Walbot, 1994). Since anthocyanins have weak absorption in the UV-B region, they are regarded as UV screens only at very high concentrations.

Other secondary metabolites such as furanocoumarins also accumulate upon exposure to UV-B radiation, which was shown to delay the development of certain insect larvae (McCloud and Berenbaum, 1994).

Chloroplast membrane lipids such as galactolipids were greatly reduced in many crop species after UV-B radiation. Under field conditions, Esser (1980) noticed an increase in total lipids in UV-B exposed bean plants. Tevini et al. (1981) reported reduction in total proteins and glycerolipids in a number of plants and suggested a correlation with the onset of senescence. Supporting evidences for UV-induced aging process resulting in damage to subcellular membranes were shown by Berjak (1974) and Roshchupkin et al. (1975).

Epicuticular waxes play an important role in minimizing cuticular transpiration and water diffusion. Enhanced UV-B radiation caused a 25% increase in total wax in cucumber, bean, and barley (Steinmüller and Tevini, 1986). The mode of UV-B action was suggested to be on both the biosynthesis of waxes and inhibition of elongase II resulting in low content of long chain homologues (Tevini and Steinmüller, 1987). Leaf soluble protein and carbohydrate content were affected in many of the crops tested for UV-B sensitivity (Esser, 1980; Tevini et al., 1981; Vu et al.; 1982a, b). Polyamines are considered to be the physiological marker for water stress and other damages that accumulate under high UV-B dosage (Kramer et al., 1992).

EFFECTS OF UV-B RADIATION ON CHLOROPLAST STRUCTURE AND FUNCTION

Although DNA and proteins are known to absorb UV radiation, considerable attention has been paid to the UV-B induced cellular membrane damage. Reports of loss of membrane integrity as a result of UV radiation are from ultrastructural (Allen et al., 1978; Bornman, 1986; Bornman et al., 1983; Brandle et al., 1977; Campbell et al., 1974; Noorudeen and Kulandaivelu, 1982; Skokut et al., 1977) and physiological studies (Doughty and Hope, 1973; Murphy, 1983). Electron micrographs of chloroplasts isolated from UV-B treated bronzed soybean leaves revealed the disintegrity of the chloroplasts (Vu et al., 1982a, b) and dilation of thylakoids (Bornman, 1986; Brandle et al., 1977; Noorudeen, 1982). UV-B impairment of membrane integrity was solely attributed to the formation of endogenous free radicals, which cause lipid peroxidation. It was Mantai et al. (1970) who correlated the effects of UV radiation on photosynthesis with ultrastructural changes in spinach chloroplasts.

Several papers have appeared considering the site of UV-B action in chloroplasts, all of them reporting not a single specific site but multiple target sites. Bornman (1989) had reviewed the sites of UV-B damage. Most of the studies focus on photosystem II (PSII) rather than photosystem I (PSI) as the major site of UV-B action. The effect of UV-B radiation is not evenly distributed between the photosystems (Bornman, 1989). UV-B radiation was shown to have different target sites in chloroplasts such as chloroplast membranes (Murphy, 1983), PSII reaction center (Nedunchezhian and Kulandaivelu, 1991; Noorudeen and Kulandaivelu, 1982), and enzymes of CO_2 fixation cycle (Vu et al., 1984). UV-B radiation induces formation of certain polypeptides that offer a protective function to light-harvesting Chl complex (Nedunchezhian and Kulandaivelu, 1991). Although PSI was shown to be insensitive to UV-B radiation (Iwanzik et al., 1983; Noorudeen and Kulandaivelu, 1982; Renger et al., 1982; Van et al., 1977), the cyclic photophosphorylation was significantly inhibited (Brandle et al., 1977; Van et al., 1977).

Besides acting at the photosystem level, UV-B radiation was shown to alter the stoichiometry of PSI to PSII (Lingakumar and Kulandaivelu, 1993b). The D1 protein of PSII RC was assumed to be the main target for UV-B (Greenberg et al., 1989; Renger et al. 1989). The D1 protein degradation assay was used as a sensitive in vivo probe for penetration of UV-B to the mesophyll tissues. The distortion of 33 kDa protein from the OEC was suggested to be the reason for inactivation of water oxidation under UV-B treatment (Renger et al., 1989). Strid et al. (1990) have shown that the PSII activity, levels of ATP synthase, and RuBPcase were drastically decreased upon exposure to enhanced UV-B radiation. Reduction in RuBPcase activity under UV-B illumination is primarily due to the damage of the polypeptides associated with the enzyme (Nedunchezhian and Kulandaivelu, 1991; Strid et al., 1990). A rapid decline in mRNA levels of both the subunits of RuBPcase was reported (Jordan et al., 1992). Fig. 2-4 shows the effect of enhanced UV-

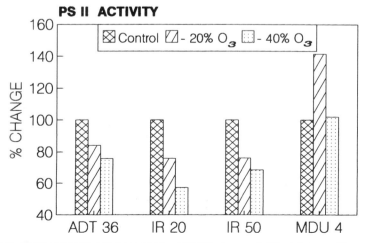

Figure 2-4. *Photosynthetic response ($H_2O \rightarrow DCPIP$) of several different rice cultivars exposed to UV-B enhanced solar radiation (simulating a 20 and 40% O_3 depletion).*

B doses on the PSII activity of four Indian rice cultivars, namely ADT 36, IR 20, IR 50, and MDU 4. With the increase in UV-B doses, a significant reduction in activity was noticed in three of the cultivars with the largest reductions in IR 20 and IR 50. Only the MDU 4 cultivar exhibited an increase in activity even under high levels of UV-B. Although rice has been reported to be a UV-B resistant species, our experiments show that the sensitivity varies with the cultivar type. Similar responses were observed for the PSI electron transport activity. When we looked for SDS-PAGE polypeptide pattern of chloroplasts, only high UV-B doses caused significant loss in the level of proteins associated with oxygen evolving complex, namely 17, 23, and 33 kDa.

Under UV-B illumination, a structural reorganization of chloroplasts with damage to PSII polypeptides was also reported (Nedunchezhian and Kulandaivelu, 1991). Although UV-B induced damage and photoinhibitory (high light) damage on PSII resemble each other, the photosensitizers mediating the damages differ from each other. Chl is the photosensitizer for photoinhibition by visible light, whereas plastoquinone directly absorbs the UV-B radition (Greenberg et al., 1989).

UV-B EFFECTS ON PHOTOSYNTHESIS

Photosynthesis and other physiological processes can be directly and significantly affected by UV-B radiation. UV-B radiation transmitted through epidermis affects photosynthesis. This has been demonstrated in plants that were grown under natural conditions as well as in plants grown under growth chamber/greenhouse conditions. Exposure to high levels of UV-B (> 1 W m^{-2}) affects photosynthesis and normal growth and development in many species of economically important crops (Bartholic et al., 1975; Biggs et al., 1975; Brandle et al., 1977; Teramura et al., 1980). Net photosynthesis was affected by high UV-B radiation and low PAR. Such reduction was thought to be due to an increase in mesophyll diffusion resistance (Brandle et al., 1977; Sisson and Caldwell, 1976; Teramura and Perry, 1982; Teramura et al., 1980).

Overall photosynthesis could also be limited by the diffusion of CO_2 entering the leaf through the stomata. Though stomatal resistance accounts for 10–15% of the total leaf resistance to CO_2, a direct correlation between increase in stomatal resitance and reduction in net photosynthetic rates could not be observed (Bennett, 1981; Teramura et al., 1980, 1983). All these resistances (residual, stomatal, and boundary layer) account for the decrease in carboxylation activity and electron transport rates (Teramura, 1983). An important step in the CO_2 incorporation in plants involves the carboxylating enzyme, namely RuBPcase in C_3 plants and PEPcase in C_4 plants.

Recent studies indicate that C_3 plants are more vulnerable to UV-B than C_4 plants (Fig. 2-5). Such differences are associated with the polypeptide composition of thylakoids (Kulandaivelu et al., 1991). Low sensitivity of C_4 plants to UV-B radiation could also be due to sclerification of leaf tissues and vertical orientation of leaves with protective leaf sheaths. The effect of UV-B radiation on carbon

PS II ACTIVITY

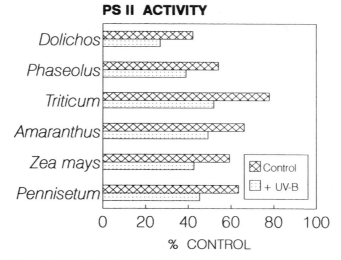

Figure 2-5. *Differential sensitivity of several C_3 and C_4 species to UV-B enhanced solar radiation. The duration of UV-B (5 W m^{-2}) exposure was 30 minutes.*

metabolism rates has been determined for a relatively large number of agronomically important crops. With a few exceptions, all plants showed detrimental effects to increased solar UV-B radiation. Takeuchi et al. (1989) presented evidence for the UV-B effects on the reduction of organic acids and soluble sugars.

UV-B DEFENSE AND GENE EXPRESSION

Plants have the capacity to develop various means of protection from the deleterious effects of UV-B radiation. One of the most important and sensitive targets of UV-B radiation is DNA. Action spectra for DNA damage in most organisms strongly suggest that DNA is the primary target (Caldwell, 1971; Setlow, 1974). Though proteins and RNA absorb at the relevant wavelengths still they are subjected to UV-B damage. The special role of DNA as the storehouse of genetic information makes any damage to it of particular consequence to organisms. Because of this, whenever DNA damage occurs, the organisms develop several mechanisms for repairing such damages caused by UV radiation. Plants have numerous biochemical mechanisms protecting cells from the harmful effects of photooxidation (Elstner, 1982). Some of them include quencher molecules and free radical scavenging enzymes such as superoxide dismutase and peroxidases. Though plant cells have enough enzyme activity to overcome photooxidation, the enzymes involved in the production of phenolics serve as an efficient protective mechanism against UV-B damage.

Beggs et al. (1986) reviewed the UV-B induced pigment formation in various plant species. Flavonoid accumulation is one of the protective measures to attenuate the harmful UV radiation. UV-B irradiation induced considerably higher levels of glutathione and ascorbic acid, presumably as an antioxidative mechanism (Rao and Ormrod, 1995). Both UV-B damage and O_3 injury were shown to have similar antioxidative mechanisms. Willekans et al. (1994) have reported that O_3, SO_2, and UV-B activate certain defense genes within the plant using a common signal transduction pathway.

Pea leaves exposed to supplemental UV-B showed severe reductions in the level of nuclear coded cab mRNA transcript level. On the contrary, chloroplast coded psb mRNA levels were not affected even after long duration of UV-B exposure. This indicates that LHCP is more vulnerable than the D1 protein of PSII to UV-B radiation (Strid et al., 1990). Greenberg et al. (1989) have shown that under UV-B treatment, the rate of degradation of D1 protein is increased rather than the synthesis of the protein being reduced. Schulze-Lefert et al. (1989) have demonstrated UV-B induced increase in transcription rates for chalcone synthase (CHS) and 4-coumarate CoA ligase (4-CL), the key enzymes of flavonoid biosynthetic pathway.

Compared to the nuclear-encoded RNA transcripts, chloroplast-encoded RNA transcripts show more stability after UV-B exposure. This is due to the fact that, in contrast to the nuclear-encoded genes, chloroplast-encoded genes are frequently controlled at the level of posttranscriptional mechanism (Jordan et al., 1991). To combat UV damage, organisms develop important repair mechanisms. Thus DNA damage by UV radiation is encountered not only by the protective molecules but also by repair systems.

UV-induced DNA damage can be repaired by several distinct mechanisms. One such mechanism is photoreactivation, in which the enzyme photolyase, a light-requiring enzyme, cleaves cyclobutane pyrimidine dimer and returns the bases to their original state. Pang and Hays (1991) demonstrated the presence of photolyase gene, which transcripts in the presence of light. The gene has been cloned from mustard by Batschaeur (1993). Photolyase activity has also been detected in maize and bean (McClennan, 1987). Another repair system that involves an excision repair mechanism is also well known. The pyrimidine (6-4) and pyrimidone dimers [(6-4) photoproducts)] are repaired by an excision repair process in which incisions are made on either side of the lesion, an oligonucleotide stretch containing the lesion is removed, and a replacement strand is resynthesized. Since plants have the potential to identify specific repair mechanisms for UV damage, it is necessary to study mutants having varied UV sensitivity. Britt et al. (1993) have developed a UV-hypersensitive *Arabidopsis* mutant called UVr1, and recently another mutant UVh1 was developed by Harlow et al. (1994). Both the mutants are UV-B sensitive. Flavonoid-defective *Arabidopsis* mutants were developed by Li et al. (1993). An attempt to quantify the UV-B-induced pyrimidine dimers was made by Quaite et al. (1992). They have also demonstrated the effective DNA repair system by the action of DNA-photolyase activity.

INTERACTION OF OTHER STRESSES WITH UV-B RADIATION

A brief summary of the studies carried out on the interaction of UV-B with other environmental factors is given in Table 2-1.

TABLE 2-1   A Summary of UV-B Interactions with Other Environmental Factors on Growth and Photosynthesis of Higher Plants

Interaction Type	Plant Species	Responses	References
	UV-B & PAR/Other Radiations		
High PAR and high UV-B	Wheat	Significant decrease in dry matter	Teramura (1980)
UV-B and UV-A/blue light	Many species	Photoreactivation of UV-B effects on growth and photosynthesis	Fernbach and Mohr (1992)
High PAR and UV-B	Cucumber	Accumulation of poly-amines as a protective mechanism	Kramer et al. (1992)
Low PAR and UV-B	Soybean	UV-B has negative effects on net photosynthesis, transpiration, diffusive resistance	Mirecki and Ter-amura (1984)
UV-B and monochromatic red light	Cowpea	Phytochrome reversal of UV-B inhibition on mor-phology and photosyn-thesis	Lingakumar and Kulandaivelu (1993a)
	UV-B and Water Stress		
Sequential treatment	Soybean	Water-stressed seedlings show greater resistance to UV-B on growth and photosynthetic activity	Sullivan and Teramura (1990)
Combined treatment	Radish, cu-cumber	Accumulation of UV-absorbing compounds as a defense mechanism	Tevini et al. (1983)
Combined treatment	Soybean	Photosynthetic recovery is high after combined stresses	Teramura et al. (1983)
Combined treatment	Soybean	Water-stressed plants show thicker leaves and accumulation of flavo-noids	Murali and Tera-mura (1986)
	UV-B and Temperature		
High temperature and UV-B	Arabidopsis	Photolyase enzyme in photoreactivating UV-B damage on DNA, en-zyme is temperature sensitive at narrower temperature 22–33°C	Pang and Hays (1991)
Low and high temperature and UV-B	Cowpea	Induction of HSPs at 40°C in chloroplasts	Nedunchezhian et al. (1992)
Low and high temperature and UV-B	Cowpea	Increased stability of thyla-koid components to UV-B at high tempera-ture (40°C)	Nedunchezhian and Kulan-daivelu (1993)
High temperature and UV-B	Sunflower, corn	Accelerates plant devel-opment	Tevini and Mark (1993)

TABLE 2-1   *(Continued)*

Interaction Type	Plant Species	Responses	References
UV-B and Nutrients			
High P supply and UV-B	Lettuce	Additive effects on photosynthesis	Bogenreider and Doute (1982)
Low P supply and UV-B	Soybean	Less UV-B sensitivity than plants grown at optimum P levels	Murali and Teramura (1985a)
UV-B and Heavy Metals			
Zinc and UV-B	Cotton	UV-B inhibition of translocation of Zn^{2+} affecting plant growth	Ambler et al. (1975)
Iron and UV-B	Cotton	UV-B reduces Fe^{3+} to Fe^{2+} for easy access to plant growth	Pushnik et al. (1987)
Cadmium and UV-B	Spruce	Affects photosynthesis, energy distribution between PSII and PSI	Dubé and Bornman (1992)
UV-B and Elevated CO_2			
Moderately high CO_2 and UV-B	Wheat, rice	UV-B induces reductions in growth and yield at 650 ppm of CO_2	Teramura et al. (1990)
Elevated CO_2 and high UV-B	Rice: IR36, Fujiyama 5	Reductions in RuBP regeneration capacities with minor differences in cultivars	Ziska and Teramura (1992)
UV-B and Diseases			
High UV-B and fungal disease *(Cercospora beticola)*	Sugarbeet	Deleterious effects on growth	Panagopoulos et al. (1992)
UV-B and viral disease	Chenopodium	UV-B reduces the viral infection	Semeniuk and Goth (1980)
UV-B and fungal disease *(Colletotrichum laganarium) (Cladosporium cucumerinum)*	Cucumber	UV-B pretreatment enhances the severity of disease	Orth et al. (1990)

UV-B and Other Longer Wavelength Radiations

The sensitivity of higher plants to UV-B radiation has been shown to increase under low visible radiation conditions, such as those in growth cabinets or in greenhouses. Moderate levels of increased UV-B radiation were found to be ineffective on leaf expansion in plants grown under high PPFD. Likewise, plants irradiated with UV-B along with high PPFD were found to be more resistant than those irradiated under low PPFD's (Mirecki and Teramura, 1984).

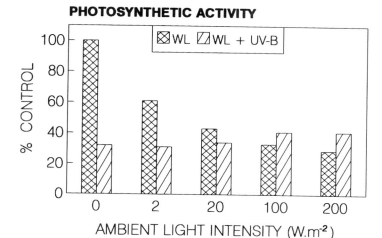

Figure 2-6. Changes in photosynthetic activity ($H_2O \rightarrow MV$) in isolated Vigna chloroplasts exposed to different white light (WL) irradiances in the absence (WL) and presence (WL + UV-B) of 5 W m^{-2} UV-B radiation.

High PPFD's affect plant sensitivity to UV-B by eliciting responses that provide UV-absorbing screens. An interaction of UV-B radiation with varying levels of PPFD was studied by Teramura (1986). The greater resistance to UV observed in plants preconditioned to high PPFD was suggested to be due to greater flavonoid concentrations and thick leaves with deeply placed sensitive organelles. Iwanzik (1986) reported that an interaction of UV-B, UV-A, and visible light resulted in partial photoreversal of some of the UV-B induced reductions in growth and photosynthesis. Fig. 2-6 shows the effect of different irradiances of supplemental white light on UV-B induced changes in photosynthetic acitivity of *Vigna* chloroplasts. UV-B irradiation (5 W m^{-2} for 60 min) in the absence of any ambient white light drastically inhibited the photosynthetic activity. Supplementation of white light reversed the UV-B changes. High irradiance (200 W m^{-2}) not only nullified the UV-B inhibition but also promoted the photosynthetic activities. It should, however, be noted that increasing the level of white light also gradually reduced the photosynthetic electron transport activity.

Water Stress and UV-B Responses

Most of the data obtained from field experiments suggest that environmental variables modify a plant's response to UV-B. One of the important environmental factors is water stress. Water stress in combination with enhanced UV-B (simulating a 12% O_3 depletion) adversely affected water loss in cucumber seedlings. A less pronounced effect was observed in radishes grown under both UV-B and water stress conditions (Tevini et al., 1983). Photosynthetic recovery after water stress was shown to be greater and more rapid in UV-B irradiated soybean plants. Since water stress produces effects similar to UV-B, the inhibitory effects of the latter are very often masked by the former (Caldwell et al., 1995).

Temperature and UV-B

In addition to water stress, an increase in ambient temperature also produces adverse effects in UV-B grown plants. High temperature regimes ameliorate UV-B effects, at least in some plant species (Teramura et al., 1991b). Tevini and Mark (1993) have found that UV-B induced changes on photosynthesis of seedlings grown at 28°C were generally overcome in plants grown under same UV-B exposure at 32°C. UV-B enhanced radiation caused a significant reduction in photosynthetic electron transport and RuBPcase activity at ambient temperatures in *Vigna* seedlings (Kulandaivelu and Nedunchezhian, 1993). Furthermore, an increased stability of thylakoid components to UV-B radiation in plants grown under elevated temperatures was shown to be due to the synthesis of new polypeptides (Nedunchezhian and Kulandaivelu, 1993). Moreover, induction of certain high and low molecular weight proteins under UV-B treatment resembled the induction patterns produced after heat shock treatments (Nedunchezhian et al., 1992). Chloroplasts isolated from plants grown at different temperatures, when exposed to UV-B radiation, demonstrated variations in their sensitivity. Chloroplasts from plants grown at high temperatures showed increased stability to UV-B compared to those from plants grown at ambient temperature (Fig. 2-7). Thus it appears that, to a certain extent, UV-B radiation mimics the response of heat shock, at least at the organelle level.

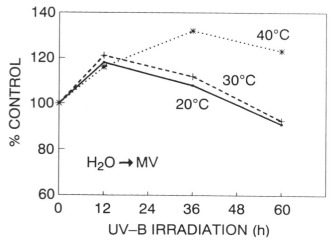

Figure 2-7. Responses of chloroplasts to UV-B radiation isolated from Vigna seedlings grown under different temperatures.

UV-B and Herbicides

Though very few papers have come out on the interaction of UV-B and herbicides in higher plants, it should be studied in detail because both UV-B and herbicides have D_1/D_2 of the PSII complex as their target site. Renger et al. (1989) have investigated the modification of the D_1/D_2 complex under UV-B treatment. They have reported that UV-B markedly reduced the number of ^{14}C-atrazine binding sites on the D1/D2 complex. A similar observation was made by Kulandaivelu and Annamalainathan (1991) in wheat seedlings pretreated with DCMU.

UV-B and Disease Severity

Certain diseases become severe in plants exposed to enhanced UV-B radiation. This was shown in a study where sugar beet grown under elevated levels of UV-B radiation was infected with *Cercospora beticola,* which led to large reductions in leaf Chl content and dry weight of total biomass (Panagopoulos et al., 1992). Furthermore, the time of infection critically affected the severity of the disease under elevated levels of UV-B. Preinfection treatment with UV-B radiation was more successful in leading to the disease development than postinfection treatment (Orth et al., 1990).

UV-B and Nutrients

In addition to these stresses, shortage of nutrients in the field poses a great problem to plants undergoing UV-B stress. Bogenreider and Doute (1982) have reported that under conditions of low nutrient supply, the effects of UV-B are not very severe when compared to those under high or optimal nutrient conditions. A similar study was conducted by Murali and Teramura (1985a); soybean plants provided with low phosphorus were less sensitive to UV-B radiation than those grown under optimal phosphorous supply. Although UV-B radiation and low phosphorus supply had deleterious effects on various parameters, these effects were nonadditive (Murali and Teramura, 1985b). UV-B radiation has a direct effect on the uptake of nutrients through anion-cation balancing (Murali and Teramura 1985b). UV-B radiation has been shown to reduce the mobilization of essential trace element like Zn^{++} (Ambler et al., 1975). Plants exposed simultaneously to UV-B radiation and cadmium showed low photosynthetic activity and reduced Chl levels, compared to those exposed under either stress alone (Dubé and Bornman, 1992).

UV-B and Elevated CO_2

The present and future threat to the whole global community is atmospheric CO_2 enrichment and global warming. Current atmospheric levels of CO_2 may double, from 340 ppm to 700 ppm, by the middle of twenty-first century. Only a few studies have been made on the combined effects of increased CO_2 and UV-B radiation. At the present-day level of CO_2, UV-B caused more shoot production than roots, while the same dose induced more roots at an elevated CO_2 level. Although Rozema et al. (1990) found no interaction in the response of pea, tomato, and aster when exposed to enhanced UV-B and elevated CO_2, the interaction seems to be species specific. It appears that plants are most responsive to UV-B only when they are grown under favorable conditions of temperature, CO_2, and soil nutrients (Caldwell and Flint, 1994). Thus when the environmental parameters vary from time to time, it is impossible to fix one environmental variable and relate its effects over UV-B. A study involving multiple stress effects combined with elevated levels of UV-B would yield the actual result. Such a study would require a large experimental area.

REFERENCES

Allen, L. H., C. V. Vu, R. H. Berg, and L. A. Garrard (1978), in *UV-B biological and climatic effects research (BACER)*, Final Report, EPA-IAG-D6-0168, USDA-EPA, Washington, DC, p. 134.

Ambler, J. E., D. T. Krizek, and P. Semeniuk (1975), *Physiol. Plant., 34*, 177–181.

Ballaré, C. L., P. W. Barnes, and S. D. Flint (1995), *Physiol. Plant., 93*, 584–592.

Barnes, P. W., S. Maggard, S. R. Holman, and B. Vergara (1993), *Crop. Sci., 33*, 1041–1046.

Bartholic, J. F., L. H. Halsey, and L. A. Garrard (1975), in *Climatic impact assessment program (CIAP)*, Monograph 5 (D. S. Nachtway, M. M. Caldwell, and R. H. Biggs, Eds.), US Dept. of Transportation Report No. DOT-TST-75-55, Natl. Tech. Info. Serv., Springfield, VA, pp. 67–71.

Batschaeur, A. (1993), *Plant J., 4*, 705–709.

Beggs, C. J. and E. Wellmann (1994), in *Photomorphogenesis in plants* (R. E. Kendrick and G. H. M. Kronenberg, Eds.), Vol. 2, Kluwer Academic, Dordrecht, Netherlands, pp. 733–750.

Beggs, C. J., U. Schneider-Zeibert, and E. Wellmann (1986), in *Stratospheric ozone reduction, solar ultraviolet radiation and plant life* (R. C. Worrest and M. M. Caldwell, Eds.), NATO ASI Series, Vol. 8, Springer-Verlag, Berlin, pp. 235–240.

Bennett, J. H. (1981), *J. Environ. Qual., 10*, 271–275.

Berjak, P. (1974), *Proc. Electron Microsc. Soc. SA, 4*, 31–32.

Biggs, R. H. and S. V. Kossuth (1978), in *UV-B Biological and climatic effects research (BACER)*, Final Report, EPA-IAG-D6-0168, USDA-EPA, Washington, DC, p. 134.

Biggs, R. H., W. B. Sisson, and M. M. Caldwell (1975), in *Climatic impact assessment program* (CIAP), Monograph 5 (D. S. Nachtway, M. M. Caldwell, and R. H. Biggs, Eds.), US Dept. of Transportation Report No. DOT-TST-75-55, Natl. Tech. Info. Serv., Springfield, VA, pp. 434–450.

Biggs, R. H., S. V. Kossuth, and A. H. Teramura (1981), *Physiol. Plant., 53*, 19–26.

Bogenreider, A. and Doute, Y. (1982), in *Biological effects of UV-B radiation* (H. Bauer, M. M. Caldwell, M. Tevini, and R. C. Worrest, Eds.), Gesellschaft fuer Strahlen-und Umweltforschung mbH, Munich, Germany, pp. 164–168.

Bornman, J. F. (1986), *Photobiochem. Photobiophy., 11*, 9–17.

Bornman, J. F. (1989), *J. Photochem. Photobiol.*, *4*, 145–158.

Bornman, J. F., R. F. Evert, and R. J. Mierzwa (1983), *Protoplasma*, *117*, 7–16.

Brandle, J. R., W. F. Campbell, W. B. Sisson, and M. M. Caldwell (1977), *Plant Physiol.*, *60*, 165–169.

Britt, A. B., J. J. Chen, D. Wykoff, and D. Mitchell (1993), *Science*, *261*, 1571–1574.

Caldwell, M. M. (1971), in *Photophysiology*, Vol. VI (A. C. Giese, Ed.), Academic Press, New York, pp. 131–177.

Caldwell, M. M. and S. D. Flint (1994), *Climat. Change*, *28*, 375–394.

Caldwell, M. M., A. H. Teramura, M. Tevini, J. F. Bornman, L. O. Björn, and G. Kulandaivelu (1995), *Ambio.*, *24*, 166–173.

Campbell, W. F., M. M. Caldwell, and J. B. Sisson (1974), *Am. J. Bot.* (Suppl.), *61*, 27–30.

Chappel, J. and K. Hahlbrock (1984), *Nature*, *311*, 76–78.

Doughty, C. J. and A. B. Hope (1973), *J. Mem. Biol.*, *13*, 185–198.

Dubé, S. L. and J. F. Bornman (1992), *Plant Physiol. Biochem.*, *30*, 761–767.

Elstner, E. F. (1982), *Annu. Rev. Plant Physiol.*, *33*, 73–96.

Ensminger, P. A. and E. Schäfer (1992), *Photochem. Photobiol.*, *55*, 437–447.

Esser, G. (1980), Einfluss einer nach Schadstoffimmission vermehrten Einstrahlung Von UV-B Licht auf Kulturpflanzen, Z. Versuchjahr, Bericht Battele Institut E.V., Frankfurt, Germany.

Fernbach, E. and H. Mohr (1992), *Trees*, *6*, 232–235.

Garrard, L. A., T. K. Van, and A. H. West (1976), *Soil Crop Sci. Soc. Fla. Pro.*, *36*, 184–188.

Gausman, H. W., R. R. Rodriguez, and D. E. Escobar (1975), *Agron. J.*, *67*, 720–724.

Greenberg, B. M., V. Gaba, A. K. Mattoo, and M. Edelman (1989), *Z. Naturforsch.*, *44c*, 450–452.

Harlow, G. R., M. W. Jenkins, T. S. Pittalwale, and D. W. Mount (1994), *Plant Cell*, *6*, 227–235.

Iwanzik, W. (1986), in *Stratospheric ozone reduction, solar ultraviolet radiation and plant life* (R. C. Worrest and M. M. Caldwell, Eds.), NATO ASI Series, Vol. 8, Springer-Verlag, Berlin, pp. 287–301.

Iwanzik, W., M. Tevini, G. Dohnt, M. Voss, W. Weiss, O. Gräber, and G. Renger (1983), *Physiol. Plant*, *58*, 401–407.

Jordan, B. R., W. S. Chow, A. Strid, and M. J. Anderson (1991), *FEBS Lett.*, *234*, 5–8.

Jordan, B. R., J. He, W. S. Chow, and M. J. Anderson (1992), *Plant Cell Environ.*, *15*, 91–98.

Kramer, G. F. D., T. Krizek, and R. M. Mirecki (1992), *Phytochemistry*, *31*, 1119–1125.

Kulandaivelu, G. and K. Annamalainathan (1991), in *Global climatic changes on photosynthesis and plant productivity* (A. P. Abrol, Govindjee, P. N. Wattal, A. Gnanam, and A. H. Teramura, Eds.), Oxford & IBH Publishers, New Delhi, pp. 59–74.

Kulandaivelu, G. and G. Nedunchezhian (1993), *Photosynthetica*, *29*, 377–383.

Kulandaivelu, G. and A. M. Noorudeen (1983), *Physiol. Plant*, *58*, 389–394.

Kulandaivelu, G., S. Maragatham, and N. Nedunchezhian (1989), *Physiol. Plant*, *76*, 398–404.

Kulandaivelu, G., N. Nedunchezhian, and K. Annamalainathan (1991), *Photosynthetica, 25*, 333–339.

Li, J., T. M. Ou-Lee, R. Raba, R. G. Amundson, and R. L. Last (1993), *Plant Cell, 5*, 171–179.

Lingakumar, K. and G. Kulandaivelu (1993a), *Photosynthetica, 29*, 341–351.

Lingakumar, K. and G. Kulandaivelu (1993b), *Aust. J. Plant Physiol., 20*, 299–308.

Mantai, K. E., J. Wong, and N. I. Bishop (1970), *Biochim. Biophys. Acta, 197*, 257–266.

McClennan, A. G. (1987), *Mutat. Res., 181*, 1–7.

McCloud, E. S. and M. R. Berenbaum (1994), *J. Chem. Ecol., 20*, 525–539.

McClure, J. W. (1975), in *The flavonoids* (J. B. Harborne, T. J. Marby, and H. Mabry, Eds.), Vol. 2, Academic Press, New York, pp. 970–1055.

Middleton, E. M. and A. H. Teramura (1993), *Plant Physiol., 105*, 741–752.

Mirecki, R. M. and A. H. Teramura (1984), *Plant Physiol., 74*, 475–480.

Murali, N. S. and A. H. Teramura (1985a), *J. Plant Nutrit., 8*, 177–192.

Murali, N. S. and A. H. Teramura (1985b), *Physiol. Plant, 63*, 413–416.

Murali, N. S. and A. H. Teramura (1986), *Photochem. Photobiol., 44*, 215–219.

Murphy, T. M. (1983), *Physiol. Plant, 58*, 381–388.

Nedunchezhian, N. and G. Kulandaivelu (1991), *Photosynthetica, 25*, 431–435.

Nedunchezhian, N. and G. Kulandaivelu (1993), *Photosynthetica, 29*, 369–375.

Nedunchezhian, N., K. Annamalainathan, and G. Kulandaivelu (1992), *Physiol. Plant, 85*, 503–506.

Noorudeen, A. M. (1982), Ph.D. Thesis, Madurai Kamaraj University, Madurai, India.

Noorudeen, A. M. and G. Kulandaivelu (1982), *Physiol. Plant, 55*, 161–166.

Orth, A. B., A. H. Teramura, and H. D. Sisler (1990), *Am. J. Bot., 77*, 1188–1192.

Panagopoulos, I., J. F. Bornman, and L. O. Björn (1992), *Physiol. Plant, 34*, 177–181.

Pang, Q. and J. B. Hays (1991), *Plant Physiol., 95*, 536–543.

Pushnik, J. C., G. W. Miller, D. Von Jolley, J. C. Brown, T. D. Davis, and A. M. Barnes (1987), *J. Plant Nutrit., 19*, 2283–2297.

Quaite, F. E., B. M. Sutherland, and J. C. Sutherland (1992), *Nature, 358*, 576–578.

Rao, M. V. and D. P. Ormrod (1995), *Photochem. Photobiol., 61*, 71–78.

Renger, G., P. Graeber, G. Dohnt, R. Hagemann, W. Weiss, and R. Voss (1982), in *Biological effects of UV-B radiation* (H. Bauer, M. M. Caldwell, M. Tevini, and R. C. Worrest, Eds.), Geselleschaft fuer Strahlen and Umweltforschung mbH, Munich, Germany, pp. 100–116.

Renger, G., H. Voelker, J. Eckert, R. Fromme, S. Hohm-Veit, and P. Graeber (1989), *Photochem. Photobiol., 49*, 97–105.

Robberecht, R. and M. M. Caldwell (1978), *Oecologia, 32*, 277–287.

Ros, J. (1990), *Karlsr. Beitr. Entw. Okophysiol., 8*, 1–157.

Roschchupkin, D. I., A. B. Pelenisyn, A. Potapenko, V. V. Talitsky, and Y. A. Vladimirov (1975), *Photochem. Photobiol., 21*, 63–69.

Rozema, J., G. M. Lenssen, and J. W. M. Van de Staaij (1990), in *The greenhouse effect and primary productivity in European agroecosystems* (J. Gadriaan, H. Van Keulen, and H. H. Van Laar, Eds.), Pudoc, Wageningen, The Netherlands pp. 68–71.

Schulze-Lefert, P., J. L. Daugle, M. Becker-Andre, K. Hahlbrock, and W. Schulz (1989), *Embo. J.*, *8*, 651–656.

Semeniuk, P. and R. W. Goth (1980), *Environ. Exp. Bot.*, *20*, 95–98.

Setlow, R. B. (1974), *Proc. Natl. Acad. Sci.*, *71*, 3363–3366.

Sisson, W. B. and M. M. Caldwell (1976), *Plant Physiol.*, *58*, 563–568.

Sisson, W. B. and M. M. Caldwell (1977), *J. Exp. Bot.*, *28*, 691–705.

Skokut, T. A., J. H. Wu, and R. S. Daniel (1977), *Photochem. Photobiol.*, *25*, 109–118.

Stapleton, A. E., and V. Walbot (1994), *Plant Physiol.*, *105*, 881–889.

Steinmüller, D. (1986), *Karlsr. Beitr. Entw. Okophysiol.*, *6*, 1–174.

Steinmüller, D. and M. Tevini (1986), in *Stratospheric ozone reduction, solar ultraviolet radiation and plant life* (R. C. Worrest and M. M. Caldwell, Eds.), NATO ASI Series, Vol. 8, Springer-Verlag, Berlin, pp. 261–269.

Strid, A. and R. J. Porra (1992), *Plant Cell Physiol.*, *33*, 1015–1023.

Strid, A., W. S. Chow, and J. M. Anderson (1990), *Biochim. Biophys. Acta*, *1020*, 260–268.

Sullivan, J. H. and A. H. Teramura (1990), *Plant Physiol.*, *92*, 141–146.

Takeuchi, Y., H. Akizuki, H. Shimizu, N. Kondo, and J. K. Sugahara (1989), *Physiol. Plant*, *76*, 425–430.

Teramura, A. H. (1980), *Physiol. Plant*, *48*, 333–339.

Teramura, A. H. (1983), *Physiol. Plant*, *58*, 415–427.

Teramura, A. H. (1986), in *Stratospheric ozone reduction, solar ultraviolet radiation and plant life* (R. C. Worrest and M. M. Caldwell, Eds.), NATO ASI Series, Vol. 8, Springer-Verlag, Berlin, pp. 327–343.

Teramura, A. H. and M. C. Perry (1982), in *Biological effects of UV-B radiation* (H. Bauer, M. M. Caldwell, M. Tevini, and R. C. Worrest, Eds.), Geselleschaft fuer Strahlen and Umweltforschung mbH, Munich, Germany, pp. 192–202.

Teramura, A. H., R. H. Biggs, and S. Kossuth (1980), *Plant Physiol.*, *65*, 483–488.

Teramura, A. H., M. Tevini, and W. Iwanzik (1983), *Physiol. Plant*, *57*, 175–180.

Teramura, A. H., J. H. Sullivan, and L. H. Ziska (1990), *Plant Physiol.*, *93*, 470–475.

Teramura, A. H., L. H. Ziska, and A. E. Sztein (1991a), *Physiol. Plant*, *83*, 373–380.

Teramura, A. H., M. Tevini, J. F. Bornman, M. M. Caldwell, G. Kulandaivelu, and L. O. Björn (1991b), *UNEP Rep.*, pp. 25–32.

Tevini, M. and W. Iwanzik (1986), in *Stratospheric ozone reduction, solar ultraviolet radiation and plant life* (R. C. Worrest, and M. M. Caldwell, Eds.), NATO ASI Series, Vol. 8, Springer-Verlag, Berlin, pp. 271–285.

Tevini, M. and U. Mark (1993), in *Frontiers of photobiology, Proceedings of 11th international congress on photobiology* (A. Shima, M. Ichahashi, Y. Fujiwar, and H. Takebe, Eds.), Elsevier, Amsterdam, pp. 541–546.

Tevini, M. and D. Steinmüller (1987), *J. Plant Physiol.*, *13*, 111–121.

Tevini, M. and A. H. Teramura (1989), *Photochem. Photobiol.*, *50*, 479–487.

Tevini, M., W. Iwanzik, and U. Thoma (1981), *Planta*, *153*, 388–394.

Tevini, M., W. Iwanzik, and U. Thoma (1983), *Z. Pflanzenphysiol.*, *110*, 459–467.

Tevini, M., J. Braun, and G. Fieser (1991), *Photochem. Photobiol.*, *53*, 329–333.

Van, T. K., L. A. Garrard, and S. H. West (1977), *Environ. Exp. Bot.*, *17*, 107–112.

Vu, C. V., L. H. Allen, and L. A. Garrard (1978), *Soil Crop Sci. Soc. Fla. Proc.*, *38*, 59–63.

Vu, C. V., L. H. Allen, Jr., and L. A. Garrard (1982a), *Physiol. Plant.*, *55*, 11–16.

Vu, C. V., L. H. Allen, Jr., and L. A. Garrard (1982b), *Environ. Exp. Bot.*, *22*, 465–473.

Vu, C. V., L. H. Allen, and L. A. Garrard (1984), *Environ. Exp. Bot.*, *24*, 131–143.

Wellmann, E. (1983), in *Encyclopedia of Plant Physiology, New Series* (L. O. Lange, P. S. Novel, C. B. Osmond, and H. Ziegler, Eds.), Vol. 16, Springer-Verlag, Berlin, pp. 745–756.

Willekens, H., M. Van Montagu, D. Inze, C. Langebartels, and H. Sandermann, Jr. (1994), *Plant Physiol.*, *106*, 1007–1014.

Ziska, L. H. and A. H. Teramura (1992), *Plant Physiol.*, *84*, 269–276.

3

Chilling and Freezing

Louis-Philippe Vézina, Jean-Marc Ferullo, Gilles Laliberté,
Serge Laberge, and Claude Willemot

INTRODUCTION

Plants have colonized almost all of the earth's surface. They have developed adaptative traits that allow them to grow and reproduce in regions that are remarkably stressful. Low temperature occurs episodically or seasonally over approximately 90% of earth's dry land. Although equatorial biomes possess a wide variety of plant genera and species, and represent the cradle of evolution for land plants, the plants that feed the world today are mostly grown in regions where low temperature is an environmental constraint. Chilling and freezing resistance in plants has captured the interest of scientists for a long time. Extraordinary improvements have been achieved in the selection and breeding for cold resistance in crop plants like wheat, rye, and alfalfa. Most plant breeders now face the problem of genetic diversity in their effort to push back the limits imposed by low temperature. Plant physiology and plant molecular biology have brought a tremendous amount of information on the cellular mechanisms that seem to operate when a plant undergoes the transition between its cold-sensitive state and its cold-tolerant state to acquire protection against cold injury. How these mechanisms operate at the whole plant level in natural environmental conditions has rarely been addressed; neither has it been demonstrated that the capacity to perform any of these processes is an absolute requirement for tolerance. Plant ecology has provided models of the strategies adopted by plants to tolerate low temperatures in different climatic conditions, but these models or the data from which they are derived have sometimes brought confusion rather than enlightenment to researchers seeking more universally applicable tolerance concepts. In writing this chapter we have tried to look as objectively as possible at the numerous reports available in the literature, in an effort to draw a realistic picture of the means by which poikilothermic organisms like plants succeed in tolerating low temperature stress. The comprehensive study by Sakai and Larcher (1987) was extensively consulted by the authors for the sections dealing with biogeography; the readers should refer to this work for additional information. We made the effort to bring together apparently divergent approaches in the study of cold tolerance, and we emphasized, where pertinent, the need for more integrated research approaches.

BIOGEOGRAPHICAL AND EVOLUTIONARY ASPECTS OF LOW-TEMPERATURE TOLERANCE IN PLANTS

The Occurrence of Low Temperature

In January, mean sea level temperature of continental areas varies from 30°C at the Equator to −40°C in the Arctic. This uneven mean temperature distribution is a direct result of extreme differences in the incidence angle of sunlight as a function of latitude. Poles receive far less solar energy per unit area than equatorial regions. In addition, outside of the tropics of Cancer and Capricorn, temperature fluctuates seasonally in response to variations in day-length and sun elevation. Cold air masses will form as a result of radiation deficit in winter, and the atmospheric movement of these masses can provoke severe drops in temperature, which are called advective freezes. Although more frequent in winter, advective freezes can also occur during the warmer seasons, and thus affect plant productivity (Bowers, 1994). In exceptional instances, advective freezes will strike subtropical regions, where they can suddenly threaten the survival of whole plant populations (Yelenosky, 1985). Formation of cold air masses can also be the result of radiation heat loss at ground level. These radiation freezes are responsible for extreme drops in temperature during clear winter nights. In a given region, mean air temperature fluctuates with altitude (about 0.64°C per 100 m of altitude), and with distance from the coasts, since the land gains and loses heat more rapidly than the sea. Temperature variations are less pronounced in the western parts of continents, outside the tropics, because prevalent winds blow from west to east, and thus bring warm humid air to the western sides of continents.

In the oceans, surface temperature varies along a latitudinal gradient and follows seasonal variations in air temperature. In January, mean surface temperature of seawater is slightly under 0°C at the North Pole and around 30°C at the Equator. However, because only the upper part of the water column is heated, and because cold water sinks to the bottom due to its greater density, the temperature of approximately 75% of seawater is below 4°C (Ross, 1995).

Plant Distribution and the History of Temperature Changes in Recent Eras

The present global distribution of plants reflects both their evolutionary history and the effects of past and recent climatic conditions. The origin of the angiosperms has been traced as far back as the early Cretaceous, some 120 million years ago (Cox and Moore, 1993). Angiosperms were differentiated into modern families about 95 million years ago in the equatorial band (0–20° latitude). Although the description of the world's past climate regimes and their effect on plant distribution goes beyond the scope of the present chapter, it is noteworthy to acknowledge the influence of the dramatic climatic fluctuations observed through earth's history.

Two recent lines of evidence now demonstrate that green algae from the Charophyceae are on the phylogenic branch leading to land plants. This has been demonstrated through the comparison of chloroplastic DNA sequences encoding for two tRNA species (tRNA^{Ala} and tRNA^{Ile}) in land plants and in eubacteria and algae. It was found that this DNA fragment is characteristically interrupted by introns in land plants, while it is uninterrupted in most green algae. Distribution of these introns in members of the Charophyceae confirms that these taxa are part of the lineage that gave rise to land plants (Manhart and Palmer, 1990). Identical conclusions were drawn from molecular comparisons of the nuclear-encoded small subunit of ribosomal RNA throughout members of several algae and land plants genera (Kranz et al., 1995).

In the late Mesozoic and early Cenozoic (65–100 million years ago), forests extended close to the poles, and latitudinal changes in mean temperature were far less pronounced than they are today (Cox and Moore 1993; Huggett 1991). This period of global warmth was followed by a gradual cooling interspersed with periodic rewarming, which increased seasonality of the climates, especially in the higher latitudes (Fig. 3-1). Numerous data indicate that the adaptation of plants to cooling and seasonality was relatively rapid; for example, in North America, following a brief low temperature incursion at the Cretaceous–Tertiary boundary, mesothermal evergreens, which prefer a mean annual temperature above 20°C, disappeared, leaving a niche that was filled with deciduous vegetation apparently more tolerant to low temperature (Wolfe and Upchurch, 1986). More recently, plant distribution was influenced by periods of drastic climatic cooling, which resulted in extensive freezing and glacier formation during the Quaternary era. Four major glacier advances and withdrawals occurred in North America during the Pleistocene epoch. The most recent glaciation, which ended approximately 18 000 years ago, covered most of Canada and regions of north and central United States with ice, snow, and glaciers. These ice ages eradicated virtually all tropical and subtropical plants from North America and Eurasia. Upon rewarming of the climate, many cold-adapted species disappeared from regions of lower latitudes and only survived there in the coldest areas, most often at high altitudes. The Norwegian mugwort, *Artemisia norvegica,* an alpine plant now found in the Ural Mountains and in two isolated localities in Scotland, is an example of such climatic relics (Cox and Moore, 1993).

While the limits of plant expansion in higher latitudes and altitudes are mostly controlled by temperature, their expansion in lower latitudes and altitudes is limited by other factors. Woodward (1988) suggested that the energy put into acclimation might be at the expense of competitive energy. The potential for expansion of a plant into areas of higher latitudes is determined by the sensitivity of each step of

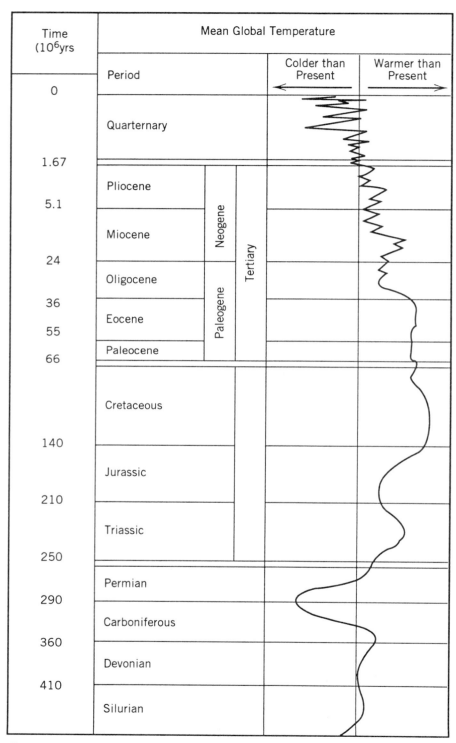

Figure 3-1. Mean temperature variations over earth's history. (From R. A. Spicer and J. L. Chapman, Trends Eco., 5, Sept. 1990, 279).

its life cycle (germination, establishment, reproduction, dispersal) to low temperature, and its capacity to complete its cycle in a short growing season. For example, it has been demonstrated that the distribution limits of the perennial herb *Verbena officinalis* are due to its limiting capacity to flower when mean air temperature is below 16°C (Woodward, 1990).

The Effects of Low Temperature on Plants

Plants face three major problems when exposed to low temperature: an alteration in the spatial organization of cell membranes, a slowing down of their chemical and biochemical reactions, and under freezing conditions, changes in water status and availability.

Plants are poikilotherms; that is, their temperature is variable and follows that of the surrounding environment. However, plant temperature may deviate substantially from the mean temperature of the surrounding area, and accurate estimation of plant temperature can only be achieved by direct measurements or by calculations of energy budgets. Such calculations are based on the simple assumption that the energy "in" minus the energy "out" is equal to the stored energy, and involve six variable parameters, namely shortwave and longwave radiations, heat conduction and convection, latent heat, and heat storage (Nobel, 1988). Unless these measurements or calculations are made, it is impossible to accurately determine the temperature stress experienced by plants or plant parts. An additional difficulty in measuring plant temperature is that not all plant parts experience the same temperature stress at a given moment.

Some plants or parts of plants will be injured by exposure to cool above-zero temperatures, and these are described as chilling sensitive. Others are chilling survivors but may be injured when exposed to freezing temperatures; they are freezing sensitive. Some plants will withstand extremely harsh low-temperature treatments, both in natural environments and in controlled conditions; they are freezing survivors. All sorts of nuances exist within this rough classification, as will be described below, but by studying members of each of these broad categories, one can envision the concept of low-temperature tolerance in its whole range.

Current Biome Boundaries Set by Low Temperature

Climatic regions are described by a series of parameters including mean temperature, rainfall, prevailing wind conditions, and humidity. However, several studies have suggested that the main limitation for plant life in a given region is the minimum temperature (Parker, 1963; Prentice et al., 1992; Sakai and Larcher, 1987; Sakai and Weiser, 1973; Woodward, 1987, 1988). Based on these observations, Woodward (1987) was the first to develop a prediction model for geographical distribution of major types of vegetation based on minimum temperatures. Woodward's biome boundaries are based on the approximate point of failure of each strategy used for cold tolerance; they are summarized in Table 3-1.

It is implicit throughout this chapter that biome boundaries are not set by temper-

TABLE 3-1   Minimum Temperature Encountered and Expected Physiognomy

Expected Physiognomy	Minimum Temperature	Phenomenon
Broad-leaved raingreen	0 to 10	Chilling sensitive
Broad-leaved evergreen	0	Frost sensitive
Broad-leaved evergreen	−15	Frost resistant, supercooling
Broad-leaved summergreen	−40	Frost resistant, deep supercooling
Broad-leaved summergreen	No limit	Dehydration resistant
Needle-leaved evergreen	−15	Frost resistant, supercooling
Needle-leaved evergreen	−45	Frost resistant, deep supercooling
Needle-leaved evergreen	−60	Dehydration resistant
Needle-leaved summergreen	No limit	Dehydration resistant

Source: From Woodward, 1987; Prentice et al., 1992.

ature alone. How different environmental parameters act together with low temperature to limit a habitat is hardly predictable; the best example to illustrate this is the timberline. The timberline, either in alpine or arctic areas, where continuous forest gradually gives way to treeless vegetation, is probably the easiest biome boundary to recognize. The timberline is characterized by low temperature, high winds, great accumulation of snow (at alpine timberline), and surfacing of permafrost (at the arctic timberline). Although it forms a sharp temperature-dependent boundary, roughly corresponding to the 10°C isotherm during the warmest month, low temperature minimum per se is not the primary factor controlling the limit of tree growth. Instead, this limit is correlated with degree-days over 5°C (Prentice et al., 1992). The mechanism behind this requirement for heat by arborescent species is still a matter of debate, but it appears that the conditions required for cellulose synthesis and deposition are not met beyond the timberline.

LOW-TEMPERATURE OCCURRENCE AND TOLERANCE STRATEGIES IN PLANTS

The Humid Lowland Regions of the Tropics

In the humid lowlands of northern South America, central Africa, and southeast Asia, the temperature rarely drops below 15°C and seasonal temperature fluctuations are minimal. These regions are the first regions where plants emerged as land organisms in earth history. These plants did not need to develop tolerance to low temperature, and therefore, most plants that thrive in these regions exhibit a strong sensitivity to chilling temperatures (+12 to +15°C), like the marine algae from which they have evolved (Biebl, 1964). Typically, the vegetation is dominated by broad-leaved evergreens such as the evergreen trees of the genus *Guarea* and members of the Meliaceae, shrubs, lianas, and herbaceous forest undergrowth of the genera *Anthurium* and *Episcia,* by members of the Araceae and Gesneriaceae, and by palms (Fig. 3-2).

When exposed to low above-zero temperatures for various periods of time, these

Figure 3-2. *Evergreen tropical forest (South America, Argentina). (Courtesy of M. M. Grantner.)*

plants will suffer damages that may become readily apparent, or that will become apparent upon returning to normal growth conditions. They are, but for a few exceptions, chilling sensitive.

Very little information is available on the chilling sensitivity of tropical plants in their natural environment. However, sensitivity to chilling has become a key factor for marketing and agronomic exploitation of tropical plants outside of their original ecological niche. Many ornamental species from the tropics are shipped or stored at low temperature, and many tropical crops are grown in temperate regions. The disorder caused by chilling temperatures on these plants has been thoroughly reviewed in the volume *Chilling Injury of Horticultural Crops,* edited by C. Y. Wang (1990). In these plants, chilling sensitivity will develop into symptoms if the low-temperature stress imposed is beyond a certain threshold. In most cases, both duration and temperature are important parameters of the stress.

It is generally recognized that the deleterious action of chilling is a two-phase phenomenon and that the cellular membrane is the primary site of damage. First, the cellular membranes of sensitive cells will undergo a phase transition, a process through which the functionality of the membrane is temporarily or permanently impaired (Raison and Orr, 1990). Membrane phase transition is a phenomenon by which the lipid bilayer goes from a liquid-crystalline state to a gel-crystalline state. The concept of phase transition and its molecular bases will be dealt with in more details in the second half of this chapter. When the phase transition has occurred to the gel-crystalline state, the membrane is leaky and compartmentation of cellular content is reduced. Membrane-bound enzymes that need a definite steric environment to perform their catalytic activity will also be affected by changes in membrane physical characteristics.

Typically, tropical plants have membranes with a phase transition temperature as high as $+15°C$; that is, their membrane lipids have a higher degree of saturation than those of plants from temperate climates. They will thus experience physical damage at the very first decrease in ambient temperature. This generally triggers changes in vital metabolic functions, which represents a shift to a second phase of damage. This sequence of events is reversible when the low-temperature stress remains below a plant-specific threshold. It leads to irreversible alterations of cell functions when the stress goes beyond this limit. Not all parts of a plant are equally susceptible to low temperature, nor are plants at all development stages; for example, pollen grain formation in *Anthurium* sp. is disrupted by milder temperatures than that causing lesions in the leaves (Patterson and Reid, 1990).

Chilling-sensitive species from the tropics are restricted in their range of habitat because they did not evolve in environments where chilling is a seasonally recurring event and they do not possess the genetic characteristics to develop less chilling-sensitive ecotypes (Sakai and Larcher, 1987). Because they are never exposed to seasonal variations in temperatures, they do not have the capacity to go through adaptative changes that could prevent subsequent injuries; for example, they do not have the capacity to lower the stress threshold. This inability to acclimate when exposed to low noninjurious temperature stresses has been exemplified many times under controlled conditions (Wilson, 1979).

Similarly, tropical seaweeds are extremely sensitive to low temperature. For example, the seaweeds *Botryocladia spinulifera* and *Haloplegma duperreyi*, which grow in the tropical western Atlantic, and whose distribution is limited by a 20°C seawater isotherm, show extensive injuries after 2 weeks at 18°C, and die after 2 weeks at 15°C (Pakker et al., 1995).

The Subtropical and Warm-Temperate Regions

In most of the areas between the tropics of Cancer and Capricorn, South and Central America, Africa, India, southeast Asia, and the northern and western coasts of Australia, temperature and precipitation fluctuate seasonally and the life-forms that colonize these regions must endure prolonged exposure to low above-zero temperatures and drought. The vegetation there is varied: some broad-leaved evergreen trees and evergreen herbaceous species in the forests (Fig. 3-3*a*), C_4 grasses, deciduous trees, hemicryptophytes, and palms in regions with a dry season (Fig. 3-3,*b*). Adaptation to low temperature has also evolved in an altitudinal series of vegetation. The best examples of plant groups that have developed ecotypes spreading from the tropics through regions of seasonal temperature variations are the orchids, the palms, and the mangroves. Examples of genera that comprise ecotypes spreading through altitudinal series are *Lycopersicon* and *Passiflora* (Patterson and Reid, 1990).

These frost-free regions represent the limit of sustainable life for a large number of genera. The plants in these regions are all resistant to chilling, although to different degrees. Phylogeneticists believe that during evolution, the ability to acclimate came with the endogenous ability of a species to alternate between periods of active growth and periods of quiescence (Sakai and Larcher, 1987). Although the molecular basis for acclimation to chilling is probably different from the basis for acclimation to freezing or drought, it is believed that the ability to cease or refrain from growth is a prerequisite for the colonization of regions where conditions can be temporarily unfavorable.

As mentioned in the preceding section on tropical plants, it appears that adaptation to life under chilling environments required lowering of the critical phase transition temperature of the cell membrane. Confidence in this concept has come from extensive comparisons between membrane lipid composition, phase transition temperature and resistance to chilling of various plants. These studies show that a close correlation exists between the relative abundance of high-melting point phosphatidylglycerol species in the membrane and the threshold temperature of chilling sensitivity (Kenrick and Bishop, 1986; Murata, 1983; Roughan, 1985). However logical this theory might seem, some exceptions exist, and experimental data gathered through the use of an *Arabidopsis* mutant demonstrate that the phase transition temperature of membrane systems is not the only parameter involved in chilling tolerance (Wu and Browse, 1995). Recent studies have demonstrated that the ability to repair damages resulting from exposure to an injurious temperature stress might be as important as the ability to avoid damages (Moon et al., 1995).

In orchids chilling resistance is inherited as a dominant trait through crosses between tolerant and sensitive ecotypes (Patterson and Reid, 1990).

The main limitations to the spreading of these species at higher altitudes or latitudes appear to be the occurrence of episodic or seasonal frost. Resistance to freezing stress seem to be fundamentally different from resistance to chilling stress, and thus not an extension of the strategies required to acclimate to low above-zero temperatures. Although light frosts rarely bring plant water to the freezing point,

Figure 3-3. (a) Subtropical evergreen forest (Asia, Japan). (b) Savanna woodland (Polynesia, Australia). (Courtesy of M. M. Grantner.)

this 0°C limit, that is, the freezing of external water, seems to put an end to the physiological tolerance established through time by many genera.

When tropical plants are exposed to a deep frost, injury will often occur following the formation of extracellular ice in the tissues and overdehydration of the cells. In extreme conditions, ice crystals will also grow through the extracellular space, leading to a maceration of the tissue. It appears that even in mild frost conditions, palms and other tropical plants will never recover from the injuries caused by cell dehydration or membrane dysfunction (Sakai and Larcher, 1987).

The Humid and Semiarid Regions with Mild Winters

Outside the intertropical band, episodic frosts are a seasonal recurrence. Two fairly different types of climates impose this kind of stress on living organisms: the semiarid climates of the Mediterranean type and the coastal humid type. The semiarid regions are more or less confined to a relatively narrow subtropical band between the 30th and 50th parallels, while the humid regions with mild winters spread to higher latitudes on western oceanic coasts such as in California, Chile, and the European Atlantic coast (Fig. 3-4). Winter mean temperatures in these regions are between 0 and 8°C, with mean minima between −5 and 3°C. Episodic frosts also occur in the mountainous regions during clear winter and spring nights, when intense radiative cooling and movement of cold air masses combine to increase the negative temperature balance. Freezing temperatures can then reach −10 to −15°C, and even occasionally −20 to −25°C. However, this temperature stress is restricted

Figure 3-4. *Seasonal rainforest (North America, United States). (Courtesy of D. Gagnon.)*

in time and rarely lasts more than a few hours. All the plants living under these climatic conditions are survivors of chilling environments (Bramlage and Meir, 1990). It seems that these plant genera must have first developed chilling resistance to expand from their initial range in tropical lowlands before they spread into regions where seasonal chilling occurs together with recurrent frost. However, as noted later in this chapter, some frost-resistant plants have apparently lost the ability to withstand prolonged exposure to low above-zero temperatures.

In the temperate rainforest of the Southern Hemisphere the mean winter temperature is between 5 and 12°C. These forests stand south of the 53rd parallel in South America and south of the 45th parallel in Australia and New Zealand. The vegetation in these regions is typically composed of conifers and broad-leaved evergreen arborescent plants of the genera *Podocarpus, Fitzroya,* several species of *Dacrydium, Nothofagus,* and in Australia, *Eucalyptus.* These species can withstand sudden temperature drops between −8 and −15°C, down to −20°C for species that grow in higher altitude like *Podocarpus nivali, Nothofagus solandri,* and *Eucalyptus gunii* (Sakai et al., 1981).

Between the 30th and the 45th parallel, on the western side of continents, a Mediterranean climate prevails, with its hot, dry summers and short, mild winters. Because drought is a major limiting factor for growth in spring and summer, these plants have their active growth period during the winter and remain quiescent during the summer. Thus in the Mediterranean climate, frost spells occur when the plants are in active growth (maquis, garrigue). The Mediterranean flora is mainly represented by shrubs, geophytes, and annual herbs. Xeromorphism is also a general feature of the species there; leaves with little ratio of surface to volume, sclerenchymatous deposit in leaves, thick cuticles, and pubescence. Among the few trees that can whitstand these conditions, *Olea europea* and *Quercus ilex* are the most representative species (Flahaut, 1937). *Q. ilex,* as well as *Phillirea latifolia, Cupressus sempervirens,* and some *Erica* species, can withstand momentary drops of temperature down to −15 to −20°C and are by far the most tolerant species. *O. europea,* like most other Mediterranean plants including *Pinus pinea, Laurus nobilis, Q. coccifera, Q. suber,* and *Pistacia lenticscus,* will suffer damage from episodic frosts around −10°C (Larcher, 1970).

Altitudes between 3000 and 4500 m in equatorial South America and east Africa will create conditions of frost occurrence similar to those of the Mediterranean climate. This is the place were caulescent genera such as *Espeletia, Dendrosenecio,* and *Lobelia* dominate the surrounding vegetation by their stature. These giant rosette plants stand several meters tall and can withstand night frosts down to −13°C at any time of the year (Beck, 1986).

Genuine freezing tolerance is generally lacking within plants from these regions. Some of them succeed in surviving frost spells by avoiding temperature drops within their tissue. The means by which they accomplish this avoidance are varied, and they are commonly designated "frost-mitigation mechanisms." The underground organs, the parts covered by humus, the inner tissue of plants with a thick bark, and the inner tissue of cushion plants are all shielded from transient cooling of ambient air by thermal inertia (Hedberg and Hedberg, 1979). In giant arborescent rosettes, this inertia is amplified by the accumulation of dead leaves that remain attached to the part of the stem most sensitive to freezing. Pubescent organs are transiently protected from temperature drops by a layer of immobilized air, and reproductive organs in many species are also thermally insulated by the nyctinastic movement of leaves or flower parts (Enright, 1982; Schwintzer, 1971).

Thermal inertia of large organs also comes from the increased capacity to store calories during the day (Mooney et al. 1977; Nobel, 1980; Ruthsatz, 1978). Plants bearing aerial parts that are close to the ground benefit from the release of ground heat during the night (Clements and Ludlow, 1977). Finally, in some species, heat is generated internally by overactive metabolic pathways during temperature drops. In some cases, this mechanism is so efficient that the plant temperature will remain far above zero when ambient air temperature falls to $-8°C$ (Knutson, 1974; Meeuse, 1975; Rees et al., 1981).

Some plants will also lower the temperature at which tissue water will freeze during an acclimation phase; this is accomplished either by accumulating metabolites that will lower the freezing point or through the supercooling of water. Accumulation of osmotically active compounds like sugars or polyols will contribute to a $3-5°C$ decrease in the freezing point of the tissue water (Alden and Hermann, 1971; Levitt, 1980; Parker, 1963; Siminovitch, 1981). However, the accumulation of such compounds will also increase the osmolarity of the cytosol and affect the biological activity of compounds within the cell. Plants or tissues that accumulate osmolytes to avoid freezing must therefore endure a transient water stress.

When frost is so intense that mitigation fails to keep the plant tissue above the freezing point, a supplementary delay in freezing is provided in some cases by supercooling (Marcellos and Single, 1979). Supercooling is an unstable, temporary state during which solutions are maintained as liquids, below their theoretical crystallization point. Morphological and cytological adaptations, such as narrow extracellular spaces or reduced size of mesophyll cells, seem to favor supercooling of the water in certain species (Sakai and Larcher, 1987). It has been demonstrated that in *Olea europea* and in species from genera *Espeletia, Polylepsis,* and bamboos, leaf water can remain in a supercooled state at temperatures of -6 to $-12°C$ (Ishikawa, 1984; Larcher, 1963, 1970, 1975; Larcher and Wagner, 1976; Larcher and Winter, 1981).

Whether all these plants have to go through acclimation before they exhibit frost avoidance is debatable. Examples of plants that go through acclimation include *Olea europea,* where the freezing point drops to $-5°C$ during acclimation, and *Polylepsis sericea,* in which the supercooling point of water drops from -9 to $-20°C$ following a prolonged exposure to $0°C$ (Larcher, 1963, Rada et al., 1985). In some species like the giant arborescent rosettes, the ability to withstand transient drops of temperature is equal throughout the year.

The mechanisms of freezing mitigation and freezing avoidance allow these organisms to remain in a state of thermodynamic imbalance. Unless this state is maintained by an expensive source of heat, as in homeothermic animals, freezing of tissue water will inevitably occur if the temperature stress lasts for more than a few hours per day. In order to survive to more prolonged periods of low temperature, plants had to develop true frost tolerance strategies, that is, the ability to withstand ice formation inside tissues.

The Cool Temperate Climates

In the mean latitudes, seasonal variations in temperature increase in amplitude, getting extreme in continental areas. What is called the temperate climate zone is a mixture of several climatic conditions ranging from humid oceanic climates to dry-lands (Fig. 3-5). The typical vegetation types found in these regions are the prairies with an abundance of perennial grasses, the steppes with an abundance of shrubs, the temperate deciduous forests, the wetlands, and, in the Northern Hemisphere, the mixed forest (deciduous to boreal conifers). At high altitudes the deciduous forest is also replaced by conifers, which are more adapted to short growing seasons, high wind velocity, recurrent soil moisture stress, and low nutrient availability (Barnes, 1991).

Most of the plants that live within zones of temperate climates have favored means by which they can avoid exposing the organs that are the most active for growth and reproduction to deep freezes. Geophylls, for example, have a prostrate or creeping type of growth, which allows them to benefit from snow cover, thus alleviating the mixed effect of low temperature extremes and wind. Perennial herbs will lose all of their aerial parts and leave only roots and crowns for overwintering; as for geophylls, the actual stress experienced by these remaining organs is far less

Figure 3-5. *Temperate deciduous and evergreens (North America, Canada). (Courtesy of M. M. Grantner.)*

than that of exposed organs. Annuals will overwinter by leaving only the dehydrated embryo-bearing seeds on the ground. These seeds will remain dehydrated, although surrounded by moisture, until favorable germinating conditions are encountered and the integuments allow imbibition to proceed. Trees have developed a different strategy to protect their actively growing parts and reproductive organs from freezing; for most angiosperm trees it seems that leaf shedding and formation of an overwintering bud is a successful adaptive trend. Numerous studies suggest that in the Cretaceous, leaf abscission was a strategy adapted by most plants in response to periods of severe drought, and became an advantageous characteristic with the advent of seasonal frost as the latitudinal range in temperature increased (Barnes, 1991; Spicer, 1989). Whatever environmental pressures led to the development of deciduousness, it is one of the most remarkable adaptations of plants to low temperature.

Most but not all of the plants that live in cool temperate climates undergo adaptive changes at the onset of the cool seasons. Some of these changes are obvious for nonoverwintering organs, like the leaves of deciduous trees, but others are more subtle, and they allow overwintering plant parts to go from a frost-sensitive to a frost-tolerant state. Cold acclimation is a broad concept that describes the process by which an organism acquires cold tolerance. By extension, it refers to the cellular, physiological, morphological and biochemical events that are essential for the development of cold tolerance. In deciduous trees, leaf abscission is but one of these changes. However, in these organisms, as in most plants living in the cool temperate climates, cold acclimation must succeed in providing the overwintering plant parts with protection against prolonged exposure to frost.

In many plants, cold acclimation occurs as a two-step process. The first step occurs in the fall in response to changing day-length and recurrent drops to chilling temperatures during the night. It is associated with important changes in the concentration of amino acids, carbohydrates, nucleic acids, lipids, and plant hormones (Li, 1984). It is also associated with changes in gene expression (Castonguay et al., 1993; Guy et al., 1985; Johnson-Flanagan and Singh, 1987; Mohapatra et al., 1987), isozymic patterns, and enzyme activity (Levitt, 1980). The extent of these changes, their physiological significance, and the genetic mechanisms underlying these biochemical modifications are dealt with in the following section of this chapter. Whether cold acclimation can be dissociated from dormancy is not clear in most cases studied up to now. However, it is clear that most of the processes associated with cold acclimation generally occur when plants go into dormancy, independently of temperature shifts. Only in rare instances has dormancy been separated from cold acclimation (Heichel and Henjum, 1990). It is generally accepted that neither cold temperature nor decreasing day-length alone will allow full acclimation (McKersie and Leshem, 1994).

What is known of the second step of cold acclimation is far less documented, but it has been clearly demonstrated that exposure to below-freezing temperatures will increase tolerance in many cold-tolerant species (Olien and Lester, 1985). This stage of acclimation has also been associated with changes in gene expression (Castonguay et al., 1993). Since the buildup of extracellular ice results in a withdrawal of cell water, it is believed that cold acclimation will also bring most cells to a state where they can withstand prolonged dehydration, and rehydrate without damage upon rewarming. However, many types of cells will not endure dehydration. This is true for ray parenchyma cells and for flower bud cells in some deciduous trees. In tissues where these types of cells are abundant, extracellular ice formation is a rare event, and cells will not usually lose their water to the extracellular matrix. Rather, the water will undergo deep supercooling into a more stable state than the transitory supercooled state of water in plants like *Olea* sp. The probability of ice nucleation increases with decreasing temperature, and deep supercooled water will rarely be stable under −40°C, which is the homogeneous ice nucleation point of pure water (Ashworth, 1992). In nature, where ice nucleation is heterogeneous, that is, initiated by surfaces or particles foreign to water, it usually happens at much warmer temperatures.

Deep supercooling has been reported in more than 240 species belonging to 33 families in the Angiospermae, and one in the Gymnospermae. In these plants water in the xylem parenchyma and in the buds can remain supercooled for months, unless the drops in temperature exceed the plant's limits. There is a strong correlation among trees between ice nucleation point and geographical distribution. Trees with ice nucleation point at −40°C will rarely be found in regions where the mean minimum temperature is below −40°C (Quamme, 1985). Nucleation avoidance is usually acquired through acclimation; thus the timing of frost spells is crucial for most species. Freezing of deep supercooled water releases heat of fusion, which can be detected by thermal analysis and is observed as a low-temperature exotherm (LTE) (Quamme, 1991). Recent data obtained through comparisons between true hardiness and LTEs determined that in natural hardening conditions the LTE varies seasonally, but that lethal temperatures are often milder than a LTE for a given species (Lindstrom et al., 1995). Thus in species that exhibit deep supercooling, factors other than the freeze-avoidance ability of xylem are likely to limit survival at temperatures above the LTE. How water is maintained in a supercooled state in a certain tissue, and not in adjacent tissues, is still unclear (Ishikawa and Sakai, 1985; Quamme, 1995). It has been proposed that discontinuities in the vascular system might act as a physical barrier to ice propagation from one tissue to another (Ashworth et al 1989, 1992; Kader and Proebsting, 1992; Quamme et al., 1995). It has been demonstrated, however, that restriction to ice propagation does not proceed through the same strategy in all cases (Flinn and Ashworth, 1994).

Acclimation does not bring all plants into a state of true dormancy. For trees living in the milder parts of the temperate climate zones, and for many herbaceous plants that loose their aerial parts during acclimation, dehardening will occur in a few days upon return to normal growth conditions. Likewise, hardening can occur over a relatively short period for some species.

Because relatively mild below-zero temperatures can be lethal even for the more hardy species in the nonacclimated state, timing of acclimation and deacclimation is crucial for plant survival in a given area, sometimes independent of the tolerance level to be acquired. A good example was provided in a study performed by Weiser

(1970). Northern ecotypes of red osier dogwood *(Cornus stolonifera)* went into acclimation earlier than more southern ecotypes, even if they were planted in southern latitudes. In this study, it was demonstrated that both ecotypes eventually reached the same level of resistance, but that inadequate timing in the acclimation resulted in the death of plants from the southern ecotype when they were displaced at northern latitudes.

The Subarctic and Subalpine Climates

North of the 50th parallel in North America, and north of the 60th parallel in Europe and Asia, the climatic conditions become harsh for plant life. In these regions the growing season gets shorter and is often restricted to two months in midsummer, day-length variations are extreme, and mean minimum temperatures in winter are typically below −40°C. The same conditions are found at high elevations in most continents. Vegetation formations in these climates are the boreal forests (taïga), the forest-tundra (Fig. 3-6a), which is a transition between tundra and the true boreal forest, and the cordilleran or mountain vegetation (Scott, 1995).

All plants that use deep supercooling as an avoidance mechanism for low-temperature survival cannot live in these regions. Typical trees found under such inhospitable climatic conditions are aspens, alders, poplars, birches, larches, and pines (Fig. 3-6b). Most of them can survive frost spells at −80°C.

North American boreal forests are dominated by trees from the genera *Picea* (spruce), *Pinus* (pine), and *Abies* (fir), by shrubs of the genera *Betula, Salix, Populus, Ledum,* and *Vaccinium,* and by mosses and lichens (Scott, 1995). Numerous conifers of the Taxodiaceae were abundant in these latitudes during the Cretaceous, but they retreated southwards during global cooling. At the same time, conifers of the Pinaceae became dominant, due to their hardiness and ability to survive long winters and short growing seasons (Sakai and Larcher, 1987). The shoot primordia of these extremely hardy conifers avoid freezing injury even at −70°C, by tolerating dehydration following extracellular ice formation.

A trait peculiar to plants found in cold mountainous regions areas is the ability to respond rapidly to changing environmental conditions. The variability of microclimatic conditions is greater at higher altitudes than in geographic areas with fewer natural disturbances, and these have a drastic effect on plant distribution. For example (extremely) hardy species from genera *Loiseleuria, Silene,* and *Saxifraga* are found in wind-exposed sites and they do not benefit from snow cover. In contrast, rhododendron will live at the same altitude, but in the few areas were snow accumulates. Bryophytes and some lichens benefit from ground heat release so that their temperature usually exceeds air temperature. Furthermore, many bryophytes and lichens lack an effective cuticle and are poikilohydric. They are used to extreme changes in water availability. When dry, they become inactive, but they will resume normal growth upon rehydration. This ability to withstand recurrent dryness is probably the means by which they also withstand loss of cellular water during extracellular ice formation (Longton, 1992). Mosses are also peculiar among plants in regards

Figure 3-6. (a) *Subalpine forest (North America, Canada). (Courtesy of M. M. Grantner.)* (b) *Timberline (North America, Alaska). (From* Physical Geography of the global environment, *1st ed., H. J. de Blij and O. Muller, Eds., Wiley, New York, 1993, p. 231.)*

to cold tolerance in that within a given species, there is no significant differences in cold-tolerance level between ecotypes found in northern or southern latitudes (Biebl, 1964).

The Arctic and Alpine Climates

In arctic regions, the soil remains frozen throughout the year and in restricted areas its surface will thaw during summer. The growing season is so short that it would be more accurate to say that plants there are adapted to episodic thaw rather than to episodic frost. In the northern latitudes where plants can still be found, only four months of the year have a mean temperature above 0°C, and none above 10°C (Scott, 1995). The Antarctic is a polar desert, with mean temperatures around −30°C near the coast, and low continental annual precipitation.

Plants that live in the northern latitudes are of the tundra type, which is characterized by the absence of trees. Only two plant species are endogenous of the Antarctic, the grass *Deschampsia antarctica,* and the cushion-forming pearlwort *Colobanthus crassifolius.*

In response to the extremely low temperatures that prevail in such regions and the combined dehydrating effect of wind, which is amplified by the absence of natural barriers, plants have developed extreme morphological and physiological characteristics (Fig. 3-7). They exhibit a prostrate type of growth to benefit from snow cover and minimize wind dessication, their buds are protected by old leaf material, they typically have enzymes with low temperature optima, they are resis-

Figure 3-7. Alpine tundra (North America, Canada). (Courtey of M. M. Grantner.)

tant to extreme dehydration, and they have small xeromorphic leaves to minimize evapotranspiration (Spicer and Chapman, 1990). Many of these plants survive as perennials and they are evergreens (Körner and Larcher, 1988), thus minimizing dependence on seasonal timing. It is not clear whether these plants exist in acclimated or nonacclimated states, or if they modulate their growth by spontaneous responses to changing temperature conditions.

In extreme altitudes, plants are often exposed to low temperatures equivalent to those experienced by arctic plants. However, the main differences between the two ecosystems is that alpine plants grow in locations where the growth season is longer (e.g., mountains in the tropics), maximum temperature is higher, radiation is more intense, and the soil is not permanently frozen. Plants there, as in the arctic, often benefit from favorable microclimatic conditions that allow plants to grow above the snow line, which is 6000 m in subtropical mountains and 4000 m in temperate regions (Grabherr et al., 1995).

THE MOLECULAR BASIS OF LOW-TEMPERATURE TOLERANCE

Historical Survey

Scientists have long searched for unified concepts that could describe the means by which plants survive low-temperature stress. In the first part of this chapter, we have attempted to demonstrate that oversimplification of the cold-resistance concept can be misleading. It is clear that plants have adapted various means to survive exposure to low temperature, and that the means that have been selected through evolution and history in a definite area are most often suited for the cold stress imposed by a given set of climatic conditions, and rarely suited for survival in a different environment. However, as in many fields of biological research, scientists have looked for model organisms in which the mechanism of cold resistance could be studied, independently of peripheral parameters such as morphological and anatomical adaptative features. Plants like alfalfa and many species within the Brassicaceae and the Graminaceae, for example, possess ecotypes or ecovariants that exhibit astonishing differences in cold tolerance, with no apparent morphological adaptation. Preliminary comparative studies of these ecotypes demonstrated that there was no simple means to differentiate the tolerant from the sensitive genotypes, aside from submitting them to a survival test at low temperature. Then came the idea that these plants should exhibit different responses during cold acclimation, and that these differential responses could be used to identify mechanisms that lead either to superior or inferior cold acclimation. A tremendous amount of research has been done using this basic approach. Essentially, plants have been compared in their hardened and nonhardened states, and plants of differing hardiness have been compared in their fully hardened state.

It was found that almost all metabolic pathways are affected by acclimation, and consequently the accumulation, disappearance, or transformation of most major (or minor) metabolites was monitored during acclimation in different plant systems.

Various molecular species were at times singled out as essential for cold resistance, on the basis of the tight correlation between their accumulation and the acquisition of resistance. However, most of these have been temporarily set aside since they did not fulfill one of the basic prerequisites for essentiality, that is, being essential for acclimation in all plants. Another major limitation of this approach became evident in that, however precise and acurate the measurements were, they could hardly prove whether the accumulation of a compound was physiologically essential for the acquisition of resistance, or if it was a result of alterations of metabolic functions by low temperature.

A lot of information also arose from the analysis of the mechanisms by which cells or tissues are injured by frost. These observations confirmed the general concepts that had been hypothesized earlier, that the protoplasts of resistant cells have characteristics that allow them to survive freezing, either by avoiding lethal dehydration, by withstanding severe dehydration, or by recovering functionality after dehydration-induced contraction (Uemura et al., 1995). It has been shown that this protection is acquired through acclimation, that it reaches different levels during acclimation, and that it is not acquired by cold-sensitive plants.

Current research is still focused on the search for a unified concept for frost survival. It now seems obvious to most researchers in this field that the main processes that occur during hardening are the acquisition of resistance to dehydration by the protoplast, the acquisition of resistance to plasma membrane dysfunction, and the promotion of extracellular ice formation. It is also apparent that there are no compelling reasons to think that the means by which various plants accomplish these processes are identical at the molecular level, or that all plants must accomplish all three of these processes.

It should be pointed out, though, that studying the molecular basis of cold acclimation should not be considered as the only approach to the understanding the hardiness concept. A recent study by Stone et al. (1993) demonstrated that in *Solanum* spp., two components of winter survival, freezing tolerance in the nonacclimated state and cold-acclimation ability, were not correlated in segregated populations. The mode of inheritance of these two traits was investigated in F_1 and backcross populations of two wild diploid species of *Solanum* exhibiting extremes of freezing tolerance and acclimation capacity. The study also demonstrated that inheritance of cold acclimation ability is relatively simple in *Solanum* sp. These results could provide new avenues to investigate the link between molecular and genetic aspects of cold tolerance.

The Molecular Basis of Chilling Resistance

The existence of an acclimation process that would lead to chilling resistance has long been debated. Some of the species exhibiting chilling resistance and freezing sensitivity do acquire a higher degree of resistance to chilling when exposed to noninjurious temperatures (Patterson and Reid, 1990; Saltveit and Morris, 1990; Wilson, 1979). However, many studies have shown that in most plant families, the main exception being the Solanaceae, membrane composition prior to acclimation is critical for chilling resistance (Kenrick and Bishop, 1986; Murata, 1983; Roughan, 1985).

Signaling The nature of the cold sensor in the plants that acclimate to chilling is still unknown. The only available evidence on the nature of a cold sensor demonstrates that the plasma membrane of a cyanobacteria (PCC6803) performs this task through changes in fluidity under different temperature regimes (Vigh et al., 1993). Using catalytic hydrogenation, which performs in situ hydrogenation of plasma membrane phospholipids as a means to alter fluidity of the membrane, Murata and Vigh and colleagues have demonstrated that chemically controlled changes in membrane fluidity specifically trigger the expression of a mRNA encoding for a desaturase enzyme. They also demonstrated that this specific response is similar to that triggered by lowering the temperature. These results, therefore, provide direct evidence that in *Synechocystis* the membrane acts as a sensor for low temperature and affects nuclear transcriptional activity (Maresca and Cossins, 1993).

How this signal is transduced is still unknown. It is believed that most of the injuries caused by chilling in sensitive plants are linked to chloroplast membranes (Hariyadi and Parken, 1993), and since cyanobacteria are thought to be the ancestors of modern plant chloroplasts, comparisons between their reaction to chilling and that of chloroplasts could provide a deeper understanding of the mechanisms underlying chilling resistance in higher plants.

Most of the data on transcriptional regulation by chilling temperatures come from studies made with frost-resistant plants during acclimation at above-zero temperatures. To our knowledge, there is only one report on the regulation of gene expression during acclimation to chilling using a chilling-tolerant, frost-sensitive species (Binh and Oono 1992), apart from studies on detached organs in postharvest technology (Schaffer and Fischer, 1988; Yu et al., 1995).

Acclimation and Resistance to Chilling Most plants react to chilling temperatures by changing the relative abundance of unsaturated phospholipids in their membranes (Williams et al., 1995). A lipid layer made of identical phospholipids with saturated acyl moieties will be rigid and cohesive; that is, it will turn into the gel-crystalline state at a relatively high temperature. In comparison, unsaturated aliphatic chains in a lipid layer will create disorder, increase fluidity, and contribute to lowering the phase transition temperature of membranes. How and where the fatty acids or glycerolipids are desaturated has long been a mystery. It had long been known that most lipid desaturating enzymes are bound to membranes, and this characteristic had made their purification impossible. Only by using complementation of mutants did Somerville and coworkers succeed in isolating the genes encoding for desaturase genes (Arondel et al., 1992; Browse et al., 1992). Labeling studies have shown that the first double bond is always introduced on a 18–0 fatty acyl-ACP by desaturation of 18:0 (stearoyl)-ACP to 18:1 (oleoyl)-ACP by a Δ9-stearoyl-ACP desaturase in the plastid, which is the only known hydrosoluble desaturase in higher plants. All subsequent desaturations are catalyzed by membrane-bound enzymes and use glycerolipids as substrates. Oleoyl phosphatidylcholine

(PC) is desaturated to $\Delta 9,12-18:2$ in the endoplasmic reticulum. Some desaturation to $\Delta 9,12,15$ also occurs in the endoplasmic reticulum. The fatty acids of galactolipids are desaturated to $18:2$ and $18:3$ at *sn-1* or *sn-2,* and $16:1$, $16:2$, and $16:3$ at *sn-2* in the plastid envelope. $16:0$ fatty acyl moieties at *sn-2* of phosphatidylglycerol (PG) can also be desaturated to $16:1$ $\Delta 3$-*trans* in the plastid. Whether the adjustment of membrane lipid saturation is a protective mechanism against chilling-induced injuries is debated, but many reports now show that the organisms that use this strategy will experience injuries if refrained from doing so. Murata and Yamaya (1984) showed that, with the higher proportion of 16:0/16:0 (with the 16:0 fatty acid in both positions *sn-1* and *sn-2* of the glycerol backbone) plus 16:0 (trans) lipids, the phase transition temperature is higher. In two elegant demonstrations using transgenic plant material (Murata et al., 1992; Wolter et al. 1992), it was shown that low temperature stability of the thylakoid membranes can be modulated by altering the degree of saturation of acyl moieties in phosphatidylglycerol.

How this reaction is controlled at the whole plant level is also unclear. Apart from the work reported with a cyanobacterium by Murata (Los et al., 1993), data are lacking on the description of the initial response of the plant to chilling. Gibson et al. (1994) demonstrated that in Arabidopsis, of the six desaturase genes used as probes, only one, a plastid Δ-15 desaturase, was up-regulated by low temperature, suggesting that changes in desaturation activity could be controlled by posttranscriptional processes in plant cells.

Sekiya (1992) has proposed that in *Citrus junos* desaturation could proceed through expression of a novel isozymic form of $18:1$-specific glycerol-3-phosphate acyltransferase. In addition, unlike bacterial and chloroplast membranes, plant membranes also contain sterols, which modulate fluidity of the structure and could thus provide additional resistance to changes in phase transition, and postpone the need for desaturation of membranes outside the chloroplast (Williams et al., 1995).

Experiments by Murata et al. (1992) and Wolter et al. (1992) showed that genetically engineered alterations in the saturation level of PG modulates chilling tolerance in tobacco. However, the role of PG in chilling sensitivity is probably limited to photosynthetic tissues. Different mechanisms must be involved in chilling injury in nonphotosynthetic tissues and in tissues chilled in the dark. Saturated species of PG were shown not to be a determining factor for tolerance in the leaves of Solanaceae (Kenrick and Bishop, 1986; Murata, 1983; Roughan, 1985). Kodama et al. (1994) have shown that transgenic tobacco plants with high levels of trienoic fatty acyl moieties exhibit superior cold tolerance.

Since chloroplasts are the prime site of disorganization by chilling temperatures in sensitive plants, and since fatty acid synthesis in plants is restricted to the plastids, it is probable that initial destabilization of the chloroplastic membrane will result in decompartmentation at the cellular level (Yu et al., 1995). Recent studies with detached tomato pericarp disks indicate that initial injury to the chloroplasts might originate from the dysfunction of the galactolipid synthesis pathway which involves several enzymes both inside and outside the chloroplasts (Yu and Willemot, 1996; Yu et al., 1995).

Membrane lipid peroxidation is another possible cause for chilling-induced disorders. It is not clear, however, if peroxidation is a primary cause or a result of initial membrane dysfunction (Shewfelt and Purvis, 1995). It has been shown that low temperatures will induce accumulation of free radicals, and peroxidation of membrane lipids by these radicals. The products of lipid peroxidation will destabilize the membrane and induce further lipid peroxidation, making it an autocatalytic process (Shewfelt and Purvis, 1995).

The Molecular Basis of Freezing Resistance

Cold Signal Transduction The exact nature of the primary events that take place when a plant cell is exposed to low above-zero temperatures, that is during cold acclimation, is still unclear. These primary events are designated as signal transduction mechanisms, since they are thought to occur at the onset of temperature change and to be responsible for a cascade of subsequent events that result in acclimation. Signal transduction must then be extremely specific, since its effect is generally irreversible and manifold. Three interesting lines of study have brought some light into the understanding of this intriguing and crucial phenomenon.

As was demonstrated in *Synechocystis,* membranes seem able to both sense subtle changes in temperature and signal this change to the cell (see the preceding section on the molecular basis of chilling resistance). What remains to be shown is whether the plasma membrane and other important membrane systems in higher plant cells can sense temperature shifts by changes in fluidity and whether membrane signaling can trigger processes that are essential for the acquisition of freezing tolerance.

Other groups have worked on the hypothesis that since abscisic acid (ABA) can increase frost tolerance in nonacclimated plants, it must act as a signal transducer. Exogenous applications of ABA trigger transcriptional activity that is somewhat similar to that induced by cold acclimation (Mohapatra et al., 1988; Robertson and Gusta, 1986). However, the level of acclimation gained through exposure to exogenous ABA is always lower than that attained through exposure to cold. Studies have demonstrated that ABA-responsive elements are present in the 5′-upstream sequences of many genes whose transcription is up-regulated during cold acclimation. Holappa and Walker-Simmons (1995) have shown that transcription of a cold-induced gene encoding for a protein kinase is also up-regulated by ABA.

Studies with ABA-deficient mutants have brought new information on the role of ABA as a signal transducer. Nordin et al. (1991) showed that expression of cold-regulated genes during cold acclimation can occur without a concomittant rise in endogenous ABA level, and more recently, studies by Dallaire et al. (1994), Lång et al. (1994), and Mäntylä et al. (1995) suggest that transduction of low temperature and ABA signaling proceed through different routes, but that their triggering effect can be convergent under certain conditions.

Temperature declines, like many other environmental stimuli, will increase the influx of calcium in cells. Calcium is considered by many as a secondary messenger since it regulates many metabolic processes, either through binding to modulating proteins like calmodulin, or directly through the activation of catalytic enzymes with regulatory functions like protein kinases and phosphatases (Poovaiah and Reddy, 1993). Studies performed with isolated alfalfa cells suggest that when calcium influx is blocked during acclimation, the cells will not harden, and some cold-regulated genes will not express. When nonacclimated cells are treated with ionophores or agonists of calcium channels, some cold-induced genes will be expressed (Monroy and Dhindsa, 1995). However, no causal link has yet been established between expression of these cold-induced genes and acquisition of tolerance, and it has not been shown that these treatments could trigger acclimation. Subsequent studies by the same group also showed that transcription of a messenger encoding for a calcium-dependent protein kinase is induced by exposure to hardening conditions.

It seems likely that calcium and ABA can act as transducers of a low-temperature signal. However, since ABA is involved in transduction of other environmental signals, and calcium influx rates are modulated by a host of environmental cues, they hardly seem specific enough to be a primary signal. Moreover, none of these signals can fully replace the cold signal. The hypothesis that membrane fluidity is the ideal prospect for sensing temperature shifts is still currently reinforced by experimental results. The model proposed by Monroy and Dhindsa (1995), which describes the putative sequence of events from changes in membrane fluidity to transcriptional regulation of cold acclimation-specific genes, takes into account most of the recent findings in both signal transduction and cold-specific gene expression. Molecular tools now exist to produce data that will challenge this model, and to demonstrate whether, for example, hydrogenation of the plasma membrane, or any other treatments, that can specifically alter plasma membrane fluidity, can also modulate extracellular calcium influx, endogenous ABA levels, expression of cold-induced genes, or acclimation of isolated cells to below-zero temperatures.

Transcriptional Regulation by Cold   If the cold signal is perceived and transduced to the nucleus, it should trigger a series of molecular modifications that will result in over- or under-expression of specific genes. It is generally believed that genes encoding *trans*-acting factors will first respond to signal transduction. These factors are regulatory proteins that are synthesized in the cytosol and targeted to the nucleus, and bind to *cis*-acting elements in the 5'-upstream region of genes encoding for proteins with various nonregulatory functions. The binding of these factors, together with the binding of other less specific nuclear proteins, modify the upstream region of targeted genes and initiate a transcriptional wave. This cascade of events is described as transcriptional regulation.

Cold-induced synthesis of a *trans*-acting factor has yet to be demonstrated, and thus it is still unclear if cold-induced signal transduction is followed by such a highly organized set of molecular interactions. However, since the genomic response during cold acclimation seems both complex and tightly controlled, it is probable that cold modulation of gene expression operates through a cascade of molecular events, and that *trans*-acting factors play an important role in initiating a transcription wave that results in such a cascade.

A *cis*-acting element that confers cold inducibility has been isolated and sequenced in *Arabidopsis thaliana* by Yamaguchi-Schinozaki and Shinozaki (1994). It is a nine base pair fragment present in the promoting sequence of at least two genes that are overexpressed during cold acclimation. However, this motif also confers transcriptional modulation during dehydration. There are also other reports showing that specific regions of promotors confer transcriptional regulation by cold. Dolferus et al (1994) have demonstrated that the promotor of the *AdH* gene in *A. thaliana* contains four regions harboring *cis*-acting elements that interact to control transcription under various stress conditions, including low-temperature stress.

Other studies by Horvath et al. (1993), Baker et al. (1994), and Wang et al. (1995) have also demonstrated that once the cold signal is transduced, specific regulatory elements act to modulate transcription. *Trans*-acting factors that bind to these *cis*-acting elements can be isolated from the pool of nuclear proteins. Since some regulatory elements have now been characterized, the identification of cold-specific *trans*-acting factors should arise shortly, at least for *Arabidopsis thaliana*. It remains to be seen if the molecular machinery used by *A. thaliana* to sense and react to cold bears similarities with that of plants exhibiting superior hardiness, or that are true perennials. It is, however, an ideal model plant for studies at the molecular level. It also bears close phylogenetic similarities to *Brassica napus,* an important crop species cultivated in cool temperate regions, where cold tolerance is an important agronomic trait.

Cold-Induced Genes There is no doubt now that low temperature modulates the expression of nuclear genes. The experimental evidence for this phenomenon comes from two experimental approaches: differential hybridization and in vitro translation. In the first approach, mRNAs are isolated from acclimated and nonacclimated plants. The pool of mRNAs obtained from acclimated plants is used to build a cDNA library that is then probed with cDNA synthesized from both types of mRNAs. Any difference between the hybridization patterns of the RNA populations obtained from these two stages of acclimation indicates the presence of mRNAs synthesized in response to acclimation, and allows the isolation of cold-acclimation specific (CAS) cDNAs. When isolated, these cDNAs are used as probes to monitor the expression level in various acclimation conditions.

In the second approach, mRNA species from acclimated and nonacclimated plants are translated in vitro, and the products separated for analysis by two-dimensional-electrophoresis. Comparisons of the translation patterns will exhibit the differences between the two mRNA populations and give preliminary information on the size and isoelectric point of the encoded peptides.

These two lines of evidence have shown that response to low nonfreezing temperature acts at least partly at the transcriptional level, and that expression of numerous genes is modulated by low temperature. Such responses have been shown to occur in wheat (Danyluk and Sarhan, 1990), in alfalfa (Castonguay et al., 1993; Mohapatra et al., 1987), in *Arabidopsis thaliana* (Lin et al., 1990), and in spinach (Guy et al., 1985). In most cases, cold regulation of expression is correlated with increased cold tolerance (for a review see Li and Christersson, 1993) and in one study on alfalfa, it was also demonstrated that specific modulation of gene expression also occurs at subfreezing temperatures (Castonguay et al., 1993).

Table 3-2 lists genes that have been isolated by differential hybridization in model plants. Although not complete, the list exemplifies that most of the genes isolated by these methods bear little similarity to genes with known metabolic or structural functions, and that some of the proteins encoded by up-regulated genes following a low-temperature treatment are species specific. Recent studies also indicate that some of these genes are cold induced in more that one species, and that cold-acclimation specific genes are in fact rare. Many of the genes that are induced by cold treatments are also expressed in response to various stresses involving water availability, for example, high salt and drought. In alfalfa, of the five genes isolated by differential hybrization using cold as a stress signal, only one is not expressed in reaction to salt and dehydration stress (Castonguay et al., unpublished results). In alfalfa, a thorough examination of in vitro translation patterns from plants maintained in natural acclimation conditions revealed that expression of a few specific genes occurred in the heart of winter, long after the ground is frozen. This pattern of expression was correlated with the highest level of resistance of the plant, and seems specific to the hardiest cultivars (Castonguay et al., unpublished results). An important reality thus emerges from gene expression studies conducted in natural conditions, in that even though it is convenient to study cold acclimation under controlled conditions using low above-zero temperatures to trigger the plant response, such studies might lead to the characterization of mechanisms that bear little resemblance to those of complete acclimation. Studies under natural environmental conditions revealed that, although the response of a species to low temperature involves expression of numerous genes, achieving full acclimation potential of each ecotype within a species might be related to the expression, under more extreme temperature conditions, of a limited number of genes.

Homologies have been found, though, between cold-induced genes and other types of genes with unknown functions. A number of cold-induced genes encode proteins harboring glycine-rich repeated motifs. Many of these proteins are synthesized in an immature form, with a leader amino acid sequence at the N-terminal, which promotes cell wall targeting. Experimental evidence shows that some glycine-rich proteins are found in their mature form in the cell wall, and they are thought to play a structural role (Showalter, 1993). Glycine-rich domains provide a relative softness to the secondary structure of the mature protein; how this characteristic is related to the putative structural role of the protein is still hypothetical, as is its relation to cold acclimation. Another type of glycine-rich protein has been isolated from alfalfa in the hardened state, a protein with no apparent leader sequence, but with a putative nuclear targeting signal at the carboxy terminal. This protein has only a few repeated motifs, and the glycine residues are distributed more evenly throughout the amino acid chain. The function of this protein is still unknown, but its putative association with the nucleus suggests that it might play a regulatory role.

Cold-induced genes bearing similarities to dehydrin-like proteins have also been

TABLE 3-2 Cold-Induced cDNAs from *Medicago sativa* (alfalfa), *Arabidopsis thaliana*, and *Triticum aestivum* (Wheat)

Name	Transcript Size (Kb)	Amino Acids	Regulation[a]	Class and/or Sequence Information	Reference
				Medicago sativa	
cas18	1.6, 1.4, 1.0	168	C	(Gly-Thr)$_n$ repeats Dehydrin-like	Wolfraim et al. (1993)
cas17	0.8	160	C	(Gly-Thr)$_a$ repeats Dehydrin-like	Wolfraim and Dhindsa (1993)
psm2075	0.70	159	C, ABA	7 (GGYNHGGGYN) repeats Glycine-rich	Luo et al. (1991)
msaCIA	0.90	204	C, D	(G$_4$HG$_2$H)$_n$ and (GGGYNH)$_n$ repeats Glycine-rich	Laberge et al. (1993)
pUM90-1	1.1	>192	C, D, S, W, ABA	Gly-rich (35—40%) Gly-rich tandem repeats	Luo et al. (1992)
pUM90-2	>0.54	>133	ND	Gly-rich (35—40%) Gly-rich tandem repeats	Luo et al. (1992)
pUM91-4	>0.68	>157	ND	Gly-rich (35—40%) Gly-rich tandem repeats	Luo et al. (1992)
msaCIC	0.90	166	C, W	Proline-rich and hydrophobic domains Biomodular protein	Castonguay et al. (1994)
cas15 or masCIB	0.80	136	C, D	Putative bipartite nuclear targeting signal	Monroy et al. (1993)
msaCID	0.70	158	C, D	85% identity with ABR17 (Iturriaga et al., 1994) Pathogenesis-related protein	Laberge et al. unpublished results
msaCIE	1.2	ND	C	Homologous to Glyceraldehyde-3P dehydrogenase Central metabolism	Laberge et al. unpublished results
psm2358	1.2, 0.9	ND	C	ND Unclassified	Mohapatra et al. (1989)
				Arabidopsis thaliana	
cor15	0.7	139	C	10^8 times more effective than sucrose in protecting LDH against freeze inactivation in vitro	Lin and Tomashow (1992)

Clone	kb	aa	Induction[a]	Comments	Reference
cor47	1.4	294	C	Homology with LEA proteins (group II)	Gilmour et al. (1991)
cor160	2.5	>354	C	Contain at least 3 types of repeat (incomplete clone)	Hajela et al. (1990)
cor6-6	0.6	66	C	24% Ala, 14% Lys, 11% Gly	Kurkela and Frank (1990); Gilmour et al. (1991)
Triticum aestivum					
pWG1 (cor39)	3.3, 1.5, 0.8	391	C, D Seeds, ABA	Relate to cor47 of Arabidopsis; Hydrophylic peptide; One sequence Lys rich is repeated 6X and is related to group II LEA proteins; Another sequence repeated 6X is Gly rich; Homologous to Wcs120	Guo et al. (1992)
Wcs120	1.65	390	C	Boiling stable; Gly 26.7%, thr 16.7% his 10.8%; Domain A repeat 6X, Domain B 11X	Houde et al. (1992a)
				Related to cor39 but regulation is different; The antibody is specific for freezing-tolerant cereals; Many proteins are recognized	Houde et al. (1992b)
Wcs19	1.0	190	C	Leaf specific, light stimulated C-terminal half has a high content of acidic amino acids with a net negative charge of -7; N-terminal half is rich in pro, lys, arg with a net positive charge of $+10$; 21% alanine, not boiling stable	Chauvin et al. (1993)
Wcs200	3.5 (2.0, 1.7)	ND	C specific	Homology to Wcs120	Ouellet et al. (1993)
Wcs66	1.6	491	C	Homology to Wcs120	Chauvin et al. (1994)
pkaba1	1.3	>332	C (ABA, S, D)	Protein kinase	Holoppa and Walker-Simmons (1995)

[a] C = cold, D = dehydration, S = salt, W = wounding, ND = not determined.

89

isolated. These proteins are induced by dehydration in many plant species, and are characterized by a lysine-rich amino acid consensus sequence (Close and Bray, 1993). Their general structure suggests that they might act as stabilizers for proteins and membranes in dehydrated cells. These proteins are also related to proteins that accumulate in the later stages of embryogenesis.

One protein encoded by a cold-induced gene in *Arabidopsis thaliana* has been studied in much more detail. This protein has been produced in a prokaryotic expression system and purified. Lin and Thomashow (1992) then demonstrated that this chimeric protein is 10^8 times more efficient than sucrose in protecting lactate dehydrogenase against freeze inactivation in vitro. In *Brassica napus* and *Spinacea oleracea,* proteins homologous to heat-shock proteins accumulate during cold acclimation (Krishna et al., 1995; Neven et al., 1992). These proteins are thought to participate in protein transport, folding, assembly, and disassembly processes.

In the last few years, several proteins specifically expressed under low temperature have been isolated and characterized (Table 3-3). All of these proteins share common features, especially a strong bias in their amino acid composition, a strong hydrophobicity and solubility at 100°C. Whether these proteins confer superior cold tolerance to the plants from which they were isolated has not yet been demonstrated. One family of proteins from wheat, immunologically related to WCS120 (Dallaire et al., 1994), accumulates in the vascular transition zone during acclimation (Houde et al., 1995). Another protein, PCA 60, is more abundant in hardy deciduous varieties than in sensitive evergreen varieties during acclimation (Arora and Wisniewski, 1994). Five proteins of approximately 19, 26, 32, 34, and 36 kDa, which accumulate in the apoplast of acclimated rye leaves, exhibit significant antifreeze activity in vitro (Hon et al., 1994). Recombinant WCS120 produced in *E.coli* prevents freeze denaturation of proteins (Houde et al., 1995).

In most instances, the evidence for up-regulation has been obtained from mRNA expression studies using cold-induced cDNAs as probes. This kind of information does not imply per se that the protein for which the cDNAs encode is synthesized at a higher rate, or that it accumulates in the tissue during acclimation. A better understanding of the role of these proteins during the acclimation process will arise from the characterization of their structure, from the monitoring of their accumulation and localization in the tissues and cells, and from the utilization of antisense technology to prove that their presence is a prerequisite for the acquisition of resistance.

Nothing is known of the proteins encoded by genes that are down-regulated. Although it seems probable that most of these genes encode for proteins playing a role in cellular functions that are nonspecifically inhibited by low temperature, there is no reason to believe that down-regulation is not as important and significant as up-regulation during acclimation. The study of proteins that disappear during acclimation should also be taken in account since they participate in the physicochemical adjustment of the plants. Furthermore, as is shown in the following section, there is still no obvious link between the putative function of proteins encoded by cold-induced genes and most of the cellular or biochemical functions that are altered during acclimation.

TABLE 3-3 Cold-Induced Proteins

Protein	Species	mw	Solubility	Amino Acid Content	Cellular Localization	Putative Role	Reference
WCS120	Wheat	50 kDa	Boiling soluble	Glycine-rich		Enzyme cryo-protective agent	Houde et al. (1992b)
WSC120-related	Wheat	25-200 kDa	Boiling soluble		Cytoplasmic/nucleoplasmic		Dallaire et al. (1994)
WCS200	Wheat	200 kDa	Boiling soluble	Glycine-rich			Ouellet et al. (1993)
PCA60	Peach	60 kDa	Boiling soluble	Glycine-rich			Arora and Wisniewski (1994)
BN28	Rapeseed	<6.6 kDa	Boiling soluble	Glycine/alanine-rich			Boothe et al. (1995)
CAP85	Spinach	85 kDa	Boiling soluble	Lysine-rich	Cytosolic		Neven et al. (1993)
CAP160	Spinach	160 kDa					Guy et al. (1992)
RAB18	Arabidopsis	22 kDa					Mäntylä et al. (1995)
LT178	Arabidopsis	140 kDa					Mäntylä et al. (1995)
Rye AFPs	Rye	19-36 kDa				Antifreeze protein	Hon et al. (1944)
MsaciA	Alfalfa	32 kDa		Glycine-rich			Ferullo et al. (1995)
MsaciB	Alfalfa	23 kDa		Glycine-rich			Ferullo et al. (1995)

Cellular and Metabolic Functions Altered by Low Temperature

Resistance to Dehydration   Resistance to dehydration of the protoplasts is thought to occur through the accumulation of osmotically active compounds that will lower the difference in vapor pressure between extracellular ice and free cell water. Small hydrophilic molecules are best suited for this purpose. Mono- or disaccharides, amino acids, small peptides, polyamines, and quaternary ammonium compounds have all been shown to accumulate in certain plant model systems (Hanson and Burnet, 1994).

It has long been demonstrated that perennial plants will accumulate carbohydrate reserves in the fall. As has been shown in frost-sensitive tubers (Levitt, 1980) this change in overall physiology bears little relation to the acquisition of frost tolerance. However it should be remembered that most overwintering plants rely on these reserves to resume growth in the spring.

Carbohydrate metabolism can also respond to decreases in temperature in a different manner. During acclimation, starch reserves decrease in some organs at the expense of sucrose accumulation. This decrease in reserves occurs in many plants, and recent experimental data suggest that metabolism of sucrose, leading to the synthesis of oligosaccharides like stachyose and raffinose, is related to the acquisition of tolerance (Castonguay et al., 1995; Hinesley et al., 1992). In vitro studies have demonstrated that these molecular species might be active in cryoprotection rather than in the prevention of water loss (Caffrey et al., 1988; Hincha, 1990). Furthermore, since the level of sucrose in the cell generally exceeds that of stachyose and raffinose, it is doubtful these metabolic transformations of sucrose are aimed at increasing the ability to retain water in the cell.

Whatever the role of raffinose and stachyose in the acquisition of tolerance, it has been hypothesized that synthesis of these compounds is crucial during acclimation and that it is closely related to the level of tolerance of the plant. Since the gene encoding for galactinol synthase, the enzyme catalyzing the first committed step in the biosynthetic pathway of raffinose, has been cloned and sequenced, antisense technology may now provide information on the causal relationship between acclimation and the accumulation of these oligosaccharides.

As for carbohydrates, free amino acids have long been identified both as potential reserve materials for overwintering organs and as osmolytes. Model algal systems provide evidence that accumulation of proline is a rapid, reversible response to changes in external osmotic potential. Recent studies demonstrate that Δ-pyrroline-5-carboxylate synthetase has a key regulatory role in controlling the synthesis and accumulation of proline, and that overexpression of this enzyme in transgenic plants confers additional osmotolerance (Kavi Kishor et al., 1995).

Although correlations exist between the accumulation of other potential osmolytes and cold tolerance, the evidence that determines whether they are physiologically active as cryoprotectants is generally lacking.

Protective Mechanisms   The evidence produced through the last two decades has indicated that destabilization of the plasma membrane is the primary cause of freezing injury in plants (Steponkus et al., 1993). Cold acclimation induces changes in the composition of the plasma membrane (Lynch and Steponkus, 1987; Uemura and Steponkus, 1994), and these changes are associated with alterations in the cryo-behavior of the membrane. The main changes observed in the membrane lipid composition are: an increase in the proportion of diunsaturated phospholipids, and a decrease in cerebrosides. The cryobehavior of the plasma membrane was first assessed by the combined use of isolated protoplasts and liposomes, and eventually, by the use of protoplasts in which the composition of the membrane was engineered through artificial enrichment of mono- and disaturated species of phosphatidylglycerol. Briefly, in rye, it was observed that injuries are caused by contraction of the protoplasts following freeze-induced dehydration in nonacclimated protoplasts, as a result of the formation of endocytotic vesicles (Dowgert and Steponkus, 1984). Upon rehydration, this irreversible membrane condition will result in lysis as the protoplasts expand. Acclimated protoplasts will form exocytotic extrusions during contraction, a condition in which membrane surface area is conserved such that expansion lysis does not occur (Gordon-Kamm and Steponkus, 1984). In addition, it was shown that membranes from acclimated protoplasts do not undergo transition from the lamellar to H_{II} phase in the region of close apposition with the subtending plasma membrane. As a whole, this series of experiments demonstrated that cold-induced alterations in the lipid composition of plasma membranes are causally related to the increase in their cryostability. Recent studies have shown that similar mechanisms might be acting in *Arabidopsis thaliana* (Uemura et al., 1995).

Apart from the compelling evidence showing that membranes protect themselves from cold-induced injuries by altering their composition, other studies indicated that several putative protective mechanisms are triggered during acclimation. Some proteins produced during acclimation are potent cryoprotectants for thylakoid membranes (Hincha et al., 1989) or for enzymes in vivo (Lin and Thomashow, 1992). It has also been observed that plants acquire protection from the activated oxygen that is produced during frost damage. This was established by comparing the damage caused by free oxygen radicals and by frost to the plasma membrane, and by demonstrating the increase in the capacity to scavenge radicals under these conditions (Kendall and McKersie, 1989; Smirnoff and Colombé, 1988). It was also demonstrated that transgenic alfalfa expressing bacterial Mn-dependent superoxide dismutase (an enzymatic scavenger) have increased regrowth potential after freezing (McKersie et al., 1993).

Promotion of Extracellular Ice Formation   Intracellular ice formation is a cause or an indication of cell death in plants and most organisms exposed to frost. It has long been debated whether plants acquire the ability to control extracellular ice formation during acclimation, or whether this trait is intrinsic to cold-resistant plants or plant parts. Several studies have shown that a particular type of proteins, called antifreeze proteins (AFP), are synthesized and accumulate in the extracellular space during acclimation (Griffith et al., 1993). These proteins bind to ice crystals in formation and restrict their growth in the leaves of *Secale cereale*. Another type of protein, ice nucleators, seem active in promoting heterogeneous ice nucleation.

The combined effect of these proteins is thought to promote extracellular ice formation at a much higher temperature than the theoretical ice nucleation point, and to restrict crystal expansion. Both of these proteins are present in the extracellular space of cold-acclimated leaves of winter rye, and seem to play an essential role in freezing tolerance (Brush et al., 1994; Hon et al., 1994). The genes encoding for a particular AFP have been cloned and expressed in transgenic plants (Hightower et al., 1991). Although they seem to provide the transgenic plants with the ability to refrain from intracellular crystallization, it is still not clear whether this trait confers increased cold tolerance.

CONCLUDING REMARKS

Willows and poplars are nonclimaxic species that have high relative growth rates. They grow in various climatic conditions over wide latitudinal and altitudinal ranges. They are the most cold tolerant of all deciduous tree species, and in boreal regions, they grow side by side with *Picea mariana,* one of the slowest-growing tree species. *Astragalus alpinus* and *Oxytropis maydelliena* are two arctic species of the Fabaceae family that are phylogenetically related to sainfoin, a perennial forage with moderate cold-tolerance potential. Like many arctic plants they will stop or resume growth in response to sudden variations in ambient air temperature, and thus seem to have lost the need to acclimate or undergo dormancy to withstand frost. However hard it is to reconcile these observations with the belief that cold tolerance is an evolutionary trait linked with low relative growth rates and early fall dormancy, these are biological facts that should be given due attention in the definition of cold-tolerance models, at the ecological or cellular level.

There is still an apparent gap between the picture drawn by the analysis of cold-specific gene expression and that drawn by the analysis of other cellular processes. For example, even if membranes seem the primary site for cold perception and the primary site for cold injury, differential hybridization has singled out genes encoding for functions apparently remote from membrane metabolism. It is possible that the methods used to isolate cold-specific genes up to now are not sensitive enough, and that substraction libraries made at various stages of acclimation would give a more reliable picture of the genomic response.

Recent advances in nuclear magnetic resonance technology, molecular biology, and cell biology will help to define the strategies involved in plant resistance to cold stress. The use of mutants and, even more so, the use of transgenic plant material now offers the means to challenge the biological models that were first proposed by physiology. These technologies will most probably unveil mechanisms that have not yet been associated with cold tolerance and in this way, contribute to draw a more integrated perspective. It is our hope that the new tools that are being offered by emerging technologies will be used to increase our understanding of this important phenomenon, rather than generating additional isolated lines of research.

REFERENCES

Alden, J. and R. K. Hermann (1971), *Bot. Rev., 37,* 37–142.

Arondel, V., B. Lemieux, I. Hwang, S. Gibson, H. M. Goodman, and C. R. Somerville (1992), *Science, 258,* 1353–1355.

Arora, R. and M. E. Wisniewski (1994), *Plant Physiol., 105,* 95–101.

Ashworth, E. N. (1992), *Hort. Rev., 13,* 215–255.

Ashworth, E. N., G. A. Davis, and M. E. Wisniewski (1989), *Plant Cell Environ., 12,* 521–528.

Ashworth, E. N., T. J. Willard, and S. R. Malone (1992), *Plant Cell Environ., 15,* 507–512.

Baker, S. S., K. S. Wilhelm, and M. F. Thomashow (1994), *Plant Mol. Biol., 24,* 701–713.

Barnes, B. V. (1991), *Ecosystems of the world, Vol 7, Temperate deciduous forests* (E. Röhrig and B. Ulrich, Eds.), Elsevier, New York, pp. 219–344.

Beck, E. (1984), in *Tropical alpine environments: Plant form and function* (P. W. Rundel, A. P. Smith, F. C. Meinzer, Eds.), Cambridge University Press, pp. 77–110.

Biebl, R. (1964), *Protoplasma, 59,* 133–156.

Binh, L. T. and K. Oono (1992), *Plant Physiol., 99,* 1146–1150.

Boothe, J. G., M. D. deBeus, and A. M. Johnson-Flanagan (1995), *Plant Physiol., 108,* 795–803.

Bowers, M. C. (1994), in *Plant-environment interactions* (R. E. Wilkinson, Ed.), Marcel Dekker, New York, pp. 391–411.

Bramlage, W. J. and S. Meir (1990), in *Chilling injury of horticultural crops* (C. Y. Wang, Ed.), CRC Press, Boca Raton, FL, pp. 37–49.

Browse, J., M. Miquel, J. Lightner, and J. Okuley (1992), in *Biochemistry and molecular biology of membrane and storage lipids of plants* (N. Murata and C. Somerville, Eds.), American Society of Plant Physiologists, Rockville, MD, pp. 50–59.

Brush, R. A., M. Griffith, and A. Nlynarz (1994), *Plant Physiol., 104,* 725–735.

Caffrey, M., V. Fonseca, and A. C. Leopold (1988), *Plant Physiol., 86,* 754–758.

Castonguay, Y., S. Laberge, P. Nadeau, and L.-P. Vézina (unpublished data).

Castonguay, Y., P. Nadeau, and S. Laberge (1993), *Plant Cell Physiol., 34,* 31–38.

Castonguay, Y., S. Laberge, P. Nadeau, and L.-P. Vézina (1994), *Plant Mol. Biol., 24,* 799–804.

Castonguay, Y., P. Nadeau, P. Lechasseur, and L. Chouinard (1995), *Crop Sci., 35,* 509–516.

Chauvin, L. P., M. Houde, and F. Sarhan (1993), *Plant Mol. Biol., 23,* 255–265.

Chauvin, L. P., M. Houde, and F. Sarhan (1994), *Plant Physiol., 105,* 1017–1018.

Clements, R. J. and M. M. Ludlow (1977), *J. Appl. Ecol., 14,* 551–556.

Close, T. J. and E. A. Bray (1993), *Plant responses to cellular dehydration during environmental stress,* American Society of Plant Physiologists, Rockville, MD.

Cox, C. B. and P. D. Moore (1993), *Biogeography. An ecological and evolutionary approach.* Blackwell Scientific Publications, London.

Dallaire, S., M. Houde, Y. Gagné, S. S. Hargurdeep, S. Boileau, N. Chevrier, and F. Sarhan (1994), *Plant Cell Physiol., 35,* 1–9.

Danyluk, J. and F. Sarhan (1990), *Plant Cell Physiol., 31,* 609–619.

Dolferus, R., M. Jacobs, W. J. Peacock, and E. S. Dennis (1994), *Plant Physiol., 105,* 1075–1087.

Dowgert, M. F. and P. L. Steponkus (1984), *Plant Physiol., 75,* 1139–1151.

Enright, J. T. (1982), *Oecologia, 54,*253–259.

Ferullo, J.-M., L.-P. Vézina, Y. Castonguay, P. Nadeau, and S. Laberge (1995), *Plant Physiol. Suppl., 108,* abst. No. 512.

Flahaut, C. (1937), *Distribution géographique des végétaux dans la région méditerranéenne française.* Lechevalier, Paris.

Flinn, C. L. and E. N. Ashworth (1994), *J. Am. Soc. Hort. Sci., 119,* 295–298.

Gibson, S. V. Arondel, K. Iba, and C. Somerville (1994), *Plant Physiol., 106,* 1615–1621.

Gilmour, S. J., N. N. Artris, and M. F. Thomashow (1991), *Plant Mol. Biol., 18,* 13–21.

Gordon-Kamm, W. J. and P. L. Steponkus (1984), *Proc. Natl. Acad. Sci. USA, 81,* 6373–6377.

Grabherr, G., M. Gottfried, A. Gruber, and H. Pauli (1995), in *Arctic and alpine biodiversity: Patterns, causes and ecosystem consequences* (F. S. Chapin, III, and C. Körner, Eds.), Springer-Verlag, New York, pp. 167–181.

Griffith, M., E. Marentes, P. Ala, and D. S. Yang (1993). in *Advances in plant cold hardiness* (P. H. Li and L. Christersson, Eds.), CRC Press, Boca Raton, FL, pp. 177–186.

Guo, W., R. W. Ward, and M. F. Thomashow (1992), *Plant Physiol., 100,* 915–922.

Guy, C. L., K. J. Niemi, and R. Brambl (1985), *Proc. Natl. Acad. Sci. USA, 82,* 3673–3677.

Guy, C. L., D. Haskell, L. Neven, P. Klein, and C. Smelser (1992), *Planta, 188,* 265–270.

Hajela, R. K., D. P. Horvath, S. J. Gilmour, and M. F. Thomashow (1990), *Plant Physiol., 93,* 1246–1252.

Hanson, A. D., and M. Burnet (1994), in *Biochemical and cellular mechanisms of stress tolerance in higher plants* (J. C. Ed, Ed.), Springer-Verlag, New York; pp. 291–302.

Hariyadi, P. and K. Parken (1993), *J. Plant Physiol., 141,* 733–738.

Hedberg, I. and O. Hedberg (1979), *Oikos, 33,* 297–307.

Heichel, G. H. and K. I. Henjum (1990), *Crop Sci., 30,* 1123–1127.

Hightower, R., C. Baden, E. Penzes, P. Lund, and P. Dunsmuir (1991), *Plant Mol. Biol., 17,* 1013–1021.

Hincha, D. H. (1990), *Cryo-Letters,* 11, 437–444.

Hincha, D., U. Heber, and J. M. Schmitt (1989), *Plant Physiol. Biochem., 27,* 795–801.

Hinesley, L. E., D. M. Pharr, L. K. Snelling, and S. R. Funderburk (1992), *J. Am. Soc. Hort. Sci., 117,* 852–855.

Holappa, L. and M. K. Walker-Simons (1995), *Plant Physiol., 108,* 1203–1210.

Hon, W.-C., M. Griffith, P. Chong, and D.S.C. Yang (1994), *Plant Physiol., 104,* 971–980.

Horvath, D. P., B. K. McLarney, and M. F. Thomashow (1993), *Plant Physiol., 103,* 1047–1053.

Houde, M., R. S. Dhindsa, and F. Sarhan (1992a), *Mol. Gen. Genet., 234,* 43–48.

Houde, N., J. Danyluk, J. F. Laliberté, E. Rassart, R. S. Dhindsa, and F. Sarhan (1992b), *Plant Physiol., 99,* 1381–1387.

Houde, M., C. Daniel, M. Lachapelle, F. Allard, S. Laliberté, and F. Sarhan (1995), *Plant J.*, *8*, 583–593.

Huggett, R. J. (1991), *Climate, earth processes and earth history*, Springer-Verlag, New York.

Ishikawa, M. (1984), *Plant Physiol.*, *75*, 196–202.

Ishikawa, M. and A. Sakai (1985), *Plant Cell Environ.*, *8*, 333–338.

Iturriaga, E. A., M. J. Leech, D.H.P. Barratt, and T. L. Wang (1994), *Pl. Mol. Biol.*, *24*, 235–240.

Johnson-Flanagan, A. M. and J. Singh (1987), *Plant Physiol.*, *85*, 699–705.

Kader, S. A. and E. L. Proebsting (1992), *J. Am. Soc. Hort. Sci.*, *117*, 955–960.

Karkela, S. and M. Franck (1990), *Plant Mol. Biol.*, 15, 137–144.

Kavi Kishor, P. B., Z. Hong, G. H. Miao, C.A.A. Hu, and D.P.S. Verma (1995), *Plant Physiol.*, *108*, 1387–1394.

Kendall, E. J. and B. D. McKersie (1989), *Physiol. Plant*, *76*, 86–94.

Kenrick, J. R. and D. G. Bishop (1986), *Plant Physiol.*, *81*, 946–949.

Knutson, R. N. (1974), *Science, 186*, 746–747.

Kodama, H., T. Hamada, G. Horiguchi, M. Nishimura, and K. Iba (1994), *Plant Physiol.*, *105*, 601–605.

Körner, C. and W. Larcher (1988), in *Plants and temperature* (S. P. Long and F. I. Woodward, Eds.), The Company of Biologists Ltd. Cambridge, UK, pp. 25–57.

Kranz, H. D., D. Miks, M. L. Siegler, I. Capesius, C. W. Sensen, and V.A.R. Huss, (1995), *J. Mol. Evol.*, *41*, 74–84.

Krishna, P., M. Sacco, J. F. Cherutti, and S. Hill (1995), *Plant Physiol.*, *107*(915–923).

Kurkela, S., and M. Franck (1990), *Plant Mol. Biol.*, *15*, 137–144.

Laberge, S., L.-P. Vézina, P. Nadeau, and Y. Castonguay (unpublished data).

Laberge, S., Y. Castonguay and L.-P. Vézina (1993), *Plant Physiol.*, *101*, 1411–1412.

Lång, V., E. Mäntylä, B. Welin, B. Sundberg, and E. T. Palva (1994), *Plant Physiol.*, *104*, 1341–1349.

Larcher, W. (1963), *Protoplasma*, *57*, 569–587.

Larcher, W. (1970), *Oecol. Plant*, *5*, 267–286.

Larcher, W. (1975), *Anz Math.-Naturewiss Klösterr Akad Wiss*, *11*, 1–20.

Larcher, W. J. Wagner (1976), *Oecol. Plant*, *11*, 361–374.

Larcher, W. and A. Winter (1981), *Principles*, *25*, 143–152.

Levitt, J. (1980), *Responses of plants to environmental stresses, Vol. I. Chilling, freezing, and high temperature stresses*, 2nd ed., Academic Press, London.

Li, P. H. (1984), *Hort. Rev.*, *6*, 373–416.

Li, P. H. and Christersson, L. (1993), *Advances in plant cold hardiness*. CRC Press, Boca Raton, FL.

Lin, C. and M. F. Thomashow (1992), *Plant Physiol.*, *92*, 519–525.

Lin, C., W. W. Guo, E. Everson, and M. F. Thomashow (1990), *Plant Physiol.*, *94*, 1078–1083.

Lindstrom, O. M., T. Anisko, and M. A. Dirr (1995), *J. Am. Soc. Hort. Sci.*, *120*, 830–834.

Longton, R. E. (1992), in *Bryophytes and lichens in a changing environment* (J. W. Bathes and A. M. Farmer, Eds.), Clarendon Press, Oxford, UK, pp. 32–78.

Los, D., I. Horvath, L. Vigh, and N. Murata (1993), *FEBS Lett., 318*, 57–60.

Luo, M., L. Lin, R. D. Hill and S. S. Mohapatra (1991), *Plant Mol. Biol., 17*, 1267–1269.

Luo, M., J.-H. Liu, S. Mohapatra, R. D. Hill and S. S. Mohapatra (1992), *J. Biol. Chem., 267*, 15367–15374.

Lynch, D. V. and P. L. Steponkus (1987), *Plant Physiol., 83*, 761–767.

Manhart, J. R. and J. D. Palmer (1990), *Nature, 345*, 268–270.

Mäntylä, E., V. Lång, and E. T. Palva (1995), *Plant Physiol., 107*, 141–148.

Marcellos, H. and W. V. Single (1979), *Cryobiology, 16*, 74–77.

Maresca, B. and A. R. Cossins (1993), *Nature, 365*, 606–607.

McKersie, B. D. and Y. Y. Leshem (1994), in *Stress and stress coping in cultivated plants*, Kluwer Academic Publishers, Dordrecht, The Netherlands.

McKersie, B. D., Y. Chen, M. deBeus, S. R. Bowley, C. Bowler, D. Inze, K. D'Halluin, and J. Botterman (1993), *Plant Physiol., 103*, 1150–1163.

McKersie, B. D., and Y. Y. Leshem (1994), *Stress and stress coping in cultivated plants*, Kluwer Academic Publishers, Dordrecht, The Netherlands.

Meeuse, B.J.D. (1975), *Annu. Rev. Plant Physiol., 26*, 117–126.

Mohapatra, S. S., R. J. Poole, and R. S. Dhindsa (1987), *Plant Physiol., 84*, 1172–1176.

Mohapatra, S. S., L. Wolfraim, R. J. Poole, and R. S. Dhindsa (1989), *Plant Physiol., 89*, 375–380.

Mohapatra, S. S., R. J. Poole, and R. S. Dhindsa (1988), *Plant Physiol., 87*, 468–473.

Monroy, A. F., Y. Castonguay, S. Laberge, F. Sarhan, L.-P Vézina and R. S. Dhindsa (1993), *Plant Physiol., 102*, 873–879.

Monroy, A. F. and R. F. Dhindsa (1995), *Plant Cell, 7*, 321–331.

Moon, B. Y., S.-I. Higashi, Z. Gombos, and N. Murata (1995), *Proc. Natl. Acad. Sci. USA, 92*, 6219–6223.

Mooney, H. A., P. J. Weisser, and S. L. Gulmon (1977), *Flora, 166*, 117–124.

Murata, N. (1983), *Plant Cell Physiol., 24*, 81–86.

Murata, N. and J. Yamaya (1984), *Plant Physiol., 74*, 1016–1024.

Murata, N., O. Ishizaki-Nishizawa, S. Higashi, H. Hayashi, Y. Tasaka and I. Nishida (1992), *Nature, 356*, 710–713.

Neven, L. G., D. W. Haskell, C. L. Guy, N. Denslow, P. A. Klein, L. G. Green, and A. Silverman (1992), *Plant Physiol., 99*, 1362–1369.

Neven, L. G., D. W. Haskell, A. Hofig, Q. B. Li, and C. L. Guy (1993) *Plant Mol. Biol., 21*, 291–305.

Nobel, P. S. (1980), *Ecology, 61*, 1–7.

Nobel, P. S. (1988), in *Plants and temperature* (S. P. Long and F. I. Woodward, Eds.), The Company of Biologists Ltd. Cambridge, UK, pp. 1–23.

Nordin, K., P. Heino, and E. T. Palva (1991), *Plant Mol. Biol., 16*, 1061–1071.

Olien, C. R. and G. E. Lester (1985), *Crop Sci., 25*, 288–290.

Ouellet, F., M. Houde, and F. Sarhan (1993), *Plant Cell Physiol., 34*, 59–65.

Pakker, H., A. M. Breeman, W. F. Prud'homme van Reine, and C. van den Hoek (1995), *J. Phycol., 31*, 499–507.

Parker, J. (1963), *Bot. Rev.*, *29*, 123–201.

Patterson, B. D. and M. S. Reid (1990), in *Chilling injury of horticultural crops* (C. Y. Wang, Ed.), Boca Raton, FL, CRC Press, pp. 87–112.

Poovaiah, B. W. and A.S.N. Reddy (1993), *Crit. Rev. Plant Sci.*, *12*, 185–211.

Prentice, I. C., W. Kramer, S. P. Harrison, R. Leemans, R. A. Monserud, and A. M. Solomon (1992), *J. Biogeogr.*, *19*, 117–134.

Quamme, H. A. (1978), *J. Am. Soc. Hort. Sci.*, *103*, 59–61.

Quamme, H. A. (1985), *Acta Hort.*, *168*, 11–30.

Quamme, H. A. (1991), *Hortsci.*, *26*, 513–517.

Quamme, H. A., W. A. Su, and L. J. Veto (1995), *Am. Soc. Hort. Sci.*, *120*, 814–822.

Rada, F. G. Goldstein, A. Azocar, and F. Meinzer (1985), *J. Exp. Bot.*, *36*, 989–1000.

Raison, J. K. and G. R. Orr (1990), in *Chilling injury of horticultural crops* (C. Y. Wang, Ed.), CRC Press, Boca Raton, FL, pp. 145–164.

Rees ap, T. W. A. Fuller, and J. H. Green (1981), *Planta*, *152*, 79–86.

Robertson, A. J. and L. V. Gusta (1986), *Can. J. Bot.*, *64*, 2758–2763.

Ross, D. A. (1995), *Introduction to oceanography*, HarperCollins, New York.

Roughan, P. G. (1985), *Plant Physiol.*, *77*, 740–746.

Ruthsatz, B. (1978), *Darwiniana*, *21*, 492–539.

Sakai, A. and W. Larcher (1987), *Frost survival of plants.* Springer Verlag, New York.

Sakai, A. and C. J. Weiser (1973), *Ecology*, *54*, 118–126.

Sakai, A., D. M. Paton, and P. Wardle (1981), *Ecology*, *62*, 563–570.

Saltveit, Jr., M. E. and L. L. Morris (1990), in *Chilling injury of horticultural crops* (C. Y. Wang, Ed.), CRC Press, Boca Raton, FL, pp. 3–15.

Schaffer, M. A. and R. L. Fischer (1988), *Plant Physiol.*, *87*, 431–436.

Schwintzer, C. R. (1971), *Plant Physiol.*, *48*, 203–207.

Scott, A. J. (1995), *Canada's vegetation. A world perspective.* McGill-Queen's University Press, Montreal, Kingston.

Sekiya, J., K. Nagai, and T. Miyata (1992), in *Biochemistry and molecular biology of membrane and storage lipids of plants* (N. Murata and C. Somerville, Eds.) The American Society of Plant Physiologists, Rockville Maryland, pp. 89–95.

Shewfelt, R. L. and A. C. Purvis (1995), *Hortci.*, *30*, 213–218.

Showalter, A. M. (1993), *Plant Cell*, *5*, 9–23.

Siminovitch, D. (1981), *Cryobiology*, *18*, 166–185.

Smirnoff, N. and S. V. Colombé (1988), *J. Exp. Bot.*, *39*, 1097–1108.

Spicer, R. A. (1989), *Trans. Roy. Soc. Edinburgh: Earth Sci.*, *80*, 321–329.

Spicer, R. A. and J. L. Chapman (1990), *Tree*, *5*, 279–284.

Steponkus, P. L., M. Uemura and M. S. Webb (1993), in *Advances in low temperature biology* (P. L. Steponkus, Ed.), J.A.I. Press, Greenwich, CT, pp. 183–210.

Stone, J. M., J. P. Palta, J. B. Bamberg, L. S. Weiss and J. F. Harbage (1993), *Proc. Natl. Acad. Sci. USA*, *90*, 7869–7873.

Uemura, M. and P. L. Steponkus (1994), *Plant Physiol.*, *104*, 479–496.

Uemura, M., R. A. Joseph and P. L. Steponkus (1995), *Plant Physiol.*, *109*, 15–30.

Vigh, L., D. A. Los, I. Horvath and N. Murata (1993), *Proc. Natl. Acad. Sci. USA, 90*, 9090–9094.

Wang, C. Y. (1990), *Chilling injury of horticultural crops*. CRC Press, Boca Raton, FL.

Wang, H., R. Datla, F. Georges, M. Lowen and A. J. Cutler (1995), *Plant Mol. Biol., 28*, 605–617.

Weiser, C. J. (1970), *Science, 169*, 1269–1278.

Williams, J. P., U. K. Mobasher, and D. Wong (1995), *Plant Lipid Metab.*, 372–377.

Wilson, J. M. (1979), in *Low temperature stress in crop plants* (J. M. Lyons, D. Graham, and J. K. Raison, Eds.), Academic Press, New York, pp. 47–66.

Wolfe, J. A. and J.G.R. Upchurch (1986), *Nature, 324*, 148–152.

Wolfraim, L. A. and R. S. Dhindsa (1993), *Plant Physiol., 103*, 667–668.

Wolfraim, L. A., R. Langis, H. Tyson and R. S. Dhindsa (1993), *Plant Physiol., 101*, 1275–1282.

Wolter, F. P., R. Schmidt and E. Heinz (1992), *Embo. J., 11*, 4685–4692.

Woodward, F. I. (1987), *Climate and plant distribution*, Cambridge University Press, Cambridge, UK.

Woodward, F. I. (1988), in *Plants and temperature* (S. P. Long and F. I. Woodward, Eds.), The Company of Biologists, Cambridge, UK, pp. 59–75.

Woodward, F. I. (1990), in *Phil. Trans. Soc. Hond.* B. 326, pp. 585–593.

Wu, J. and J. Browse (1995), *Plant Cell, 7*, 17–27.

Yamaguchi-Shinozaki, K. and K. Shinozaki (1994), *Plant Cell, 6*, 251–264.

Yelenosky, G. (1985), *Hort. Rev., 7*, 201–238.

Yu, H. L and C. Willemot (1996), *Plant Sci., 113*, 33–41.

Yu, H. L., C. Willemot, P. Nadeau, S. Yelle, and Y. Castonguay (1995), *Postharvest Biol. Technol., 7*, 231–241.

4

High Temperature

Sneh Lata Singla, Ashwani Pareek, and Anil Grover

INTRODUCTION

All living organisms are adapted to grow in a narrow window of temperature regime. Unlike homeothermic animals, plants are, by and large, incapable of maintaining the optimum temperature crucial for their growth, reproduction, and development. A slight increase in temperature, even transiently, may affect the physiological and biochemical processes of plants. Adverse effects of high-temperature stress have been noted during both vegetative and reproductive growth stages in various crop plants. Processes leading to floral development and quality of seeds are critically affected by abnormally high temperatures (Helm et al., 1989; Rao et al., 1992, 1995). Information on agronomic and crop physiological aspects of high-temperature stress response is covered in several reviews (Brodl, 1990; Burke, 1990; Howarth, 1991; Howarth and Ougham, 1993; McKersie and Leshem, 1994; Sage and Reid, 1994). The damaging effects of high temperature on some of the economically important crop plants are summarized in Table 4-1.

In this chapter, responses of plants, mainly the crops, to high-temperature stress are examined. The molecular aspects of the high-temperature stress response have been emphasized here as the information on these aspects is rapidly expanding, thanks to the advent of plant molecular biology and genetic engineering techniques. Discussion in this chapter is focused on:

1. How the rise in temperature may affect various physiological and biochemical reactions
2. Characteristics of heat shock genes and proteins
3. Physiological roles of the heat shock proteins

TABLE 4-1 Adverse Effects of High Temperature on Selected Crop Plants

Plant	Damaging Effect	Reference
Barley	Grain size and yield	MacNicol et al. (1993)
Brassica	Pollen viability	Rao et al. (1992)
Cowpea	Male sterility and pollen viability	Ahmed et al. (1992)
Potato	Tuber initiation and small-sized tubers	Sterrett et al. (1991)
Rice	Root and shoot growth	Seetraraman (1992)
	Pistil hyperplasia, stamen hypoplasia in spikelets	
	Grain-filling	
Summer rape	Pollen and female fertility	Morrison (1993)
Tobacco	Pollen viability and seed set	Rao et al. (1995)
Wheat	Grain weight and yield	Blum (1988)

CELLULAR RESPONSES TO HIGH TEMPERATURE

There is hardly any physiological activity/metabolic process in plants that is not adversely affected by high-temperature stress. High-temperature-induced symptoms of impairment are noted on activities associated with seedling growth and vigor, root growth, nutrient uptake, water relations of the cell, solute transport, photosynthesis, respiration, general metabolism, fertilization, and maturation of fruits. The sum total of metabolic changes elicited when living cells are subjected to a sudden and transient increase in temperature is referred to as heat shock (HS) response. Altered patterns of enzymes, membrane structure, photosynthetic activities, and protein metabolism represent principal components of the HS response.

While general observations on HS response are made in this chapter, it is important to appreciate the fact that the genetic constitution of the organism makes an inherent difference to the physiological, biochemical, and molecular alterations exhibited in response to high-temperature stress. A temperate plant like potato may exhibit heat stress at 30°C, while the same temperature may be optimal for a tropical crop species. Brassica and pea exhibit HS response (in terms of altered protein metabolism) at lower temperature values than do rice, sorghum, and maize, possibly for the same reason (Pareek et al., 1995). For antarctic algae, a mere 5°C causes high-temperature stress with regard to their molecular responses (Vayda and Yuan, 1994).

Enzyme Activities

For cells to survive during periods of high temperature, it is essential that their metabolic activities be finely modulated. Altered enzymatic activities provide an effective way to carry out such modulations (Burke, 1990; Burke et al., 1988). The alterations in enzyme activity can potentially be brought about by induction/repression of isozymes, changes in concentration of enzyme molecules (by increased or decreased rate of synthesis/degradation), or modifications of enzyme activities by substrates or effectors (Hochachka and Somero, 1984). Representative examples in support of the above alterations in enzyme activity are covered in a chapter by Burke (1990). A significant body of literature exists on isozyme changes in response to temperature stress. One such example is that of spinach glutathione reductase, which has different isozymic forms in response to cold acclimation, exhibiting different Km* values for the oxidized glutathione, which is a substrate of this enzyme (Guy and Carter, 1984). In *E. coli*, high-temperature stress causes induction of manganese-containing superoxide dismutase (MnSOD), possibly in response to a heat-mediated increase in O^-2 production (Privalle and Fridovich, 1987).

Burke et al. (1988) have put forward the concept of thermal kinetic windows (TKW) to highlight the difference in average body temperature needed to bring about selective changes in enzyme molecules. The TKW of optimal enzyme function is defined as the temperature range in which the value of the apparent Km was within 200% of the minimum apparent Km value observed for that enzyme. From the observation that the temperature ranges comprising the TKWs for wheat and cotton are 17.5–23°C and 23.5–32°C, respectively, it is evident that these plants are within the optimal range of temperature for only some fraction of the growing season (Burke et al., 1988). In such cases, molecular engineering can be employed to assist in broadening of the temperature characteristics of limiting enzymes (Burke, 1995).

Photosynthesis

Extreme temperatures cause protein denaturation, loss of membrane integrity, photoinhibition, and ion imbalance (Sage and Reid, 1994). The rate of photosynthesis in most species declines above 35°C (Sage and Reid, 1994). As each of the constituent activities of photosynthetic machinery has a unique thermal dependence, one needs to examine how the HS affects individual partial reactions of photosynthesis in order to understand how precisely temperature affects this process.

High temperature induces stomatal limitations to photosynthesis in natural environment through stomatal closure (Berry and Bjorkman, 1980). Chloroplast biogenesis in barley has been shown to be predominantly affected when plants are grown at high temperature (Smillie et al., 1978). In wheat seedlings, leaf senescence process was appreciably hastened (in terms of decline in chlorophyll content) at 35°C as compared to 25°C (Grover et al., 1986). In the same study, high temperature caused increased damage to chloroplast lamellar structure during leaf senescence. Temperature above 30°C has been shown to cause disorientation of the lamellar system in barley chloroplasts (Smillie et al., 1978). Under high light intensities, which normally induce elevated leaf temperatures, lateral migration of the light harvesting complex of photosystem II (LHCII) has been noted. This event is probably important in preventing overexcitation of photosystem II (PSII) (Sundby and Andersson, 1985). In bean chloroplasts, high-temperature stress caused disintegra-

---

*Substrate concentration at which an enzyme-catalyzed reaction proceeds at one-half its maximum velocity.

tion of grana and disruption of membrane protein-lipid interactions with a tendency to form nonbilayer structures, as well as bringing about other changes in fatty acyl residues of membrane lipids such that the functional roles of the photosynthetic membranes could be preserved under prevailing thermal conditions (Gounaris et al., 1983). Raison et al. (1982) have shown that the saturation of lipid molecules stabilizes the photosynthetic evolution of oxygen against heat inactivation. Gombos et al. (1994) addressed the question of whether fatty acid unsaturation plays a role in the heat stability of photosynthesis in *Synechocystis* PCC 6803 (desaturation of fatty acids was altered through transformation of fatty acid desaturases in this cyanobacterium). In this study, unsaturation of membrane lipids was found to stabilize photosynthesis against heat inactivation.

High-temperature-induced reduction in electron transport activities (at rate-saturating light intensities) has been documented in several plant species. Grover et al. (1986) showed that the decline in PSII-catalyzed electron transport activity ($H_2O \rightarrow$ methyl viologen) was greatly accentuated at 35°C as compared to 25°C, in wheat seedlings. The photosystem I (PSI) catalyzed electron transport activity (ascorbate + 2,6-dichlorophenolindophenol $\rightarrow$ methyl viologen) was also affected to a greater extent at 35°C as compared to 25°C, in the same study. Photosynthetic CO_2 fixation rates in leaves and intact chloroplasts of spinach (measured at 18–20°C) are substantially decreased by pretreatment at temperatures exceeding 20°C; this is probably because the ribulose-1,5-bisphosphate carboxylase oxygenase (rubisco) enzyme requires a form of activation that is temperature-sensitive (Weis, 1981). Monson et al. (1982) have reported that heat stress induces alterations in rubisco structure in such a way that its affinity for CO_2 is hampered. In wheat leaves, both inactivation as well as the per se loss of enzyme protein are possibly the cause for thermal loss of rubisco activity (Grover et al., 1986). According to Vierling and Key (1985), reduction in rubisco small subunit synthesis is correlated with the lowered steady-state level of its mRNA. In contrast, changes in rubisco large subunit synthesis show little relationship to the corresponding mRNA. The oxygenation to carboxylation ratio of rubisco increases with an increase in temperature, which results in increased sensitivity of photosynthesis to O_2 (Sage and Reid, 1994). Reduced ribulose-1,5-bisphosphate (RuBP) regeneration capacity at high temperatures may also limit photosynthesis at high temperatures (Sage and Reid, 1994).

Increased temperature affects carbohydrate metabolism through reduction in leaf starch levels, which in turn is either due to hydrolysis of starch or inhibition of starch formation (Burke, 1990; MacLeod and Duffus, 1988). High-temperature stress adversely affects UDP sucrose synthase enzyme in developing ears of barley, thereby causing an irreversible reduction in the capacity of the endosperm to convert sucrose to starch (MacLeod and Duffus, 1988). Changes in sucrose-starch balance have been noted in several plant species in response to elevated temperatures (Burke, 1990; MacLeod and Duffus, 1988).

Ultrastructural Effects

High-temperature stress causes considerable changes in the ultrastructure of cells (Ciamporova and Mistrik, 1993). Studies on high-temperature-associated membrane perturbations are considered important, as membranes represent an important barrier separating cellular activities from outside temperatures and control crucial requirements of compartmentation within the cell. Prominent ultrastructural changes in membranes accompanying HS have been reported for the nucleus, endoplasmic reticulum (ER), mitochondria, and plastids (Ciamporova and Mistrik, 1993; Collins et al., 1995). In secretory cells of barley aleurone, which have a notably high network of the ER lamellae, a principal response to high-temperature stress is the dissociation of the lamellar structure—which causes arrest of α-amylase mRNA translation (Belanger et al., 1986). Similar observations were made upon subjecting carrot root tissues to HS, in which case there was a concomitant reduction in mRNA translation for extensin protein (Brodl et al., 1987). Likewise, high humidity and heat stress cause dissociation of ER in tobacco pollen (Ciampoloni et al., 1991). High-temperature-induced dissociation of the ER lamellae may result either from declined levels of the enzymes involved in the formation of these lamellae or from altered fluidity of the lamellae (Brodl, 1990). During HS, the fatty acids associated with phosphotidylcholine (primary phospholipids of the ER) shift from longer chain unsaturated species to shorter chain saturated species (Brodl, 1990). It is suggested that this change is possibly an adaptive shift by which a membrane with decreased degree of fluidity is formed that would be able to both sustain secretion and maintain integrity at high temperatures (Brodl, 1990).

High-temperature stress causes programmed changes in the ultrastructural appearance of nucleoli. The nucleoli of heat-shocked root cells of soybean appear less granular, with reduced heterochromatin content (Mansfield et al., 1988). Heat stress (45°C) in *Allium cepa* caused loosened nucleolus with changed ratio of fibrillar and granular components (Ciamporova and Mistrik, 1993). Associated with these changes, HS induces inhibition of the RNA processing and ribosome assembly (Neumann et al., 1984).

Formation of electron-dense cytoplasmic granules is promoted by elevated temperatures. Such granules have been noted in tomato leaves (Neumann et al., 1994; Nover et al., 1983) and yeast cells (Parsell et al., 1994). In tomato, these granular complexes are mainly comprised of heat shock proteins (HSPs) and mRNA species (Nover et al., 1989). It is postulated that these structures represent transient sites for mRNA: during recovery from HS, these granules become dispersed and closely associate with polysomes to facilitate translation (Nover et al., 1989).

Maize root cells show dispersed nuclei with conspicuous stress granules, decreased density of mitochondria with decreased number of cristae, as well as dense inclusions both in mitochondria and plastids, in response to high-temperature stress (Ciamporova and Mistrik, 1993).

MOLECULAR RESPONSES TO HIGH TEMPERATURE

Heat Shock Proteins

The molecular basis of the HS response was revealed for the first time when Ritossa (1962), employing *Drosophila* as a system, found that even one minute of temperature elevation brings about an altered puffing pattern of the polytene chromosomes. Subsequently, Tissieres et al. (1974) showed that the HS condition results in an altered protein profile in the *Drosophila* cells. Further research has established that nearly all organisms, ranging from bacteria to man, respond to HS by synthesizing a new set of proteins called heat shock proteins (HSPs). In case of plants, the first report on HS-induced alterations in protein profile appeared with a gap of six years from that of the animal systems (Barnett et al., 1980). In subsequent years, HSPs have been detected in a number of crop species (Table 4-2). Fig. 4-1 shows several HSPs, examined by one- and two-dimensional protein gel electrophoresis tech-

TABLE 4-2 Selected Examples of the Detection of HSPs in Plants

Plant	HSPs	Reference
Brassica oleracea	90, 88, 86, 74, 69, 66, 47, 43, 42, 27, 23, 21, 19, 18 kDa	Fabijanski et al. (1987)
Cucurbita sp.	76, 73 kDa	Strzalka et al. (1994)
Glycine max	69–70, 24, 22, 15–18	Lin et al. (1984)
	15–18 kDa	Hsieh et al. (1992)
	101 kDa	Lee et al. (1994)
Gossypium hirsutum	100, 94, 89, 75, 60, 58, 37, 21 kDa	Burke et al. (1985)
Hordeum vulgare	94, 85, 76, 71, 39, 32, 24	Clarke and Critchley (1992)
Lycopersicon esculentum	Seven LMW HSPs (15–25 kDa)	Kato et al. (1993)
Oryza sativa	33 kDa	Fourre and Lhoest (1989)
	110 kDa	Singla and Grover (1993)
	104 kDa	Singla and Grover (1994)
	90 kDa	Pareek et al. (1995)
Phaseolus aureus	70 kDa	Wu et al. (1993)
Sorghum bicolor	94, 73, 60, 34, 30, 24	Clarke and Critchley (1994)
Triticum aestivum	16–18 kDa, 27–94 kDa	Hendershot et al. (1992)
	Twelve new HSPs (99–83, 69–35, 15, 14 kDa)	Necchi et al. (1987)
	60, 58, 46, 40, 14 kDa	Helm et al. (1989)
	118, 90, 70, 18 kDa	Kraus et al. (1995)
Vigna radiata	114, 79, 73, 70, 60, 56, 51, 46, 18 kDa	Collins et al. (1995)
Zea mays	60 kDa	Sinibaldi and Turpen (1985)
	108, 89, 84, 76, 73, 30, 23, 18 kDa	Atkinson et al. (1989)

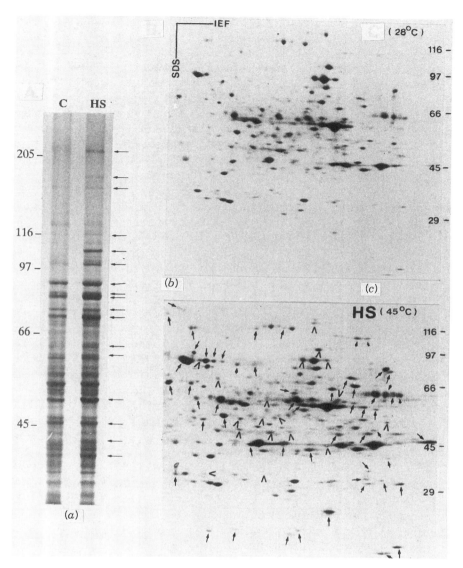

Figure 4-1. *Analysis of the electrophoretic profiles of the soluble proteins of rice shoots in response to HS treatment given to 5-day-old seedlings (Oryza sativa L. cultivar Pusa-169). (a) Proteins are resolved on 5–20% linear gradient polyacrylamide sodium dodecylsulphate gel and stained with silver nitrate. (b) Proteins are resolved on two-dimensional gel (first dimension, isoelectric focusing using 4 vols of 5–8 pH and 1 vol of 3–10 pH ampholines; second dimension, 5–20% linear gradient polyacrylamide sodium dodecylsulphate gel) followed by silver staining. C: Control; HS: Heat shock (45°C, 8 hours). Arrows mark the polypeptides that accumulate in response to HS while arrowheads mark the proteins that show a decline during HS. Details of the treatments and techniques are published elsewhere (Pareek et al., 1995; Singla and Grover, 1994).*

niques followed by highly sensitive silver staining (see legends and Pareek et al., 1995, for more details on techniques employed in this study), which accumulate in rice shoots when 5-day-old seedlings are subjected to HS (Singla and Grover, unpublished results). Detailed information on HSPs has been reviewed in several chapters/papers (Becker and Craig, 1994; Hightower, 1991; Lindquist, 1986; Lindquist and Craig, 1988; Morimoto et al., 1990; Parsell and Lindquist, 1993; Schlesinger, 1990; Schlesinger et al., 1982; Welch, 1993; Yura et al., 1993). Major biochemical and molecular aspects of plant HSPs have specifically also been covered (Gurley and Key, 1991; Harrington et al., 1994; Howarth, 1991; Howarth and Ougham, 1993; Nagao and Key, 1989; Vierling, 1990, 1991). We describe below some of the salient features of HSPs.

Molecular Weights   HSPs are either high (80–100 kDa, HMW-HSPs), intermediate (68–73 kDa, IMW-HSPs), or low molecular weight (15–20 kDa, LMW-HSPs) proteins (Lin et al., 1991; Schoffl et al., 1988). Based on their molecular mass and homology of the amino acid sequence, Neumann et al. (1989) have classified these proteins into six distinct families: HSP 110 (95–110 kDa), HSP 90 (80–95 kDa), HSP 70 (63–79 kDa), HSP 60 (53–62 kDa), HSP 20 (10–30 kDa), and HSP 8.5 (ubiquitin).

Other Inducers   Apart from HS, HSPs are synthesized and accumulated in response to a large number of factors (such as arsenite, ethanol, cadmium chloride) both in plant and animal cells (Ashburner and Bonner, 1979; Nover, 1989; Sanchez et al., 1992). Specifically in the case of plants, HSPs are also induced by water stress, abscisic acid, wounding, heavy metals, an insecticide-nematicide (mathomyl), excess NaCl, chilling, and anoxic conditions (Borkird et al., 1991; Cabane et al., 1993; Czarnecka et al., 1984, 1988; Edelman et al., 1988; Heikkila et al., 1984; Kapoor, 1986; Lafuente et al., 1991; Mocquot et al., 1987; Neven et al., 1992; Pareek et al., 1995; Rees et al., 1989).

Cellular Locations   The majority of HSPs are localized in cytoplasm and encoded by nuclear genes (Nover et al., 1989; Vierling, 1991). During HS, some HSPs are transported from the cytoplasm toward the interior of certain organelles (Nover et al., 1989; Vierling, 1990). For instance, specific HSPs have been shown to become stably localized with purified nuclei, mitochondria, and ribosomes in soybean and bean seedlings (Lin et al., 1984; Vidal et al., 1993) and with plasma membrane, mitochondria, and glyoxysomal fractions in maize (Cooper and Ho, 1987). The nuclear-encoded HSPs are also transported to the chloroplasts (Clarke and Critchley, 1992; Kloppstech et al., 1985; Marshall et al., 1990; Vierling et al., 1988). In tomato cells, HSP 70 localizes primarily in the nucleolus during HS and then redistributes to the cytoplasm during recovery (Neumann et al., 1987). HSP 70/HSC 70 (HSC means heat shock cognate; these proteins are present constitutively and show a marked resemblance to HSPs in terms of their amino acid sequence) in maize and spinach, as well as a LMW-HSP (22 kDa) in *Arabidopsis*, have been shown to be localized in the ER during HS (Anderson et al., 1994; Cooper and Ho, 1987; Helm et al., 1995).

On the other hand, certain HSPs are synthesized within the organelles under the control of organeller DNA. For instance, maize HSP 60 is shown to be synthesized in the mitochondria (Sinibaldi and Turpen, 1985).

Transient Synthesis   HSPs are mostly transiently synthesized in cells (Ashburner and Bonner, 1979; Howarth, 1991; Nagao et al., 1990). Further, the synthesis and accumulation of the HSPs exhibit both time and temperature dependence (Ashburner and Bonner, 1979; Hsieh et al., 1992; Lin et al., 1984; Singla and Grover, 1994). In *Drosophila* as well as in soybean, the presence of HSP mRNA has been detected within 10–15 minutes of the HS (Ashburner and Bonner, 1979; Kimpel and Key, 1985). Interestingly, mRNA for soybean HSPs are more stable at high temperatures than at control growth temperatures (Kimpel et al., 1990).

Coordinated and Noncoordinated Synthesis   The accumulation of HSPs can either be coordinated or noncoordinated. This essentially means that the whole spectrum of HSPs might or might not appear at the same time in the heat-shocked cells. In rice seedlings, accumulation of HSP 104 and HSP 90 was found coordinated in several aspects such as cross-inducibility by different stresses and their spatial and temporal distributions (Pareek et al., 1995). However, HSP 82 and HSP 70 (adjudged on northern blots employing specific gene probes) in rice system are shown to be noncoregulated with respect to the inducibility in response to different stresses (Breusegem et al., 1994).

Synthesis Under Field Conditions   Do HSPs accumulate in field-grown plants? This question is relevant because, in most experimental situations, induction of HSPs is accomplished by a rapid exposure to HS. However, field-grown plants are exposed to a gradual increase in temperature (diurnally or seasonally). The experimental results with laboratory-grown plants, therefore, may not be fully applicable to plants of the same species growing in the field. Furthermore, field stress conditions reflect a multitude of factors, and thus the cellular response may not be due to a single factor (high temperature) alone.

Accumulation of HSPs has been reported in field-grown cotton (Burke et al., 1985) and wheat plants (Nguyen et al., 1994). Induction of HSP mRNA has been shown in field-grown soybean (Kimpel and Key, 1985). The HSP mRNA was detectable in response to drought stress (imposed by withdrawing water for irrigation) in the later study, thus indicating a close relation between high temperature and drought stress in the field. In rice seedlings, 104 and 90 kDa HSPs are accumulated when the seedlings are subjected to water stress in the field (Pareek et al., 1995). According to Almoguera et al. (1993), the LMW-HSPs accumulated in sunflower (growing under field conditions) in response to HS and to water stress are different with respect to molecular weights and isoelectric points. Differential tissue-specificity for the water-stress-induced and HS-induced LMW-HSPs was noted in this study. Thus response to water stress and HS in context of HSPs may not be similar on finer analysis.

HSPs are constitutively synthesized in flowers, pollen, embryos, pods, and seeds of several plant species (Duck and Folk, 1994; Hernandez and Vierling, 1993).

These observations show that the HSPs are developmentally regulated proteins. Detailed understanding of the mechanisms controlling the developmental and genetic regulation of HSPs (Frova and Gorla, 1993; Jorgensen and Nguyen, 1995; Nguyen et al., 1994) could be of help in modulating the HSP synthesis at desired growth stages.

Cellular Levels   It is usually necessary to employ radiolabeling techniques for detecting HSPs synthesized in low or moderate amounts (Yucel et al., 1992). In all biological systems, HSP 70 proteins are found to be the most predominant, and the members of HSP 90 family stand next to HSP 70 in terms of their predominance (Parsell and Lindquist, 1993). The amounts of these proteins during HS are so massive that often these proteins can be seen by employing silver and Coomassie-stained sodium dodecylsulphate (SDS)-polyacrylamide gels. HSP 100 family of proteins are also visible on stained gels (Singla and Grover, 1994). Nearly 0.4% of the total soluble protein fraction in heat-shocked rice seedlings is constituted by HSP 104 (Singla and Grover, 1994). In contrast, 15–18 kDa HSPs accumulate to the extent of 0.76–0.98% of total protein in soybean, while the increase in the 18.1 kDa HSP at 40°C HS in soybean can be as high as 1000–2000 fold (DeRocher et al., 1991; Hsieh et al., 1992).

Conservation   Most HSPs are highly conserved proteins, as related proteins from diverse species show a marked homology in their structure and function. Structural resemblance of the related HSPs is noted in terms of their immunological cross-reactivity, homology of amino acid sequence of the proteins, or the homology of nucleotide sequence of the concerned genes (see the section on heat-shock genes below).

Immunological cross-reaction has proven a valuable tool in such studies. Antibodies raised against HSP 70 of yeast cross-react with wheat HSP 70 proteins (Giorini and Galili, 1991). We employed this technique to show that rice HSP 110 is immunologically related to yeast HSP 104 (Singla and Grover, 1993). Cross-reactivity of antiyeast HSP 104 antibodies has been further checked against several plant species (Lee et al., 1994; Schirmer et al., 1994). Comparison of the amino acid sequences of related HSPs has been carried out employing amino acid sequences obtained from N-terminus or tryptic peptide sequencing or from deduced amino acid sequence determined from cDNA or genomic DNA clones of the HS genes. Significant homology has been noted amongst HSP 70 as well as HSP 90 on the basis of deduced amino acid sequence (Lindquist, 1986; Parsell and Lindquist, 1993; Vierling, 1991). Functional resemblance of the HSPs has been elegantly shown in two recent studies in which mutation of yeast HSP 104 has been shown to be complementable by related genes from *Arabidopsis* (Schirmer et al., 1994) and soybean (Lee et al., 1994).

Heat-Shock Genes

Genes encoding HSPs are referred to as hs genes or hsp genes. *Drosophila*, yeast, and *E. coli* are the best studied organisms with respect to characterization of hs genes. In recent past, several plant HS genes have been cloned and sequenced (Table 4-3). In plants, more thoroughly characterized hs genes are those encoding the LMW-HSPs and HSP 70, while some work has also been carried out toward characterizing genes encoding HMW-HSPs (Nagao and Key, 1989; Vierling, 1991).

HSP 70 proteins in most eukaryotes are represented by a multigene family. Yeast has nine different hsp 70-like genes organized into five gene families (SSA 1 to SSA 4, SSB 1 and SSB 2, SSC 1, SSD 1, and KAR 2). In this case, SSA 3 and SSA 4 are heat-inducible genes, whereas SSA 1 and SSA 2 are constitutively expressed (and are thus considered HSCs). The SSA 1 and SSA 2 genes show 80% sequence identity (Craig et al., 1985). Likewise, plant hsp 70 gene family also has several members (Nagao et al., 1990; Wu et al., 1988), and not all members are heat inducible (Nagao et al., 1990). The nucleotide sequences of various plant hsp 70 genes are remarkably similar among themselves (Vierling, 1991). The identity of plant hsp 70 genes can also be extended across the biological kingdom, as is revealed by experiments in which utilization of *Drosophila* hsp 70 gene as a probe has enabled isolation of the corresponding plant genes from maize (Rochester et

TABLE 4-3 Selected Examples of the Characterization of HSP-Coding Genes in Plants

Plant	Gene/cDNA Clone	Reference
Arabidopsis thaliana	LMW-HSP-coding genes (hsp 17.4, 18.2)	Takahashi and Komeda (1989)
	Athsp 17.6	Helm and Vierling (1989)
	Athsp 22	Helm et al. (1995)
Catharanthus roseus	hsp 90	Schroder et al. (1993)
Daucus carota	hsp 70 (DChsp 70)	Lin et al. (1991)
Glycine max	LMW-HSP-coding genes (hsp 17.5, 17.6)	Nagao et al. (1985)
	Gmhsp 22	Helm et al. (1993)
	hsp 101	Lee et al. (1994)
	Ubiquitin	Xia et al. (1994)
Helianthus annus	Tetraubiquitin (HaUbiS)	Almoguera et al. (1995)
Lycopersicon esculentum	Polyubiquitin (Ubq 1-1)	Rollfinke and Pfitzner (1994)
Nicotiana tabacum	Calmodulin-binding hsp (pTC B48)	Lu et al. (1995)
Oryza sativa	hsp 70, ubiquitin	Borkird et al. (1991)
	hsp 16.9	Tzeng et al. (1992)
	Class I LMW-HSP-coding genes (pTS1, pTS3)	Tseng et al. (1993)
	hsp 82 B	Breusegem et al. (1994)
	Ubiquitin (pRMA630)	Kim et al. (1995)
Pharbitis nil	hsp 83 A	Felsheim and Das (1992)
Phaseolus vulgaris	hsp 70	Vidal et al. (1993)
Pisum sativum	LMW-HSP-coding genes (Pshsp 22.7)	Helm et al. (1993)
Spinacia oleracea	hsp 70	Anderson et al. (1994)
Zea mays	hsp 70	Bates et al. (1994)

al., 1986), *Arabidopsis* (Wu et al., 1988), and soybean (Roberts and Key, 1991). As against higher eukaryotic systems, *E. coli* has a single heat-inducible hsp 70 related gene *(dnaK)*, which shows 48% sequence homology to the *Drosophila* hsp 70 gene (Bardwell and Craig, 1984).

The structural features of the hs genes are conserved too. This can be illustrated by taking position of introns as an example. The coding sequence of the maize hsp 70 gene is interrupted by an intron (Rochester et al., 1986). The position of this intron coincides with the position of the intron located in one of the hsc 70 genes (hsc 1) of *Drosophila* (Rochester et al., 1986). A putative intron in three genes of *Arabidopsis* hsp 70 family has been shown at the same position as that in maize hsp 70 and petunia hsp 70, as well as in *Drosophila* hsc 70 genes (Nagao and Key, 1989). Recent studies have shown that there exist at least three highly conserved hsp 70 genes in maize and a conserved intron is present in all three (Bates et al., 1994).

The hs genes of eukaryotes are present at various chromosomal locations. Initial indications of the location of HS genes in *Drosophila* were based on the appearance of puffs on polytene chromosomes (Ritossa, 1962). Individual coding sequences were subsequently mapped to these sites by in situ hybridization of the heat-induced mRNAs (Mckenzie and Meselson, 1977). Further work has shown that several of the hs genes are clustered, while a few others are interspersed with independently regulated genes. There is not much information on the chromosomal location of the plant HS genes. It is expected that the current work on construction of restriction fragment length polymorphism (RFLP) maps in crop species such as rice, wheat, and tomato, using cloned cDNAs as markers, would furnish data on this aspect.

Heat-Shock Elements

Though translational regulation of plant HSPs has been reported in some isolated cases (Apuya and Zimmerman, 1992), the bulk of evidence suggests that the expression of HS genes at higher temperatures is controlled primarily at the level of transcription (Gurley and Key, 1991; Howarth and Ougham, 1993).

The transcription-level regulation of HS genes is mediated by a core DNA sequence called the heat-shock element (HSE), located in the promoter region of the HS genes (to the 5' side of the TATA box). Multiple copies of HSEs may occur clustered within about 150 nucleotides upstream from the TATA box sequence (Howarth and Ougham, 1993; Nagao et al., 1990; Raschke et al., 1988; Schoffl et al., 1988). Comparison of different HS genes has indicated that the key feature of the HSE is the presence of at least three 5 base pair (bp) modules (nGAAn) arranged as contiguous inverted repeats—nGAAnnTTCnnGAAn (Lis and Wu, 1993). The HSE sequences are highly conserved, as the 5 bp repeat 5'-nGAAn-3' motif has been found in various organisms such as yeast, slime molds, amphibians, nematodes, and mammals (Lindquist, 1986). The HSEs have been identified in various plant species including soybean, maize, and *Arabidopsis,* with configuration similar to that reported for animal cells above (Nagao et al., 1990; Gurley and Key, 1991; Rochester et al., 1986; Wu et al., 1988). Importantly, almost all plant hs genes sequenced to date contain partly overlapping multiple HSEs proximal to TATA motif (Raschke et al., 1988; Schoffl et al., 1988). Far upstream HSEs (located upstream of approximately -120 bp) seem to be of minor importance in plants (Czarnecka et al., 1989).

Gene transfer experiments have been performed in order to assess the biological relevance of the HSEs in regulating the expression of hs genes. In one such experiment, a 457 bp upstream sequence derived from the *Drosophila* hsp 70 gene was fused to a reporter gene npt II (bacterial gene that encodes for enzyme neomycin phosphotransferase, thus enabling the cells to grow on kanamycin—an antibiotic), and the resulting chimeric gene construct was introduced into tobacco cells (Spena and Schell, 1987; Spena et al., 1985). The results showed that the *Drosophila* upstream sequence was capable of expressing the NPT II protein in a heat-regulated manner in the regenerated tobacco plants. The important finding of this experiment is that the *cis*-acting elements present in *Drosophila* DNA are correctly recognized by the tobacco plant, indicating that the transcription machinery of the animal and plant HS genes is homologous. In a recent study, β-glucuronidase (gus) gene from *E. coli* under the control of promoter of hsp 18.2 gene from *Arabidopsis thaliana* has been successfully expressed in yeast in response to HS, indicating that higher plant *cis*-acting elements can function in yeast (Takahashi et al., 1993).

Plant genetic engineering technologies are of much help in understanding the signal transduction components involved in HS response also (Harrington et al., 1994). The HS treatment, being a physical parameter, is not expected to affect the transcriptional activity itself, at least not with the present understanding of the biological systems. Severin and Schoffl (1990) transformed tobacco plants with hygromycin resistance gene linked to a HS promoter. The transformed tobacco showed sensitivity to hygromycin at normal growth temperature, while resistance was seen when temperature was elevated. Any cell that will show resistance to hygromycin even at control temperature will be expected to have altered signal transduction component in this strategy. On a similar line, a promoter of hsp 18.2 of *Arabidopsis* was fused with the reporter gene gus, and this chimeric construct was transferred into *Petunia* and *Arabidopsis* (Takahashi and Komeda, 1989; Takahashi et al., 1992). The hsp-gus fusion gene showed heat-inducible gus expression in this analysis. This approach can also be used for the selection of mutants expressing hs genes constitutively (Takahashi et al., 1992). In another analysis of this kind, fusion of an HSE-containing promoter fragment from soybean with the upstream region of light-inducible gene (rbcS-3A) from pea resulted in a light-dependent, organ-specific heat-inducibility of the linked reporter gene (Strittmatter and Chua, 1987). This approach can also aid in learning how, in molecular terms, different stimuli can affect the expression of a single gene, and also how different *cis*-acting elements interact with each other to bring about proper coordination.

Recent evidence suggests that the induction of hsp genes may be regulated by a variety of signals in addition to HS (synthesis of HSPs in response to stimuli apart from HS has been mentioned in the section on other inducers, above). Thus one can ask how multiple signals influence *cis*-acting regulatory sequences controlling expression of a given HS gene. It will be of interest to finely dissect the structural features of HS promoters with this viewpoint.

Heat-Shock Factors

For the regulation of HS promoter, a positively acting transcription factor has been identified that binds specifically to the HSEs. This protein, termed heat shock factor (HSF), has been purified from several organisms including *Drosophila,* human, and yeast systems (Kingston et al., 1987; Parker and Topol, 1984; Wiederrecht et al., 1987). Genes encoding HSF have been cloned in several organisms (Lis and Wu, 1993; Wiederrecht et al., 1988). In contrast to microbial and animal systems, not much information is available on HSF of higher plants (Table 4-4). Nuclear extract of soybean plumules contain components that bind the HSE region of plant hs genes and provide protection to HSEs in DNA footprint analysis (Czarnecka et al., 1990; 1992). Three different HSF clones have been isolated from tomato cell cultures (Scharf et al., 1990). One of the potential HSF-coding genes was shown to be constitutively expressed, whereas the other two were induced by HS. Sequence comparison defined a single domain of approximately 90 amino acid residues common to all three genes and to the HSE-binding domain of yeast HSF (Scharf et al., 1990; 1993). We now consider details on the characterization of HSF, which have come mainly through employing yeast, *Drosophila,* and human cells.

Binding to Heat-Shock Elements In unstressed cells, HSF is present in both the cytoplasm and nucleus in a monomeric form and has no DNA binding activity (Morimoto, 1993). In response to HS, HSF assembles into a trimer, in which form it can bind to DNA (Morimoto, 1993). The response to HS is rapid; activation and binding of HSF to HSE is detected within minutes of temperature elevation (Sorger and Nelson, 1989). In yeast, constitutively trimeric HSF remains bound to HSEs under normal and HS conditions (Kingston et al., 1987; Sorger et al., 1987; Wu, 1984; Zimarino and Wu, 1987). The ability of HSF to promote transcription in yeast is modulated by a heat-induced change in its phosphorylation state (Sorger and Pelham, 1988; Sorger et al., 1987). Nucleotide sequences involved in oligomerization of HSF have been mapped with fine experiments (Lis and Wu, 1993). The kinetics and magnitude of DNA binding activity during HS are often proportional to the transcriptional response (Abravaya et al., 1991; Jurivich et al., 1992; Morimoto, 1993).

The transition of monomer to trimer form of HSF involves leucine zipper interactions (Rabindran et al., 1993; Westwood et al., 1991). Although this mode of regu-

TABLE 4-4 Selected Examples of the Characterization of HSF-Coding Genes in Different Organisms

System	Molecular Weight of the HSF	Reference
Arabidopsis thaliana	54 kDa	Hubel and Schoffl (1994)
Drosophila	77.3 kDa	Clos et al. (1990)
Human (He La cells)	79 kDa	Schuetz et al. (1991)
Lycopersicon peruvianum	33, 35, 50 kDa	Scharf et al. (1990, 1993)
Saccharomyces cerevisiae	93.2 kDa	Wiederrecht et al. (1988)

lation is shared among many eukaryotic species, there is a variation in the temperature at which DNA binding activity of HSF is induced. The basis of this variation has been experimentally analyzed by studying the response of a human HSF expressed in *Drosophila* cells and *Drosophila* HSF expressed in human cells (Clos et al., 1993). It was found that the temperature that induces DNA binding and trimerization of human HSF in *Drosophila* was decreased by nearly 10°C to the induction temperature for the host cell, whereas *Drosophila* HSF expressed in human cells was constitutively active. This clearly indicates that the activity of HSF in vivo is not a simple function of the absolute environmental temperature (Clos et al., 1993).

Regulation Through Heat-Shock Proteins It has recently come to light that HSPs have a role to play in transcriptional regulation of their corresponding genes. Excess cellular levels of HSPs have a negative effect on their transcriptional induction, possibly mediated by binding of HSPs with their corresponding HSFs (Abravaya et al., 1992; Morimoto, 1993). Complexes containing HSP 70 and the active trimeric form of HSF have been detected in extracts of heat-shocked cells (Abravaya et al., 1992). The association between the HSF trimer and HSP 70 may be important in the conversion of active HSF trimer to the monomer, a key event in the attenuation of HS transcriptional response (Morimoto, 1993).

Requirement for Growth Under Normal Conditions Notably, HSF is required for growth of yeast cells at normal temperature also (Sorger and Pelham, 1988). Hence, the role of HSF is not obligatory for viability under stress conditions alone. The need for HSF at normal growth temperatures has not been thoroughly investigated in other systems.

CELLULAR ROLES OF HEAT-SHOCK PROTEINS

Protein biosynthesis shows a general decline during most stress conditions (Cooper and Ho, 1983). HS-associated reduction in protein biosynthesis rate has been well documented (Bewley et al., 1983; Singla and Grover, 1994). Despite an overall reduction in protein synthesis activity, it is notable that cells preferentially synthesize HSPs. It is therefore logical to think that HSPs may be of some importance to cells in order to sustain them during stress and to recover when the stress is over.

Heat-Shock Proteins and High-Temperature Tolerance

Accumulation of HSPs has been related with acquisition of induced thermotolerance in both animal and plant systems (Carper et al., 1987; Howarth and Ougham, 1993; Nagao et al., 1990; Petersen, 1990). This contention has basically emerged from the fact that the course of the HSPs synthesis and the development of acquired thermotolerance are temporally correlated (Blumenthal et al., 1994; Lin et al., 1984). Acquired or induced thermotolerance refers to a phenomenon by which plants subjected to a pretreatment with sublethal heat stress exhibit increased tolerance to the ensuing lethal stress levels. This is illustrated in Fig. 4-2, which shows that 5-day-old rice (*Oryza sativa L.*, cultivar Pusa-169, grown at 28°C) seedlings pretreated at 40°C for 4 hours and then subjected to 50°C (4 hours) show better growth after return to 28°C, as compared to seedlings that were taken directly to 50°C from 28°C (S. L. Singla and A. Grover, unpublished results).

Further support for the hypothesis that HSPs have a role to play in thermotolerance has come from the following evidence:

1. It has come to light that the chemical agents that induce HSPs (such as cadmium, arsenite, malonate, and ethanol) also induce thermotolerance (Howarth, 1990; Lindquist, 1986).
2. Experimental systems in which HSPs are not synthesized by selective mutagenesis of hs genes (Sanchez and Lindquist, 1990), or where HSPs are inacti-

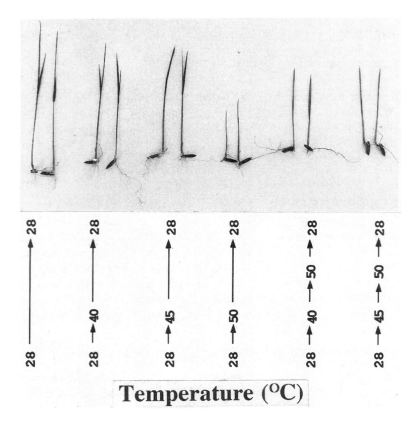

Figure 4-2. *Effect of various HS treatments on the development of induced thermotolerance in rice. 5-day-old rice (Oryza sativa L. cultivar Pusa-169) seedlings grown at 28°C were subjected to HS treatments (4 hours in each case) as shown. The analysis of growth in terms of increase in shoot length was carried out after 63 hours of recovery at 28°C following the HS.*

vated by antibody binding, are not capable of developing thermotolerance (Riabowol et al., 1988).

3. Comparison of heat-tolerant and heat-susceptible ecotypes that show natural variations in developing thermotolerance has established a high level of correlation between the extent of accumulation of HSPs and thermotolerance (DiMascio et al., 1994; Weng and Nguyen, 1992).

4. Still more support for this role of HSPs was shown by specifically inhibiting the expression of HS genes by creating a competition of HSF binding between the native HS gene promoter and that of the introduced HS gene promoter (Johnston and Kucey, 1988).

The search for the precise biochemical/molecular reactions in the cells for which HSPs provide protection is continuing. We now consider the details of this subject by taking specific HSPs into consideration (Fig. 4-3).

HSP 60   Another major class of HSPs that act as the molecular chaperons is the HSP 60 family of proteins. These proteins are found in all bacteria, mitochondria, and plastids (rubisco subunit binding protein) of eukaryotic cells (Becker and Craig, 1994). HSP 60 members differ from HSP 70 proteins structurally in that, while HSP 70 proteins are functional as monomers or dimers, HSP 60 proteins (chaperonins) are large oligomeric structures usually consisting of 14 subunits of approximately 60 kDa each arranged in two heptameric rings stacked on top of each other (Becker and Craig, 1994). HSP 60 proteins bind unfolded polypeptides and play a crucial role in catalyzing the folding of unfolded proteins and assembling higher order protein structures.

HSP 70   Physiological roles of the members of the hsp 70 gene family are beginning to emerge. It has come to light that members of this family are involved in the transport of cytoplasmically synthesized proteins to the cell organelles such as ER (Chirico et al., 1988; Deshaies et al., 1988). Transient binding of HSP 70 proteins to a wide array of newly synthesized proteins has been noticed in He La cells (Beckmann et al., 1990). This binding is considered important for the proper folding of newly synthesized polypeptides. According to the current model, many precursor polypeptides that are destined to be translocated to specific cell organelles accumulate in the cell so long as the synthesis of HSP 70 proteins is not keeping pace with its increased requirement under stress conditions. Once the HSP 70 has been synthesized in sufficient amounts, precursor proteins may be transported and processed in their normal way (Beckman et al., 1990). The targeting of proteins to lysosomes for their eventual degradation is also shown to be mediated by HSP 70 proteins in chicken, hamster, and human lung fibroblast cells (Chiang et al., 1989).

The above examples indicate that HSP 70 performs the role of molecular chaperon. The important aspect of this molecular chaperon function is to prevent the aggregation of partially folded nascent and newly completed polypeptides and to

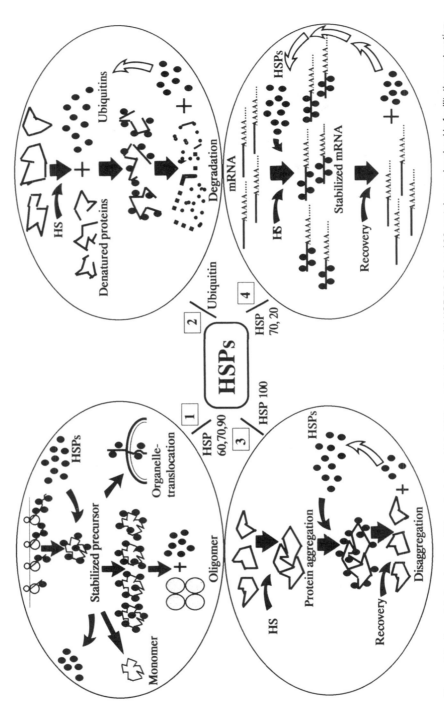

Figure 4-3. A model showing four different roles in which HSPs are implicated. (1) HSP 70, 60, and 90 proteins are involved in facilitating maturation of newly synthesized proteins in their capacity as molecular chaperons. (2) Ubiquitins help in the proteolysis of denatured proteins. (3) Yeast HSP 104 and equivalent plant proteins from Arabidopsis and soybean help in disaggregation of protein aggregates formed during heat shock. (4) HSP 70 and HSP 20 are suggested to stabilize mRNA molecules in the form of HS granules during HS conditions.

ensure their proper folding (Ellis, 1990; Ellis and van der Vies, 1991). All HSP 70 members bind ATP with high affinity and possess a weak ATPase activity that can be stimulated by binding to unfolded proteins (Rothman, 1989). In a recent study, Anderson and Guy (1995) have shown that spinach HSC 70 stabilizes the enzymatic activity of bovine adrenal glucose-6-phosphate dehydrogenase. It is possible that HSP 70 members may have a general protective role in maintaining functions of various cellular proteins.

HSP 90   Another class of HSPs that function as molecular chaperons is the HSP 90 family (Becker and Craig, 1994; Gething and Sambrook, 1992). In yeast, HSP 90-related proteins are important for the activation of steroid hormone receptors toward their binding to hormonal signal (Picard et al., 1988). HSP 90 binds ATP, undergoes a conformational change upon ATP binding, and has been implicated in facilitating the folding of denatured citrate synthase in vitro (Becker and Craig, 1994). HSPs in the range of 90 kDa have been detected in several plant species, and the corresponding genes have been cloned, sequenced, and characterized in a few instances. Interestingly, expression of HSP 90 proteins has been correlated to diverse functions such as flowering and pathogenesis, apart from induction by abiotic stresses like high and low temperatures, salinity, and drought (Breusegem et al., 1994; Felsheim and Das, 1992; Pareek et al., 1995; Schroder et al., 1993; Walther-Larsen et al., 1993).

HSP 100   HSP 104 is a major HSP in yeast whose role in thermotolerance has been unequivocally shown (Sanchez and Lindquist, 1990). Mutant yeast cells harboring a defective hsp 104 gene do not show thermotolerance under conditions in which the normal wild type cells do. This thermotolerance defect could be rescued by transformation with the wild type gene of yeast (Sanchez and Lindquist, 1990). Seeking to understand the basic mechanism of HSP 104's action, researchers hypothesized that either it protects cellular proteins from heat denaturation (much like molecular chaperons) or it aids in proteolysis of heat-damaged proteins—much like ubiquitins (Parsell et al., 1994). However, it has been found that HSP 104 plays neither of these roles; rather, it functions in a manner not previously described for any other HSP as it mediates the resolubilization of heat-inactivated proteins from insoluble aggregates (Parsell et al., 1994).

Homologues of yeast HSP 104 are noted in rice, soybean, *Arabidopsis,* wheat, maize, pea, and so on (Lee et al., 1994; Schirmer et al., 1994; Singla and Grover, 1993). It has been shown that *Arabidopsis* Athsp101 gene (homologue of yeast hsp 104) can partially substitute for the function of hsp 104 in yeast, restoring induced thermotolerance in strains carrying mutation of their own hsp 104 gene (Schirmer et al., 1994). Similarly, the soybean homologue of yeast HSP 104 (HSP 101) has been found to provide partial complementation of the thermotolerance defect of the HSP 104 mutant of yeast (Lee et al., 1994). These studies demonstrate that yeast HSP 104 and related proteins from plant systems are structurally and functionally homologous. Importantly, these reports genetically implicate the involvement of plant HSPs in thermotolerance (Schirmer et al., 1994).

Ubiquitins   Ubiquitin is a highly conserved 76 amino acid protein that is present either free or covalently bound to a variety of cytoplasmic, nuclear, and integral membrane proteins (Vierstra, 1993). During heat stress, synthesis of ubiquitin increases; hence it is considered to be an HSP (Borkird et al., 1991; Lindquist and Craig, 1988). Conjugation of ubiquitin to various acceptor proteins (translationally or posttranslationally) in eukaryotic cells regulates a number of cellular processes (Finley and Chau, 1991). It has been proposed that increased synthesis of ubiquitin during heat stress probably reflects an increased demand for removal of proteins damaged by stress. Ubiquitin has protease activity and aids in removal of denatured or nonfunctional proteins, thereby preventing their buildup to toxic levels, and recycles them as peptides or amino acids (Vierstra, 1993).

HSPs and Cold Tolerance

In several instances, accumulation of HSPs has been found in response to low temperatures; for example, accumulation of HSP 90 and 104 proteins has been noted in rice seedlings in response to low temperatures (Pareek et al., 1995). In soybean also, one of the chilling-acclimation-related proteins has been identified as a member of the HSP 70 family (Cabane et al., 1993). In a recent study, Collins et al. (1995) have shown that specific HSPs play a role in determining HS-induced chilling tolerance in mung bean hypocotyl tissues. These facts highlight that the HSPs might have a general effect in controlling tolerance to temperature stress in plants. According to Neven et al. (1992), increased synthesis of HSC 70 proteins by exposure of spinach seedlings to cold stress could result from an influence of low temperature on protein folding and/or assembly processes. Apart from HS and cold stress, accumulation of HSPs has been noted in response to water stress (Almoguera et al., 1995; Heikkila et al., 1984), salt stress (Borkird et al., 1991), and hypoxic stress conditions (see the discussion of other inducers in the section on molecular responses to high temperature, above). It is possible that a part of the response (in which HSPs are implicated) is common to all these stresses.

CONCLUSIONS AND FUTURE WORK

It is a challenge for plant scientists to evolve strategies for breeding high-temperature-tolerant crop plants, by either sexual methods or genetic engineering techniques. Any such endeavor needs a thorough understanding of the molecular basis of the high-temperature stress response. It is quite apparent from the preceding discussion that the progress made in understanding the molecular aspects of the high-temperature response is no doubt remarkable. Specific HSPs and genes have been detected and characterized, and the mechanisms of their transcriptional regulation have been understood to an extent. The recent hypothesis that the regulation of HSFs is controlled by the HSPs has attracted a lot of attention. However, several lacunae are still left to be removed. The full spectrum of HSPs from a number of plant species is yet to be documented. There are only limited instances in which HS genes have been cloned. Most information on transcriptional regulation of HS genes has come through work on limited species of these proteins (particularly 70 kDa HSP family). Most details on the roles of HSPs have been obtained employing microbial and animal systems, but the relationship of these proteins to plant processes is not adequately established.

Plant genetic engineering techniques, through which specific modulations of the desired genes can be achieved, may be an important approach in discovering the roles of HSPs. Recently, transgenic *Arabidopsis* plants have been raised, plants that constitutively express high levels of HSPs. The level of basic thermotolerance was significantly enhanced in these transgenic plants (Lee et al., 1995). However, there are multiple HSPs, and it is possible that experiments involving modulation of a single gene may not reflect the complete picture. At this point in time, it appears that the plant biotechnologists are not totally convinced whether the enhanced expression levels of heat shock genes would positively lead to successful manipulation of the major crop plants with respect to their high-temperature responses.

ACKNOWLEDGMENTS

We thank the Department of Science and Technology (DST), Government of India, for the financial support (SP/SO/A-04/93). SLS and AP are thankful to University Grants Commission and Council of Scientific and Industrial Research (CSIR), respectively, for the award of research fellowships.

REFERENCES

Abravaya, K., B. Phillips, and R. I. Morimoto (1991), *Genes Dev., 5*, 2117–2127.

Abravaya, K., M. P. Myers, S. P. Murphy, and R. I. Morimoto (1992), *Genes Dev., 6*, 1153–1164.

Ahmed, F. E., A. E. Hall, and D. A. DeMason (1992), *Am. J. Bot., 79*, 784–791.

Almoguera, C., M. A. Coca, and J. Jordano (1993), *Plant J., 4*, 947–958.

Almoguera, C., M. A. Coca, and J. Jordano (1995), *Plant Physiol., 107*, 765–773.

Anderson, J. V. and C. L. Guy (1995), *Planta, 196*, 303–310.

Anderson, J. V., L. G. Neven, Q. B. Li, D. W. Haskell, and C. L. Guy (1994), *Plant Physiol., 104*, 303–304.

Apuya, N. R. and J. L. Zimmerman (1992), *Plant Cell, 4*, 657–665.

Ashburner, M. and J. J. Bonner (1979), *Cell, 17*, 241–254.

Atkinson, B. G., L. Liu, and I. S. Goping (1989), *Genome, 31*, 698–704.

Bardwell, J. C. A. and E. A. Craig (1984), *Proc. Natl. Acad. Sci. USA, 81*, 848–852.

Barnett, T., M. Altschuler, C. N. McDaniel, and J. P. Mascarenhas (1980), *Dev. Genet., 1*, 331–340.

Bates, E. E. M., P. Vergne, and C. Dumas (1994), *Plant Mol. Biol., 25*, 909–916.

Becker, J. and E. A. Craig (1994), *Eur. J. Biochem., 219*, 11–23.

Beckmann, R. P., L. A. Mizzen, and W. J. Welch (1990), *Science, 248,* 850–854.

Belanger, F. C., M. R. Brodl, and T. H.-D. Ho (1986), *Proc. Natl Acad. Sci. USA, 83,* 1354–1358.

Berry, J. A and O. Bjorkman (1980), *Annu. Rev. Plant Physiol., 31,* 491–543.

Bewley, J. D., K. M. Larson, and J. E. T. Papp (1983), *J. Exp. Bot., 34,* 1126–1133.

Blum, A. (1988), *Plant breeding for stress environments,* CRC Press, Boca Raton, FL.

Blumenthal, C., C. W. Wrigley, I. L. Batey, and E. W. R. Barlow (1994), *Aust. J. Plant Physiol., 21,* 901–909.

Borkird, C., C. Simeons, R. Villarroel, and M. V. Montagu (1991), *Physiol. Plant, 82,* 449–457.

Breusegem, F. V., R. Dekeyser, A. B. Garcia, B. Claes, J. Gielen, and M. V. Montagu (1994), *Planta, 193,* 57–66.

Brodl, M. R. (1990), in *Environmental injury to plants* (F. Katterman, Ed.), Academic Press, San Diego, CA, pp. 113–135.

Brodl, M. R., M. Tierney, T. H.-D. Ho, and J. E. Varner (1987), *Plant Physiol., 83* (suppl.), 78.

Burke, J. J. (1990), in *Stress responses in plants: Adaptation and acclimation mechanisms* (R. G. Alscher and J. R. Cumming, Eds.), Wiley-Liss, New York, pp. 295–309.

Burke, J. J. (1995), in *Environment and plant metabolism* (N. Smirnoff, Ed.), BIOS Scientific Publishers, Oxford, UK, pp. 63–78.

Burke, J. J., J. L. Hatfield, R. R. Klein, and J. E. Mullet (1985), *Plant Physiol., 78,* 394–398.

Burke, J. J., J. R. Mahan, and J. L. Hatfield (1988), *Agron. J., 80,* 553–556.

Cabane, M., P. Calvet, P. Vincens, and A. M. Boudet (1993), *Planta, 190,* 346–353.

Carper, S. W., J. J. Duffy, and E. W. Gerner (1987), *Cancer Res., 47,* 5249–5255.

Chiang, H.-L., S. R. Terlecky, C. P. Plant, and J. F. Dice (1989), *Science, 246,* 382–385.

Chirico, W. J., M. G. Waters, and G. Blobel (1988), *Nature, 332,* 805–810.

Ciampolini, F., K. R. Shivanna, and M. Cresti (1991), *Bot. Acta, 104,* 110–116.

Ciamporova, M. and I. Mistrik (1993), *Environ. Exp. Bot., 33,* 11–26.

Clarke, A. K. and C. Critchley (1992), *Plant Physiol., 100,* 2081–2089.

Clarke, A. K. and C. Critchley (1994), *Physiol. Plant., 92,* 118–130.

Clos, J., J. T. Westwood, P. B. Becker, S. Wilson, K. Lambert, and C. Wu (1990), *Cell, 63,* 1085–1097.

Clos, J., S. Rabindran, J. Wisniewski, and C. Wu (1993), *Nature, 364,* 252–255.

Collins, G. G., X. L. Nie, and M. E. Saltveit (1995), *J. Exp. Bot., 46,* 795–802.

Cooper, P. and T.-H. D. Ho (1983), *Plant Physiol., 71,* 215–222.

Cooper, P. and T.-H. D. Ho (1987), *Plant Physiol., 84,* 1197–1203.

Craig, E. A., M. R. Slater, W. R. Boorstein, and K. Palter (1985), in *Sequence specificity in transcription and translation* (R. Calendar and L. Gold, Eds.), Liss, New York, pp. 659–667.

Czarnecka, E., L. Edelman, F. Schoffl, and J. L. Key (1984), *Plant Mol. Biol., 3,* 45–58.

Czarnecka, E., R. T. Nagao, J. L. Key, and W. B. Gurley (1988), *Mol. Cell Biol., 8,* 1113–1122.

Czarnecka, E., J. L. Key, and W. B. Gurley (1989), *Mol. Cell Biol., 9,* 3457–3463.

Czarnecka, E., P. C. Fox, and W. B. Gurley (1990), *Plant Physiol.*, *94*, 935–943.

Czarnecka, E., J. C. Ingersoll, and W. B. Gurley (1992), *Plant Mol. Biol.*, *19*, 985–1000.

DeRocher, A. E., K. W. Helm, L. M. Lauzon, and E. Vierling (1991), *Plant Physiol.*, *96*, 1038–1047.

Deshaies, R. J., B. D. Koch, M. Werner-Washburne, E. A. Craig, and R. Schekman (1988), *Nature*, *332*, 800–805.

DiMascio, J. A., P. M. Sweeney, T. K. Danneberger, and J. C. Kamalay (1994), *Crop Sci.*, *34*, 798–804.

Duck, N. B. and W. R. Folk (1994), *Plant Mol. Biol.*, *26*, 1031–1039.

Edelman, L., E. Czarnecka, and J. L. Key (1988), *Plant Physiol.*, *86*, 1048–1056.

Ellis, R. J. (1990), *Nature*, *346*, 710.

Ellis, R. J. and S. M. van der Vies (1991), *Annu. Rev. Biochem.*, *60*, 321–347.

Fabijanski, S., I. Altosaar, and P. G. Arnison (1987), *J. Plant Physiol.*, *128*, 29–38.

Felsheim, R. F. and A. Das (1992), *Plant Physiol.*, *100*, 1764–1771.

Finley, D. and V. Chau (1991), *Annu. Rev. Cell Biol.*, *7*, 25–69.

Fourre, J. L. and J. Lhoest (1989), *Plant Sci.*, *61*, 69–74.

Frova, C. and M. S. Gorla (1993), *Theor. Appl. Genet.*, *86*, 213–220.

Gething, M. J. and J. Sambrook (1992), *Nature*, *355*, 33–45.

Giorini, S. and G. Galili (1991), *Theor. Appl. Genet.*, *82*, 615–620.

Gombos, Z., H. Wada, E. Hideg, and N. Murata (1994), *Plant Physiol.*, *104*, 563–567.

Gounaris, K., W. P. Williams, and P. J. Quinn (1983), *Biochem. Soc. Trans.*, *11*, 388–389.

Grover, A., S. C. Sabat, and P. Mohanty (1986), *Plant Cell Physiol.*, *27*, 117–126.

Gurley, W. B. and J. L. Key (1991), *Biochemistry*, *30*, 1–11.

Guy, C. L. and J. V. Carter (1984), *Cryobiology*, *21*, 454–464.

Harrington, H. M., S. Dash, N. Dharmasiri, and S. Dharmasiri (1994), *Aust. J. Plant Physiol.*, *21*, 843–855.

Heikkila, J. J., J. E. T. Papp, G. A. Schultz, and D. Bewley (1984), *Plant Physiol.*, *76*, 270–274.

Helm, K. W. and E. Vierling (1989), *Nucleic Acids Res.*, *17*, 7995.

Helm, K. W., N. S. Petersen, and R. H. Abernethy (1989), *Plant Physiol.*, *90*, 598–605.

Helm, K. W., P. R. LaFayette, R. T. Nagao, J. L. Key, and E. Vierling (1993), *Mol. Cell Biol.*, *13*, 238–247.

Helm, K. W., J. Schmeits, and E. Vierling (1995), *Plant Physiol.*, *107*, 287–288.

Hendershot, K. L., J. Weng, and H. T. Nguyen (1992), *Crop Sci.*, *32*, 256–261.

Hernandez, L. D. and E. Vierling (1993), *Plant Physiol.*, *101*, 1209–1216.

Hightower, L. E. (1991), *Cell*, *66*, 191–197.

Hochachka, P. W. and G. N. Somero (1984), *Biochemical adaptation*, Princeton University Press, Princeton, NJ.

Howarth, C. J. (1990), *J. Exp. Bot.*, *41*, 877–883.

Howarth, C. J. (1991), *Plant Cell Environ.*, *14*, 831–841.

Howarth, C. J. and H. J. Ougham (1993), *New Phytol.*, *125*, 1–26.

Hsieh, M.-H., J.-T., Chen, T.-L. Jinn, Y.-M. Chen, and C.-Y. Lin, (1992), *Plant Physiol.*, *99*, 1279–1284.

Hubel, A. and F. Schoffl (1994), *Plant Mol. Biol.*, *26*, 353–362.

Johnston, R. N. and B. L. Kucey (1988), *Science*, *242*, 1551–1554.

Jorgensen, J. A. and H. T. Nguyen (1995), *Theor. Appl. Genet.*, *91*, 38–46.

Jurivich, D. A., L. Sistonen, R. A. Kroes, and R. I. Morimoto (1992), *Science*, *255*, 1243–1245.

Kapoor, M. (1986), *Int. J. Biochem.*, *18*, 15–29.

Kato, S., K. Yamagishi, F. Tatsuzawa, K. Suzuki, S. Takano, M. Eguchi, and T. Hasegawa (1993), *Plant Cell Physiol.*, *34*, 367–370.

Kim, H. U., C. H. Yun, W. S. Cho, S. K. Kang, and T. Y. Chung (1995), *Plant Physiol.*, *108*, 865.

Kimpel, J. A. and J. L. Key (1985), *Plant Physiol.*, *79*, 672–678.

Kimpel, J. A., R. T. Nagao, V. Goekjian, and J. L. Key (1990), *Plant Physiol.*, *94*, 988–995.

Kingston, R. E., T. J. Schuetz, and Z. Larin (1987), *Mol. Cell Biol.*, *7*, 1530–1534.

Kloppstech, K., G. Meyer, G. Schuster, and I. Ohad (1985), *EMBO J.*, *4*, 1901–1909.

Kraus, T. E., K. P. Pauls, and R. A. Fletcher (1995), *Plant Cell Physiol.*, *36*, 59–67.

Lafuente, M. T., A. Bewer, M. G. Guye, and M. E. Saltviet, Jr. (1991), *Plant Physiol.*, *95*, 443–449.

Lee, Y. R. J., R. T. Nagao, and J. L. Key (1994), *Plant Cell*, *6*, 1889–1897.

Lee, J. H., A. Hubel, and F. Schoffl (1995), *Plant J.*, *8*, 603–612.

Lin, X., M.-S. Chern, and J. L. Zimmerman (1991), *Plant Mol. Biol.*, *17*, 1245–1249.

Lin, C.-Y., J. K. Roberts, and J. L. Key (1984), *Plant Physiol.*, *74*, 152–160.

Lindquist, S. (1986), *Annu. Rev. Biochem.*, *55*, 1151–1191.

Lindquist, S. and E. A. Craig (1988), *Annu. Rev. Genet.*, *22*, 631–677.

Lis, J. and C. Wu (1993), *Cell*, *74*, 1–4.

Lu, Y.-T., M. A. N. Dharmasiri, and H. M. Harrington (1995), *Plant Physiol.*, *108*, 1197–1202.

MacLeod, L. C. and C. M. Duffus (1988), *Aust. J. Plant Physiol.*, *15*, 367–375.

MacNicol, P. K., J. V. Jacobsen, M. M. Keys and I. M. Stuart (1993), *J. Cereal Sci.*, *18*, 61–68.

Mansfield, M. A., W. L. Lingle, and J. L. Key (1988), *J. Ultrastruct. Mol. Struct. Res.*, *99*, 96–105.

Marshall, J. S., A. E. DeRocher, K. Keegastra, and E. Vierling (1990), *Proc. Natl. Acad. Sci. USA*, *87*, 374–378.

McKenzie, S. L. and M. Meselson (1977), *J. Mol. Biol.*, *117*, 279–283.

McKersie, B. D. and Y. Y. Leshem (1994), *Stress and stress coping in cultivated plants*, Kluwer Academic, Dordrecht, The Netherlands, p. 256.

Mocquot, B., B. Richard, and A. Pradet (1987), *Biochimie*, *69*, 677–681.

Monson, R. K., M. A. Stidham, G. J. Williams, III, G. E. Edwards, and E. G. Usibe (1982), *Plant Physiol.*, *69*, 921–928.

Morimoto, R. I. (1993), *Science*, *259*, 1409–1410.

Morimoto, R. I., A. Tissieres, and C. Georgopoulos (1990), *Stress proteins in biology and medicine*, Coldspring Harbor Laboratory Press, Coldspring Harbor, NY.

Morrison, M. J. (1993), *Can. J. Bot., 71,* 303–308.

Nagao, R. T. and J. L. Key (1989), in *Cell culture and somatic cell genetics of plants* (J. Schell and I. K. Vasil, Eds.), Vol. 6, Academic Press, New York, pp. 297–328.

Nagao, R. T., E. Czarnecka, W. B. Gurley, F. Schoffl, and J. L. Key, (1985), *Mol. Cell Biol., 5,* 3417–3428.

Nagao, R. T., J. A. Kimpel, and J. L. Key (1990), *Adv. Genet., 28,* 235–274.

Necchi, A., N. E. Pogna, and S. Mapelli (1987), *Plant Physiol., 84,* 1378–1384.

Neumann, D., K.-D. Scharf, and L. Nover (1984), *Eur. J. Cell Biol., 34,* 254–264.

Neumann, D., U. Z. Nieden, R. Manteuffel, G. Walter, K.-D. Scharf, and L. Nover (1987), *Eur. J. Cell Biol., 43,* 71–81.

Neumann, D., L. Nover, B. Parthier, R. Rieger, K.-D. Scharf, R. Wollgiehn, and V. Z. Nieden (1989), *Biol. Zentralbl., 108,* 1–156.

Neumann, D., O. Lichtenberger, D. Gunther, K. Tschiersch, and L. Nover (1994), *Planta, 194,* 360–367.

Neven, L. G., D. W. Haskell, C. L. Guy, N. Denslow, P. A. Klein, L. G. Green, and A. Silverman (1992), *Plant Physiol., 99,* 1362–1369.

Nguyen, H. T., C. P. Joshi, N. Klueva, J. Weng, K. L. Hendershot, and A. Blum (1994), *Aust. J. Plant Physiol., 21,* 857–867.

Nover, L. (1989), in *Heat shock and other stress response systems of plants* (L. Nover, D. Neumann, and K.-D. Scharf, Eds.), Springer Verlag, New York, pp. 30–43.

Nover, L., K.-D. Scharf, and D. Neumann (1983), *Mol. Cell Biol., 3,* 1648–1655.

Nover, L., K.-D. Scharf, and D. Neumann (1989), *Mol. Cell Biol., 9,* 1298–1308.

Pareek, A., S. L. Singla, and A. Grover (1995), *Plant Mol. Biol., 29,* 293–301.

Parker, C. S. and J. Topol (1984), *Cell, 37,* 273–283.

Parsell, D. A. and S. Lindquist (1993), *Annu. Rev. Genet., 27,* 437–496.

Parsell, D. A., A. S. Kowal, M. A. Singer, and S. Lindquist (1994), *Nature, 372,* 475–478.

Petersen, N. S. (1990), *Adv. Genet., 28,* 275–296.

Picard, D., S. J. Salser, and K. R. Yamamoto (1988), *Cell, 54,* 1073–1080.

Privalle, C. T. and I. Fridovich (1987), *Proc. Natl. Acad. Sci. USA, 84,* 2723–2726.

Rabindran, S. K., R. I. Haroun, J. Clos, J. Wisniewski, and C. Wu (1993), *Science, 259,* 230–234.

Raison, J. K., C. S. Pike, and J. A. Berry (1982), *Plant Physiol., 70,* 215–218.

Rao, G. U., A. Jain, and K. R. Shivanna (1992), *Ann. Bot., 68,* 193–198.

Rao, G. U., K. R. Shivanna, and V. K. Sawhney (1995), *Current Sci., 69,* 351–355.

Raschke, E., G. Baumann, and F. Schoffl (1988), *J. Mol. Biol., 199,* 549–557.

Rees, C. A. B., A. M. Gullons, and D. B. Walden (1989), *Plant Physiol., 90,* 1256–1261.

Riabowol, K. T., L. A. Mizzen, and W. J. Welch (1988), *Science, 242,* 433–436.

Ritossa, F. (1962), *Experientia, 18,* 571–573.

Roberts, J. K. and J. L. Key (1991), *Plant Mol. Biol., 16,* 671–683.

Rochester, D. E., J. A. Winter, and D. M. Shah (1986), *EMBO J., 5,* 451–458.

Rollfinke, I. K. and U. M. Pfitzner (1994), *Plant Physiol., 104,* 299–300.

Rothman, J. (1989), *Cell, 59,* 591–601.

Sage, R. F. and C. D. Reid (1994), in *Plant environment interactions* (R. E. Wilkinson, Ed.), Marcel Dekker, New York, pp. 413–449.

Sanchez, Y. and S. L. Lindquist (1990), *Science, 248,* 1112–1115.

Sanchez, Y., J. Taulien, K. A. Barkovich, and S. Lindquist (1992), *EMBO J., 11,* 2357–2364.

Scharf, K.-D., S. Rose, W. Zott, F. Schoffl, and L. Nover (1990), *EMBO J., 9,* 4495–4501.

Scharf, K.-D., S. Rose, J. Thierfelder, and L. Nover (1993), *Plant Physiol., 102,* 1355–1356.

Schirmer, E. C., S. Lindquist, and E. Vierling (1994), *Plant Cell, 6,* 1899–1909.

Schlesinger, M. J. (1990), *J. Biol. Chem., 265,* 12111–12114.

Schlesinger, M. J., M. Ashburner, and A. Tissieres (1982), *Heat shock: From bacteria to man,* Coldspring Harbor Laboratory Press, Coldspring Harbor, NY.

Schoffl, F., G. Baumann, and E. Raschke (1988), in *Plant gene research* (D. P. S. Verma and R. B. Goldberg, Eds.), Springer-Verlag, Vienna, Austria, pp. 253–273.

Schroder, G., M. Beck, J. Eichel, H.-P. Vetter, and J. Schroder (1993), *Plant Mol. Biol., 23,* 583–594.

Schuetz, T. J., G. J. Gallo, L. Sheldon, P. Tempst, and R. E. Kingston (1991), *Proc. Natl. Acad. Sci. USA, 88,* 6911–6915.

Seetraraman, R. (1992), in *Handbook of agriculture* (A. M. Wadhwani, Ed.), Indian Council of Agricultural Research, New Delhi, pp. 760–790.

Severin, K. and F. Schoffl (1990), *Plant Mol. Biol., 15,* 827–833.

Singla, S. L. and A. Grover (1993), *Plant Mol. Biol., 22,* 1177–1180.

Singla, S. L. and A. Grover (1994), Plant Sci., 97, 23–30.

Sinibaldi, R. M. and T. Turpen (1985), *J. Biol. Chem., 260,* 15382–15385.

Smillie, R. M., C. Critchley, J. M. Bain, and R. Nott (1978), *Plant Physiol., 62,* 191–196.

Sorger, P. K. and H. C. M. Nelson (1989), *Cell, 59,* 807–813.

Sorger, P. K. and H. R. B. Pelham (1988), *Cell, 54,* 855–864.

Sorger, P. K., M. J. Lewis, and H. R. B. Pelham (1987), *Nature, 329,* 81–85.

Spena, A. and J. Schell (1987), *Mol. Gen. Genet., 206,* 436–440.

Spena, A., R. Hain, U. Zeirvogel, H. Saedler, and J. Schell (1985), *EMBO J., 4,* 2739–2743.

Sterrett, S. B., G. S. Lee, M. R. Henninger, and M. Lentner (1991), *J. Am. Soc. Hort. Sci., 116,* 701–705.

Strittmatter, G. and N. H. Chua (1987), *Proc. Natl. Acad. Sci. USA, 84,* 8986–8990.

Strzalka, K., R. Tsugeki, and M. Nishimura (1994), *Folia Histochem. Cytobiol., 32,* 45–49.

Sundby, C. and B. Andersson (1985), *FEBS Lett., 191,* 24–27.

Takahashi, T. and Y. Komeda (1989), *Mol. Gen. Genet., 219,* 365–372.

Takahashi, T., S. Naito, and Y. Komeda (1992), *Plant Physiol., 99,* 383–390.

Takahashi, T., N. Yabe, and Y. Komeda (1993), *Plant Cell Physiol., 34,* 161–164.

Tissieres, A., H. K. Mitchell, and U. M. Tracy (1974), *J. Mol. Biol., 84,* 389–398.

Tseng, T. S., S. S. Tzeng, K. W. Yeh, F. C. Chang, Y. M. Chen, and C. Y. Lin (1993), *Plant Cell Physiol., 34,* 165–168.

Tzeng, S. S., K. W. Yeh, Y. M. Chen, and C. Y. Lin (1992), *Plant Physiol., 99,* 1723–1725.

Vayda, M. E. and M. L. Yuan (1994), *Plant Mol. Biol., 24,* 229–233.

Vidal, V., B. Ranty, M. Dillen Scheider, M. Charpenteau, and R. Ranjeva (1993), *Plant J.*, *3*, 143–150.

Vierling, E. (1990), in *Stress responses in plants: Adaptation and acclimation mechanisms* (R. G. Alscher and J. R. Cumming, Eds.), Wiley-Liss, New York, pp. 357–375.

Vierling, E. (1991), *Annu. Rev. Plant Physiol. Plant Mol. Biol.*, *42*, 579–620.

Vierling, E. and J. L. Key (1985), *Plant Physiol.*, *78*, 155–162.

Vierling, E., R. T. Nagao, A. E. DeRocher, and L. M. Harris (1988), *EMBO J.*, *7*, 575–581.

Vierstra, R. D. (1993), *Annu. Rev. Plant Physiol. Plant Mol. Biol.*, *44*, 385–410.

Walther-Larsen, H., J. Brandt, D. B. Collinge, and H. Thordal-Christensen (1993), *Plant Mol. Biol.*, *21*, 1097–1108.

Weis, E. (1981), *Planta*, *151*, 33–39.

Welch, W. J. (1993), *Sci. Am.*, *268*, 34–41.

Weng, J. and H. T. Nguyen (1992), *Theor. Appl. Genet.*, *84*, 941–946.

Westwood, J. L., J. Clos, and C. Wu (1991), *Nature*, *353*, 822–827.

Wiederrecht, G., D. J. Shuey, W. A. Kibbe, and C. S. Parker (1987), *Cell*, *48*, 507–515.

Wiederrecht, G., D. Seto, and C. S. Parker (1988), *Cell*, *54*, 841–853.

Wu, C. (1984), *Nature*, *311*, 81–84.

Wu, C. H., T. Caspar, J. Browse, S. Lindquist, and C. Somerville, (1988), *Plant Physiol.*, *88*, 731–740.

Wu, D. H., D. L. Laidman, and C. J. Smith (1993), *J. Exp. Bot.*, *44*, 457–461.

Xia, B. S., R. N. Waterhouse, Y. Watanabe, H. Kajiwara, S. Komatsu, and H. Hirano (1994), *Plant Physiol.*, *104*, 805–806.

Yucel, M., J. J. Burke, and H. T. Nguyen (1992), *Environ. Exp. Bot.*, *32*, 125–135.

Yura, T., H. Nagai, and H. Mori (1993), *Annu. Rev. Microbiol.*, *47*, 321–350.

Zimarino, V. and C. Wu (1987), *Nature*, *327*, 727–730.

5

Drought

Helena M. O. Freitas

THE PROBLEM OF DEFINITION

It is not easy to define drought. In fact, it is an imprecise term with both popular and technical meanings. To some, it indicates a long dry spell, usually associated with insufficient precipitation, when crops decline and reservoirs shrink. To others, it is simply a combination of multiple meteorological elements.

Drought can be recognized as a human concept, and most of the current studies of drought are related to its impact on human activities, particularly those involving agriculture. Agricultural drought is defined in terms of crop growth or development. Drought can also be defined in meteorological terms, where moisture deficiency is measured against normal or average conditions, established through long-term observation (Katz and Glantz, 1977).

ARIDITY AND DROUGHT

Distinction must be made between aridity and drought. Aridity is usually considered to be the result of low average rainfall, and it is a permanent feature of the climatology of a region. The deserts of the world, for example, are permanently arid, with rainfall amounts of less than 100 mm yr^{-1}. In contrast, drought is a temporary feature, occurring when precipitation falls below normal or when near normal rainfall is made less effective by other weather conditions such as high temperature, low humidity, and strong winds (Felch, 1978).

Aridity is not required for drought to occur. Even areas normally considered humid may suffer drought from time to time, and some of the worst droughts ever experienced have occurred in areas that include some degree of aridity in their climatological makeup (Kemp, 1990). The problem lies not in the small amount of precipitation but rather in its variability. Rainfall variability is recognized as a major factor in the occurrence of drought (Oguntoyinbo, 1986).

THE DESERTIFICATION THREAT

The problems that arise in most areas where seasonal or contingent drought occur are seen as transitory, disappearing when the rain returns. If this does not happen, the land becomes progressively more arid until, eventually, desert conditions prevail. This process is called desertification.

Though there is no widely accepted definition of desertification, most modern approaches recognize this process as the result of the combined impact of adverse climatic conditions and the stress originated by human activity (Verstraete, 1986).

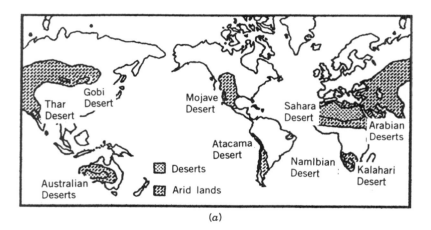

(a)

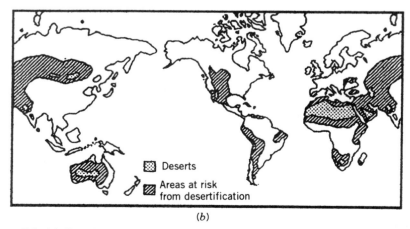

(b)

Figure 5-1. (a) *The deserts and arid lands of the world.* (b) *Desertification-prone areas.* *(Kemp, 1990.)*

According to Kemp (1990), there are many causes that together bring desertlike conditions to perhaps as much as 60,000 km^2 of the earth's surface every year and threaten 30 million km^2 more. At least 50 million people are directly at risk of losing life or livelihood in these regions (Fig. 5-1).

By the end of this century, probably one-third of earth's arable lands will be lost due to desertification and 14% of the human population living in arid zones in the world will be directly threatened (United Nations, 1977).

DROUGHT-PRONE OR SEMIARID REGIONS

The Mediterranean-Climate Regions

Areas of mediterranean climate are found between latitudes 30° and 40° north and south of the equator on the west coasts of continents. They may extend into higher latitudes on the north side of the Mediterranean Sea, and into lower latitudes in western Australia (Aschmann, 1973a). These regions closely resemble the lands bordering the Mediterranean Sea not only in climate, but also in the gross aspect of their vegetation; hence, both climate and vegetation are said to be mediterranean (Raven, 1973). Mediterranean-climate systems generally occur at these latitudes on the western or southwestern edges of the continent, whereas toward the more equatorial latitudes, and also to the continental interior, there is a transition of vegetation types from mediterranean-climate types to desert forms (di Castri and Mooney, 1973; Ehleringer and Mooney, 1983). The rarity of this climatic type, as well as its wide distribution, is apparent (Fig. 5-2).

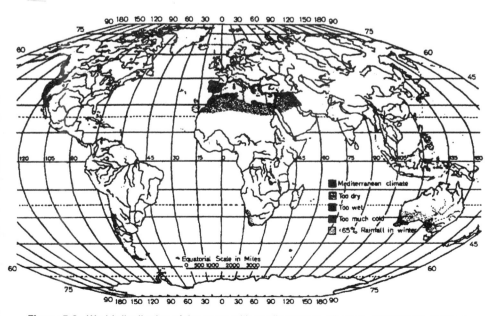

Figure 5-2. *World distribution of the areas with mediterranean climates. (Aschmann, 1973a.)*

Mediterranean-climate and desert ecosystems are characterized by low precipitation. Typically, precipitation for mediterranean-climate systems is in the range of 250–900 mm annually, and occurs almost exclusively in winter (November through April in the northern hemisphere and May through October in the southern) (Aschmann, 1973a). Desert ecosystems receive 0–250 mm annually, and seasonal precipitation may occur predominantly during either the winter months or the summer months, or in some regions during both seasons (Ehleringer and Mooney, 1983).

The mediterranean climate is severe. During the summer months temperatures are favorable for growth but moisture is limiting and atmospheric and soil drought prevails. During the winter the rain returns and water is available, but temperatures are low and growth is limited. As temperatures increase in late winter and early spring and while moisture is still available, annuals that have survived the drought period as seeds greatly accelerate growth. Although frost may occur in almost all areas of mediterranean climate, it is relatively rare and normally not severe.

Life-Form Composition of the Mediterranean Regions' Vegetation

One of the classical observations in plant geography is the phenomenon of convergence of plant life-forms in areas of similar climate (Dunn et al., 1976). This convergence is perhaps most dramatic in the mediterranean-climatic areas of the world.

Mediterranean-climate and desert ecosystems have a high diversity of life-forms. The dominant life-form of both ecosystem types is the shrub, although subshrub, herbaceous, and ephemeral life-forms constitute a significant fraction of the vegetation, and under some conditions may actually become the dominant life-forms (Ehleringer and Mooney, 1983).

As indicated before, the mediterranean-climate type occurs in California, South Africa, central Chile, and southern Australia, as well as in the mediterranean region. The lands that lie around the Mediterranean Sea represent more than half of the world area of mediterranean climate. In all of these areas the native vegetation has a similar appearance: a dense shrub dominated by woody evergreen sclerophyllous species. In deserts the tree form is generally absent, and the shrubs, for the most part, lose all or a portion of their leaves during the summer drought.

The general life strategies of plants found in representative mediterranean climate areas are similar. Drought-evading annuals predominate (Mooney and Dunn, 1970). Herbaceous perennials, which die back to the ground surface during the drought period, are also common. Similar proportions of these life forms are characteristic of the desert.

Perennial species may be distinguished into evergreen sclerophyll and drought deciduous shrubs or subshrubs. Evergreen sclerophylls are regarded as one of the most typical components of the mediterranean-type vegetation. Sclerophylly is interpreted as an adaptation to drought (Levit, 1980; Mooney, 1983; Salleo and Lo Gullo, 1990).

Ecological Significance of Sclerophylly

The compact, small, and heavily cutinized evergreen leaf is one of the most characteristic features of mediterranean-climate vegetation (Lange, 1988). Sclerophylls invest a significant amount of their carbon economy in an evergreen leaf structure with the potential for photosynthesis all year round. One obvious benefit of this thick leaf structure is the ability to avoid or delay drought stress (Dunn et al., 1976; Kummerow, 1973). The significance of the sclerophyll leaf in this regard is reflected in the relationship between tissue water potential and relative water content. As a drought adaptation, this characteristic response provides low water potentials to maintain gradients for water movement from the soil to the leaf with relatively small changes in relative cell volumes (water content) and therefore small structural changes at the sites of metabolic activity (Fig. 5-3).

There are some detrimental aspects of the sclerophyll leaf structure, related to restrictions on CO_2 uptake rates. The thick cuticle and effective stomatal responses have been shown to restrict rates of water loss and CO_2 uptake (Dunn et al., 1976; Morrow and Mooney, 1974).

Evergreen, sclerophyllous species, exploiting soil water at depth and enduring drought periods of unpredictable but not excessive duration, can extend their period of net CO_2 uptake through the entire year. Species with these characteristics effectively exploit all the environmental resources of temperature, light, and moisture. This adaptative mode is a more competitive strategy than other growth-form strategies that utilize some of these environmental resources for only limited periods of time (Dunn et al., 1976).

Examples of Mediterranean Sclerophylly

Holm oak *(Quercus ilex* ssp. *ilex)* is usually considered as a model of mediterranean sclerophylly. It is also the dominant species of most mature communities over large areas of the Mediterranean basin. Holm oak is limited in its southern range by increased summer drought and in altitude by factors associated with low temperature (Terradas and Savé, 1992). According to these authors, dense holm oak forests require over 440 mm of annual rainfall to balance the losses due to evapotranspiration and interception.

Although *Quercus ilex* cannot be regarded as a true drought avoider, it succeeds in reducing the effects of drought by recovering at least partially from water loss. Like the leaves of other sclerophyllous plants, holm oak leaves have a thick cutinized cuticle on the upper surface and a dense layer of white hairs hiding the small and abundant stomata on the lower surface. Stomatal control is efficient and cuticular transpiration is low, due to an epidermic surface with xerophytic characters and low angles of exposure protecting the rest of the canopy during evaporative peaks (Salleo and Lo Gullo, 1990; Terradas and Savé, 1992).

Another evergreen oak, very important in the mediterranean landscape of the Iberian Peninsula, is the cork-oak, *Quercus suber*. This species, one of the most important woody species occurring in Portugal, is drought resistant, and it can grow

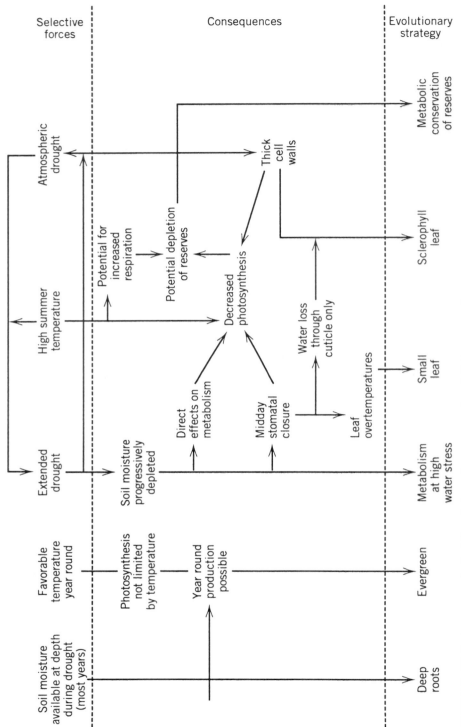

Figure 5-3. *Plant responses and adaptation strategy of evergreen sclerophylly in response to environmental stresses characteristic of mediterranean-type climates. (Dunn et al., 1976.)*

on very poor soils, due to its structural and physiological xerophytic adaptations (Oliveira et al., 1992).

The South of Portugal—An Example of a Drought-Prone Region

It is generally agreed that the Mediterranean region has suffered more than other regions in the world from landscape decadence and aridity as a result of human misuse of this landscape.

In Portugal, for many years the people of the southern region of Alentejo did not react to the increased aridity of the land. They continued to cultivate the same crops, sometimes even increasing the area under cultivation in order to obtain higher yields. In recent years the drought has been prolonged, the crops have died, and the bare earth is now being exposed to soil erosion. If the topsoil is removed, nothing will retain the rain and water quickly will run off the surface, causing further erosion. Colonization of this barren land will be difficult even for the most drought-resistant plants.

Another factor that contributed to desertification in this region was the removal of the native forest cover, often followed by substitution of introduced species that provide wood for the paper industry, like *Eucalyptus* spp. In Alentejo, the combination of drought and unsuitable agricultural practices is thus creating desertlike conditions.

Eucalyptus have a remarkable worldwide distribution and are abundant in all areas with a mediterranean-type climate. In the last 150 years they have been planted deliberately, and often hopes for their commercial utilization as lumber was unrealistic (Aschmann, 1973b). These species survive without care, growing slowly on dry slopes and quickly where their roots have access to groundwater. As Aschmann (1973b) predicted, eucalyptus is today a "prominent feature in the man-modified landscapes of all the mediterranean lands."

Desertification of large areas in Portugal is also the result of intense and constant summer fires. Although fire is an integral part of the mediterranean ecosystems and the vegetation shows a remarkable capacity to regenerate following a burn that may destroy all living tissue above ground, the increasing frequency of the fires, together with aridity, is not permitting vegetation to recover.

VEGETATION OF DRY HABITATS

General Characterization

Large territories of the mediterranean climatic areas are characterized by a physiognomically uniform vegetation. The predominant vegetation of these regions is termed "xerophytic," or plants of dry habitats. Walter and Breckle 1991 state that xerophytes are those plants growing in arid zones and on dry habitats without access to groundwater. Excluded from this group are the water-accumulating species, the succulents. The remaining group of xerophytes is further subdivided in poikilohydrous, malakophyllous, sclerophyllous, aphyllous, and stenohydrous.

Distinction has been made between drought-evading and drought-resistant species, but in arid areas all plants are exposed to drought in that their uptake of water from the soil is made difficult. The critical question is the state in which they survive the drought period. Some species, the ephemerals, survive as seeds; some, the geophytes, survive as underground organs. Others go through the drought period as living shoots in a certain state of dormancy (xerophytes, which do not store water), or as shoots in an active state as a result of stored water reserves (succulents). According to Walter and Breckle (1985) the xerophytes, not including the water-storing succulents, can be divided in three types, according to increasing drought resistance: (1) malacophyllous xerophytes, (2) sclerophyllous xerophytes, and (3) stenohydric xerophytes.

The malacophyllous xerophytes have soft and often very hairy leaves. At the start of the dry period, the leaves are relatively large; they have a high rate of transpiration (they show loosely arranged mesophyll cells with large intercellular spaces and a thin cuticle). As drought continues, the newly formed leaves are smaller and more hairy. The larger leaves die and in a long-lasting drought, so do the smaller leaves, so that only the shoot tips with small leaf primordia remain and water losses are then small. This group includes many herbaceous steppe species (Lamiaceae, Asteraceae), the semishrub of the semidesert of the Mediterranean area *Artemisia* spp., *Cistus* spp., and *Phlomis fruticosa* (Grammatikopoulos et al., 1995), and *Encelia farinosa* of the Sonoran desert.

The sclerophyllous xerophytes include both woody plants with leaves hardened by the presence of woody tissue that gives mechanical support, and broom species, which have green photosynthetic axial organs, sprouts, or twigs and cast off their leaves at the onset of drought. These plants are active during the drought, but they reduce the transpiration when water uptake becomes difficult by closing their stomata.

The stenohydric xerophytes are characterized by an almost constant cell-sap concentration. They are typical of extreme deserts where rain is rare. At the beginning of drought, they close their stomata; leaves, if present, turn yellow, and photosynthesis ceases. Cuticular transpiration is extremely low, as a result of the xeromorphic anatomy of the axial organs. These plants need minimal amounts of water and with the often small available reserves they can survive for years. However, they do not survive complete drying-out of the soil.

According to Walter and Breckle (1985) the eusucculents—to be distinguished from the halosucculents or succulent halophytes—form a special group; they differ from the other three xerophytic groups in their ability to survive for a long time on completely dry soil. When conditions are favorable, they store large amounts of water and then use it during the drought period. The root system is very shallow, lying in the upper soil layers that dry out during drought. The small absorbing roots die, and only robust, nonabsorbing roots remain alive. As soon as sufficient rain falls to wet the upper soil layers, new absorbing roots are formed, often within 24 hours, in order to refill the water-storage organs. Photosynthesis is maintained even during times of drought by means of the crassulacean acid metabolism (CAM). In succulent plants the leaves of the axial organs serve as storage organs for water. Accordingly, there is a distinction between leaf succulents—Crassulaceae, Bromeliaceae, epiphytic orchids, *Agave* and *Aloe* species—and stem succulents—Cactaceae, *Euphorbia* spp.

ASPECTS OF DROUGHT TOLERANCE IN MEDITERRANEAN SPECIES

The droughts of the mediterranean climatic regions are somewhat unpredictable as to their time of initiation and their duration (Mooney and Dunn, 1970). However, drought periods are frequent, which makes them predictable in a certain manner. The type of fluctuations in the soil moisture in these regions, with frequent natural wetting and drying, provides an opportunity for studying drought stress effects on physiological regulatory processes, canopy conductance, carbon fixation, and production (Mooney, 1987; Tenhunen et al., 1994).

Drought can be avoided by leaf drop. However, it would not be efficient to couple leaf drop or subsequent leaf initiation to an anticipatory environmental trigger, such as photoperiod, because of the unpredictability of the timing of the drought. The alternate strategy would be to hold the leaves year round and tolerate the drought in some manner. Evergreeness would solve the problem of unpredictability and would be a competitively advantageous system in terms of production. When conditions are suitable carbon fixation would occur without the energy expense and loss of time of building a production apparatus. Deep root systems could reach water at depth and shorten or eliminate the duration of the drought.

Summer drought in mediterranean areas can be so severe that perennial plants have to survive for several months without any photosynthetically active lamina. Environmental conditions are such that evergreeness is an appropriate strategy in mediterranean-type climates. The cost of maintaining evergreen leaves that can withstand periodic environmental stress is less than that of producing a new photosynthetic system annually, which would only be present during completely favorable periods (Mooney and Dunn, 1970).

Although evergreeness solves certain environmental problems, it creates others. When water stress periods arise, stomatal control can provide temporal drought-escape mechanisms. Effective cuticular resistance to water loss is also necessary. Restricting water loss in this way also restricts the influx of CO_2 and hence production. This occurs during periods of high temperatures when potential loss of energy through respiration would be greatest. Some energy-conserving mechanisms are thus required unless production has been so great as to compensate for this loss. Further, reduction of water loss by stomatal closure also results in the reduction of evaporative cooling and potential overheating of the leaf. This overheating effect is greater the larger the leaf surface. This could lead to a selection for smaller leaf sizes. Another advantage of small leaves is that they have a low ratio of blade tissue to water conduction tissue and thus are less subject to desiccation injury.

Mediterranean sclerophyll shrubs respond to seasonal drought by adjusting the amount of leaf area exposed and by reducing gas exchange via stomatal closure mechanisms. Tenhunen et al. (1990) have shown that midday stomatal closure may contribute to drought avoidance, increase water-use efficiency, and strongly alter physiological efficiency in the conversion of intercepted light energy to photoproducts.

Pistacia lentiscus is a dioecious evergreen sclerophyllous growing as a shrub or a small tree on the Serra da Arrábida, west coast of Portugal, and an important component of the *Oleo-lentiscetum* scrub—late successional stage. *Cistus* spp. are drought deciduous shrubs from early stages of succession (Correia et al., 1987). They are frequent in open and disturbed stands, but they are progressively eliminated under canopies of evergreen sclerophyllous species of the late successional stages. Correia and Catarino (1994) showed that water relations on these species are closely related with their rooting habit. *Pistacia lentiscus* has deeper and more extensive roots than *Cistus* spp., which has a less extensive root system, restricted to the upper and drier soil layers, thus tolerating lower xylem potentials. According to these authors, the root system development should be in agreement with an hydraulic resistivity.

HOW DO PLANTS COPE WITH WATER STRESS?

Over thousands of years, certain plants have adapted to life with limited moisture. Theoretically the needs of these plants are met and thus no drought exists. In reality, it is a much more complex situation because although the flora may exist in a state of equilibrium with other elements in the environment, it is a dynamic equilibrium, and the balance can be disturbed. Changes in precipitation patterns may cause dry areas to become drier. If water availability becomes very reduced, plants will suffer drought stress. Depending upon the extent of the change, plants might die from lack of moisture, they might be forced out of the area by competition with species more adapted to the new conditions, or they might survive by reducing productivity.

Most higher plants are exposed to varying degrees of water stress at some stage in ontogeny. The type of water stress may vary from small fluctuations in atmospheric humidity to extreme soil water deficits and low humidity in arid environments. With the exception of plants that tolerate extreme cellular dehydration (Gaff, 1989), adaptation to water stress involves the reduction of cell dehydration by either avoidance or tolerance of stress. Some examples of avoidance of water stress are rapid completion of ontogeny, leaf shedding, leaf rolling, and low stomatal conductance. Tolerance of water stress usually involves osmotic adjustment (Morgan, 1984).

Lack of water has been a major selective force in plant evolution, and ability to cope with water deficits is an important determinant of natural distribution of plants and of crop productivity. Plants exposed to water shortage undergo a series of physiological and biochemical, as well as molecular, changes. These changes include increase and reallocation of abscisic acid (ABA), the closure of stomata, decrease in photosynthesis, and alterations in gene expression. Metabolic or biochemical adaptations to water stress have been the subject of several studies (Hanson and Hitz, 1982).

Strategies to Reduce Water Stress

Water stress induces many morphological, phenological, anatomical, and physiological responses in higher plants, and various characteristics of plants influence these responses (Table 5-1). Ludlow (1980a, b) and Ludlow et al. (1983) proposed that responses such as leaf movements and characteristics such as deep rootedness should not be treated separately because they do not occur independently, but coincide in combinations or groupings, which Ludlow called strategies. Ludlow (1989) defined strategy as a combination or grouping of mechanistically linked responses and characteristics that comprise a particular type of behavior during periods of

TABLE 5-1 Characteristics of Plants with Different Strategies to Reduce Water Stress (LWP—leaf water potential; RWC—relative water content)

Responses and Characteristics	Consequences
Escape	
Short phenology	Low yield potential
Developmental plasticity	Good survival
Photoperiod sensitivity in crop plants	
Avoidance	
Tissues sensitive to dehydration (lethal LWP −1.5 to −2.5 MPa; lethal RWC >50%)	No penalty for growth if water uptake is maximized
Maximize water uptake	
Deep roots	Short- but no long-term cost for growth if water loss is reduced by elastic responses
Minimize water loss	Short, but no long-term cost for growth if water loss is reduced by plastic responses
Stomatal control	
Leaf movements	
Smaller leaves	
Shedding older leaves	
Low osmotic adjustment and little stomatal adjustment and photosynthetic adjustment	Good short- but poor long-term survival of water stress
Tolerance	
Tissues withstand severe dehydration (lethal LWP < −13 MPa; lethal RWC <25% RWC)	Potential for carbon acquisition and growth during water stress
Poor to moderate avoidance of dehydration (leaf rolling, deep roots)	No known metabolic costs
Moderate to high osmotic adjustment	Good long-term leaf and plant survival

Source: Ludlow (1989). By permission of SPB Academic Publishing, The Hague, Netherlands.

water stress. Such a use of this word is commonly applied in ecology after Grime (1979).

Three basic types of strategy are recognized: escape, avoidance, and tolerance (Levitt, 1972).

Escape   The life cycle is regulated in such a way that plants rarely experience water shortage. Plants survive the dry period as seeds, a particularly dehydration-tolerant form. After rain seeds germinate and plants grow, flower, and produce seed very quickly, before the water supply is exhausted. Desert annuals and short-season, annual crop and pasture are examples of this strategy. Plants with this strategy need to ensure some seed is produced, independently of the amount of rainfall.

Avoidance   Plants with this strategy have tissues that are very sensitive to dehydration and therefore must avoid water deficit situations. They must have responses and characteristics that maximize water uptake, and minimize water loss. This strategy is found in several plants (Ludlow, 1989). The highest development is found in epiphytes and desert succulents.

Tolerance   Plants with this strategy are mostly perennials and have tissues that tolerate dehydration. Most tropical C_4 grasses, many tropical C_3 forage legumes, and some crop plants belong to this group (Ludlow, 1989).

Resurrection Plants: The Extreme Tolerance to Water Stress

The most extreme examples of tolerance strategy are found in a small group of angiosperms termed poikilohydric or resurrection plants, which survive desiccation to air-dryness (Gaff, 1977; Walter, 1955).

Spores of microorganisms and plant seeds can survive periods of severe desiccation. During the germination of seeds, the acquisition of desiccation tolerance is part of the normal development program. The onset of germination disturbs tolerance and the emerging seedling rapidly loses the ability to survive desiccation. Although tolerance to dehydration in the majority of higher plants is restricted to mature seeds, there is a small group of plants—the resurrection plants—in which all vegetative organs can withstand desiccation and regain full function again upon rehydration.

One resurrection plant, *Craterostigma plantagineum,* can withstand a water loss greater than 90% for long periods and yet recover within 24 hours after rewatering (Bartels et al., 1990). Desiccation tolerance is induced in resurrection plants at moderate to severe water deficits as their protoplasm becomes air-dry. Membranes of chloroplasts and mitochondria degrade, photosynthesis and respiration decline, chlorophyll often degrades, while starch and water insoluble proteins are hydrolyzed (Gaff, 1989).

The majority of desiccation-tolerant angiosperms have been reported only in the last two decades, mainly from seasonally arid subtropical and tropical regions (Table 5-2). Little is known of their metabolic characteristics. No desiccation-tolerant

TABLE 5-2   Vascular Plant Families with Species Reported to Have Desication-Tolerant Foliage

Family	Number of Genera with Tolerant Species	Number of Species Dessication Tolerant	Number of Species in Group	Proportion of Tolerant Species (per 100,000)
		Filicopsida		
Actiniopteridaceae	1	2		
Adiantaceae	1	1		
Aspleniaceae	3	7		
Grammitidaceae	1	1		
Heminiotidaceae	2	21		
Lomariopsidaceae	1	1		
Oleandraceae	1	1		
Polypodiaceae	2	3		
Schizaceae	2	2		
Sinopteridaceae	4	41		
Group total	18	61	11,000	550
		Lycopsidae		
Selaginellaceae	1	12		
Group total	1	12	1,230	975
		Monocotyledonea		
Cyperaceae	7	12		
Liliaceae	1	6		
Poaceae	9	36		
Velloziaceae	6	34		
Group total	23	88	34,000	259
		Dicotyledonea		
Lamiaceae	1	1		
Gesneriaceae	5	8		
Myrothamnaceae	1	2		
Scrophulariaceae	3	7		
Group total	10	18	166,000	11

Source: Gaff (1989). By permission of SPB Academic Publishing, The Hague, Netherlands.

gymnosperms have been reported, and lower plants have poorly developed drought avoidance mechanisms (Gaff, 1989).

Stomatal Regulation Under Water Stress

It has long been recognized that soil drying can greatly affect both the rate at which plants grow and the rate and pattern of plant development. To a large extent, this happens because both the growth processes and the regulation of the fixation of CO_2 are sensitive to water stress. The traditional view of how plants might sense a reduction in soil water availability is that water uptake by the plant is reduced and this results in a reduction in the water content, water potential, or turgor of the leaves. This is a common situation but it is not always the case because unwatered plants can show similar or even higher shoot water potentials than plants that have been watered regularly. Fine-tuning of the stomata to leaf water status can explain how water potentials of watered and unwatered plants can be comparable. An alternative explanation is that drying of the soil can promote stomatal closure and reduction in leaf growth rate even when the plant is in shoot.

Plants under a progressive drought stress decrease their transpirational water loss by stomatal control. During long periods of water stress stomatal closure is very important to maintain a favorable water balance. It is during these times that effective means of controlling cuticular water loss are important.

Drying of the soil may result in the collapse of the water potential gradient between the xylem and the growing cells. This change would prevent water movement into the growing zone, limiting growth of plants with high leaf water potentials. The main reason for stomatal closure under water stress could be to conserve soil water, to improve water-use efficiency, to avoid damaging water deficits, or a combination of these.

If the prime reason for damage by low water potentials is through xylem cavitation, rather than through the direct effects on physiological processes that have traditionally been assumed, it might be that stomatal responses act mainly to avoid catastrophic collapse, while allowing maximal production. The physical ability of xylem tissues to resist cavitation under severe water stress is probably also a significant component of drought tolerance (Jones and Sutherland, 1991; Kolb and Davis, 1994).

In fact, low water potentials (high tensions on xylem water) are known to cause severe embolism formation in the xylem vessels of woody plants, blocking water transport and potentially causing shoot dieback. So plants like the mediterranean shrubs, surviving water potentials as low as -9 MPa, have to be extremely resistant to water-stress induced embolism (Kolb and Davis, 1994). According to these authors, susceptibility to xylem embolism cannot be used as a sole index of water stress tolerance but must be evaluated together with other traits that also impart drought resistance.

Photosynthesis and Water Stress

In general, desert plants can respond immediately to precipitation events by producing leaves that have a high photosynthetic capacity. The highest photosynthetic rates and leaf conductances for desert plants are found in ephemerals (Ehleringer et al., 1979; Mooney and Ehleringer, 1978; Mooney et al., 1976). Both C_3 and C_4 photosynthetic pathways are found in desert ephemerals. The winter ephemerals use the C_3 pathway, whereas both C_3 and C_4 pathways occur in summer ephemerals (Mulroy and Rundel, 1977).

Drought deciduous shrubs and herbaceous perennials have peak photosynthetic rates approaching those of the ephemerals (Bjorkman et al., 1972; Ehleringer and Bjorkman, 1978). Succulent subshrubs with CAM and facultative CAM metabolism and ferns have the lowest photosynthetic rates of all the life-forms (Ehleringer and Mooney, 1983). CAM plants and ferns have the greatest capacity to survive extended periods of drought.

In mediterranean-climate systems the C_3 pathway predominates in all life-forms; only in the coastal zones and at mediterranean-climate-desert ecotones can CAM be important (Ehleringer and Mooney, 1983). According to these authors, within desert ecosystems trees and shrubs possess the C_3 pathway, except for halophytic shrubs, which often possess the C_4 pathway. Succulent subshrubs are usually CAM or CAM facultative.

Comparing mediterranean-climate life-forms, evergreen species generally have lower (about one-half) photosynthetic capacities than drought-deciduous species on a leaf area basis (Gignon, 1979; Mooney et al., 1977). In a survey of the net photosynthesis rates of mediterranean-climate growth measured in field plants, Mooney (1981) gives an average rate of 8 μmol CO_2 m^{-2} s^{-1} for evergreen shrubs and trees and 15 μmol CO_2m^{-2}s^{-1} for drought-deciduous shrubs.

ROOTS: SHOOT SENSORS OF WATER STRESS

Plant water status in many arid and semiarid areas is dependent on rainfall for replenishment of near-surface soil water. These upper soil layers generally contain the greatest concentrations of both roots and nutrients, but are subjected to severe dehydration from the combined effects of root uptake and direct evaporative loss to the atmosphere. Plants have to compete with the evaporating soil surface for the water that has been added to the system, particularly when light or moderate rains occur and only the upper few centimeters of the profile may rewet (Wraith et al., 1995). Aboveground measures of plant response during and following soil drought provide only indirect indications of the associated root system's responses or of the actual soil water uptake, especially under normal conditions of spatially variable soil wetness and root system density. Improved understanding of root function during and immediately following soil drought is fundamental (Baker et al., 1992; Kosola and Eissenstat, 1994; Passioura, 1983; 1988; Wraith et al., 1995).

Plants must "sense" the drying of the soil around the root and communicate this information to the shoot by some means other than a reduction in the flux of water to the shoots. Little information is available on water relations profiles along roots, or even on root-to-root variations in water relations. Dehydration of only a few roots may generate a chemical response that moves to the shoot and is effective, even if the flux of water has not yet been reduced (Davies and Zhang, 1991). According to these authors, another possible sensing mechanism is that the plant may "measure" the increase in mechanical impedance as the soil dries, as growing roots encounter increased soil strength at reduced soil water content (Fig. 5-4). High mechanical impedance can restrict plant growth under water stress, and careful attention should be given to this aspect when assessing water stress both in field and laboratory conditions (Freitas and Mooney, 1996).

The root can communicate to the shoot perturbations in the soil environment. It

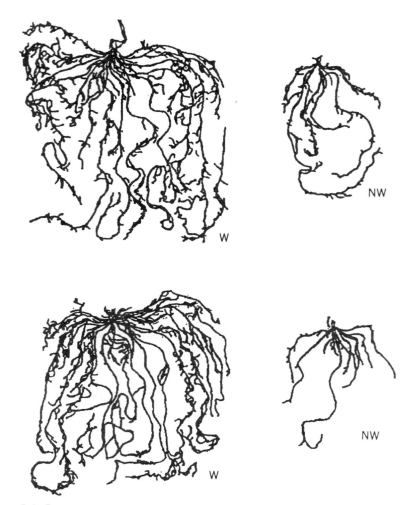

Figure 5-4. *Root branching patterns of* Bromus *plants after 12 weeks of growth in sandstone soil, watered (W) and under water stress (NW). Source:* Freitas and Mooney, 1996.

seems possible that if root activity declines for any reason the effects of increased or reduced transport of some chemical compound will be felt in the shoots (Blackman and Davies, 1985). Root signals not only influence stomatal behavior, but also regulate leaf initiation, leaf expansion, and other developmental processes (Davies and Zhang, 1991). Roots seem to be the primary sensors of water deficits, and biochemical signals transported from the roots cause many of the physiological perturbations observed in the shoots of plants subjected to drought (Schulze, 1986a, b; Turner, 1986). Signals are probably related to changes in the kind and concentration of hormones exported to the shoots, possibly in the transpiration stream. Water stress reduces cytokinin concentration in the xylem sap and increases ABA concentration (Davies and Zhang, 1991; Zhang et al., 1987).

Soil drying causes some type of chemical signaling between roots and shoots. ABA seems to be involved in signaling between different plant organs, and at transmembrane and genome levels (Davies et al., 1990). At the site of action ABA-induced stomatal closure improves water relations and increases resistance to water stress.

The Key Role of ABA

Phytohormones, especially ABA and ethylene, play a predominant role in the conversion of stressful environmental signals into changes in plant gene expression. ABA is involved in the desiccation tolerance of established plants and seeds (Voesenek and Van der Veen, 1994).

The key role of ABA as a plant hormone regulating metabolism and stomatal behavior under conditions of water stress is well established (Davies and Mansfield, 1988; Davies et al., 1990; Voesenek and Van der Veen, 1994). Attention was focused on the role of ABA in drought stress physiology, because unwatered plants show substantially enhanced ABA content (Wright and Hiron, 1969) or redistribution of existing ABA (Hartung and Davies, 1991). Also, detached leaves that are subjected to water deficit rapidly accumulate ABA within 30 minutes after applying the stress (Cohen and Bray, 1990). Moreover, ABA application reduces water loss by restricting leaf conductance (Trejo et al., 1993).

Tenhunen et al. (1994) have studied the effect of drought and the role of ABA in the shrub *Ceanothus thyrsiflorus,* under natural conditions, in a chaparral region of Southern California. Concentration of ABA in the xylem sap increased as water potential decreased in plants under drought conditions.

DROUGHT EFFECTS ON ECOSYSTEMS

What Happens to Biodiversity?

The most diverse terrestrial plant communities in the world occur on nutrient-poor soils. The association of high plant diversity with nutrient-poor habitats is not limited to mediterranean and tropical climates, nor to terrestrial plant communities. In temperate and tropical regions, in grasslands, forests, lakes, rivers, estuaries, and oceans, plant species diversity is higher in habitats relatively poor in nutrients than in nutrient-rich sites or in sites with very few nutrients (Huston, 1979; Tilman, 1983).

Extreme climate conditions may periodically lower population densities and thus increase the probability of extinction of rare species. The instability caused by environmental fluctuations may impact the biodiversity of an otherwise undisturbed environment. Considering the environmental fluctuations as a consequence of the global climatic change that leads to a more variable or extreme climate, increasing rates of species extinction may be expected.

Tilman and Haddi (1992) conducted a field study to evaluate the effect of drought on species richness and on rates of species loss and gain. Each year for nine years (1982–1990) they collected data on species composition, abundance, and richness in 46 grassland plots. A severe drought in 1988 (such severe droughts occur every 50 years in Minnesota) caused the local extinction of many rare species, and few of them recolonized the plots even after two years of normal precipitation.

The results of this experiment support the hypothesized mechanism whereby environmental variation may limit biodiversity. During the drought of 1988, when aboveground living plant biomass fell to about half of its previous level, average species richness per plot was 37% lower than during predrought conditions (1982– 1986). Despite the return to normal precipitation during the following two years, there was no significant recovery in species richness in the plots. One may argue that the species in the plots are not lost but simply dormant in the soil, in the seed bank, "waiting" for normal conditions to return. However, the seeds remained in the dormant state in 1989 and 1990, despite normal precipitation. The slow recovery after the drought suggests that species richness was recruitment limited.

CONCLUDING REMARKS

All over the world there are areas where seasonal or unpredictable droughts occur, and eventually desert conditions may predominate. It is widely accepted that this process of desertification is the result of the combined impact of adverse climatic conditions and stress originated by human activity.

Climatic change may bring about increased aridity to extended areas of the world. Thus it is important to evaluate the effects of elevated CO_2 in association with other stress factors, namely high temperature and water deficits.

Drylands and marginal zones bounding desert areas with a sparse and fragile rain-fed vegetation cover have to be managed properly in order to prevent desertification. Field projects to protect desert-prone/desertified areas against spreading of desertification by rehabilitation of degraded lands have been performed in several countries during the last few years, as for example in China (Berndtsson and Chen, 1994) and in some areas of northern Africa (Ourcival et al., 1994)

From a human perspective, drought is seen as a technological problem, an economic problem, a political problem, a cultural problem, or sometimes a multifaceted problem involving all of these (Kemp, 1990). Whatever the case, it has to be faced as an environmental problem, and fundamental to any comprehension of the situation is the relationship between society and environment in drought-prone areas.

REFERENCES

Aschmann, H. (1973a), in *Mediterranean type ecosystems: Origin and structure* (F. di Castri and H. A. Mooney, Eds.), Ecol. Stud. Vol. 7, Springer-Verlag, Berlin, pp. 11–19.

Aschmann, H. (1973b), in *Mediterranean type ecosystems: Origin and structure* (F. di Castri and H. A. Mooney, Eds.), Ecol. Stud. Vol. 7, Springer-Verlag, Berlin, pp. 363–371.

Baker, J. M., J. M. Wraith, and F. N. Dalton (1992), *Adv. Soil Sci., 19*, 53–72.

Bartels, D., K. Schneider, G. Terstappen, D. Piatkowski, and F. Salamini (1990), *Planta, 181*, 27–34.

Berndtsson, R. and H. Chen (1994), *J. Arid Environ., 27*, 127–139.

Bjorkman, O., R. W. Pearcy, A. T. Harrison, and H. A. Mooney (1972), *Science, 175*, 786–789.

Blackman, P. G. and W. J. Davies (1985), *J. Exp. Bot., 36* (162), 39–48.

Cohen, A. and E. A. Bray (1990), *Planta, 182*, 27–33.

Correia, O. A. and F. M. Catarino (1994), *Acta Oecol., 15*, 289–300.

Correia, O., F. M. Catarino, J. D. Tenhunen, and O. L. Lange (1987), in *Plant response to stress*, (J. D. Tenhunen et al., Eds.), NATO ASI Series, Vol. G15, Springer-Verlag, Berlin.

Davies, W. J. and T. A. Mansfield (1988), *ISI atlas of science—Plants & animals*, Vol. 1, pp. 263–269.

Davies, W. J. and J. Zhang (1991), *Annu. Rev. Plant Physiol. Plant Mol. Biol., 42*, 55–76.

Davies, W. J., T. A. Mansfield and A. M. Hetherington (1990), *Plant Cell Environ., 13*, 709–719.

di Castri, F. and H. A. Mooney (1973), *Mediterranean type ecosystems: Origin and structure*, Ecol. Stud. Vol. 7, Springer-Verlag, Berlin.

Dunn, E. L., F. M. Shropshire, L. C. Song, and H. A. Mooney (1976), in *The water factor and convergent evolution in mediterranean-type vegetation* (O. L. Lange, L. Kappen, and E. D. Schulze, Eds.), Springer-Verlag, Berlin, pp. 492–505.

Ehleringer, J. R. and O. Bjorkman (1978), *Plant Physiol. 62*, 185–190.

Ehleringer, J. and H. A. Mooney (1983), in *Produtivity of desert and mediterranean-climate plants* (A. Pirson and M. H. Zimmermann, Eds.). Enc. Plant Physiol., Vol. 12D, Springer-Verlag, Berlin, pp. 205–231.

Ehleringer, J. R., H. A. Mooney, and J. A. Berry (1979), *Ecology, 60*, 280–286.

Felch, R. E. (1978), in *North American droughts* (N. J. Rosenberg, Ed.), Westview Press, Boulder, CO.

Freitas, H. and Mooney, H. (1996), *Acta OECol., 17 (in press)*.

Gaff, D. F. (1977), Oecologia, 31, 95–109.

Gaff, D. F. (1989), in *Structural and functional responses to environmental stresses* (K. H. Kreeb, H. Richter, and T. M. Hinckley, Eds.), SPB Academic, The Hague, Netherlands, pp. 255–268.

Gignon, A. (1979), *Oecol. Plant, 14*, 129–150.

Grammatikopoulos, G., A. Kyparissis, and Y. Manetas (1995), *Flora, 190*, 71–78.

Grime, J. P. (1979), *Plant strategies and vegetation processes*, Wiley Interscience, New York.

Hanson, A. D. and W. D. Hitz (1982), *Annu. Rev. Plant Physiol., 33,* 163–203.

Hartung, W. and W. J. Davies (1991), in *Abscisic acid physiology and biochemistry* (W. J. Davies and H. G. Jones, Eds), BIOS, Oxford, UK, pp. 63–79.

Huston, M. (1979), *Am. Nat., 113,* 81–101.

Jones, H. G. and R. A. Sutherland (1991), *Plant, Cell Environ., 14,* 607–612.

Katz, R. W. and M. H. Glantz (1977), in *Desertification: Environmental degradation in and around arid lands* (M. H. Glantz, Ed.), Westview Press, Boulder, CO.

Kemp, D. D. (1990), *Global environmental issues. A climatological approach.* Routledge, New York.

Kolb, K. J. and S. D. Davis (1994), *Ecology, 75* (3), 648–659.

Kosola, K. R. and D. M. Eissenstat (1994), *J. Exp. Bot., 45* (280), 1639–1645.

Kummerow, J. (1973), in *Mediterranean type ecosystems: Origin and structure* (F. di Castri and H. A. Mooney, Eds.), Ecol. Stud. Vol. 7, Springer-Verlag, Berlin, pp. 157–170.

Lange, O. L. (1988), *J. Ecol., 76,* 915–937.

Levitt, J. V. (1972), *Responses of plants to environmental stresses,* Academic Press, New York.

Levitt, J. V. (1980), *Responses of plants to environmental stresses,* Vol. 2, Academic Press, New York.

Ludlow, M. M. (1980a), in *Adaptation of plants to water and high temperature stress* (N. C. Turner and P. J. Kramer, Eds.), Wiley Interscience, New York, pp. 123–138.

Ludlow, M. M. (1980b), *Trop. Grass, 14,* 136–145.

Ludlow, M. M. (1989), in *Structural and functional responses to environmental stresses* (K. H. Kreeb, H. Richter, and T. M. Hinckley, Eds.), SPB Academic, The Hague, Netherlands, pp. 269–281.

Ludlow, M. M. A. P. C. Chu, R. J. Clements, and R. G. Kerslake (1983), *Aust. J. Plant Physiol., 10,* 119–130.

Mooney, H. A. (1981), in *Ecosystems of the world, vol. 11, Mediterranean-type shrublands* (F. Di Castri, D. W. Goodall, R. L. Specht, Eds.), Elsevier, Amsterdam, pp. 249–255.

Mooney, H. A. (1983), in *Mediterranean-type ecosystems. The role of nutrients* (F. J. Kruger, D. T. Mitchell, and J. U. M. Jarvis, Eds.), Springer-Verlag, Berlin, pp. 103–119.

Mooney, H. A. (1987), in *Plant response to stress-Functional analysis in mediterranean ecosystems* (J. D. Tenhunen, F. M. Catarino, O. L. Lange, and W. C. Oechel, Eds.), Springer-Verlag, Berlin, pp. 661–668.

Mooney, H. A. and E. L. Dunn (1970), *Evolution, 24,* 292–303.

Mooney, H. A. and J. Ehleringer, (1978), *Plant Cell Environ., 1,* 307–311.

Mooney, H. A., J. Ehleringer, and J. A. Berry (1976), *Science, 194,* 322–324.

Mooney, H. A., J. Ehleringer, and O. Bjorkman (1977), *Oecologia, 29,* 301–310.

Morgan, J. M., (1984), *Annu. Rev. Plant Physiol., 35,* 299–319.

Morrow, P. A. and H. A. Mooney (1974), *Oecologia* (Berl.), *15,* 205–222.

Mulroy, T. W. and P. W. Rundel (1977), *Bioscience, 27,* 109–114.

Oguntoyinbo, J. (1986), *Climat. Change, 9,* 79–90.

Oliveira, G., O. A. Correia, M. A. Martins-Loução, and F. M. Catarino (1992), *Vegetatio, 99–100,* 199–208.

Ourcival, J. M., C. Floret, E. Le Floc'h, and R. Pontanier (1994), *J. Arid Environ., 28,* 333–350.

Passioura, J. B. (1983), *Agric. Water Manag., 7,* 265–280.

Passioura, J. B. (1988), *Annu. Rev. Plant Physiol. Plant Mol. Biol., 39,* 245–265.

Raven, P. H. (1973), in *Mediterranean type ecosystems: Origin and structure* (F. di Castri and H. A. Mooney, Eds.), Ecol. Stud. Vol. 7, Springer-Verlag, Berlin, pp. 213–224.

Salleo, S. and M. A. Lo Gullo (1990), *Ann. Bot., 65,* 259–270.

Schulze, E.-D. (1986a), *Annu. Rev. Plant Physiol., 37,* 247–274.

Schulze, E.-D. (1986b), *Aust. J. Plant Physiol., 13,* 127–141.

Tenhunen, J. D., A. Sala Serra, P. C. Harley, R. L. Dougherty, and J. F. Reynolds (1990), *Oecologia, 82,* 381–393.

Tenhunen, J. D., R. Hanano, M. Abril, E. W. Weiler, and W. Hartung (1994), *Oecologia, 99,* 306–314.

Terradas, J. and R. Savé (1992), *Vegetatio, 99–100,* 137–145.

Tilman, D. (1983), in *Mediterranean-type ecosystems. The role of nutrients* (F. J. Kruger, D. T. Mitchell, and J. U. M. Jarvis, Eds.), Springer-Verlag, Berlin, pp. 322–336.

Tilman, D. and A. El Haddi (1992), *Oecologia, 89,* 257–264.

Trejo, C. L., W. J. Davies, and L. M. P. Ruiz (1993), *Plant Physiol., 102,* 497–502.

Turner, N. C. (1986), *Adv. Agron., 39,* 1–51.

United Nations (1977), *UN Conference on desertification control.* United Nations, Nairobi, New York, 49 pp.

Verstraete, M. M. (1986), *Climat. Change, 9,* 5–18.

Voesenek, L. A. C. J. and R. Van der Veen (1994), *Acta Bot. Neerl., 43* (2), 91–127.

Walter, H. (1955), *Annu. Rev. Plant Physiol., 6,* 239–252.

Walter, H. and S.-W. Breckle (1985), *Ecological systems of the geobiosphere. 1. Ecological principles in global perspective,* Springer-Verlag, Berlin.

Walter, H. and S.-W. Breckle (1991), in *Ökologie der Erde, Band 1,* Gustav Fischer Verlag, Stuttgart, p. 238.

Wraith, J. M., J. M. Baker, and T. K. Blake (1995), *J. Exp. Bot., 46* (288), 873–880.

Wright, S. T. C. and R. W. P. Hiron (1969), *Nature, 224,* 719–720.

Zhang, J., V. Schuur, and W. J. Davies (1987), *J. Exp. Bot., 38,* 1171–1181.

6

Flooding

Richard N. Arteca

INTRODUCTION

Vascular plants originated on earth approximately 400 million years ago. Since that time terrestrial plants have gone through many adaptive changes in order to survive and eventually to flourish in a gaseous atmosphere rather than an aquatic environment. Once adapted to growth above the water surface, vigorous growth occurred due to an abundance of light and oxygen. In addition to rapid growth, living on land promoted genetic diversity by favoring cross-pollination. Wetlands contain many ecological niches between a strict aquatic environment and growth on land in well-drained aerated soils. It is estimated that 6% of the earth's surface is covered by wetlands (Maltby, 1991). They are considered by many to be one of the worlds most important environmental resources. However, they are one of the most abused and endangered ecosystems in the world. Wetlands comprise fresh, brackish, and saltwater marshes, inland and coastal swamps, flood plains, low and upland mires (fen, bogs, moor, muskeg), plus agricultural wetlands such as rice paddies. Although there is no universal definition for a wetland, they are characterized by permanent or long-term soil flooding where vegetation is wholly or partially submerged. They are found worldwide, and the flooding regimes vary dramatically, leading to a wide variety of vascular plants found in wetlands. The physical, chemical, biological, hydrological, and ecological processes that occur in wetland ecosystems are complex and are very often difficult to measure and evaluate. However, they contain a wealth of information that has been used to produce artificial wetlands on a smaller scale to purify domestic, agricultural, and industrial effluents. In addition, they contain the key to better understanding of how organisms respond and adapt to the anaerobosis that results from flooding, plus they provide a habitat for many forms of plant and animal life. In many areas throughout the world flooding commonly occurs (Kozlowski, 1984) and can have adverse effects on terrestrial plant growth and development (Blom, 1990; Maltby, 1991). Therefore information gained by better understanding of processes occurring in wetlands could be used to better manage them and to provide additional benefits to solve other ecological and agricultural problems. In this chapter the morphological, physiological, biochemical, and ecological reasons why some plants flourish under low oxygen conditions, while others cannot survive, are discussed.

PLANT RESPONSES TO FLOODING OR PARTIAL SUBMERGENCE

Aerenchyma Development

It has been shown that flooding or partial submergence induces longitudinal, interconnected gas spaces called aerenchyma in both old and new roots (Fig. 6-1) (Jackson, 1985). Aerenchyma may develop in roots, rhizomes, stems, petioles, and leaves (Jackson, 1989). There are two types of aerenchyma: *schizogenous aerenchyma,* which develops through a separation of cells in the cortex or pericycle, and *lysogenous aerenchyma,* which results from the breakdown or partial lysis of these cells. Aerenchyma development varies greatly depending on the tissue's type, amount of porosity, and its specific anatomy (Crawford, 1989; Justin and Armstrong, 1987). It is now generally accepted that aerenchyma tissue aids in root growth and survival under low oxygen conditions (ApRees and Wilson, 1984; Armstrong, 1979). This is accomplished by decreasing resistance encountered by oxygen and other gases when moving throughout the plant (Crawford, 1982; Konclova, 1990; Laan et al., 1989a, b) and increases the oxygen storage capacity while decreasing the total oxygen demand. It has also been suggested that in low-oxygen

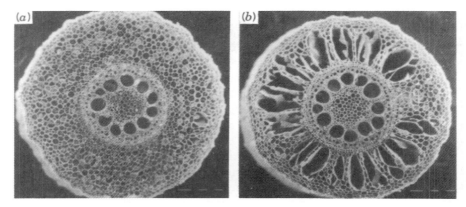

Figure 6-1. *Effect of ethylene treatment on aerenchyma formation in the cortex of well-aerated Zea mays roots. Scanning electron micrographs of 5-day-old root sections. (a) From controls given no additional ethylene. (b) From roots supplied with 5 $\mu l l^{-1}$ ethylene. (From M. B. Jackson,* Growth regulators in root development, Monogr. 10 *(M. B. Jackson and A. D. Stead, Eds.), British Plant Growth Regulator Group, Wantage, 1983, p. 103. With permission.)*

soil conditions, toxins are oxidized and rendered inactive (Ernst, 1990; Mendelsohn et al., 1981; Ponnamperuma, 1984). Many hydrophytes contain a large amount of aerenchyma when grown in well-drained soils, whereas, in flood-tolerant species of mesophytes such as maize (Norris, 1913; Trought and Drew, 1980), wheat, barley, and oats (Larsen et al., 1986; McPherson, 1939; Trought and Drew, 1980), willow and sunflower (Kawase and Whitmoyer, 1980), and others (Justin and Armstrong, 1987), there is little aerenchyma present under well-aerated soil conditions, but under low-oxygen tensions such tissue is formed very rapidly.

The mechanism of aerenchyma formation is unknown at the present time (Justin and Armstrong, 1987). However, it has been shown that the plant hormone ethylene is involved in lysogenous aerenchyma development. In maize roots exogenous applications of ethylene at a concentration of 5 $\mu l \cdot l^{-1}$ (Drew et al., 1979) has the ability to cause the breakdown of cortical cell walls in the same manner as low-oxygen conditions. It has been shown that oxygen partial pressures between those found in air and zero cause a stimulation in ethylene production, while the absence of oxygen blocks the last step of the ethylene biosynthetic pathway (Fig. 6-2) in many plant species (Abeles et al., 1992). Ethylene induced by low oxygen is thought to promote aerenchyma formation (Brailsford et al., 1993). The increase in

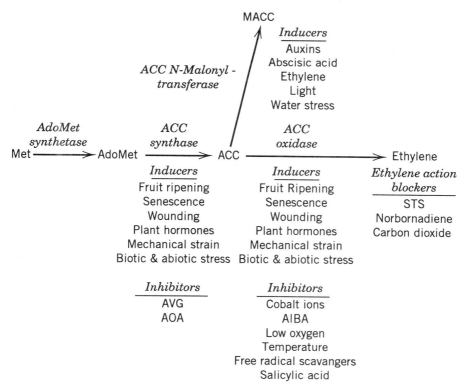

Figure 6-2. *Ethylene biosynthetic pathway: AIBA = α-amino isobutyric acid; MACC = malonyl-ACC (ACC is used in text several times); STS = Silver thiosulfate.*

ethylene formation promoted by reduced levels of oxygen are due to increased levels of 1-aminocyclopropane-1-carboxylic acid (ACC) resulting from enhanced ACC synthase activity (Wang and Arteca, 1992). The ACC produced can then move radially to better aerated tissues for its conversion to ethylene (Jackson, 1994). It has also been shown that corresponding with the increase in ethylene production there is a buildup of putrescine, the diamine precursor of spermidine and spermine (Jackson and Hall, 1993).

Recently three genes encoding for ACC synthase have been cloned in rice and two have been shown to be induced by anaerobiosis. Early studies showed that there was no change in ACC oxidase activity in response to anaerobic conditions (Atwell et al., 1988). However, it has recently been observed that changes in ACC oxidase activity in shoot tissue caused by soil flooding or as a result of the introduction of the antisense ACC oxidase gene construct can influence rates of ethylene production to a physiologically significant extent (English et al., 1996). Low oxygen tensions and ethylene-induced aerenchyma formation correlate with a large increase in cellulase activity (Grinieva and Bragina, 1993; Morgan et al., 1993). There are also a number of inhibitor studies supporting the involvement of ethylene in aerenchyma formation. Ethylene action inhibitors, such as silver ions (Drew et al., 1981) or norbornadiene, and biosynthesis inhibitors, such as cobalt ions, aminooxyacetic acid (AOA), or aminoethoxyvinylglycine (AVG), inhibit aerenchyma formation (Jackson, 1985; Konings, 1982). ACC, together with ethylene biosynthesis inhibitors, overcomes their inhibitory effect on aerenchyma formation. In addition to ethylene produced by the plant, it has been shown that under flooding conditions ethylene is produced by the soil. The physiological significance of ethylene produced by the soil remains unclear, but it can be taken up by the plant and utilized under flooding conditions (Jackson, 1985; Jackson and Campbell, 1975). Other biochemical and molecular events involved in aerenchyma formation are largely unknown.

At the present time the relative importance of plant hormones other than ethylene remains unclear. It has been shown that naphthaleneacetic acid (NAA), abscisic acid (ABA), and kinetin can inhibit aerenchyma formation, while gibberellins promote it (Konings and deWolf, 1984). However, concentrations required to inhibit activity are typically high, leading to distortion of root morphology (Justin and Armstrong, 1991). There is also evidence that nutrient status in the roots has an effect on aerenchyma formation. Deficiencies in nitrogen and phosphorous have been shown to induce aerenchyma formation in maize (Konings and Verschuren, 1980) and in *Nardus stricta* (Smirnoff and Crawford, 1983).

Schizogenous aerenchyma formation is a characteristic of many wetland plants and is formed by cell separation rather than lysogeny (Laan et al., 1989a); however, little is known in this area.

Root Extension

There is a considerable amount of information in the literature showing that root extension is sensitive to ethylene, suggesting that low-oxygen-induced ethylene production caused by flooding may have a role in controlling extension growth.

Exogenously applied ethylene at low concentrations has the ability to stimulate elongation of roots in tomato, mustard, and rice (Konings and Jackson, 1979), rye (Smith and Robertson, 1971), and broadbean (Kays et al., 1974). At concentrations above a certain threshold, however, ethylene has an inhibitory effect on root extension (Konings and Jackson, 1979; Menegus et al., 1992). Ethylene is readily entrapped by water; in fact, diffusive loss of ethylene imposed by water is 10000 times greater than by air (Jackson, 1985). Therefore, faster ethylene production, deeper water, and thicker tissue increase the ability of plant organs to entrap or retain ethylene (Jackson, 1985). Ethylene is produced at different rates depending upon the species and a variety of other factors (Abeles et al., 1992). *Oryza sativa* (rice) produces ethylene at a very slow rate (Jackson and Pearce, 1991) under flooding conditions, enabling only small amounts to accumulate in the tissue, which has a stimulatory effect on root extension (Konings and Jackson, 1979), while in mustard roots large amounts of ethylene are produced and accumulated, causing an inhibition of root extension (Konings and Jackson, 1979). Therefore, it is possible that plants such as rice produce minimal rates of ethylene in response to low-oxygen conditions encountered during flooding as a survival mechanism (Yamaguchi, 1976). However, the significance of ethylene in root extension under low-oxygen conditions remains unclear.

Reorientation of Leaves and Stems

The ability of flooding to induce ethylene production in the shoots of plants has been shown in *Lycopersicon esculentum* (Bradford and Yang, 1980a, b), *Vicia faba* (El-Beltagy and Hall, 1974), *Phaseolus vulgaris* (Wadman-van Schravendijk and Van Andel, 1986), and *Rumex paulustris* (Van der Sman et al., 1991). Under flooding conditions there is a depletion of oxygen in the root zone, leading to the production of a variety of compounds, including ACC (Bradford and Yang, 1980a, b), which is the immediate precursor to ethylene. In roots growing under low-oxygen conditions, ACC cannot be converted to ethylene (Fig. 6-2); however, it can be transported to the shoot via the transpiration stream. When ACC reaches the aerobic conditions found in the shoots it is converted to ethylene, which in turn causes an epinastic response (Bradford and Yang, 1980a, b; Abeles et al., 1992). It has been suggested that petiole epinasty in response to flooding may be an adaptive response (Armstrong et al., 1994) enabling plants to cope with stress (Abeles et al., 1992). Epinasty in tomato plants results from accelerated growth on the upper side of the petiole resulting in downward bending (Fig. 6-3). Epinastic curvature induced by flooding occurs 6–12 hours following the initiation of flooding (Jackson and Campbell, 1975) and is caused by ethylene produced and accumulated in shoot tissues (Abeles et al., 1992). When ethylene biosynthesis is blocked at the step between AdoMet and ACC by AVG or AOA or from ACC to ethylene by cobalt chloride, ethylene production and epinasty in shoots of flooded plants is decreased (Bradford et al., 1982). It has also been shown that low-oxygen conditions cause increases in gene transcription and activity of ACC synthase in tomato plants (Wang and Arteca, 1992; Zarembinski and Theologis, 1993). ACC oxidase that catalyzes the last step of the ethylene biosynthetic pathway has generally been thought to be in excess,

Figure 6-3. *The effects of root zone applications of air (left) or N_2 (right) on 4-week-old hydroponically grown tomato plants.*

and thus does not inhibit ethylene production rates. However, it has recently been shown that flooded tomato plant shoots contain higher levels of ACC oxidase than well-drained shoots, indicating that it is responsible for the faster ethylene production rates that result in epinasty in tomato plants (English et al., 1993). The enhanced expression of ACC oxidase in the leaves caused by flooding is thought to be due to systemic chemical or hydraulic signaling (English et al., 1996). Petioles from flooded transgenic tomato plants containing the ACC oxidase *(ACO1)* antisense construct were shown to produce less ethylene and to have lower ACC oxidase activity than the wild type (English et al., 1996). Flooding-induced ACC production in the roots of tomato plants is the same in transgenic and wild type plants. However, flooding promoted epinastic curvature to a lesser degree in transgenic than in wild type plants, further supporting that the induction ACC oxidase activity in the leaves of plants subjected to flooding is at least partially responsible for increased ethylene production and epinastic curvature (English et al., 1996).

Hypertrophy and Adventitious Rooting

When a plant is partially submerged during flooding hypertrophy (swelling) occurs at the base of the stem; this has been shown in tomato (Kramer, 1951), maize (Kuznetsova et al., 1981), sunflower (Phillips, 1964), and a variety of woody plant species (Imaseki and Pjon, 1970). Hypertrophy is caused by accelerated lateral cell expansion and is thought to be associated with increased intercellular space, cell lysis resulting in aerenchyma formation and adventitious rooting (Jackson, 1985). Increased ethylene and cellulase production caused by flooding is thought to be responsible for hypertrophy in sunflower hypocotyls (Kawase, 1974, 1981; Wample and Reid, 1979). Indole-3-acetic acid (IAA) has also been shown to stimulate hypertrophy in sunflower hypocotyls and under flooding conditions endogenous levels of IAA increase in the region where hypertrophy occurs (Wample and Reid, 1979). IAA has also been reported to stimulate cellulase and growth activity alone (Ridge and Osborne, 1969). However, IAA also induces ethylene production in vegetative tissues (Arteca, 1990) suggesting that ethylene may, in fact, be responsible for the induction of hypertrophy. It is thought that hypertrophy acts as a survival mechanism in plants subjected to flooding by increasing the porosity in the region at the base of the stem to enhance aeration for adventitious root formation (Armstrong et al., 1994; Justin and Armstrong, 1987). Adventitious root formation is a more complex process then hypertrophy and its regulation at the present time is poorly understood. Most of the literature on adventitious root formation deals with cuttings; comparatively speaking there is little work on flooding-induced adventitious root formation. It has been shown that flooding or partial submergence induces the initiation and/or outgrowth of adventitious roots in many plant species (Jackson, 1985; Jackson and Drew, 1984; Kawase, 1974; Visser et al., 1995). Within a few days following exposure to flooding conditions adventitious roots typically become visible (Blom et al., 1994; McNamara and Mitchell, 1990), resulting from preformed primordia or due to de novo initiation (Kozlowski, 1992). Adventitious roots replace old roots under flooding conditions and act as a survival mechanism (Blom et al., 1990; Kozlowski, 1992). Evidence supporting the physiological importance of flooding-induced adventitious root formation has been summarized by Vosenek and Van der Veen (1994) as follows:

1. There is a very good correlation between flooding resistance in plants and adventitious root formation (Blom et al., 1990, 1993; Kozlowski, 1992; Laan et al., 1989a). *Rumex* has been used as a model system to study the relationship between flooding resistance and plant distribution (Fig. 6-4) (Voesenek et al., 1992a, b). It has been shown that *Rumex* species such as *R. palustris,* typically grown under wet conditions, produce a large number of adventitious roots, whereas *Rumex* species such as *R. thyrsiflorus* and *R. acetosa,* which are usually found under dry conditions, produce only a limited number of adventitious roots; this is further illustrated with six *Rumex* species shown in Table 6-1.

2. When adventitious roots are pruned, there is a reduction in their survival rates when subjected to flooding (Jackson and Palmer, 1981).

3. Adventitious roots induced by flooding aid in water and nutrient uptake in flood-resistant plants (Sena Gomas and Kozlowski, 1980).

4. In flooding-resistant species such as *R. palustris* there is an increase in nitrate availability to plants under flooding stress due to the large number of porous adventitious roots formed, thereby providing nitrifying bacteria with oxygen

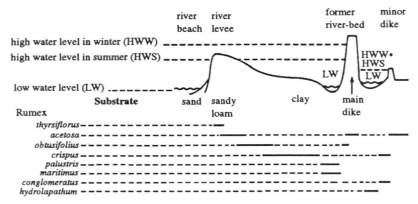

Figure 6-4. The zonation of eight Rumex species in the river flood plains of the Rhine in the Netherlands. (From Voesenek et al., 1992, J. Plant Growth Regul., 11, 171–188. With permission.)

necessary to produce nitrate. In species such as *R. thyrsiflorus*, which produce few adventitious roots in response to flooding, nitrate availability is limiting (Blom et al., 1994; Else et al., 1995; Engelaar, 1994).

5. Flooding-resistant species also have the ability to detoxify soil-borne toxins produced during flooding (Ernst, 1990; Laan et al., 1989b; Laanbroek, 1990).

The mechanism by which adventitious roots occur during flooding remains unclear; however, ethylene and auxins have been implicated in their formation. Earlier correlative work in the literature was contradictory, with some studies showing a relationship between ethylene production and adventitious roots (Bleecker et al., 1987; Tang and Kozlowski, 1984a, b; Zimmerman and Hitchcock, 1933) and some

TABLE 6-1   Adventitious Root Formation by Six *Rumex* Species[a]

		Number of Adventitious Roots per Plant	
Natural Habitat	Species	After 4 d	After 16 d
Dry	*R. thyrsiflorus*	8.0 ± 2.3	37.3 ± 6.0
	R. acetosa	22.8 ± 5.8	58.3 ± 5.0
	R. obtusifolius	16.3 ± 1.7	61.3 ± 1.7
	R. crispus	17.8 ± 2.8	61.3 ± 4.2
	R. sanguineus	23.3 ± 3.4	121.0 ± 6.1
Wet	*R. palustris*	49.8 ± 1.7	130.5 ± 9.8

[a]Root formation after 4 and 16 d of soil waterlogging (mean of three replications ± s.e.). The species occur in the field along a flooding-frequency gradient from seldom flooded (*R. thyrsiflorus*) to regularly waterlogged or submerged (*R. palustris*).

Source: Data from Blom et al. (1994).

showing no relationship (Wample and Reid, 1979; Yamamoto and Kozlowski, 1987). It has since been shown that ethylene is not directly involved in adventitious root formation; rather, it acts by increasing the sensitivity of tissue to auxin (Liu and Reid, 1992). In addition, recent reports have shown that auxin is the major factor regulating flooding-induced adventitious root formation in *Rumex* and that auxin transport appears to be a key factor (Visser et al., 1995).

Fast Underwater Shoot Extension   As previously mentioned in this chapter, many dryland species have the ability to physiologically adapt to low-oxygen tensions resulting from flooding by developing aerenchyma and adventitious roots (Jackson, 1985). Another adaptation of higher plants to low-oxygen tensions is found in deepwater rice, which responds to submergence with an enhancement in internodal growth (Jackson, 1985; Kende, 1987). Survival of deepwater rice depends on its ability to keep its foliage above the water to facilitate the movement of air to submerged parts (Raskin and Kende, 1985). The elongation response caused by flooding has been studied in deepwater rice and has been the subject of several reviews (Armstrong et al., 1994; Jackson and Pearce, 1991; Kende, 1987).

Partial Submergence of Shoots   Deepwater rice can germinate and grow rapidly in order to keep its shoots above the surface of the water. The increased stem length of deepwater rice enables it to grow in several meters of water with little or no loss of yield (Bekhasut et al., 1990). Due to water entrapment ethylene has been reported to accumulate in the central stem lacuna of deepwater rice (Stunzi and Kende, 1989). The enhanced ethylene production has been shown to result from enhanced ACC synthase activity (Cohen and Kende, 1987; Kende, 1993). Ethylene is a key regulator of flooding-induced adaptations (Abeles et al., 1992; Drew et al., 1979; Jackson, 1985; Kende, 1993); in fact, it is thought that enhanced ethylene production is one of the factors responsible for the increase in internodal length found in deepwater rice (Kende, 1987, 1993). Gibberellins have also been shown to increase in concentration during submergence (Azuma et al., 1990), while levels of ABA decrease (Hoffman-Benning and Kende, 1992). When the growth retardant tetcyclasis is applied to deepwater rice elongation growth is inhibited, further supporting that gibberellins are involved in the elongation response induced by flooding. It has been suggested that ethylene sensitizes extending cells to gibberellins, promoting rapid growth of these cells (Sauter and Kende, 1992a). The increase in growth is thought to be due to a reduction in lignin production caused by increases in β-1-3, and β-1-4-glucans associated with cell wall loosening (Sauter and Kende, 1992b). During submergence cell wall synthesis only occurs in the outer epidermal cells (Kutschera and Kende, 1989) which is where gibberellins and ethylene cause a reorientation of microtubules and cellulose microfibrils favoring cell elongation (Sauter et al., 1993). Ethylene also promotes osmotic adjustment by causing faster and more extensive cell growth (Kutschera and Kende, 1988). This is accomplished by the enhancement of hydrolytic enzymes causing the conversion of starch to sugar.

Submergence of Seedlings   Complete submergence of wheat, millet, and sorghum prevents germination even when oxygen levels are above zero (Pearce and Jackson, 1991). Certain aquatic and semiaquatic plant species are able to dramatically increase their growth rate in response to submergence (Voesenek and Van derVeen, 1994). Shoot elongation under water can be broken down into two groups: (1) elongation is enhanced under low-oxygen conditions and (2) elongation occurs in the complete absence of oxygen. Rice will germinate under low-oxygen tensions caused by submergence with the root emerging first. Ethylene and carbon dioxide that are entrapped in the coleoptile promote its elongation (Ishizawa and Esashi, 1984; Pearce and Jackson, 1991; Raskin and Kende, 1983b), and the buoyant forces of these gases gives the coleoptile a sense of direction (Kutschera et al., 1990). It has been reported that the promotive effects of low oxygen and carbon dioxide on elongation growth are not mediated by increased ethylene production (Pearce et al., 1992; Raskin and Kende, 1983a, b) rather they are probably dependent on ethylene action, since norbornadiene (an ethylene action inhibitor) inhibits the promotive effect of carbon dioxide (Ishizawa et al., 1988). The plant species and organ in which elongation is promoted by ethylene under flooding conditions are outlined in Table 6-2 (Voesenek and Van der Veen, 1994). The promotive effect of submer-

TABLE 6-2   Plant Species and Organ in Which Ethylene Is Involved in Stimulated Shoot Elongation Under Water

Plant Species	Organ	References
Apium nodiflorum	Petiole	Ridge (1987)
Callitriche platycarpa	Internode	Musgrave et al. (1972)
Daltha palustris	Petiole	Ridge (1987)
Epilobium hirsutum	Internode	Ridge (1987)
Geum rivale	Petiole	Ridge (1987)
Hydrocharis morsus-ranae	Petiole	Cookson and Osborne (1978)
Nasturtium officinale	Internode	Cookson (1976)
		Schwegler and Brandle (1991)
Nymphoides peltata	Petiole	Ridge and Amarasinghe (1984)
Oenanthe crocata	Petiole	Ridge (1987)
Oenanthe fistulosa	Petiole	Ridge (1987)
Oryza sativa	Coleoptile	Ku et al. (1970)
Oryza sativa	Internode	Metraux and Kende (1983)
Plantago major	Petiole	Ridge (1987)
Potamogeton distinctus	Stem	Suge and Kusanagi (1975)
Ranunculus flammula	Petiole	Ridge (1987)
Ranunculus lingua	Petiole	Ridge (1987)
Ranunculus repens	Petiole	Ridge (1987)
Ranunculus sceleratus	Petiole	Musgrave and Walters (1973)
Ranunculus pygmaeus	Petiole	Horton (1992)
Regnellidium diphyllum	Petiole	Musgrave and Walters (1974)
Rumex crispus	Petiole	Voesenek and Blom (1989)
Rumex maritimus	Internode	Van der Sman et al. (1991)
Rumex palustris	Petiole	Voesenek and Blom (1989)
Sagittaria pygmaea	Internode	Suge and Kusanagi (1975)

Source: Data from Voesenek and Van der Veen (1994).

gence on elongation appears to be due to increased cell osmolarity (Kutschera et al., 1990) and wall extensibility (Ishizawa and Eshashi, 1984). In addition to extension growth seen in dark-grown rice coleoptiles, there is also extension by the true leaves of rice growing in the light (Jackson et al., 1987).

Shoot elongation in the absence of oxygen has been shown in rice coleoptiles (Taylor, 1942), *Echinochloa oryzoides* mesocotyls (Pearce and Jackson, 1991; VanderZee and Kennedy, 1981), *Schoenoplectus, Scirpus, Typha,* and *Potamogeton filiformis* (Barclay and Crawford, 1982), and *Potamogeton pectinatus* (Summers and Jackson, 1993). Comparatively speaking there is a limited amount of research on the mechanism of shoot elongation in the absence of oxygen. Carbon dioxide and the plant hormones ethylene, auxin, and gibberellins have been reported to have no effect on elongation of rice coleoptiles grown in the absence of oxygen (Horton, 1991; Pegoraro et al., 1988; Satler and Kende, 1985). ABA has been shown to inhibit elongation in rice coleoptiles under anaerobic conditions (Sanders et al., 1990). It has been shown that putrescine is involved in the elongation of rice coleoptiles grown without oxygen (Reggiani et al., 1989); however, more research is necessary in order to establish a role.

Root-to-Shoot Communication in Flooded Plants

When roots are subjected to stress, levels of plant hormones, their precursors, and other solutes that enter the shoots via the transpirational stream are altered. Changes that occur in response to stress can act as physiologically active messages that modify the aboveground portion of the plant (Bradford and Yang, 1980a, b; Davies et al., 1994; Jackson, 1985; Jackson and Campbell, 1975). These modifications are stomatal closure, slower leaf expansion, epinasty, and leaf senescence.

In many plant species such as *Lycopersicon esculentum, Pisum sativum,* and *Phaseolus vulgaris,* there is an increase in ABA concentration resulting from flooding, which is associated with stomatal closure (Armstrong et al., 1994). It has previously been suggested that in roots under low-oxygen tensions, ABA is transported to the shoots via the transpirational stream (Neuman and Smit, 1991; Zhang and Davies, 1987). However, recent reports have shown that earlier work may not have provided a complete picture (Else et al., 1994, 1995). In order to more accurately estimate the amounts of a given compound that is transported it is important to take the following precautions: (1) avoid temporary distortions caused by wounding tissues when sampling the xylem sap; and (2) account for the effects of dilution when sap flow or rates of transpiration change as a result of sampling method or due to stresses such as low-oxygen tensions caused by flooding (Jackson, 1994). When these precautions were taken it was shown that there was a decrease in ABA transported from the roots to the shoots of plants subjected to flooding (Else et al., 1995). Therefore, it has been suggested that the increase in ABA found in the leaves of plants subjected to flooding stress is due to its synthesis in the leaves and slower export rates (Armstrong et al., 1994).

It has also been shown that the delivery of nitrate from the roots to the shoots is decreased by soil flooding (Else et al., 1995). The reduction in the ability of plants to take up nitrate is probably due to a reduced capacity for nitrate uptake coupled with decreased nitrate availability (Gambrell et al., 1991).

In contrast to ABA and nitrate uptake, both ACC and phosphate were shown to be transported from the roots to the shoots at a greater rate when plants were subjected to flooding. The delivery rate of ACC was 3.1-fold greater in plants flooded for 24 hours than in well-drained plants and delivery of phosphate was 2.3-fold greater (Else et al., 1995). The levels of ACC increase rapidly following the induction of flooding prior to the beginning of epinasty and slowing down of leaf expansion, suggesting a relationship between the two (Else et al., 1995; Jackson, 1994). The amount of ethylene formed in the leaves from ACC transported from the roots determines the degree of epinasty that occurs. Both epinasty and stomatal closure give the plant short-term protection against stress caused by flooding (Bradford and Yang, 1980a, b; Jackson, 1994).

Survival of plants for longer periods of time requires some roots to function. In some plant species long-term survival is due to internal transfer of oxygen from the shoots to the roots, which is facilitated by high internal root porosity found in aerenchyma. Prolonged exposure to ethylene produced from ACC transported to the leaves leads to senescence and eventual death of flooding sensitive species (Armstrong et al., 1994). It has also been shown that there is an increase in phosphate transferred from the roots to the shoots of flooded plants (Else et al., 1995). Then increased delivery of phosphate in the xylem sap has been suggested to be due to the release of stored phosphorous from root cell vacuoles (Lee and Radcliff, 1983; Zhang and Greenway, 1994), resulting in phosphate accumulation in the leaves of flooded plants (Jackson, 1979). Further research in this area is necessary in order to determine the physiological significance of phosphorous accumulation.

BIOCHEMICAL AND MOLECULAR CHANGES INDUCED BY OXYGEN DEPRIVATION CAUSED BY FLOODING

Biochemical Aspects

At the present time the two most important theories to explain flooding tolerance are: (1) Crawford's metabolic theory for flooding tolerance; and (2) Davies–Roberts pH stat hypothesis (Kennedy et al., 1992). In 1971 McManmon and Crawford introduced their metabolic theory on flooding tolerance. Basically this theory stated that flood tolerance in plants was dependent on the reduction in ethanol production caused by low alcohol dehydrogenase (ADH) levels, which reduced the presumed toxic effects of ethanol. Also contained within this theory was that some flooding-tolerant plants had the ability to reroute glycolytic intermediates to alternate end products such as malate, lactate, or other organic acids. More recently the major components of this theory have been shown to be incorrect but the major features contained within it, including ethanol, ADH activity, and shifts in alternative fermentative end products, are still important factors when studying flood tolerance (Kennedy et al., 1992).

The Davies–Roberts pH stat hypothesis was first reported by Davies in 1980. This hypothesis requires the tight regulation pH stat in order to prevent cytoplasm acidosis. The relative rate of synthesis of lactate versus ethanol depends on the pH found in the cytoplasm. When plants are subjected to low-oxygen conditions, pyruvate is first converted to lactate; however as the pH in the cytoplasm begins to decrease there is also a reduction of lactate dehydrogenase (LDH), together with an increase in pyruvate decarboxylase (PDC), which results in greater ethanol synthesis. Using ^{31}P and $^{13}CNMR$ Roberts (1989) has presented experimental evidence supporting this theory.

Even though shifts in anaerobic metabolism involving changes in pH have been reported to occur in several plant species (Menegus et al., 1989; Roberts et al., 1984), it does not appear that this can serve as a unifying theory for flood tolerance since plants such as rice and *E. phyllopogon,* which are flooding tolerant, do not act in the same manner (Kennedy et al., 1992).

Until recently it was generally accepted that metabolic responses of plants to low or no oxygen caused by flooding stress involved only glycolysis and the accumulation of toxic levels of ethanol. Today it is generally accepted that anaerobic metabolism involves more than just glycolysis. However, most of the studies reported in the literature to date involve the classical reactions associated with glycolysis and utilize agronomic crops such as maize, sorghum, and pea. At the present time our knowledge of the actual mechanism(s) is limited; however, there are still a number of studies that have been summarized by Kennedy et al. (1992) on biochemical adaptations of plants to low-oxygen tensions caused by flooding stress.

Molecular Aspects

The plant hormone ethylene can be induced by a variety of factors such as fruit ripening, senescence, wounding, auxins, brassinosteroids, cytokinins, and a variety of biotic and abiotic stresses. The rate-limiting enzyme in the ethylene biosynthetic pathway is 1-aminocyclopropane-1-carboxylate (ACC) synthase, which catalyzes the step between S-adenosylmethionine (AdoMet) and ACC (Abeles et al., 1992). ACC synthase has been cloned in a variety of experimental systems, which has led to a better understanding of its induction at the molecular level (see reviews by Kende, 1993; Wang and Arteca, 1995). However, at the present time the expression patterns of these genes in response to a wide variety of inducing agents has not been determined.

The physiological responses of plants exposed to low-oxygen tensions in their root zone caused by flooding have been studied in detail (Abeles et al., 1992). When tomato plants are subjected to low-oxygen conditions in the roots the following changes are observed: epinasty, leaf chlorosis, senescence, reduced stem elongation, and enhanced adventitious root formation (Abeles et al., 1992). Early studies showed that flooding causes an increase in ethylene production in the leaves and petioles of tomato plants, resulting in the flooding damage symptoms previously mentioned (Jackson, 1985). It has since been shown that when tomato roots are subjected to low-oxygen tensions caused by flooding there is an increase in ACC production in the roots. ACC produced in the roots cannot be converted to ethylene there, due to low oxygen levels blocking the final step of the ethylene biosynthetic pathway. However, it has been shown that ACC produced in the roots is transported to the top of the plant via the transpirational stream, where it is subsequently converted to ethylene, which causes flooding symptoms (see references within Abeles et al., 1992).

More recently it has been confirmed that the step between AdoMet and ACC is being stimulated in response to flooding stress. This was shown by demonstrating that when tomato roots are subjected to low-oxygen levels there is a dramatic increase in extractable ACC synthase levels found in the roots (Wang and Arteca, 1992). It has also been shown that when tomato plants are subjected to flooding conditions there is an increase in ACC oxidase activity in the leaves and petioles, which is partially responsible for the conversion of ACC to ethylene in the top portion of the plant (English et al., 1996).

Deepwater rice responds to partial submergence with an increase in internodal growth (Jackson, 1985; Kende, 1987). In fact, its survival depends on its ability to keep part of its foliage above water. Ethylene has been shown to be responsible for enhanced internodal growth of deepwater rice (Kende, 1987). In rice, low-oxygen tensions in the roots promotes a dramatic stimulation of ACC synthase (Cohen and Kende, 1987; Kende, 1987, 1993).

In order to understand the molecular basis for ethylene production in response to low-oxygen tensions caused by flooding, the genes for ACC synthase (which catalyzes the rate-limiting step in the ethylene biosynthetic pathway) induced by flooding needed to be identified and characterized. Zarembinski and Theologis (1993) were the first to identify two genes *OS-ACS-1* and *OS-ACS-3*, which are induced in rice plants subjected to flooding conditions. A third ACC synthase gene in rice, *OS-ACS-2*, was also identified; however, its expression was repressed in response to low-oxygen conditions (Zarembinski and Theologis, 1993). In addition to being induced by low-oxygen tensions, *OS-ACS-1* and *OS-ACS-3* are also induced by indole-3-acetic acid (IAA), benzyladenine (BA) and LiCl treatment. The induction of these genes in response to low-oxygen conditions was shown to be differential and tissue specific; *OS-ACS-1* is induced in the shoots, while *OS-ACS-3* is induced in the roots. Induction of these genes is insensitive to protein synthesis inhibitors, suggesting that they are primary responses due to the inducing agent. All three genes are induced when protein synthesis is inhibited, suggesting that this may be due to negative control or due to mRNA instability (Zarembinski and Theologis, 1993). The *OS-ACS-1* gene was structurally characterized and confirmed to be ACC synthase by expression studies. The amino acid sequence for *OS-ACS-1* shares significant sequence identity with other ACC synthases and is more similar to sequences found in other plant species. For example, *OS-ACS-1* shows 69% sequence identity with the tomato *LE-ACS3* gene (Rottmann et al., 1991), while only 44–50% identity with the other rice ACC synthases. Further work in rice may lead to a better understanding of the molecular events and signal transduction pathways involved in flooding-induced elongation in rice and ultimately to genetically engineered high-yielding rice varieties for growth in deepwater areas.

Recently, an ACC synthase gene *(LE-ACS3),* initially identified and character-ized by Rottmann et al. (1991), has been shown to be induced shortly following the initiation of flooding in the roots of tomato plants (Olson et al., 1995). It was also shown that a second ACC synthase *(LE-ACS2)* previously identified and character-ized by Rottmann et al. (1991) was induced in response to flooding; however, the induction was at a slower rate than *LE-ACS3* (Olson et al., 1995). Further work on the ACC synthase gene *LE-ACS3* at the molecular level may lead to a better understanding on how flooding-induced ACC and ethylene production are involved in the plant's ability to cope with flooding stress.

SUMMARY AND CONCLUSIONS

Flooding is a common occurrence throughout the world. It can occur naturally, as in wetlands, or it may be man-made in river areas due to canalization, drainage from agricultural areas, woodland cutting, or urbanization. At the present time the mechanism by which plants survive flooding stress remains unknown. However, there are a number of reports on morphological, physiological, biochemical, and molecular changes that occur in response to low-oxygen tensions caused by flooding stress. Unfortunately, there is still comparatively speaking little information bridg-ing the gap between these changes and plant ecology.

Plant responses to flooding stress may be summarized as follows:

1. In most plant parts aerenchyma, which are longitudinal interconnected gas spaces, are produced in response to flooding. These spaces are thought to aid root growth and survival under low-oxygen conditions. This is probably accomplished by decreasing the resistance encountered by oxygen and other gases in moving throughout the plant and by increasing oxygen storage capacity while reducing de-mand. The mechanism of aerenchyma formation is unknown; however, the plant hormone ethylene has been shown to be involved in its formation.

2. It has been suggested that root extension, which is promoted by low oxygen, is due to increased ethylene production under these conditions; however, the sig-nificance of ethylene-induced root extension remains unclear at the present time.

3. ACC is produced in the roots under low-oxygen conditions and transported to the shoots, where under aerobic conditions it is converted to ethylene and results in the reorientation of leaves and stems. This epinasty of the leaves is thought to act as a survival mechanism when plants are subjected to short-term stress caused by flooding.

4. When plants are subjected to flooding conditions, the submerged portion of the stem exhibits hypertrophy (swelling). In the area where swelling occurs, there is an increase in lateral cell expansion, aerenchyma, and adventitious root forma-tion, which are thought to act as survival mechanisms for plants under flooding stress.

5. Survival of deepwater rice is dependent on its ability to keep its foliage above

water. Ethylene and gibberellins have been shown to be responsible for the promotion of internodal elongation growth of deepwater rice. It has been suggested that ethylene sensitizes extending cells to gibberellin, which promotes rapid growth of these cells, resulting in internodal elongation.

6. When plants are completely submerged they are subjected to low-oxygen or zero-oxygen conditions. Under low-oxygen conditions elongation growth is thought to be dependent upon ethylene action, whereas under zero oxygen ethylene and other plant hormones have no effect. Evidence has been presented that putrescine is involved in the elongation of rice coleoptiles in the absence of oxygen; however, more research is necessary.

7. When roots are subjected to flooding stress it is thought that they communicate with the shoots by altering levels of plant hormones, their precursors, and other solutes, which are transported to the shoots via the transpirational stream. These signals result in stomatal closure, slower leaf expansion, epinasty, and leaf senescence. It has been shown that in response to flooding stress ABA and nitrate transport are reduced while ACC (the immediate precursor to ethylene) and phosphate transport are dramatically increased.

8. Until recently it had generally been accepted that metabolic responses of plants to low or no oxygen caused by flooding involved only glycolysis and the accumulation of toxic levels of ethanol. Today it is no longer accepted that anaerobic metabolism involves just glycolysis; however, the actual mechanism on plant adaptation to low oxygen tensions caused by flooding stress remains unknown.

9. Ethylene has been shown to be one of the key hormones involved in changes that occur in the plant as a result of flooding. The rate-limiting step in the ethylene biosynthetic pathway is between AdoMet and ACC, which is catalyzed by ACC synthase. Recently, genes for ACC synthase induced by flooding have been identified and characterized in rice and tomato. These findings may led to a better understanding of the molecular events and signal transduction pathways involved in flooding-induced responses in plants promoted by ethylene.

Although there has been a lot of research on morphological, physiological, biochemical, and molecular changes that occur in plants during flooding, the mechanisms that permit some plants to tolerate flooding stress while others cannot remain unknown. Further research is necessary to better understand the mechanisms involved and to bridge the gap between physiological and molecular research with plant ecology in order to clarify relationships between adaptive responses of individual plants and species distribution in the real world.

REFERENCES

Abeles, F. B., P. W. Morgan, and M. E. Saltveit, Jr. (1992), *Ethylene in plant biology*, Academic Press, London.

ApRees, T. and P. M. Wilson (1984), *Z. Pflanzen Physiol.*, *114*, 493–503.

Armstrong, W. (1979), in *Advances in botanical research* (H. W. W. Woodhouse, Ed.), Vol. 7, Academic Press, London, pp. 225–232.

Armstrong, W., R. Brandle, and M. B. Jackson (1994), *Acta Bot. Neerl., 43,* 307–358.

Arteca, R. N. (1990), in *Polyamines and ethylene: Biochemistry, physiology and interactions* (H. E. Flores, R. N. Arteca, and J. C. Shannon, Eds.), American Society of Plant Physiology, Rockville, MD, pp. 216–223.

Atwell, B. J., M. C. Drew, and M. B. Jackson (1988), *Physiol. Plant, 72,* 15–22.

Azuma, T., F. Mihara, N. Uchida, T. Yasuda, and T. Yamaguchi (1990), *Jap. J. Trop. Agric., 34,* 271–275.

Barclay, A. M. and R. M. M. Crawford (1982), *J. Exp. Bot., 33,* 541–549.

Bekhasut, P., D. W. Puchridge, A. Wiengweera, and T. Kupkanchanakul (1990), *Field Crops Res., 24,* 195–209.

Bleecker, A. B., S. Rose-John, and H. Kende (1987), *Plant Physiol., 84,* 395–398.

Blom, C. W. P. M. (1990), *Aquat. Bot., 38,* 1.

Blom, C. W. P. M., G. M. Bogemann, P. Laan, A. J. M. Van der Sman, J. M. Van der Steeg, and L. A. C. J. Voesenek (1990), *Aquat. Bot., 38,* 29–47.

Blom, C. W. P. M., L. A. C. J. Voesenek, and A. J. M. Van der Sman (1993), in *Interacting stresses on plants in a changing climate* (M. B. Jackson and C. R. Black, Eds.), Springer-Verlag, Berlin, pp. 243–266.

Blom, C. W. P. M., L. A. C. J. Voesenek, M. Banga, W. M. H. G. Engelaar, J. H. G. M. Rijnders, H. M. van de Steeg, and E. J. W. Visser (1994), *Ann. Bot., 74,* 253–263.

Bradford, K. J. and S. F. Yang (1980a), *Plant Physiol., 65,* 322–326.

Bradford, K. J. and S. F. Yang (1980b), *Plant Physiol., 65,* 327–330.

Bradford, K. J., T. C. Hsiao, and S. F. Yang (1982), *Plant Physiol., 70,* 1503–1507.

Brailsford, R. W., L. A. C. J. Voesenek, C. W. P. M. Blom, A. R. Smith, M. A. Hall, and M. B. Jackson (1993), *Plant Cell Environ., 16,* 1071–1080.

Cohen, E. and H. Kende (1987), *Plant Physiol., 84,* 282–286.

Cookson, C. (1976), *Auxin, ethylene and cell growth of water plants,* PhD. Thesis, University of Cambridge, Cambridge, UK.

Cookson, C. and D. J. Osborne (1978), *Planta, 144,* 39–47.

R. M. M. Crawford (1982), in *Encyclopedia of plant physiology* (O. L. Lange et al., Eds.), Vol. 12B, Springer-Verlag, Berlin, pp. 453–477.

Crawford, R. M. M. (1989), *Studies in plant survival. Ecological case histories of plant adaptation to adversity,* Blackwell Scientific Publications, Oxford, UK.

Davies, D. D. (1980), in *The biochemistry of plants: A comprehensive treatise* (P. K. Stumpf and E. E. Conn, Eds.), Vol, 2, Academic Press, New York, pp. 581–611.

Davies, W. J., F. Tardieu, and C. L. Trejo (1994), *Plant Physiol., 104,* 373–374.

Drew, M. C., M. B. Jackson, and S. Giffard (1979), *Planta, 147,* 83–88.

Drew, M. C., M. B. Jackson, S. C. Giffard, and R. Campbell (1981), *Planta, 153,* 217–224.

El-Beltagy, A. S. and M. A. Hall (1974), *New Phytol., 73,* 47–60.

Else, M. A., W. J. Davies, P. N. Whitford, K. C. Hall, and M. B. Jackson (1994), *J. Exp. Bot., 45,* 317–323.

Else, M. A., K. C. Hall, G. M. Arnold, W. J. Davies, and M. B. Jackson (1995), *Plant Physiol., 107,* 377–384.

Engelaar, W. M. H. G. (1994), *Roots nitrification and nitrate aquistition in waterlogged and compacted soils*, Ph.D. thesis, Catholic University, Nijmegen, Netherlands.

English, P. J., G. W. Lycett, J. A. Roberts, K. C. Hall, and M. B. Jackson (1993), in *Cellular and molecular aspects of the plant hormone ethylene* (J. C. Pech, A. Latche, and C. Balague, Eds.), Kluwer Academic, Dordrecht, Netherlands, pp. 261–262.

English, P. J., G. W. Lycett, J. A. Roberts, and M. B. Jackson (1996), *Plant Physiol.*, in press.

Ernst, W. H. O. (1990), *Aquat. Bot., 38*, 73–90.

Gambrell, R. P., R. D. Delaune, and W. H. Patrick, Jr. (1991), in *Plant life under oxygen stress* (M. B. Jackson, D. D. Davies, and H. Lambers, Eds.), SPB Academic, The Hague, Netherlands, pp. 101–117.

Grinieva, G. M. and T. V. Bragina (1993), *Russ. J. Plant Physiol., 40*, 662–667.

Hoffman-Benning, S. and H. Kende (1992), *Plant Physiol., 99*, 1156–1161.

Horton, R. F. (1991), *Plant Sci., 79*, 57–62.

Horton, R. F. (1992), *Aquat. Bot., 44*, 23–30.

Imaseki, H. and C. Pjon (1970), *Plant Cell Physiol., 11*, 827–829.

Ishizawa, K. and Y. Esashi (1984), *Plant Cell Environ., 7*, 239–245.

Ishizawa, K., M. Hoshina, K. Kawabe, and Y. Eshashi (1988), *J. Plant Growth Regul., 7*, 45–58.

Jackson, M. B. (1979), *Physiol. Plant, 46*, 347–351.

Jackson, M. B. (1985), *Annu. Rev. Plant Physiol., 36*, 145–174.

Jackson, M. B. (1989), in *Cell separation in plants. Physiology, biochemistry and molecular biology* (D. J. Osborne and M. B. Jackson, Eds.), Springer-Verlag, Berlin, pp. 263–274.

Jackson, M. B. (1994), *Agron J., 86*, 775–782.

Jackson, M. B. and D. J. Campbell (1975), *New Phytol., 74*, 397–406.

Jackson, M. B. and M. C. Drew (1984), in *Flooding and plant growth* (T. T. Kozlowski, Ed.), Academic, Orlando, FL, pp. 47–128.

Jackson, M. B. and K. C. Hall (1993), *Ann. Bot., 72*, 569–575.

Jackson, M. B. and J. H. Palmer (1981), *Plant Physiol., 67*, 58.

Jackson, M. B. and D. M. E. Pearce (1991), in *Plant life under oxygen stress* (M. B. Jackson, D. D. Davies, and H. Lambers, Eds.), SPB Academic, The Hague, Netherlands, pp. 47–67.

Jackson, M. B., I. Water, T. Setter, and H. Greenway (1987), *J. Exp. Bot., 38*, 1826–1838.

Justin, S. H. F. W. and W. Armstrong (1987), *New Phytol., 105*, 465–495.

Justin, S. H. F. W. and W. Armstrong (1991), *New Phytol., 117*, 607–618.

Kawase, M. (1974), *Physiol. Plant., 31*, 29–38.

Kawase, M. (1981), *Am. J. Bot., 68*, 651–658.

Kawase, M. and R. E. Whitmoyer (1980), *Am. J. Bot., 67*, 18–22.

Kays, S. J., C. W. Nicklow, and D. H. Simons (1974), *Plant Soil, 40*, 565–571.

Kende, H. (1987), in *Physiology of cell expansion during growth* (D. J. Cosgrove and D. P. Kneivel, Eds.), American Society of Plant Physiologists, Rockville, MD, pp. 221–238.

Kende, H. (1993), *Annu. Rev. Plant Physiol. Mol. Biol., 44*, 283–307.

Kennedy, R. A., M. E. Rumpho, and T. C. Fox (1992), *Plant Physiol., 100*, 1–6.

Konclova, H. (1990), *Aquat. Bot., 38,* 127–134.

Konings, H. (1982), *Physiol. Plant., 54,* 119–124.

Konings, H. and A. deWolf (1984), *Physiol. Plant, 60,* 309–314.

Konings, H. and M. B. Jackson (1979), Z. *Pflanzen Physiol., 92,* 385–397.

Konings, H. and G. Verschuren (1980), *Plant Physiol., 49,* 265–270.

Kozlowski, T. T. (1984), in *Flooding and plant growth* (T. T. Kozlowski, Ed.), Academic Press, New York, pp. 1–7.

Kozlowski, T. T. (1992), *Bot. Rev., 58,* 107–222.

Kramer, P. J. (1951), *Plant Physiol., 26,* 722–736.

Ku, H. S., H. Suge, L. Rappaport, and H. K. Pratt (1970), *Planta, 90,* 333–339.

Kutschera, U. and H. Kende (1988), *Plant Physiol., 88,* 361–366.

Kutschera, U. and H. Kende (1989), *Ann. Bot., 63,* 385–388.

Kutschera, U., C. Siebert, Y. Masuda, and A. Sievers (1990), *Planta, 183,* 112–119.

Kuznestova, G. A., M. G. Kuznetsova, and G. M. Grineva (1981), *Fizol. Rast., 28,* 340–348.

Laan, P., M. J. Berrovoets, S. Luthe, W. Armstrong, and C. W. P. M. Blom (1989a), *J. Ecol., 77,* 693–703.

Laan, P., A. Smolders, C. W. P. M. Blom, and W. Armstrong (1989b), *Acta Bot. Neerl., 38,* 131–145.

Laan, P., M. Tosserams, C. W. P. M. Blom, and B. W. Veen (1990), *Plant Soil, 122,* 39–46.

Laanbroek, H. J. (1990), *Aquat. Bot., 38,* 109–125.

Larsen, O., H.-G. Nilsen, and H. Aarnes (1986), *J. Plant Physiol., 122,* 269–272.

Lee, R. B. and R. G. Ratcliff (1983), *J. Exp. Bot., 34,* 441–444.

Liu, J. and D. M. Reid (1992), *J. Exp. Bot., 43,* 1191–1198.

Maltby, E. (1991), in *Plant life under oxygen deprivation* (M. B. Jackson, Ed.), SPB Academic, The Hague, Netherlands, pp. 3–21.

McManmon, M. and R. M. M. Crawford (1971), *New Phytol., 70,* 299–306.

McNamara, S. T. and C. A. Mitchell (1990), *Hortscience, 25,* 100–103.

McPherson, D. C. (1939), *New Phytol., 38,* 190–202.

Mendelsohn, I. A., K. McKee, and W. J. Patrick (1981), *Science, 214,* 439–441.

Menegus, F., L. Cattaruzza, A. Chersi, and G. Fronza (1989), *Plant Physiol., 90,* 29–32.

Menegus, F., L. Cattaruzza, and E. Ragg (1992), *Physiol. Plant., 86,* 168–172.

Metraux, J. P. and H. Kende (1983), *Plant Physiol., 72,* 441–446.

Morgan, P. W., J. I. Sarquis, C. J. He, W. R. Jordan, and M. C. Drew (1993), in *Cellular and molecular aspects of the plant hormone ethylene* (J. C. Pech, A. Latche, and C. Balague, Eds.), Kluwer Academic, Dordrecht, Netherlands, pp. 232–237.

Musgrave, A. and J. Walters (1973), *New Phytol., 72,* 783–789.

Musgrave, A. and J. Walters (1974), *Planta, 121,* 51–56.

Musgrave, A., M. B. Jackson, and E. Ling (1972), *Nature, 238,* 93–96.

Neuman, D. S. and B. A. Smit (1991), *J. Exp. Bot., 42,* 1499–1506.

Norris, F. de la M. (1913), *Proc. Bristol Naturalists Soc., 4,* 134–138.

Olson, D. C., J. H. Oetiker, and S. F. Yang (1995), *J. Bio. Chem., 270,* 14056–14061.

Pearce, D. M. E. and M. B. Jackson (1991), *Ann. Bot., 68,* 201–209.

Pearce, D. M. E., K. C. Hall, and M. B. Jackson (1992), *Ann. Bot., 69,* 441–447.

Pegoraro, R., S. Mapelli, G. Torti, and A. Bertani (1988), *J. Plant Growth Regul., 7,* 85–94.

Phillips, I. D. J. (1964), *Ann. Bot., 28,* 17–35.

Ponnamperuma, F. N. (1984), in *Flooding and plant growth* (T. T. Kozlowski, Ed.), Academic Press, New York, pp. 9–45.

Raskin, I. and H. Kende (1983a), *Plant Physiol., 72,* 447–454.

Raskin, I. and H. Kende (1983b), *J. Plant Regul., 2,* 193–203.

Raskin, I. and H. Kende (1985), *Science, 228,* 327–329.

Reggiani, R., A. Hochkoeppler, and A. Bertani (1989), *Plant Cell Physiol., 30,* 893–898.

Ridge, I. (1987), in *Plant life in aquatic and amphibious habitats* (R. M. M. Crawford, Ed.), Blackwell Scientific Publications, Oxford, UK.

Ridge, I. and I. Amarasinghe (1984), *Plant Growth Regul., 2,* 235–249.

Ridge, I. and D. J. Osborne (1969), *Nature, 223,* 318–319.

Roberts, J. K. M. (1989), in *The ecology and management of the wetlands* (D. D. Hook, Ed.), Vol. 1, Croom-Helm Press, London, pp. 392–397.

Roberts, J. K. M., J. Callis, D. Jardetsky, V. Walbot, and M. Freeling (1984), *Proc. Natl. Acad. Sci., USA, 81,* 6029–6033.

Rottmann, W. H., G. F. Peter, P. W. Oeller, J. A. Keller, N. F. Shen, B. P. Nagy, L. P. Taylor, A. D. Campbell, and A. Theologis (1991), *J. Mol. Biol., 222,* 937–964.

Sanders, I. O., N. V. J. Harpham, I. Raskin, A. R. Smith, and M. A. Hall (1990), *Plant Cell Physiol., 30,* 1091–1099.

Satler, S. O. and H. Kende (1985), *Plant Physiol., 79,* 194–198.

Sauter, M. and H. Kende (1992a), *Planta, 188,* 362–368.

Sauter, M. and H. Kende (1992b), *Plant Cell Physiol., 33,* 1089–1097.

Sauter, M., R. W. Seagull, and H. Kende (1993), *Planta, 190,* 354–362.

Schwegler, T. and R. Brandle (1991), *Bot. Helv., 101,* 135–140.

Sena Gomas, A. R. and T. T. Kozlowski (1980), *Physiol. Plant., 49,* 373–377.

Smirnoff, N. and R. M. M. Crawford (1983), *Ann. Bot., 51,* 237–249.

Smith, K. A. and P. D. Robertson (1971), *Nature, 234,* 148–149.

Stunzi, J. T. and H. Kende (1989), *Plant Cell Physiol., 30,* 49–56.

Suge, H. and T. Kusanagi (1975), *Plant Cell Physiol., 16,* 65–72.

Summers, J. E. and M. B. Jackson (1993), in *Interacting stresses on plants in a changing climate* (M. B. Jackson and C. R. Black, Eds.), Springer-Verlag, Berlin, pp. 315–325.

Tang, Z. C. and T. T. Kozlowski (1984a), *Can. J. Bot., 62,* 1659–1664.

Tang, Z. C. and T. T. Kozlowski (1984b), *Plant Soil, 77,* 183–192.

Taylor, D. L. (1942), *Am. J. Bot., 29,* 721–738.

Trought, M. C. T. and M. C. Drew (1980), *J. Exp. Bot., 31,* 1573–1585.

Van der Sman, A. J. M., L. A. C. J. Voesenek, W. C. P. M. Blom, F. J. M. Harren, and J. Reuss (1991), *Funct. Ecol., 5,* 304–313.

VanderZee, D. and R. A. Kennedy (1981), *Am. J. Bot., 68,* 1269–1277.

Visser, E. J. W., C. J. Heijink, J. J. G. M. van Hout, L. A. C. J. Voesenek, G. W. M. Barendse, and C. W. P. M. Blom (1995), *Physiol. Plant.*, *93*, 116–122.

Voesenek, L. A. C. J. and C. W. P. M. Blom (1989), *Plant Cell Environ.*, *12*, 433–439.

Voesenek, L. A. C. J. and R. Van Der Veen (1994), *Acta Bot. Neerl.*, *43*, 91–127.

Voesenek, L. A. C. J., A. J. M. van der Sman, F. J. M. Harren, and C. W. P. M. Blom (1992a), *J. Plant Growth Regul.*, *11*, 171–188.

Voesenek, L. A. C. J., M. C. C. De Graaf, and C. W. P. M. Blom (1992b), *Acta Bot. Neerl.*, *41*, 331–343.

Wadman-van Schravendijk, H. and O. M. Van Andel (1986), *Physiol. Plant.*, *66*, 257–264.

Wample, R. L. and D. M. Reid (1979), *Physiol. Plant.*, *45*, 219–226.

Wang, T. W. and R. N. Arteca (1992), *Plant Physiol.*, *135*, 631–634.

Wang, T. W. and R. N. Arteca (1995), *Plant Physiol.*, *109*, 627–636.

Yamaguchi, T. (1976), *Jap. J. Trop. Agric.*, *20*, 33–34.

Yamamoto, F. and T. T. Kozlowski (1987), *Environ. Exp. Bot.*, *27*, 329–340.

Zarembinski, T. I. and A. Theologis (1993), *Mol. Biol. Cell*, *4*, 363–373.

Zhang, J. and W. J. Davies (1990), *J. Exp. Bot.*, *38*, 1649–1659.

Zhang, Q. and H. Greenway (1994), *J. Exp. Bot.*, *45*, 567–575.

Zimmerman, P. W. and A. E. Hitchcock (1933), *Contrib. Boyce Thompson Inst.*, *5*, 351–359.

7

Salt

Jürgen Hagemeyer

SALINIZATION—A PROBLEM OF GLOBAL SCALE

Soil salinization is the enrichment of salts, mainly sodium chloride or sodium sulfate, at or near the soil surface. This can be a natural process in salt marshes or in arid environments. If the salt accumulation results from human activity, it is called "secondary salinization" (Thomas and Middleton, 1993). In regions with insufficient rainfall agricultural crops need irrigation. Whenever this is done without proper management, particularly without sufficient drainage, salts from the irrigation water accumulate in the soil. This is by no means a modern problem. Sumerian documents dating from 2400 B.C. mention anthropogeneous soil salinization in the mesopotamian area (Kreeb, 1974, pp. 122–123). Also in modern times mismanagement in irrigation causes the spreading of soil salinization in semiarid and arid regions (Flowers et al., 1986). In Africa, human-induced secondary salinization accounts for 10% of African saline soils, but it affects about one-half of the area of irrigated land (Thomas and Middleton, 1993). In Pakistan, 6.2 million hectares of land are salt-affected and losses due to reduced rice yield are more than 100 million US dollars per year (Aslam et al., 1993). Johnson et al. (1992) estimated that up to one-third of the irrigated cropland in the United States is affected by salinity. On a global scale, nearly 40% of the earth's land surface is potentially endangered by salinity problems (Cordovilla et al., 1994). With a rapidly growing human population and an increasing demand for food production, these problems will be aggravated in the future. Salinized soils are hard to recover for agricultural use. The best way to avoid salt accumulation is to apply an excess of irrigation water to flush salts out of the top soil layers (Bresler et al., 1982, pp. 193–198; Carter, 1975). However, due to shortage of irrigation water, this precautionary measure is often neglected.

For living organisms, salt is one of the oldest stressors, since all life originates from the saline sea. However, most vascular land plants are rather sensitive to increased salt levels in the soil. Specialized plants, halophytes, have evolved in various systematic groups, and have adapted to saline substrate conditions.

SALINE SOILS

Distribution of Saline Soils

Natural saline soils are found in coastal areas and salt marshes all over the world (Chapman, 1974). The salts originate either from periodical or episodical flooding with seawater with an average salt concentration of 3.5%, or from salt spray drifted inland.

Furthermore, saline soils occur in arid regions, where the potential evaporation exceeds the rainfall. Under such conditions, water evaporates rapidly from the soil surface. Most of the time, water moves upward through the soil profile toward the surface. Leached salts from lower soil horizons precipitate at the surface and can form white salt crusts. The resulting soil type is called solonchak (Walter, 1985, p. 244; Walter and Breckle, 1985, p. 37). In basins, which have no outlet, runoff water from the surrounding higher land flows to the lowest spots. Leached salts from the vicinity of the basin tend to accumulate in the trough. In this way salt pans develop. These are known, for instance, as takyrs from the Karakum desert (Walter, 1985, pp. 245–252). During dry periods salt dust from the soil surface can drift over long distances to other places, where it is deposited and causes soil salinization.

Recently, salinized soils are also found in industrialized regions along major roads, on which deicing salts have been applied in the winter (Brod, 1993). Seybold (1973) described the spreading of *Puccinellia distans* at roadsides in Germany. This plant is a facultative halophyte with a natural distribution in coastal saline habitats. The salinization of soils near roads gave rise to changes in plant communities. Tolerant plants started to colonize these new, man-made habitats (Runge, 1985). Hazardous effects of deicing salts on trees growing along roads have been repeatedly observed (Albert and Falter, 1978; Bogemans et al., 1989; Ernst and Joosse-Van Damme, 1983, pp. 56–61; Trockner and Albert, 1986).

Types of Saline Soils

Saline soils show a large variability in the degree of salinization and the kind of accumulated salts. Also, seasonal variations in soil salinity levels have been described (Young et al., 1994). Saline soils are usually classified according to the salt concentration of the soil solution and the degree of saturation of the exchange complex with Na^+ ions. Three major soil types have been distinguished (Shainberg, 1975; Waisel, 1972, p. 33):

1. Saline soils with salt concentrations in the soil solution high enough to inter-fere with the growth of most crop plants. The critical concentration limit can vary with environmental conditions and plant species. The pH of such a soil is below 8.5 and the exchangeable Na^+ percentage is less than 15%.
2. Nonsaline alkali soils (sodic soils) with low soluble salt concentrations and an exchangeable Na^+ percentage exceeding 15%.
3. Saline alkali soils with high concentrations of soluble salts and an exchange-able Na^+ percentage above 15%.

The kind of accumulated salt can vary. Enrichment of chlorides, like NaCl, is found under humid as well as arid conditions. The soil reaction is near neutral. In deserts and dry grasslands, soils can contain alkaline salts, like carbonates or sul-fates of Na, Mg, or Ca. Soils containing alkaline sodium salts are called solonez and have pH values between 8.5 and 11.

EFFECTS OF SALT ON VASCULAR PLANTS

Osmotic and Ion-Specific Effects

Salts can have either osmotic effects or ion-specific effects on various physiological processes in plants. Osmotic effects of salts in the substrate mainly impede water uptake into the roots by reducing the soil water potential. The symptoms are turgor loss and wilting. Specific effects of ion toxicity are more varied, as ions can inter-fere with physiological processes on all levels within an organism. Experimentally these two effects can be distinguished when plants are treated either with a nonab-sorbed osmoticum in the substrate, like polyethylene glycol (PEG), or with readily absorbed salt ions.

Salt-resistant cell lines from two species of orange trees *(Citrus sinensis* and *C. aurantium)* showed similar reactions to NaCl-induced stress, but differed in their response to PEG-induced osmotic stress (Ben-Hayyim, 1987). It was concluded that salt resistance of *C. sinensis* cells involved osmotic adaptation, whereas *C. aurantium* had no such adaptation.

Effects of both osmotic and salinity stress on two varieties of wheat *(Triticum aestivum)* differing in drought resistance were investigated by Nagy and Galiba (1995). Plants grown in hydroculture were treated first with PEG in the solution and then transferred to an equi-osmolal NaCl solution. During the PEG treatment, the drought-resistant variety had a higher water-use efficiency. This was inferred from a lower stomatal conductance, but similar net photosynthesis compared to the sensitive variety. However, upon direct transfer from control to NaCl solutions, both varieties showed no differences in water-use efficiency, net photosynthesis, or water content in shoots or roots. Thus it was concluded that the drought-resistant variety was just as salt sensitive as the drought-sensitive one. After transfer from PEG to NaCl treatment, the drought-resistant variety showed declining net photo-synthesis, whereas the sensitive variety increased its net photosynthesis steadily.

Therefore, the authors pointed out the disadvantages of the drought-resistant wheat variety, having a well-developed water economy but lacking the ability to cope with effects of toxic salt ions.

Reactions of Plants to Salinity

A multitude of salt effects on vascular plants growing in saline soils have been described. The reactions of plants depend largely on the specific degree of tolerance against soil salinity. Plants can be categorized according to their biomass production under salt stress (Fig. 7-1). Four main groups were distinguished:

1. Eu-halophytes, which show stimulation of productivity at moderate salinity (e.g., *Salicornea europaea, Suaeda maritima*)
2. Facultative halophytes, showing a slight growth enhancement at low salinity (e.g., *Plantago maritima, Aster tripolium*)
3. Nonhalophytes with low salt tolerance (e.g., *Hordeum* sp., *Gossypium* sp.)
4. Halophobic plants (e.g., *Phaseolus vulgaris, Glycine max*)

It should, however, be noted that plant reactions to salt vary greatly, and many plant species do not fall into one of those groups but show transitional behavior. Various other classification systems have been proposed, which are reviewed by Breckle (1990).

From an ecological viewpoint halophytes can be characterized as plants that survive to complete their life cycles at high salinities (Flowers et al., 1977, 1986).

A great deal of attention has been focused particularly on two plant groups: (1) salt-adapted halophytes (Breckle, 1990; Flowers, 1985; Flowers et al., 1986; Munns et al., 1983) and (2) crop plants of economic value, like barley, wheat, rice,

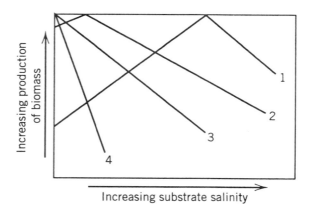

Figure 7-1. *Response of different plant groups to substrate salinity (after Greenway and Munns, 1980; Kreeb, 1974; Larcher, 1994): (1) Eu-halophytes. (2) Facultative halophytes. (3) Nonhalophytes with low salt tolerance. (4) Halophobic plants.*

or tomato (e.g., Greenway and Munns, 1980; Gorham et al., 1985, 1994; Khatun and Flowers, 1995; Mansour and Stadelmann, 1994; Perez-Alfocea et al., 1993a; Richards, 1992).

Effects on Phytohormone Levels   Phytohormones have a key role in controlling the water balance of plants (Parry and Horgan, 1991; Zeevaart and Creelman, 1988), which is affected by salts in the substrate. Abscisic acid (ABA) induces stomatal closure (Mittelheuser and Van Steveninck, 1969), whereas gibberellic acid has the opposite effect of opening of stomates (Livne and Vaadia, 1965). Phytohormone levels in plants are affected by salt stress. In leaves of grapevine plants *(Vitis vinifera)* irrigated with 50 mM NaCl solution, the ABA content increased within 6 hours (Downton and Loveys, 1981). It was suggested that ABA is involved in the perception of drought stress in plant roots, which also occurs in saline soil. Under conditions of drought, ABA production in roots is increased. With the transpiration stream the hormone is transported to the leaves, where it causes the closure of stomates (Zhang and Davies, 1989; Zhang et al., 1987). This reduces transpiration and can help to limit the water loss of plants under drought stress.

ABA also appears to control the salt tolerance of plants in some way. Photosynthesis and carbon fixation of pea plants *(Pisum sativum)* treated with 50 mM NaCl for 48 hours in hydroculture were significantly inhibited (Fedina et al., 1994). However, plants that were pretreated for 24 hours with 1 μM ABA in the culture solution showed no reduction of photosynthesis. The ABA treatment alleviated inhibitory effects of NaCl on photosynthesis and carbon fixation. The authors concluded that an ABA treatment of pea plants preceding the salt stress increased their tolerance to NaCl (Fedina et al., 1994).

Changes in the balance of growth regulators, like gibberellins, cytokinins, or ABA, can be the reason for reductions of dry matter production of salt-stressed plants (Ungar, 1991, pp. 87–93). Either synthesis or translocation of such compounds may be inhibited under salt stress. Also cell expansion growth is affected by the phytohormone balance of the plants, which is modified by salt stress (Lerner et al., 1994). The authors conclude that roots, as sensing organs for salt stress, modulate the hormonal balance in response to soil salinity. A major effect of salinity on growth seems to result from changes in phytohormone levels.

Transpiration   In general, plants under salt stress show reduced transpiration (Flowers, 1985). This can either be due to osmotic reductions in root water uptake or to stomatal closure.

In a pot culture experiment plants of the halophyte *Salsola kali* were irrigated with nutrient solutions amended with 150 mM of either NaCl or KCl (Eshel and Waisel, 1984). The salt-treated plants had only half of the transpiration rates of unsalinized control plants. Also the halophyte *Suaeda maritima* had lower transpiration rates under salt stress (Clipson and Flowers, 1987). In hydroponically grown plants without salt stress, transpiration rates were 1.3 g H_2O (g DW)$^{-1}$ h^{-1}, whereas plants treated with 400 mM NaCl reached only 0.97 g H_2O (g DW)$^{-1}$ h^{-1}.

Reduced transpiration rates under salt stress were also observed in plants of the

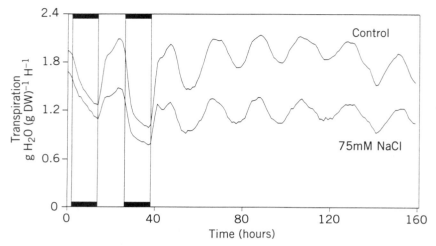

Figure 7-2. *Effect of NaCl on transpiration rates of two Tamarix aphylla plants, cultured with half-strength Hoagland solution to which 1.5 mM (control) or 75 mM NaCl were added. Transpiration was determined gravimetrically from weight loss. For 2 days plants were kept in light–dark cycles (indicated by black bars) of LD 12h: 12h, and afterwards in continuous light (see Hagemeyer, 1990; Hagemeyer and Waisel, 1987).*

halophytic tree *Tamarix aphylla* (Fig. 7-2). The plants were raised from branch cuttings in hydroculture and cultivated in growth chambers under controlled environmental conditions. Transpiration rates were calculated from weight loss of the culture vessels per g branch dry weight and per hour (Hagemeyer, 1990; Hagemeyer and Waisel, 1987, 1989a). Plants in nutrient solutions with NaCl had lower transpiration rates during the light hours of the day and also at night (Fig. 7-2). In continuous light the transpiration showed a free-running, endogenous circadian rhythm with a mean period length of 21.4 hours, which was not affected by the salt treatment (Hagemeyer, 1990). A similar rhythm of transpiration was found in *Tamarix jordanis* (Waisel, 1991a). In *T. aphylla* transpiration rates were negatively correlated with the relative humidity of the ambient air (Hagemeyer and Waisel, 1989a). This was found with control as well as with plants in 50 and 100 mM NaCl solutions. Daytime transpiration rates were about 2–3 times higher than night rates, and it was concluded that the stomates had retained their regulatory ability under salt stress.

Plant Water Balance   Plant water uptake is hampered in saline substrates, as salt ions lower the soil water potential. Halophytes overcome this problem by osmotic adjustment, which involves uptake of salt ions (Flowers et al., 1986; see the subsection on osmotic adjustment in the section on mechanisms of salt resistance, below). To a limited extent, this is also true for nonhalophytes (Greenway and Munns, 1980). Flowers et al. (1991) pointed out that salt injuries like wilting can occur in some plants, despite (1) an apparently adequate salt accumulation in leaves for osmotic adaptation and (2) too low an average leaf tissue concentration of salt to account for the observed damage. Therefore, the question arises: What caused the salt injuries in these plants? Oertli (1968) proposed that extracellular accumulation of salt ions in cell walls may produce a low water potential in the leaf apoplast. This inhibits water uptake into the cells and causes turgor loss and wilting. Therefore, salt injuries may result from the failure of plants to transport salt ions into their leaf cells and, thus to achieve osmotic adjustment of the cells (see also Greenway and Munns, 1980; Munns, 1993).

Additional evidence for Oertli's hypothesis that salt injury is caused by cell dehydration following extracellular accumulation of salt was presented by Flowers et al. (1991). Rice plants *(Oryza sativa)* were grown in hydroculture and exposed to 50 mM NaCl. After several days of treatment the plants showed symptoms of water deficit, like rolling and browning of leaves, although ion accumulation was sufficient to compensate for the drop in external water potential. NaCl concentrations in the apoplast solution reached up to 600 mM, whereas the external concentration was only 50 mM NaCl in the culture medium. X-ray microanalysis of leaf tissue revealed higher ion concentrations in the apoplast than in the vacuoles. This indicated a lower water potential in the apoplast than in the cells. Water uptake into the cells was, therefore, impeded. The authors concluded that one factor responsible for salt damage in rice plants is an extracellular accumulation of salt in the leaf tissue. Therefore, analyses of salt-ion concentrations in bulk tissue water can give no conclusive evidence for an efficient osmotic adjustment of salt-stressed plants. Ion accumulations in particular compartments within the tissue should always be considered (see the subsection on intracellular compartmentation and compatible solutes in the section on mechanisms of salt resistance, below.)

Photosynthesis   Photosynthetic performance and carbon assimilation of plants have been investigated to assess plant productivity under salt stress. Ungar (1991, p. 73) pointed out that reductions in photosynthetic activity of salt-stressed plants can be due to different causes: decreased stomatal conductance, reduced carboxylase activity, limited tissue CO_2 availability, and inhibition of the light-reaction mechanism.

Effects of Na^+ and Cl^- ions on gas exchange parameters of sweet orange *(Citrus sinensis)* seedlings were investigated by Bañuls and Primo-Millo (1992). To distinguish effects of the ions, different salts were applied (NaCl, KCl, $NaNO_3$). Accumulation of Cl^- in the leaves reduced photosynthesis and stomatal conductance, whereas these were not affected by Na^+ at concentrations up to 478 mM in tissue water. It was concluded that adverse salt effects in *Citrus* leaves were caused by Cl^- accumulation.

Seedlings of the shrub *Myrica cerifera,* which grows in coastal areas of North America under saline conditions, were cultivated in growth chambers and treated with NaCl (Sande and Young, 1992). Concentrations of NaCl above 50 mM reduced stomatal conductance to water vapor diffusion and net photosynthetic rates.

Growth and root nodulation with nitrogen-fixing actinomycorrhiza were also inhibited. However, after 5 weeks at 150 mM NaCl, plants were still alive, and thus showed the potential to survive natural periodic increases in salinity.

Salt effects on leaf gas exchange of fig trees *(Ficus carica)* were studied after short-term treatments with NaCl (Golombek and Lüdders, 1993). Plants were grown in hydroculture and exposed to 50 mM NaCl for 2 days and afterwards to 100 mM NaCl for 5 days. In different treatment groups, the root mass was varied in relation to leaf area. Within one week of treatment stomatal conductance was markedly reduced. Plants with many roots showed a transient stimulation of net photosynthetic rates during the salt treatment. Net assimilation rates of plants with few roots were only slightly reduced under salt stress. Mainly nonstomatal factors were responsible for changes in CO_2 uptake upon salt exposure. The water-use efficiency of the plants increased during the salt treatment. The authors concluded that during the first days of salt stress the salt resistance of fig plants was enhanced by the observed changes in gas exchange parameters. The reduced stomatal conductance helped to increase the water-use efficiency, but affected net CO_2 uptake only slightly. The maintenance or stimulation of net CO_2 assimilation rates under salt stress may enable fig plants to cope with salinity.

Direct salt effects on the photosynthetic apparatus of barley varieties *(Hordeum vulgare)* were investigated by Belkhodja et al. (1994). Cut leaves of different barley genotypes were immersed in water or in solutions of 100 mM NaCl, and chlorophyll fluorescence was measured in the dark and in strong light. Only in strong light were changes in rapid fluorescence kinetics found under salt stress. Such changes were attributed to delayed plastoquinone reoxidation in the dark. The observed alterations of chlorophyll fluorescence in response to salinity were significantly correlated with independent measurements of salt resistance of the tested barley varieties. Therefore, it was suggested that chlorophyll fluorescence can be used to assess the salt resistance of different plant cultivars.

Mineral Nutrient Relations   Many investigations focused on effects of substrate salinity on uptake and turnover of mineral nutrients in halophytes and non-halophytes (Munns et al., 1983; Ungar, 1991, pp. 107–138).

Ion relations of six *Brassica* species under salt stress were studied by He and Cramer (1992). Plants were grown in hydroculture with a quarter-strength Hoagland solution and exposed to seawater salinity. Salt stress reduced the growth of all species. A ranking of relative salt resistance of the species based on their dry matter production was established with *B. napus* the most resistant and *B. carinata* the most sensitive species. Mineral nutrient concentrations in shoots of the plants were analyzed to assess the relationship between relative salt resistance and nutrient imbalances under salt stress. Concentrations of Ca, Mg, K, Cl, Na, and total N were significantly affected by seawater salinity. However, only the change in calcium concentrations was significantly correlated to the relative salt resistance of the investigated *Brassica* species. It was suggested that calcium may play a regulatory role in the response of *Brassica* species to salt stress (see also the section on calcium interactions with NaCl).

Alterations in nutrient relations under salt stress were studied in five tomato *(Lycopersicon esculentum)* cultivars differing in salt resistance (Perez-Alfocea et al., 1993a). Plants were grown in sand culture and irrigated with Hoagland solutions to which up to 140 mM NaCl was added. Different responses of the nutrient balance were observed in the cultivars. One salt-resistant cultivar showed a typical halophyte response with increased shoot accumulation of sodium and chloride, but a reduced potassium content in the shoot. In contrast, another salt-resistant cultivar excluded sodium and chloride ions from the shoot and retained these ions in the roots. This cultivar maintained its potassium selectivity under salt stress. The authors concluded that the salt sensitivity of some tomato cultivars may be due to both toxic effects of sodium and chloride and imbalances of nutrients caused by salinity. Growth was inversely correlated with sodium and chloride contents, but directly correlated with potassium and calcium contents. Salt resistance of tomato plants implied physiological processes regulating sodium and chloride accumulation and maintaining potassium selectivity. A standard salt-resistance mechanism operating in all tested tomato cultivars was not found.

Biomass Production   The productivity of plants under salt stress appears to be a sensitive indicator for their salt resistance (Fig. 7-1). Dry matter production is easy to determine under all kinds of experimental conditions and has, thus, often been used to characterize plant responses to salt (Flowers et al., 1986; Greenway and Munns, 1980; Ungar, 1991, pp. 49–72). The parameter to be considered is the dry weight production, since the water content of a plant and its fresh weight can vary greatly with the salt stress conditions and the water availability (see the subsection on salt dilution by succulence in the section on mechanisms of salt resistance). In agricultural studies the productivity of crops in terms of biomass production can be a key parameter to evaluate salinity effects. These depend on the developmental stage; seedlings, for instance, are particularly sensitive (Ungar, 1991, p. 49). Therefore, the biomass production under salt stress can vary during the life cycle of a plant. This should be considered in studies of plant growth under saline conditions.

In productivity studies the dry weight increase of plants must be distinguished from the production of organic biomass (Flowers et al., 1986). As plants in saline substrates tend to accumulate considerable amounts of inorganic ions (see the subsection below on osmotic adjustment), part of the dry weight increase will be due to inorganic material. Therefore, to determine the produced organic dry matter, the total dry weight of the plant should be corrected by subtracting the amount of inorganic matter (ash content). However, in most studies such a more complicated procedure was not used.

A screening technique for salt resistance of rice plants *(Oryza sativa)* that is based on increases in fresh and dry weights under salt stress was proposed by Aslam et al. (1993). Two-week-old rice seedlings raised in silica gravel were transplanted to a nutrient solution culture and treated with NaCl. Concentrations were increased by 25 mM per day until final values of 50–200 mM were reached in different treatments. After 15 days, plants were harvested and fresh weight and dry weight of shoots and roots as well as the percentage of mortality were determined. The absolute shoot fresh weight or the relative shoot yield at 100 mM NaCl in percent of control were the best criteria for salt resistance. Based on these criteria a ranking of the salt resistance of the tested rice varieties was established. The results were consistent with those of other resistance tests in soil culture, either in containers or in the paddy field. Responses of the tested varieties were similar in the solution culture test and in the soil treatments and the ranking of salt resistance was comparable. Therefore, screening for salt resistance at the seedling stage in solution culture based on the shoot fresh weight was a rapid and reliable method.

A split-root system technique was used to investigate salt effects on growth of individual roots of the cactus *Opuntia ficus-indica* (Gersani et al., 1993). Young cactus plants were grown in sand culture irrigated with nutrient solution. Intact individual roots were placed in separate compartments of the culture container and irrigated with nutrient solutions amended with NaCl. Thus part of the root system received plain nutrient solution, whereas individual roots were exposed to 30 or 100 mM NaCl in the solution for up to 4 weeks. The dry weight of salt-treated roots decreased relative to the control by 40% in 30 mM NaCl and by 93% in 100 mM NaCl. Root respiration rates were increased under salt stress. Accumulation of carbon in the roots was reduced only in the higher salt concentration. Thus growth and respiration of small portions of the root system exposed to moderate salinity were maintained for several weeks. This may enable the roots to grow into less saline regions of the soil or to survive until soil salinity decreases. Such a strategy is of particular value in places with heterogeneous soil conditions.

The reasons for salt-induced growth reductions of plants were considered by Munns (1993). He suggested a two-phase response to salt stress. In the first phase, growth is reduced by the lowering of soil water potential, that is, by water stress. Thus in the first phase the plant responds to salt outside rather than within its organism. Such effects are not salt-specific, but occur also under any osmotic stress. This phase can last for weeks. In the following, second phase, salt-specific damage in leaves appears, when the absorbed salt ions from the soil can no longer be sequestered in the plant tissue. Concentrations rise beyond lethal levels and the respective leaves die. The loss of a few leaves does not affect the overall growth of a large plant, but if the rate of leaf loss exceeds the rate of new leaf development, the supply of assimilates to growing leaves will decrease. Furthermore, the supply of growth regulators may change and growth is further reduced. Munns (1993) pointed out that if the model is correct, rapid screening techniques for salt resistance, including tissue culture methods, may give questionable results. Such methods can test only the salt response in the first phase, that is the response of a plant to low substrate water potentials. The author suggested measuring of growth rates as a useful criterion for determining salt resistance of plants.

Seed Production   In order to colonize permanently saline habitats, plants must be able to complete their life cycle under such conditions. They must produce viable seeds under salt stress. Many crops are cultivated to produce seeds, like corn, rice, or peanuts, and the effects of salinity on seed production are of particular scientific and economic interest.

Effects of salt stress on reproductive organs of peanuts *(Arachis hypogaea)* were investigated by Silberbush and Lips (1988). Peanuts are moderately sensitive to salinity. Plants were grown in containers with calcareous sand in a greenhouse and irrigated with solutions of 0 or 50 mM NaCl as well as various nitrogen sources. Furthermore, developed gynophores were treated separately outside the main pots with NaCl solutions of up to 200 mM. The shoot dry weight of plants was not affected by treatment with 50 mM NaCl. Under salt stress the number of gynophores per plant was increased, but the mean pod weight was reduced. Gynophore vitality was reduced by treatments with 100 mM NaCl. The authors concluded that the salt sensitivity of peanut plants in the field is mainly due to the sensitivity of the reproductive organs. During their development in the field, extending gynophores touch the soil surface with its particularly high salt accumulation. Thus the pods are exposed to more severe salt stress than the whole plants.

The causes for reduced fertility of rice plants *(Oryza sativa)* under saline conditions were studied by Khatun and Flowers (1995). Rice plants were grown in sand in a greenhouse and were irrigated with nutrient solutions. In different treatment groups various amounts of an artificial seawater solution were added. The NaCl-salinity ranged from 0 to 50 mM. The treatment lasted from one month after germination to onset of flowering. Under salt stress the numbers of fertile florets and the pollen viability were reduced. Irrigation with 25 mM NaCl in artificial seawater reduced pollen germination to less than one-third of the control. Salt effects on stigmas were assessed from seed production of plants after pollination with viable pollen from unsalinized plants. Seed set was reduced by 38% in female plants grown in 10 mM NaCl in artificial seawater. Plants of the highest treatment (50 mM NaCl) produced no seeds at all. Reciprocal crossing experiments with salt-affected male plants and unsalinized female plants showed that salt effects on female plants dominated those on pollinator plants. Correlations between concentrations of sodium and chloride in pollen grains and in vitro pollen germination indicated direct ion effects on pollen viability.

Seed Germination   As mentioned above, the salt resistance of plants can change during their life cycle. Seed germination is a particularly sensitive developmental stage (Ungar, 1991, p. 49). It was generally observed that seeds of halophytes are more salt resistant than those of nonhalophytes (Ungar, 1978). However, the reasons are not as yet clear. At high salinity, halophyte seeds can remain dormant until the substrate salt level is reduced after sufficient rain has flushed the soil and germination conditions have improved (Ungar, 1978).

In salt deserts, not only salinity but also the low and rather unpredictable rainfall can restrict plant growth and seed germination. Some species apparently developed "internal water clocks" (Evenari, 1985) to detect favorable germination conditions. The seed coats of such plants contain substances, like salts, that inhibit germination.

During events of rainfall the water-soluble salts are leached. After sufficient rain has fallen and the inhibiting substances have been washed out, the seeds can start to germinate. This was described for plants of the genus *Atriplex* (Waisel, 1972, p. 186). Salt is the leachable inhibiting substance in seeds of the desert shrub *Zygophyllum dumosum* (Danin, 1983, p. 33). Thus it is ensured that germination starts only when sufficient rainwater for plant development is available. In the outer parts of fruits of *Atriplex dimorphostegia* a water-soluble substance that inhibits seed germination was found (Koller, 1957). Only after it is leached by rainwater, can germination start. The leachable substance may consist of salt incrustations. For instance, in seeds of *Atriplex triangularis* and *A. confertifolia,* highest concentrations of sodium, chloride, and of other ions were found in seed coats (Khan et al., 1987). Several plant species have developed di- or polymorphism of seeds (Khan and Ungar, 1984; Philipupillai and Ungar, 1984; Ungar, 1979, 1982). Thus amounts of leachable substances can vary in different seeds of the same plant. The seeds then require different amounts of rainfall to initiate their germination. In this way not all seeds start to grow at the same time, and competition for limited habitat resources is reduced. Furthermore, part of the seed production can remain in the soil seed bank for extended periods, thus ensuring a prolonged occupation of the habitat.

Most plant species germinate best in nonsaline conditions (Ungar, 1982). Under laboratory conditions even halophytes, like *Triglochin maritimum* or *Salicornia stricta,* had highest germination rates in pure water (Lötschert, 1970). However, seeds of the halophyte *Salicornia pacifica,* which grows on inland salt plains in Utah, were able to germinate in highly saline solutions (Khan and Weber, 1986). After 20 days of treatment with 5% NaCl solutions in a favorable temperature regime, up to one-third of the seeds had germinated.

Salt effects on the germination of seeds of *Arabidopsis thaliana* were investigated by Saleki et al. (1993). Three mutant strains had been obtained that were able to germinate in higher NaCl concentrations than the wild type. However, the later developmental stages did not differ in salt resistance from the wild type. Exposure of seeds to mannitol showed that the mutants could also germinate in lower osmotic potentials of the medium than the wild type. Reactions of mutant seeds to various potassium and lithium salts were similar to those of the wild type. Apparently, mutant seeds were mainly osmotolerant, but had also acquired a certain resistance to sodium ions. Analyses showed that NaCl-resistant seeds absorbed more ions during the imbibition than those of the salt-sensitive wild type. Such ion accumulation could contribute to osmotic adjustment of the seeds to low osmotic potentials of the media.

In germinating seeds and seedlings of rice plants *(Oryza sativa),* salt-stress-induced changes in protein profiles were observed (Radha Rani and Reddy, 1994). Rice seeds were germinated for 3 days in different NaCl concentrations ranging up to 600 mM. Two predominant polypeptides with 70 and 23 kDa were induced under salt stress. In shoots of germinated seedlings exposed to NaCl, two other polypeptides were found (15 and 26 kDa). It was suggested that these stress-responsive polypeptides may play a role in the salt-stress response of rice plants. However, the specific functions of such proteins need further investigation.

MECHANISMS OF SALT RESISTANCE

Salt resistance is the ability of plants to tolerate excess salt in their habitat without any significant impairment of their vital functions (Walter and Breckle, 1985, p. 110). It is a complex combination of various mechanisms, not a single process or adaptation, and therefore, certainly not controlled by a single gene (Munns, 1993).

Following the terminology of Levitt (1980, pp. 16–18), plants can achieve resistance to salt stress either by tolerating the stress or by avoiding it. Resistance is the more general term, including mechanisms of tolerance and of avoidance.

Tolerance

Tolerance to salt stress in the above-mentioned sense is the ability to tolerate toxic as well as osmotic effects of salt ions in the cytoplasm (Larcher, 1994, p. 319). Research in this field has focused on direct salt effects on enzymes in the cytoplasm of salt-stressed plants. Many investigated enzymes of halophytes were similar to those of other plants. They showed no particular tolerance to salt ions in in vitro studies (Flowers et al., 1977). However, high concentrations of salt ions, like Na^+ and Cl^-, have been found in the cytosol of salt-exposed plants (Eshel and Waisel, 1979). Under such conditions, the cytoplasmic enzymes have to function in the presence of salt ions. This was investigated with the enzyme phosphoenolpyruvate (PEP) carboxylase, extracted from the halophytes *Suaeda monoica* and *Chloris gayana* (Shomer-Ilan et al., 1985). It is the key enzyme in the process of CO_2 fixation in plant leaves. The in vitro salt tolerance of PEP carboxylase depended on the enzyme pretreatment as well as on the substrate (PEP) concentration in the medium. Addition of PEP to the extraction and storage medium helped to stabilize the enzyme. At low PEP levels of the assay medium, the pretreated enzyme was inhibited by NaCl, but at high PEP levels the enzyme was activated by NaCl. Salt ions changed the kinetic properties of the enzyme and were suggested to function as allosteric effectors.

The same enzyme, PEP carboxylase, was extracted from various plant species differing in salt resistance (Shomer-Ilan and Waisel, 1986). The activity of the substrate-stabilized enzymes of all tested plants was increased with 100 mM NaCl in the assay medium when 1.6 mM PEP was added. Also other agents, like betain, prolin, or glycerol, could apparently stabilize PEP carboxylase of corn plants *(Zea mays)* and affected the enzyme's salt response (Shomer-Ilan and Waisel, 1986). The authors suggested that such compounds can play a role in the regulation of enzyme activity by its substrate under saline and nonsaline conditions. The salt tolerance of plants may depend not only on salt exclusion from the cytosol, but also on changes of the microenvironment of the enzymes, for instance on whether the levels of substrate or protective agents have been increased.

A similar mechanism of enzyme protection was found with ribulose-1,5-bisphosphate carboxylase/oxygenase (rubisco) of the woody halophyte *Tamarix jordanis* (Solomon et al., 1994). It is the key enzyme in photosynthetic carbon reduction of plants. The carboxylating activity of the enzyme was inhibited by NaCl.

However, addition of N-methyl-L-proline restored the activity. The compound N-methyl-*trans*-4-hydroxy-L-proline was less effective in protecting the rubisco. Both prolin derivatives were found in some species of *Tamarix*. The salt resistance of *T. jordanis* seems to be based on two mechanisms: the increase of rubisco content and the formation of compatible solutes, like N-methyl-L-proline (see also the sub-section on intracellular compartmentation and compatible solutes, below). Such solutes enable the rubisco to function at high rates in the presence of salt ions in the cytosol.

Avoidance or Regulation

In many plants the resistance to salt stress involves a more or less efficient restriction of salt uptake. Such a restriction can concern either the whole plant, or sensitive organs, tissues, or cells. The efficiency of such avoidance mechanisms is always limited. Consequently, additional mechanisms have evolved for the regulation and stabilization of internal salt levels (Breckle, 1990; Flowers et al., 1986; Waisel, 1972, pp. 141–172, 1991a).

Restriction of Uptake or Transport   The foremost strategy to limit salt accumulation in plants is the inhibition of uptake of salt ions. On the whole plant level, this can be achieved by inhibition of root uptake, which has been described for mangroves (see the section on forests between land and sea, below). However, in most plant species such a mechanism is not efficient. Therefore, strategies have evolved to restrict salt transport into sensitive organs or tissues (Munns et al., 1983). The plants can sequester salt ions, which move with the transpiration stream, in specialized tissues, and thus prevent them from reaching sensitive parts of the organism. Such a mechanism was found in various species of the Fabaceae (Läuchli, 1984). Kramer et al. (1977) suggested that some legume species can reabsorb Na^+ ions from the upward-moving transpiration stream in mature parts of the roots. At the plasmalemma of xylem parenchyma cells the Na^+ ions are exchanged against K^+. The ion accumulation capacity of xylem parenchyma cells is limited, and thus this process is effective only at low levels of salinity. Under conditions of prolonged salt stress the retention mechanism must be supported by retranslocation of accumulated salt ions within the plants. In various crop plants, like bean, maize, and pepper, Lessani and Marschner (1978) observed a transport of Na^+ and Cl^- ions from leaves to the roots. Part of the ions were recreted back into the medium. This shows how long-distance transport processes can be involved in plant adaptations to salt stress.

The distributions of Na^+ in roots and shoots of two species of Fabaceae *(Phaseolus aureus, Trifolium alexandrinum)* and of various Chenopodiaceae *(Chenopodium schraderianum, C. album, Atriplex prostrata, Suaeda maritima)* exposed to salinity were investigated by Reimann (1992). Plants were grown in sand culture and irrigated with nutrient solutions amended with NaCl or KNO_3, so as to adjust concentrations of salt ions to 10 mM. Whereas *Phaseolus* retained most of its Na^+ in roots and hypocotyls, *Trifolium* showed no root retention of Na^+ and had highest levels in the shoots. The halophilic Chenopodiaceae, *Atriplex* and *Suaeda,* transported almost all absorbed Na^+ to the shoots, presumably for osmotic adjustment (see the subsection on that topic, below). In contrast, the salt-sensitive Chenopodiaceae, *Chenopodium schraderianum* and *C. album,* showed low Na^+ concentrations in the shoots. Uptake rates and transport rates for Na^+ from roots to shoots were low. A high proportion of Na^+ remained in roots and hypocotyls. In comparison, *Phaseolus* kept about 80% of the Na^+ in roots and hypocotyls, *C. schraderianum* 50%, *C. album* 35%, and *Trifolium* 20%. These results showed also that some species of the Chenopodiaceae can be salt-excluders.

Differences in ion relations of various members of the Chenopodiceae were also observed in another study by Reimann and Breckle (1993). Plants were grown in sand culture irrigated with 10 mM Na^+ and K^+. The halophilic *Atriplex rosea* showed highest Na^+ concentrations in the shoots, which indicated an effective Na^+ transport from roots to shoots (Fig. 7-3). This was confirmed by a low potassium/sodium ratio in shoots of only 2.4. In contrast, *Chenopodium pumilio* had highest Na^+ levels in hypocotyls and very low Na^+ concentrations in shoots. Such relations indicated Na^+ retention in roots and hypocotyls. A potassium/sodium ratio in shoots of *C. pumilio* of 42 also showed the preferential transport of K^+ and retention of Na^+ in basal parts of the shoots.

Partitioning and recirculation of salt ions in soybean plants *(Glycine max)* under moderate salt stress were studied by Durand and Lacan (1994). Plants were grown in hydroculture and treated with NaCl concentrations up to 50 mM for 8 days. The plants showed marked Na^+ retention in stems. The Na^+ transport to young leaves was restricted. Analyses of xylem sap at different stem heights showed that Na^+ concentrations of the sap were progressively lower at higher stem levels. This was attributed to retention of Na^+ by stem tissue and older leaves. Application of radioactive $^{22}Na^+$ to leaves indicated the retranslocation of this ion from leaves to roots with the phloem stream. However, recirculation rates were low and retranslocation alone was insufficient to prevent Na^+ accumulation in leaves. The salt sensitivity of soybean plants appears to depend mainly on the regulation of salt concentrations in the xylem.

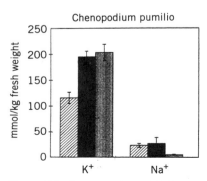 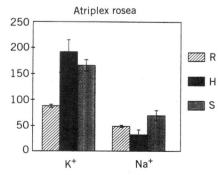

Figure 7-3. *Concentrations of Na^+ and K^+ in roots (R), hypocotyls (H), and shoots (S) of two species of the Chenopodiaceae. Plants were grown on sand and irrigated with solutions of 10 mM Na^+ and K^+. Means $\pm$ SD of 6 measurements. (After Reimann and Breckle, 1993.)*

Salt Elimination by Glands   The elimination of excess salt ions from the plant into its environment is called "recretion" (Frey-Wyssling, 1972). In the literature the terms "secretion" or "excretion" have also frequently been used (Fahn, 1988; Lüttge and Pitman, 1976). Specialized epidermal structures have evolved in various plant groups: bladder hairs and salt glands.

The salt bladder hairs are characteristic for members of the family Chenopodiaceae (Schirmer and Breckle, 1982). They are hairlike structures on leaf surfaces and consist of several stalk cells that support a large, balloon-shaped cell (Black, 1954; Reimann and Breckle, 1988). The stalk cells transport salt ions actively into the vacuole of the big bladder cell. Subsequently, the bladders of some species die and they are shed. In this way, salt is removed from the leaves (Fahn, 1988).

Salt glands are structures that can transport salt ions directly out of the plants (Hill and Hill, 1976; Thomson et al., 1988). Early descriptions were given for *Tamarix* species (Brunner, 1909; Marloth, 1887) and members of the Plumbaginaceae (Ruhland, 1915). Salt glands can have a simple structure, consisting of only 2 cells, as in grasses (Arriaga, 1992; Liphschitz and Waisel, 1982; Marcum and Murdoch, 1990; Taleisnik and Anton, 1988), or they can include up to 20 cells, as in *Limonium* species (Wiehe and Breckle, 1990). They are found in various plant families, an observation that suggests polyphyletic origin. A systematic overview was given by Liphschitz and Waisel (1982).

The efficiencies of salt glands of four haylophytic species *(Spartina anglica, Limonium vulgare, Armeria maritima,* and *Glaux maritima)* were investigated in a comparative study by Rozema et al. (1981). Plants had been collected from the field and were grown on different salts. The highest sodium recretion was found in *Spartina anglica,* and a somewhat lower recretion in *Limonium vulgare.* The other species had even lower rates. *Spartina* was able to recrete some 60% of the absorbed sodium, whereas these values were 33%, 20%, and 4%, respectively, for *Limonium, Glaux,* and *Armeria.* Such data show that salt recretion is of varying significance for the investigated species. Clearly, other mechanisms must also support the stabilization of the salt balance, like salt exclusion or dilution (see the subsection on salt dilution by succulence, below).

Salt glands of the halophytic tree *Tamarix aphylla* have been thoroughly investigated. They are found on young, green branches, and on the small, scalelike leaves. The *Tamarix* glands consist of eight cells: six outer, secretory cells and two inner, collecting cells (Bosabalidis and Thomson, 1984, 1985; Thomson and Liu, 1967). The secretory cells have a dense cytoplasm, whereas the collecting cells are vacuolated. The six secretory cells are separated from the inner cells by a cuticle. Apoplasmatic transport between inner and outer cells is inhibited. The cuticle has openings only at the common walls of innermost secretory cells and collecting cells (Bosabalidis and Thomson, 1985; Thomson and Liu, 1967). Many plasmodesmata connect the inner and outer cells, through which salt ions can be transported into the secretory cell complex (Thomson and Liu, 1967). The six secretory cells are interconnected by numerous plasmodesmata (Thomson and Platt-Aloia, 1985). Salt ions are probably stored in vesicles within the secretory cells. The membranes of such vesicles can fuse with the outer membrane of the upper secretory cell. In this way salt ions are eliminated from the cell into the outer apoplasm above the gland. The salt brine then passes through pores in the outer leaf cuticle. After drying out, salt crystals appear on the leaf surfaces (Thomson and Liu, 1967).

The composition of salts recreted by *Tamarix aphylla* depends on the composition of the root environment (Berry, 1970; Berry and Thomson, 1967). Ions like Rb^+, Li^+, or Cd^{2+} were recreted whenever these were added to culture solutions (Hagemeyer and Waisel, 1988; Thomson et al., 1969). The selectivity of the glands of *Tamarix aphylla* seems to be low (Hagemeyer, 1990; Storey and Thomson, 1994).

The ecophysiological role of the salt glands of *Tamarix aphylla* was studied by Waisel (1991b). He noted that, under conditions of high atmospheric humidity, twigs were often covered with an alkaline solution. The pH fluctuated between 8 in the dark and 10.5 in daylight. This solution resulted from precipitation of water on hygroscopic, recreted salts. It was shown that the alkaline solution can absorb CO_2 from the ambient air. The CO_2 is released during periods with a lower pH. The air surrounding the branches is thus enriched with CO_2 during prolonged periods and assimilation of *T. aphylla* is stimulated. Waisel (1991b) suggested three possible functions of salt recretion in *T. aphylla:* (1) elimination of excess salt, (2) reduction of transpiration by a moist cover of branches, and (3) stimulation of assimilation by trapping of CO_2.

Other well-studied salt glands are those of *Limonium* species within the Plumbaginaceae (Arisz et al., 1955; Faraday and Thomson, 1986; Hill, 1967; Hill and Hill, 1973a, b; Vassilyev and Stepanova, 1990; Wiehe and Breckle, 1990), those of *Glaux maritima* of the Primulaceae (Rozema and Riphagen, 1977; Rozema et al., 1977), and those of the mangrove tree *Avicennia marina* (Drennan and Pammenter, 1982; Shalom-Gordon and Dubinsky, 1990, 1993; Drennan et al., 1992; Fitzgerald et al., 1992; Ish-Shalom-Gordon and Dubinsky, 1990, 1993).

Salt Dilution by Succulence   Salinity can give rise to morphological and structural changes in some plant species. It is well known that certain plants develop thick, fleshy, succulent organs under salt stress (Waisel, 1972). This was observed only in dicotyledons, but not in monocotyledons (Flowers et al., 1986). A measure of succulence is the ratio of water content to surface area for a tissue at equilibrium with water-saturated atmosphere (Lüttge and Smith, 1984). Various plant parts can become succulent under salt stress. Fleshy leaves are observed in *Atriplex* or *Suaeda,* whereas stem succulence is found in *Salicornia, Arthrocnemum,* and others (Breckle, 1990). The increased thickness of succulent leaves is mainly due to larger mesophyll cells, which have absorbed water and increased the size of their vacuoles (Munns et al., 1983; Waisel, 1972). Succulence results, therefore, from increased water uptake of the tissue. It has been suggested that this may help to dilute absorbed salt ions (Munns et al., 1983). Consequently, succulence may be a mechanism to avoid high salt concentrations in plant organs. However, the dilution capacity of the tissue is limited, and this strategy can help plants only to cope with low levels of salinity or with short events of salt stress.

Effects of NaCl on leaf succulence of *Atriplex hastata,* a facultative halophyte of coastal salt marshes, was studied by Black (1958). Plants grown in hydroculture with added NaCl developed succulent leaves. This was explained by two observations. First, NaCl stress extended the life span of individual leaves, which by itself increased their final thickness. Secondly, the thickening process of mature leaves was accelerated at salt concentrations that did not impede the growth rate of the plants. The most rapid thickening was found when only part of the root system was exposed to 0.6 M NaCl, so as not to impede a rapid general growth rate of the whole plant.

Ion regulation strategies of halophytic *Atriplex* species were compared to those of the salt-sensitive *Phaseolus vulgaris* (Greenway et al., 1966). Plants were grown in hydrocultures and treated with radiolabeled Na^+ and Cl^- ions for various periods. From data of ion uptake and transport within plants, the authors concluded that halophytic *Atriplex* species could more efficiently regulate ion concentrations in their leaves. This was not due to ion export, but to continued growth and increased succulence. The authors suggested that plants that have no salt glands may regulate shoot ion concentrations either by limited uptake or by the diluting effects of an increased tissue volume, that is succulence.

Salt responses of *Glaux maritima,* a coastal halophyte, also involved succulence (Rozema, 1975). The degree of succulence was determined in hydroponically grown plants as water content (g) per leaf area (cm^2). Up to 60 mM NaCl in the medium succulence increased. At higher salt levels, degrees of succulence were reduced. The salt-regulating capacity of succulence in *Glaux maritima* was apparently limited to low salinity levels up to 60 mM. Under stronger salt stress other mechanisms must operate to ensure survival of the plants.

Salt-resistance mechanisms of three halophyte species from the Sudanese Red Sea salt marsh were compared by Ali and Dreyling (1994). Plants of the grass *Sporobolus spicatus,* of the perennial herb *Suaeda monoica,* and of the shrub *Zygophyllum album* were grown in sand culture in a growth chamber. The containers were irrigated with nutrient solutions amended with various concentrations of artificial seawater, which contained mainly NaCl. Vegetative growth parameters, like height, fresh and dry weights, and leaf dimensions, as well as ion uptake and recretion were determined. All three plant species showed optimal development in moderate salinity and are, thus, obligate halophytes. Only *Sporobolus* recreted substantial amounts of salt through glands. From the measured data the authors concluded that the salt resistance of *Sporobolus* was based on ion recretion, whereas succulence played a role in *Suaeda* and *Zygophyllum*. Increased succulence under salt stress was a mechanism to dilute salts in the cell sap of the halophytes.

Differences in salt resistance of two subspecies of *Salsola kali* (Chenopodiaceae) were investigated by Reimann and Breckle (1995). Plants were grown in pots and irrigated with nutrient solutions to which NaCl was added in varied concentrations. Succulence ratios were calculated as leaf water content per leaf surface area. In plants of the ssp. *tragus* dry weight was increased by treatment with 50 mM NaCl, but not in ssp. *ruthenica*. With rising salinity levels, shoot water contents increased in both subspecies, but ssp. *tragus* was more succulent than ssp. *ruthenica*. The results indicated that salt resistance of the subspecies was correlated to succulence and the determined low potassium/sodium ratios in leaves.

Osmotic Adjustment   In order to counterbalance low water potentials of saline substrates (see the section on saline soils, above), some plants use a controlled accumulation of salt ions (Flowers et al., 1977; Lerner, 1985; Waisel, 1972). This is osmotic adjustment on the whole-plant level (Flowers et al., 1986). In the cells, salt ions are compartmentalized and sequestered in vacuoles to avoid toxic effects in the cytosol (Cheeseman, 1988; Flowers, 1985; Gorham et al., 1985). At the same time, the osmotic balance between vacuole and cytosol is maintained by accumulation of compatible, organic solutes in the cytoplasm (Gorham et al., 1985; see also the subsection on intracellular compartmentation and compatible solutes, below).

Terrestrial vascular halophytes do not respond uniformly to salinity, and members of different taxa differ widely in their ion accumulation. Such specific differences in accumulated ion patterns have been used to establish groups of plants that share similar ion patterns, called "physiotypes" (Albert, 1975; Albert and Kinzel, 1973; Albert and Popp, 1977).

The process of osmotic adaptation in leaves of the semihalophytic grass *Panicum repens* under salt stress was investigated by Ramati et al. (1979). Plants were grown in nutrient solutions for two weeks, and then NaCl was added in concentrations up to 200 mM. The response to salt stress had two phases. In the first phase of 2–4 days, leaves showed dehydration and accumulation of Na^+ in cells. Chloride became dominant in the leaf cells only after 3 days. The second phase of salt adaptation lasted 4–6 days, during which time the Cl^- content of leaf cells increased. Already after 2 days, the osmotic potential of the leaves had reached lower values than that of the external medium. However, growth of the plants was inhibited until 6–8 days after transfer to salinity. Such transient growth reductions were probably caused by accumulations of Na^+ and Cl^- ions in cell walls, which prevented the full osmotic adaptation of the cytosol (see also the subsection on plant water balance in the section on the effects of salt on vascular plants, above).

Osmotic adaptation of different subspecies of the halophyte *Atriplex canescens* under salt stress was studied by Glenn et al. (1992). In a greenhouse plants were cultured for 80 days in containers with soil and irrigated with nutrient solutions amended with NaCl up to 720 mM. The three tested subspecies showed different accumulation patterns of Na^+ and K^+. At the time of harvest, *A. canescens* ssp. *canescens* had lower Na^+ and higher K^+ levels in leaves and stems than the other two subspecies *(A. canescens* ssp. *macropoda* and ssp. *linearis)* over the applied salinity range. The sodium/potassium ratios in leaves of the two high-sodium subspecies ranged from 2 in the lowest salt treatment to 10 in highest salinity. In leaves of the low-sodium subspecies, sodium/potassium ratios were between 0.4 and 2.3. Despite such differences in ion accumulation patterns, all three subspecies showed an equal salt resistance and had similar osmotic pressures in their leaves and stems. The salt resistance of this species was, therefore, not dependent on high Na^+ accumulation in leaves.

The same subspecies of *Atriplex canescens* were also studied under field conditions in Sonora, Mexico (Glenn et al., 1994). The subspecies *linearis* grows in intertidal zones of estuaries, whereas the subspecies *canescens* is found on dunes. In lysimeter pot experiments, 50% growth reduction was observed at a root zone salinity of 1160 mM NaCl for ssp. *linearis* and at 760 mM NaCl for ssp. *canescens*. The ssp. *linearis* showed adaptation to saline environments with higher net transport of Na^+ from root to shoot, higher Na^+ accumulation, as well as higher sodium/potassium ratios in the leaves. However, both subspecies showed similar osmotic adjustment in their leaves, which was equal to 2–3 times the external salinity. Even at low salt levels, both subspecies accumulated larger quantities of Na^+ than K^+ for osmotic adjustment. The authors suggested that an improvement of salt resistance of crop plants may be achieved by breeding for Na^+ accumulation rather than salt exclusion.

Osmotic adaptations were also investigated in two coastal halophytes, *Triglochin bulbosa* and *T. striata* (Naidoo, 1994). In hydroculture, plants were exposed to NaCl concentrations up to 400 mM for 9 weeks. Under salt stress, concentrations of Na^+ and Cl^- in shoots and roots of both species increased, K^+ concentrations decreased, and thus sodium/potassium ratios increased. With 200 mM NaCl in the substrate, inorganic ions in shoots of *T. striata* accounted for 83% of the osmotic adjustment and in *T. bulbosa* for 72%. At the same time, shoot concentrations of Na^+ accounted for 41% of osmotic adjustment in *T. striata* and for 22% in *T. bulbosa*. Both species accumulated proline, and, assuming that it was confined to the cytoplasm, its concentration was adequate to balance osmotic effects of salt ions (see also the next subsection). The authors suggested that the observed greater salt resistance of *T. bulbosa* was due to the more efficient salt exclusion from its roots. Growth reductions at high salinity were probably caused by interactions of Na^+ with other ions, like K^+, Ca^{2+}, Mg^{2+}, and particularly in *T. striata*, by excess Na^+ in shoots.

Intracellular Compartmentation and Compatible Solutes   For osmotic adaptation to salinity many halophytes accumulate salt (see the preceding subsection). Salt ions are compartmentalized and sequestered into cell vacuoles to protect the cytoplasm from toxic effects. This results in low osmotic potentials in the vacuoles. To prevent dehydration of the cytosol, its osmotic potential must be adjusted to the level of the vacuole. This can be achieved by accumulation of osmotically active, organic solutes in the cytosol, which do not interfere with physiological processes (Flowers et al., 1977, 1986; Munns et al., 1983). Such substances are called "compatible solutes." The outlined concept is known as the "intracellular model of solute compartmentation"; it describes osmotic adjustment on the cellular level (Gorham et al., 1985).

The chemical nature of compatible solutes varies among different plant species. They include polyoles (e.g., sorbitol or mannitol), amino acids and amides (e.g., proline), quarternary ammonium compounds (e.g., betaine), and soluble carbohydrates (sugars). Some compounds appear to have a certain association to taxonomic groups (Table 7-1; see also Gorham et al., 1985).

TABLE 7-1   Some Examples of Organic Solutes Found Enriched in Various Plant Species Under Salt Stress[a]

Solute	Plant Species and Family	Reference
Proline	*Phaseolus aureus* (Fabaceae)	Sudhakar et al. (1993)
	Mesembryanthemum crystallinum (Aizoaceae)	Thomas et al. (1992)
	Suaeda monoica (Chenopodiaceae)	Storey and Wyn Jones (1977)
	Atriplex spongiosa (Chenopodiaceae)	
	Spartina x townsendii (Poaceae)	
	Spartina anglica (Poaceae)	Van Diggelen et al. (1986)
	Sporobolus virginicus (Poaceae)	Marcum and Murdoch (1992)
	Triglochin maritima (Juncaginaceae)	Stewart and Lee (1974)
	Armeria maritima (Plumbaginaceae)	
	Lycopersicon esculentum (Solanaceae)	Alarcon et al. (1994)
		Perez-Alfocea et al. (1993b)
		Perez-Alfocea et al. (1994)
Betaine	*Spinacia oleracea* (Chenopodiaceae)	Summers and Weretilnyk (1993)
	Suaeda monoica (Chenopodiaceae)	Storey and Wyn Jones (1977)
	Atriplex spongiosa (Chenopodiaceae)	
	Spartina x townsendii (Poaceae)	
	Chloris gayana (Poaceae)	Storey and Wyn Jones (1975)
	Sporobolus virginicus (Poaceae)	Marcum and Murdoch (1992)
Sorbitol	*Plantago maritima* (Plantaginaceae)	Ahmad et al. (1979)
Sugars	*Lycopersicon esculentum* (Solanaceae)	Alarcon et al. (1994)
		Perez-Alfocea et al. (1993b)
		Perez-Alfocea et al. (1994)
	Sporobolus virginicus (Poaceae)	Marcum and Murdoch (1992)

[a] It has been suggested that such compounds function as compatible solutes for osmotic adjustment in cells.

Besides osmotic adaptation, some compatible solutes, like betaine or proline, were shown to have stabilizing effects on enzymes under salt stress (Shomer-Ilan and Waisel, 1986; Solomon et al., 1994; see the section on tolerance, above).

INTERACTIONS OF CALCIUM WITH NaCl

Calcium is an essential plant nutrient with many functions in metabolism, like stabilization of membranes, signal transduction as a second messenger, and control of enzyme activity (Hanson, 1984; Hepler and Wayne, 1985; Kirkby and Pilbeam, 1984). There are also reports stating that Ca^{2+} in the substrate can help to remediate adverse effects of salinity on plants. For instance, the salt response of bean plants *(Phaseolus vulgaris)* in hydroculture depended on the Ca^{2+} concentration of the nutrient solution (LaHaye and Epstein, 1969, 1971). When the plants were exposed to 50 mM NaCl in a nutrient solution lacking Ca^{2+}, the roots were damaged. Addition of 0.1 mM Ca^{2+} prevented this. From solutions lacking Ca^{2+}, bean plants absorbed large quantities of Na^+. With increased concentrations of Ca^{2+} in the medium, Na^+ accumulation in leaves of the plants was progressively reduced. At 3 mM Ca^{2+}, Na^+ concentrations in leaves of salt-treated plants were equal to those of control plants grown without salt. The Na$^+$ was, however, accumulated in roots and stems (see the subsection on restriction of uptake or transport in the section on mechanisms of salt resistance, above).

Additional studies have generally confirmed the ameliorating effects of Ca^{2+} on substrate salinity. Inhibitions of root elongation of pea plants *(Pisum sativum)* under salt stress were reversed by increased Ca^{2+} levels of the medium (Solomon et al., 1989). Adverse effects of NaCl on water transport of maize *(Zea mays)* root cells were partly compensated by additional Ca^{2+} (Azaizeh et al., 1992).

Effects of substrate NaCl on different *Brassica* species under varied Ca^{2+} supply were studied by Ashraf and Naqvi (1992). Supplemental Ca^{2+} improved the percentage of seed germination and shoot dry matter production of *B. napus* and *B. juncea*, but had no effect on the germination speed. In contrast, Schmidt et al. (1993) found no growth improvement of various *Brassica* species under salt stress with additional Ca^{2+}.

In barley *(Hordeum vulgare)*, a high Ca^{2+} supply mitigated NaCl-induced increases of Na$^+$ contents and decreases of K$^+$ contents of roots (Martinez and Läuchli, 1993). The effect of Ca^{2+} on Na$^+$ uptake of barley depended also on the way in which the ions were fed to the plants (Gorham et al., 1994). Soil-applied Ca^{2+} reduced accumulation of Na$^+$ in leaves of plants grown in media with NaCl. However, when the ions were sprayed on the leaves, Na$^+$ uptake was enhanced by Ca^{2+}. It was concluded that resistance to salt, either applied in the substrate or as salt spray on leaves, are different traits. This may be of ecological importance for plant species growing near the seashore, which can be affected by salt spray and/or substrate salinity.

In *Sorghum bicolor* plants, 100 mM NaCl reduced leaf elongation by 20% (Bernstein et al., 1993). An increase of Ca^{2+} concentrations from 1 to 10 mM reversed such effects. When exposed to NaCl in the medium, pH gradients across tonoplasts in root tips of *S. bicolor* were reduced. Additional Ca^{2+} reversed this reduction (Colmer et al., 1994). It was suggested that the maintenance of a pH gradient across the tonoplast plays a role in salt resistance of the plants, since this gradient might be the driving force for Na$^+$ transport from the cytoplasm into the vacuole. Thus Ca^{2+} supported the compartmentalization of Na$^+$ ions (see the subsection on intracellular compartmentation and compatible solutes in the section on mechanisms of salt resistance, above).

Effects of supplemental Ca^{2+} on the salt response of cotton *(Gossypium hirsutum)* root cells were investigated by Zhong and Läuchli (1994). At 150 mM NaCl in a nutrient solution the selectivity of K$^+$ versus Na$^+$ in the apical region of the roots was enhanced by addition of 10 mM Ca^{2+}. It was suggested that supplemental Ca^{2+} may alleviate inhibitory effects of NaCl on cotton roots by maintaining the selectivity of the plasma membrane for K$^+$ over Na$^+$.

Bean plants *(Phaseolus vulgaris)* under NaCl stress showed reductions in germination and early seedling growth (Cacharro et al., 1994). Such effects were ameliorated by supplemental Ca^{2+} in the substrate. Concentrations of Cl$^-$ were not affected by Ca^{2+} levels. However, additional Ca^{2+} reversed NaCl-induced reductions of K$^+$ and Ca^{2+} concentrations in shoots and also restored transport of these ions from roots to shoots.

A hypothesis explaining the described effects of Ca^{2+} on salt responses of plants was put forward by Rengel (1992). He suggested that Na^+ reduces the binding of Ca^{2+} to plasma membranes (see also Lynch et al., 1987), inhibits influx and increases efflux of Ca^{2+}, and causes depletion of internal Ca^{2+} stores in cell compartments. Changes in Ca^{2+} levels of cells are the primary responses to salt stress, which are perceived by root cells. The supply of Ca^{2+} to leaf cells is reduced and the activity of Ca^{2+} in leaf cells is decreased, while the Na^+ activity is increased. Therefore, the amelioration of adverse salt effects by an additional Ca^{2+} supply is probably due to preventing Na^+-induced changes in Ca^{2+} levels of the cells.

It is interesting to note that the presence of Ca^{2+} can also increase the resistance of plants to other stressors, like Cd^{2+} (Hagemeyer, 1990; Hagemeyer and Waisel, 1989b) or Ni^{2+} (Gabrielli and Pandolfini, 1984).

SALT INDUCTION OF CRASSULACEAN ACID METABOLISM IN *MESEMBRYANTHEMUM*

A remarkable response to salt stress has been discovered in members of the family Aizoaceae, particularly in the genus *Mesembryanthemum*. Grown under nonsaline conditions such plants perform C_3 photosynthesis. Winter and von Willert (1972) observed that salt-stressed plants of *M. crystallinum* absorbed CO_2 in the dark and, at the same time, malate levels of their leaves increased. These are typical features of the crassulacean acid metabolism (CAM). Apparently, under salt stress *M. crystallinum* can shift its photosynthetic pathway from C_3 to CAM.

The CAM is an adaptation of plants to stress conditions, especially drought (Lüttge, 1987a, b). In environments with limited water supply, such as in deserts or on saline substrates, it is advantageous for plants to reduce transpiration by stomatal closure during the hot daytime. As this would also hamper carbon assimilation, CAM plants have developed a way to bind CO_2 in the cooler and more humid dark hours. The CO_2 is bound to PEP by PEP carboxylase, and the formed malate is stored in cell vacuoles. Thus during the day the stomates can remain closed to limit transpiration and conserve water. At night stomates are open for CO_2 uptake. The stored malate is then broken down again during the light hours to CO_2 and pyruvate, and the carbon is introduced into the Calvin cycle. Such plants show a diurnal cycle of malate concentrations in their leaves with increasing levels during nights and decreasing levels during days.

The annual facultative halophyte *M. crystallinum* can change its photosynthetic pathway in response to environmental conditions, and is thus an inducible CAM plant (Osmond, 1978). The succulent *Mesembryanthemum* species grow in desert environments, like the southwestern African Namib, and along Mediterranean coasts (see the subsection on salt dilution by succulence in the section on mechanisms of salt resistance, above.) It is not yet clear whether the shift to CAM is a preprogrammed part of their life cycle or is only expressed in response to salt stress (Lüttge, 1993). Herppich et al. (1992) suggested that the salt induction of CAM is under strict developmental control, since CAM occurred only after plants had reached the juvenile growth phase. Perhaps plants germinate in natural habitats after rainfall and start their life with C_3 photosynthesis. During the following dry period they switch to CAM, which enables them to survive longer and produce large numbers of seeds (Lüttge, 1993; Winter et al., 1978). On the other hand, Piepenbrock and Schmitt (1991) argued that the shift to CAM in *M. crystallinum* is more dependent on environmental conditions than on endogenous programs. They found a uniform steady-state level of PEP-carboxylase activity in leaves of different ages from 7-week-old, well-watered plants. However, in salt-stressed plants, the rate of induction of PEP-carboxylase activity varied with the age of the plant and also with the age of individual leaves. Younger plants induced CAM more slowly than older ones. The activity of PEP carboxylase was down-regulated when salt-stressed leaves were cut and rehydrated, regardless of leaf age or pretreatment. From such data the authors inferred a predominantly environmental control of CAM induction in *M. crystallinum*.

Various other aspects of the CAM induction have been studied in detail. Young plants of *M. crystallinum* grown on a soil mixture under controlled conditions performed C_3 photosynthesis (Dai et al., 1994). Within one week of irrigation with 350 mM NaCl, the plants shifted to CAM, which was determined by changes in enzyme activities, nocturnal CO_2 uptake, and malate accumulation. In addition to salt, growth regulators, like ABA, supplied to the plants with the irrigation water, were also capable of inducing CAM. The authors hypothesized that ABA may also have a role in CAM induction under natural conditions. Both water stress and an external supply of ABA induced CAM. Since water stress also caused an increase in leaf ABA levels, it seemed likely that ABA is involved in CAM induction of *M. crystallinum*. Similarly, a role of ABA in the response to drought stress was described for other plant species (see the subsection on effects on phytohormone levels in the section on the effects of salt on vascular plants, above.)

The salt-stress response of *M. crystallinum* involves an increase in levels of PEP carboxylase and other enzymes associated with CAM. Furthermore, in cell suspension cultures NaCl-induced increases of proline levels were found (Thomas et al., 1992). Upon treatment with 400 mM NaCl in the medium, proline levels increased from 5% to 40% of total soluble amino acids. However, the amounts of PEP carboxylase protein in cells of suspension cultures were not affected by the NaCl treatment. Apparently, the induction of a PEP carboxylase protein, which was found in salt-treated leaves, and of other CAM enzymes is possible only in organized tissues. Thomas et al. (1992) concluded that both proline accumulation (see the subsection on intracellular compartmentation and compatible solutes in the section on mechanisms of salt resistance, above) and CAM induction contribute to the salt resistance of *M. crystallinum*. Proline accumulation is a rapid response on the cellular level, whereas induction of CAM is a slower process, which depends on tissue organization.

Regulation of gene expression in salt-stressed *M. crystallinum* plants was investigated by Vernon et al. (1993). After exposure to salinity various mRNAs encoding proteins of different biochemical pathways accumulated in leaf tissue. It was suggested that water stress triggered the coordinated induction of mRNAs involved in different aspects of the adaptive stress response of the plants. Varied stressors, like drought or salinity, caused different transcription levels of several genes. The molecular stress response of *M. crystallinum* may be triggered by multiple signals. The authors suggested a complex web of response pathways and mechanisms of gene regulation. This would enable the plants to respond to different stressors with specific increases in expression of certain genes or groups of genes.

Changes in gas exchange and photosynthetic pathways similar to those in *M. crystallinum* have also been observed in other members of the same family, for example, *Carpobrotus edulis* (Winter, 1973). Such adaptations provide instructive examples of responses to salinity or other stressors, which involve physiological shifts on various levels from gene regulation on the molecular level to the whole-plant level.

FORESTS BETWEEN LAND AND SEA: MANGROVES

Mangroves are unique forest communities in tidal zones of equatorial and tropical coasts (Ball, 1988; Walter and Breckle, 1986). Chapman (1977) used the term "mangal" for such communities and reserved "mangrove" for the specialized trees inhabiting this ecosystem. Such trees, which are permanently affected by salinity during their whole life cycle, were named "halotrees" (Waisel, 1991a). Mangrove trees have to cope with extreme conditions. Twice daily at high tide, the trees are inundated with seawater and only the crowns project to the open air. They are exposed to seawater salinity and, additionally, the roots must endure periods of low oxygen supply. At low tide, the soil between the trees falls dry. Then the substrate loses water by evaporation and transpiration of the plants, and substrate salt concentrations increase even beyond seawater values. In this way, the appearance of the mangrove community changes drastically every six hours.

Walter and Breckle (1986) distinguished three types of mangrove ecosystems:

1. Estuarine mangroves are affected both by seawater and by freshwater from rivers.
2. Coastal mangroves occur on tropical coasts sheltered from heavy surf behind coral reefs. Their only freshwater input is rain.
3. Coral reef mangroves are found on the landward side of reefs that have risen above the low-tide water level.

Tree species inhabiting mangrove ecosystems need special physiological adaptations to survive under such extreme conditions. The predominant genera, which belong to different families, are *Avicennia, Bruguiera, Ceriops, Rhizophora, Aegiceras, Aegialitis, Laguncularia, Lumnitzera, Kandelia,* and *Xylocarpus* (Walter and Breckle, 1986).

Physiological adaptations of mangrove trees to salinity have been investigated in a number of studies. Scholander et al. (1966) found that the xylem sap of several mangrove species from the genera *Rhizophora, Laguncularia,* and *Conocarpus* had very low ion concentrations. Apparently the roots, which are exposed to seawater salinity, perform an ultrafiltration at plasma membranes. This process is driven by tension in the xylem resulting from low sap pressures. Water uptake from the ascending xylem sap into the leaves is effected by accumulation of solutes in leaf parenchyma cells, which have, thus, very low osmotic pressures.

Xylem pressures measured in various mangrove plants ranged from -3 to -6 MPa (Scholander, 1968). This is sufficiently lower than the osmotic pressure of seawater (-2.5 MPa) to drive an ultrafiltration of water at root cell membranes. The author suggested that the negative sap pressure ultimately depended on the low osmotic potential of leaf cells. The filtration process continued also after chilling or poisoning of the roots. It was thus a nonmetabolic process and was even reversible by experimentally applied reverse pressure gradients (Scholander, 1968).

The salt transport in the mangrove tree *Avicennia marina* was studied in natural stands in southern Sinai (Waisel et al., 1986). Transpiration rates, xylem sap concentrations, leaf salt contents, and rates of recretion and of salt retranslocation out of leaves were continuously monitored during several days. Salt fluxes in and out of the leaves were calculated from the data. The authors concluded that the salt resistance of *A. marina* depended on three mechanisms:

1. Salt filtration by the roots was the most important factor. About 80% of the salt reaching the root surface was prevented from entering the plants.
2. Salt tolerance on the cellular level enabled the plants to survive high intracellular salt concentrations.
3. Salt retranslocation out of leaves and recretion from glands was another factor; about 40% of the salt that had entered through the roots were recreted via glands. Therefore, recretion alone did not enable the plants to cope with excess salt, but might have helped to balance mineral concentrations by selective recretion of toxic ions.

Concentrations of mineral ions and of organic solutes in leaves of 23 mangrove species from Australia were measured by Popp et al. (1985). In most species, leaf concentrations of Na^+ and Cl^-, calculated on plant water basis, were close to seawater values. Predominant organic solutes in most species were pinitol and mannitol. Proline was found in two species of *Xylocarpus*. High concentrations of methylated quarternary ammonium compounds occurred in 5 out of 23 investigated species, among them *Avicennia eucalyptifolia* and *A. marina*. Such organic compounds are known to function as compatible solutes in osmoregulation on the cellular level and can balance high concentrations of salt ions in cell vacuoles (see the subsection on intracellular compartmentation and compatible solutes in the section on mechanisms of salt resistance, above).

The long-distance water transport of the mangrove tree *Rhizophora mangle* in Hawaii was studied by Zimmermann et al. (1994). Measurements with a xylem pressure probe and staining experiments showed the occurence of viscous, mucilage- and/or protein-related compounds in the xylem fluid. Also, very low sap flow rates indicated the presence of high-molecular-weight polymeric substances in the xylem sap. Several effects of such compounds in the xylem were proposed. The increased density of viscous solutions was assumed to generate gravity-dependent convectional flows. Also, small gas bubbles occur frequently in viscous solutions, which may allow water transport by interfacial, gravity-independent streaming at the gas/water interfaces. The authors proposed that density-driven convectional flows and interfacial streaming, as well as low radial reflection coefficients of the roots to NaCl, were the traits by which *R. mangle* maintained its water transport under saline conditions. Therefore, it was suggested that high-molecular-weight polymeric substances may play an important role in long-distance water transport in this mangrove tree.

Growth of two Australian mangrove species, *Sonneratia alba* and *S. lanceolata*, in relation to salt stress was investigated by Ball and Pidsley (1995). The species showed differences in salt resistance; *S. alba* grew in salinity levels from freshwater to seawater with maximal growth between 5% and 50% seawater. In contrast, *S. lanceolata* grew only up to 50% seawater and showed best growth from 0 to 5% seawater. The growth response was compared under conditions favorable for both species, that is, 5% seawater. The less salt-resistant *S. lanceolata* showed twice the height, leaf area, and biomass of the more resistant species. The authors concluded that a higher salt resistance is achieved at the expense of growth performance and competitive ability under low salinity. The described differences in salt resistance were consistent with the distribution of the species along natural salinity gradients.

A variety of adaptations to extreme habitat conditions occurs in mangrove plants:

1. Salt exclusion by root ultrafiltration
2. Salt recretion via glands, for example, in *Avicennia, Laguncularia,* and *Conocarpus* (Atkinson et al., 1967; Roth, 1992, pp. 157–161; see also the subsection on salt elimination by glands in the section on mechanisms of salt resistance, above)
3. Ion accumulation in leaf cells for osmotic adaptation and osmotic adjustment with compatible solutes (see the subsections on osmotic adjustment and intracellular compartmentation and compatible solutes in the section on mechanisms of salt resistance, above)
4. Leaf succulence, for example, in *Rhizophora* and *Laguncularia* (Roth, 1992, pp. 130–156; see also the subsection on salt dilution by succulence in the section on mechanisms of salt resistance)

Furthermore, several species of, for example, *Sonneratia* or *Avicennia,* developed breathing roots, pneumatophores, to improve their oxygen supply in muddy, regularly inundated substrates (von Guttenberg, 1968, pp. 286–296; Hovenden and Allaway, 1994). Some mangrove trees, like *Rhizophora, Avicennia,* and *Kandelia,* have peculiarly shaped fruits with viviparous seeds, which germinate already inside the fruit (Walter and Breckle, 1986, pp. 182–183). In this way the seedling stage is shortened and they can rapidly occupy their habitat. Such characteristics are found in varying combinations in the different mangrove species. They enable mangrove trees to occupy their extreme habitats as the dominant vascular plants.

Currently many mangrove communities are endangered by human activities, like logging for timber production and clearing for human land-use, or by water pollution. For instance, in a coastal zone in eastern Thailand the estimated annual loss of mangrove forests by deforestation was 500 ha yr^{-1} in the late 1970s, but more than 2300 ha yr^{-1} at the end of the 1980s (Raine, 1994). The author warns that if the present rate of forest destruction continues, soon none will be left. Ball (1988) pointed out that there is an urgent need for development of sustainable management practices, which are based on understanding of physical and biological processes in mangrove ecosystems. Furthermore, it will be crucial to implement conservation schemes in mangrove ecosystems on all continents in order to preserve these unique plant communities for future generations.

EPILOG

Responses of many plant species to salt stress have been intensively investigated (see section "Effects of Salt on Vascular Plants"). Also, particular adaptations of specially adapted halophytes to saline conditions were described in detail (see section "Mechanisms of Salt Resistance"). What is needed now, as Munns (1993) has pointed out, is the development and experimental test of hypotheses to integrate the numerous data into a more general picture of the salt-stress response of plants (see also Lerner et al., 1994).

Promising approaches to elucidate mechanisms of salt resistance are comprehensive studies with simultaneous measurements of various ecological and physiological parameters in the same plant. Under favorable conditions, even the same plant individual can be used. In this way it is possible to assess the relative contributions of each of the different adaptational mechanisms to the fitness of that particular species. Whenever possible, such investigations should be conducted in the field, or at least under conditions that come close to the natural environment of the studied species. Instructive examples were presented by Rozema (1975) and Waisel et al. (1986).

Experimental conditions can strongly influence responses of plants to applied stress, in this respect to salinity. The discussion of calcium effects on salt resistance of plants (see section "Interactions of Calcium with NaCl") demonstrates how seemingly small changes of the nutrient supply, like increases in calcium concentrations of the medium, drastically alter the plants' response to salt ions. One may even assume that the salt resistance of a plant species is not a constant parameter, but is controlled by the calcium supply. Such effects, which are probably numerous, should be considered when results from different experiments are compared. Small variations in experimental conditions can be meaningful. The need for standardization in experimental techniques is emphasized.

A great deal of effort is spent worldwide in the improvement of salt resistance of crop plants. This seems necessary to counterbalance continual losses of agricultural area due to salinization, as discussed in the opening section of this chapter. However, from a long-term ecological point of view, it seems equally important to improve and apply advanced irrigation techniques to stop salinization and desertification processes. It seems more reasonable to avoid salt accumulation in agricultural soils than to spend resources in finding and adapting salt-resistant crop plants. Nevertheless, in order to avert the definite loss of large areas of arable land where the salinization process has already begun, the search for adapted crop plants that can be cultivated under such conditions is also required.

REFERENCES

Ahmad, I., F. Larher, and G. R. Stewart (1979), *New Phytol., 82*, 671–678.

Alarcon, J. J., M. J. Sanchez-Blanco, M. C. Bolarin, and A. Torrecillas (1994), *Plant Soil, 166*, 75–82.

Albert, R. (1975), *Oecologia, 21*, 57–71.

Albert, R. and J. Falter (1978), *Phyton, 18*, 173–197.

Albert, R. and H. Kinzel (1973), *Z. Pflanzenphysiol., 70*, 138–157.

Albert, R. and M. Popp (1977), *Oecologia, 27*, 157–170.

Ali, A. K. S. and G. Dreyling (1994), *Angew. Bot., 68*, 156–162.

Arisz, W. H., I. J. Camphuis, H. Heikens, and A. J. Van Tooren, (1955), *Acta Bot. Neerl., 4*, 322–338.

Arriaga, M. O. (1992), *Bothalia, 22*, 111–117.

Ashraf, M. and M. I. Naqvi (1992), *Z. Pflanzenernähr. Bodenk., 155*, 101–108.

Aslam, M., R. H. Qureshi, and N. Ahmed (1993), *Plant Soil, 150*, 99–107.

Atkinson, M. R., G. P. Findlay, A. B. Hope, M. G. Pitman, H. D. W. Saddler, and K. R. West (1967), *Aust. J. Biol. Sci., 20*, 589–599.

Azaizeh, H., B. Gunse, and E. Steudle (1992), *Plant Physiol., 99*, 886–894.

Ball, M. C. (1988), *Trees, 2*, 129–142.

Ball, M. C. and S. M. Pidsley (1995), *Func. Ecol., 9*, 77–85.

Bañuls, J. and E. Primo-Millo (1992), *Physiol. Plant., 86*, 115–123.

Belkhodja, R., F. Morales, A. Abadia, J. Gomez-Aparisi, and J. Abadia (1994), *Plant Physiol., 104*, 667–673.

Ben-Hayyim, G. (1987), *Plant Physiol., 85*, 430–433.

Bernstein, N., A. Läuchli, and W. K. Silk (1993), *Plant Physiol., 103*, 1107–1114.

Berry, W. L. (1970), *Am. J. Bot., 57*, 1226–1230.

Berry, W. L. and W. W. Thomson (1967), *Can. J. Bot., 45*, 1774–1775.

Black, R. F. (1954), *Aust. J. Bot., 2*, 269–286.

Black, R. F. (1958), *Aust. J. Bot., 6*, 306–321.

Bogemans, J., L. Neirinckx and J. M. Stassart (1989), *Plant Soil, 120*, 203–211.

Bosabalidis, A. M. and W. W. Thomson (1984), *Ann. Bot., 54*, 169–174.

Bosabalidis, A. M. and W. W. Thomson (1985), *J. Ultrastruct. Res., 92*, 55–62.

Breckle, S. W. (1990), in *Genetic aspects of plant mineral nutrition* (N. El Bassam et al.,

Eds.), Kluwer Academic, Dordrecht, Netherlands, pp. 167–175.

Bresler, E., B. L. McNeal, and D. L. Carter (1982), *Saline and sodic soils,* Springer-Verlag, New York.

Brod, H. G. (1993), *Ber. der Bundesanst. für Strassenwes., Heft V2,* 1–165.

Brunner, C. (1909), *Jahrb. der Hamburg Wiss. Anst. XXVI* 1908, 3. Beiheft, 89–162.

Cachorro, P., A. Ortiz, and A. Cerda (1994), *Plant Soil, 159,* 205–212.

Carter, D. L. (1975), in *Plants in saline environments* (A. Poljakoff-Mayber and J. Gale, Eds.), Springer-Verlag, New York, pp. 25–35.

Chapman, V. J. (1974), in *Ecology of halophytes* (R. J. Reimold and W. H. Queen, Eds.), Academic Press, New York, pp. 3–19.

Chapman, V. J. (1977), *Wet coastal ecosystems,* Elsevier Scientific, Amsterdam.

Cheeseman, J. M. (1988), *Plant Physiol., 87,* 547–550.

Clipson, N. J. W. and T. J. Flowers (1987), *New Phytol., 105,* 359–366.

Colmer, T. D., T. W. M. Fan, R. M. Higashi, and A. Läuchli (1994), *J. Exp. Bot., 45,* 1037–1044.

Cordovilla, M. P., F. Ligero, and C. Lluch (1994), *J. Exp. Bot., 45,* 1483–1488.

Dai, Z., M. S. B. Ku, D. Zhang, and G. E. Edwards (1994), *Planta, 192,* 287–294.

Danin, A. (1983), *Desert vegetation of Israel and Sinai,* Cana Publishing House, Jerusalem.

Downton, W. J. S. and B. R. Loveys (1981), *Aust. J. Plant Physiol., 8,* 443–452.

Drennan, P. and N. W. Pammenter (1982), *New Phytol., 91,* 597–606.

Drennan, P. M., P. Berjak, and N. W. Pammenter (1992), *S. African J. Bot., 58,* 486–490.

Durand, M. and D. Lacan (1994), *Physiol. Plant, 91,* 65–71.

Ernst, W. H. O. and E. N. G. Joosse-Van Damme (1983), *Umweltbelastung durch Mineralstoffe,* G. Fischer-Verlag, Stuttgart.

Eshel, A. and Y. Waisel (1979), *Physiol. Plant., 46,* 151–154.

Eshel, A. and Y. Waisel (1984), *Plant Cell Environ., 7,* 133–137.

Evenari, M. (1985), in *Hot deserts and arid shrublands* (M. Evenari et al., Eds.), Elsevier Scientific, Amsterdam, pp. 79–92.

Fahn, A. (1988), *New Phytol., 108,* 229–257.

Faraday, C. D. and W. W. Thomson (1986), *J. Exp. Bot., 37,* 471–481.

Fedina, I. S., T. D. Tsonev, and E. I. Guleva (1994), *J. Plant Physiol., 143,* 245–249.

Fitzgerald, M. A., Orlovich, D. A. and W. G. Allaway (1992), *New Phytol., 120,* 1–7.

Flowers, T. J. (1985), *Plant Soil, 89,* 41–56.

Flowers, T. J., P. F. Troke, and A. R. Yeo (1977), *Annu. Rev. Plant Physiol., 28,* 89–121.

Flowers, T. J., M. A. Hajibagheri, and N. J. W. Clipson (1986), *Quart. Rev. Biol., 61,* 313–337.

Flowers, T. J., M. A. Hajibagheri, and A. R. Yeo (1991), *Plant Cell Environ., 14,* 319–325.

Frey-Wyssling, A. (1972), *Saussurea, 3,* 79–90.

Gabrielli, R. and T. Pandolfini (1984), *Physiol. Plant., 62,* 540–544.

Gersani, M., E. A. Graham, and P. S. Nobel (1993), *Plant Cell Environ., 16,* 827–834.

Glenn, E. P., M. C. Watson, J. W. O'Leary, and R. D. Axelson (1992), *Plant Cell Environ., 15,* 711–718.

Glenn, E. P., M. Olsen, R. Frye, D. Moore, and S. Miyamoto (1994), *Plant Cell Environ.*, *17*, 711–719.

Golombek, S. D. and P. Lüdders (1993), *Plant Soil, 148*, 21–27.

Gorham, J., R. G. Wyn Jones, and E. McDonnell (1985), *Plant Soil, 89*, 15–40.

Gorham, J., R. Papa, and M. Aloy-Lleonart (1994), *J. Exp. Bot., 45*, 895–901.

Greenway, H. and R. Munns (1980), *Annu. Rev. Plant Physiol., 31*, 149–190.

Greenway, H., A. Gunn and D. A. Thomas (1966), *Aust. J. Biol. Sci., 19*, 741–756.

Guttenberg, H. von (1968), *Der primäre Bau der Angiospermenwurzel*, Encyclopedia of Plant Anatomy, Vol. VIII, Pt. 5, Gebr. Bornträger, Berlin.

Hagemeyer, J. (1990), *Diss. Bot., 155*, 1–194.

Hagemeyer, J. and Y. Waisel (1987), *Physiol. Plant., 70*, 133–138.

Hagemeyer, J. and Y. Waisel (1988), *Physiol. Plant., 73*, 541–546.

Hagemeyer, J. and Y. Waisel (1989a), *Physiol. Plant., 75*, 280–284.

Hagemeyer, J. and Y. Waisel (1989b), *Physiol. Plant., 77*, 247–253.

Hanson, J. B. (1984), *Adv. Plant Nutr., 1*, 149–208.

He, T. and G. R. Cramer (1992), *Plant Soil, 139*, 285–294.

Hepler, P. K. and R. O. Wayne (1985), *Annu. Rev. Plant Physiol., 36*, 397–439.

Herppich, W., M. Herppich, and D. J. von Willert (1992), *Bot. Acta, 105*, 34–40.

Hill, A. E. (1967), *Biochim. Biophys. Acta, 135*, 454–460.

Hill, A. E. and B. S. Hill (1973a), *J. Membrane Biol., 12*, 129–144.

Hill, A. E. and B. S. Hill (1973b), *Int. Rev. Cytol., 35*, 299–319.

Hill, A. E. and B. S. Hill (1976), in *Encyclopedia of plant physiology, Vol. 2 IIb, Tissues and organs* (U. Lüttge and M. G. Pitman, Eds.), Springer-Verlag, New York, pp. 225–234.

Hovenden, M. J. and W. G. Allaway (1994), *Ann. Bot., 73*, 377–383.

Ish-Shalom-Gordon, N. and Z. Dubinsky (1990), *Plant Cell Physiol., 31*, 27–32.

Ish-Shalom-Gordon, N. and Z. Dubinsky (1993), *Pacific Sci., 47*, 51–58.

Johnson, D. W., S. E. Smith, and A. K. Dobrenz (1992), *Euphytica, 60*, 27–35.

Khan, M. A. and I. A. Ungar (1984), *Am. J. Bot., 71*, 481–489.

Khan, M. A. and D. J. Weber (1986), *Am. J. Bot., 73*, 1163–1167.

Khan, M. A., D. J. Weber, and W. M. Hess (1987), *Great Basin Naturalist, 47*, 91–95.

Khatun, S. and T. J. Flowers (1995), *Plant Cell Environ., 18*, 61–67.

Kirkby, E. A. and D. J. Pilbeam (1984), *Plant Cell Environ., 7*, 397–405.

Kramer, D., A. Läuchli, A. R. Yeo, and J. Gullash (1977), *Ann. Bot., 41*, 1031–1040.

Koller, D. (1957), *Ecology, 38*, 1–13.

Kreeb, K. (1974), *Ökophysiologie der Pflanzen*, Fischer-Verlag, Jena, Germany.

LaHaye, P. A. and E. Epstein (1969), *Science, 166*, 395–396.

HaHaye, P. A. and E. Epstein (1971), *Physiol. Plant., 25*, 213–218.

Larcher, W. (1994), *Ökophysiologie der Pflanzen*, 5th ed., Ulmer-Verlag, Stuttgart.

Läuchli, A. (1984), in *Salinity tolerance in plants* (R. C. Staples and G. H. Toenniessen, Eds.), Wiley, New York, pp. 171–187.

Lerner, H. R. (1985), *Plant Soil, 89*, 3–14.

Lerner, H. R., G. N. Amzallag, Y. Friedman, and P. Goloubinoff, (1994), *Israel J. Plant Sci., 42,* 285–300.

Lessani, H. and H. Marschner (1978), *Aust. J. Plant. Physiol., 5,* 27–37.

Levitt, J. (1980), *Responses of plants to environmental stresses,* Vol. 1, 2nd ed., Academic Press, New York.

Liphschitz, N. and Y. Waisel (1982), in *Tasks for vegetation science, Vol. 2, Contributions to the ecology of halophytes* (D. N. Sen and K. S. Rajpurohit, Eds.), W. Junk, The Hague, Netherlands, pp. 197–214.

Livne, A. and Y. Vaadia (1965), *Physiol. Plant., 18,* 658–664.

Lötschert, W. (1970), *Oecol. Plant., 5,* 287–300.

Lüttge, U. (1987a), *New Phytol., 106,* 593–629.

Lüttge, U. (1987b), *Giorn. Bot. Ital., 121,* 217–227.

Lüttge, U. (1993), *New Phytol., 125,* 59–71.

Lüttge, U. and M. G. Pitman (1976), in *Encyclopedia of plant physiology, Vol. 2 IIb, Tissues and organs* (U. Lüttge and M. G. Pitman, Eds.), Springer-Verlag, New York, pp. 222–224.

Lüttge, U. and J. A. C. Smith (1984), in *Salinity tolerance in plants* (R. C. Staples and G. H. Toenniessen, Eds.), Wiley, New York, pp. 125–150.

Lynch, J., G. R. Cramer, and A. Läuchli (1987), *Plant Physiol., 83,* 390–394.

Mansour, M. M. F. and E. J. Stadelmann (1994), *Physiol. Plant., 91,* 389–394.

Marcum, K. B. and C. L. Murdoch (1990), *Ann. Bot., 66,* 1–7.

Marcum, K. B. and C. L. Murdoch (1992), *New Phytol., 120,* 281–288.

Marloth, R. (1887), *Berichte Deutsch. Bot. Ges., 5,* 319–324.

Martinez, V. and A. Läuchli (1993), *Planta, 190,* 519–524.

Mittelheuser, C. J. and R. F. M. Van Steveninck (1969), *Nature, 221,* 281–282.

Munns, R. (1993), *Plant Cell Environ., 16,* 15–24.

Munns, R., H. Greenway, and G. O. Kirst (1983), in *Encyclopedia of plant physiology, Vol. 12C, Physiological plant ecology III* (O. L. Lange, P. S. Nobel, C. B. Osmond, and H. Ziegler, Eds.), Springer-Verlag, New York, pp. 59–135.

Nagy, Z. and G. Galiba (1995), *J. Plant Physiol., 145,* 168–174.

Naidoo, G. (1994), *Environ. Exp. Bot., 34,* 419–426.

Oertli, J. J. (1968), *Agrochimica, 12,* 461–469.

Osmond, C. B. (1978), *Annu. Rev. Plant Physiol., 29,* 379–414.

Parry, A. D. and R. Horgan (1991), *Physiol. Plant., 82,* 320–326.

Perez-Alfocea, F., M. T. Estan, M. Caro, and M. C. Bolarin (1993a), *Plant Soil, 150,* 203–211.

Perez-Alfocea, F., M. T. Estan, M. Caro, and G. Guerrier (1993b), *Physiol. Plant., 87,* 493–498.

Perez-Alfocea, F., A. Santa-Cruz, G. Guerrier, and M. C. Bolarin, (1994), *J. Plant Physiol., 143,* 106–111.

Philipupillai, J. and I. A. Ungar (1984), *Am. J. Bot., 71,* 542–549.

Piepenbrock, M. and J. M. Schmitt (1991), *Plant Physiol., 97,* 998–1003.

Popp, M., F. Larher, and P. Weigel (1985), *Vegetatio, 61,* 247–253.

Radha Rani, U. and A. R. Reddy (1994), *J. Plant Physiol., 143,* 250–253.

Raine, R. M. (1994), *Biol. Conserv., 67*, 201–204.

Ramati, A., N. Liphschitz, and Y. Waisel. (1979), *Physiol. Plant., 45*, 325–331.

Reimann, C. (1992), *J. Exp. Bot., 43*, 503–510.

Reimann, C. and S. W. Breckle (1988), *Flora, 180*, 289–296.

Reimann, C. and S. W. Breckle (1993), *Plant Cell Environ., 16*, 323–328.

Reimann, C. and S. W. Breckle (1995), *New Phytol., 130*, 37–45.

Rengel, Z. (1992), *Plant Cell Environ., 15*, 625–632.

Richards, R. A. (1992), *Plant Soil, 146*, 89–98.

Roth, I. (1992), *Leaf structure: Coastal vegetation and mangroves of Venezuela*, Encyclopedia of Plant Anatomy, Vol. XIV, Pt. 2, Gebr. Bornträger, Berlin.

Rozema, J. (1975), *Acta Bot. Neerl., 24*, 407–416.

Rozema, J. and I. Riphagen (1977), *Oecologia, 29*, 349–357.

Rozema, J., I. Riphagen and T. Sminia (1977), *New Phytol., 79*, 665–671.

Rozema, J., H. Gude, and G. Pollak (1981), *New Phytol., 89*, 201–217.

Ruhland, W. (1915), *Jahrbücher für Wissenschaftl. Botanik, 55*, 409–498.

Runge, F. (1985), *Decheniana, 138*, 60–65.

Saleki, R., P. G. Young, and D. D. Lefebvre (1993), *Plant Physiol., 101*, 839–845.

Sande, E. and D. R. Young (1992), *New Phytol., 120*, 345–350.

Schirmer, U. and S. W. Breckle (1982), in *Tasks for vegetation science, Vol. 2, Contributions to the ecology of halophytes* (D. N. Sen and K. S. Rajpurohit, Eds.), W. Junk, The Hague, Netherlands, pp. 215–231.

Schmidt, C., T. He, and G. R. Cramer (1993), *Plant Soil 155/156*, 415–418.

Scholander, P. F. (1968), *Physiol. Plant., 21*, 251–261.

Scholander, P. F., E. D. Bradstreet, H. T. Hammel, and E. A. Hemmingsen (1966), *Plant Physiol., 41*, 529–532.

Seybold, S. (1973), *Göttinger Floristische Rundbriefe, 7*, 70–72.

Shainberg, I. (1975), in *Plants in saline environments* (A. Poljakoff-Mayber and J. Gale, Eds.), Springer-Verlag, New York, pp. 39–55.

Shomer-Ilan, A. and Y. Waisel (1986), *Physiol. Plant., 67*, 408–414.

Shomer-Ilan, A., D. Moualem-Beno, and Y. Waisel (1985), *Physiol. Plant, 65*, 72–78.

Silberbush, M. and S. H. Lips (1988), *Physiol. Plant., 74*, 493–498.

Solomon, M., R. Ariel, A. M. Mayer, and A. Poljakoff-Mayber (1989), *Israel J. Bot., 38*, 65–69.

Solomon, A., S. Beer, Y. Waisel, G. P. Jones, and L. G. Paleg, (1994), *Physiol. Plant., 90*, 198–204.

Stewart, G. R. and J. A. Lee (1974), *Planta, 120*, 279–289.

Storey, R. and W. W. Thomson (1994), *Ann. Bot., 73*, 307–313.

Storey, R. and R. G. Wyn Jones (1975), *Plant Sci. Lett., 4*, 161–168.

Storey, R. and R. G. Wyn Jones (1977), *Phytochemistry, 16*, 447–453.

Sudhakar, C., P. S. Reddy, and K. Veeranjaneyulu (1993), *J. Plant Physiol., 141*, 621–623.

Summers, P. S. and E. A. Weretilnyk (1993), *Plant Physiol., 103*, 1269–1276.

Taleisnik, E. L. and A. M. Anton (1988), *Ann. Bot., 62*, 383–388.

Thomas, D. S. G. and N. J. Middleton (1993), *J. Arid Environ., 24*, 95–105.

Thomas, J. C., R. L. De Armond, and H. J. Bohnert (1992), *Plant Physiol.*, *98*, 626–631.

Thomson, W. W. and L. L. Liu (1967), *Planta*, *73*, 201–220.

Thomson, W. W. and K. Platt-Aloia (1985), *Protoplasma*, *125*, 13–23.

Thomson, W. W., W. L. Berry, and L. L. Liu (1969), *Proc. Natl. Acad. Sci.*, *63*, 310–317.

Thomson, W. W., C. D. Faraday, and J. W. Oross (1988), in *Solute transport in plant cells and tissues* (D. A. Baker and J. L. Hall, Eds.), Longman Scientific & Technical, Harlow, UK, pp. 498–537.

Trockner, V. and R. Albert (1986), *Flora*, *178*, 391–408.

Ungar, I. A. (1978), *Bot. Rev.*, *44*, 233–264.

Ungar, I. A. (1979), *Bot. Gaz.*, *140*, 102–108.

Ungar, I. A. (1982), in *Tasks for vegetation science, Vol. 2, Contributions to the ecology of halophytes*, (D. N. Sen and K. S. Rajpurohit, Eds.), W. Junk The Hague, Netherlands, pp. 143–154.

Ungar, I. A. (1991), *Ecophysiology of vascular halophytes*, CRC Press, Boca Raton, FL.

Van Diggelen, J., J. Rozema, D. M. J. Dickson, and R. Broekman (1986), *New Phytol.*, *103*, 573–586.

Vassilyev, A. E. and A. A. Stepanova (1990), *J. Exp. Bot.*, *41*, 41–46.

Vernon, D. M., J. A. Ostrem, and H. J. Bohnert (1993), *Plant Cell Environ.*, *16*, 437–444.

Waisel, Y. (1972), *Biology of halophytes*, Academic Press, New York.

Waisel, Y. (1991a), in *Physiology of trees* (A. S. Raghavendra, Ed.), Wiley, New York.

Waisel, Y. (1991b), *Physiol. Plant.*, *83*, 506–510.

Waisel, Y., A. Eshel, and M. Agami (1986), *Physiol. Plant.*, *67*, 67–72.

Walter, H. (1985), *Vegetation of the earth and ecological systems of the geo-biosphere*, 3rd ed., Springer-Verlag, New York.

Walter, H. and S. W. Breckle (1985), *Ecological systems of the geobiosphere*, Vol. 1, Springer-Verlag, New York.

Walter, H. and S. W. Breckle (1986), *Ecological systems of the geobiosphere*, Vol. 2, Springer-Verlag, New York.

Wiehe, W. and S. W. Breckle (1990), *Bot. Acta*, *103*, 107–110.

Winter, K. (1973), *Planta*, *115*, 187–188.

Winter, K. and D. J. von Willert (1972), *Z. Pflanzenphysiol.*, *67*, 166–170.

Winter, K., U. Lüttge, and E. Winter (1978), *Oecologia*, *34*, 225–237.

Young, D. R., D. L. Erickson, and S. W. Semones (1994), *Can. J. Bot.*, *72*, 1365–1372.

Zeevaart, J. A. D. and R. A. Creelman (1988), *Annu. Rev. Plant Physiol. Plant Mol. Biol.*, *39*, 439–473.

Zhang, J. and W. J. Davies (1989), *Plant Cell Environ.*, *12*, 73–81.

Zhang, J., U. Schurr, and W. J. Davies (1987), *J. Exp. Bot.*, *38*, 1174–1181.

Zhong, H. and A. Läuchli (1994), *Planta*, *194*, 34–41.

Zimmermann, U., J. J. Zhu, F. C. Meinzer, G. Goldstein, H. Schneider, G. Zimmermann, R. Benkert, F. Thürmer, P. Melcher, D. Webb, and A. Haase (1994), *Bot. Acta*, *107*, 218–229.

8

Trace Metals

M. N. V. Prasad

INTRODUCTION

Trace metals, namely, Al, As, Cd, Co, Cr, Cu, Hg, Mn, Ni, Pb, Se, and Zn, have been considered to be major environmental pollutants and their phytotoxicity is well established (Ross, 1994). Bioconcentrations of trace metals in agricultural, horticultural, and silvicultural plants and weeds, and in aquatic macrophytes are of special concern in the interest of human welfare.

Trace metal toxicity reduces vigor and growth, causes death in extreme cases, interferes with photosynthesis, respiration, water relations, and reproduction, and causes changes in certain organelles and disruption of membrane structure and functions. Trace metal availability, toxicity, and adaptive strategies of plants have been discussed in a number of articles (Barcelo and Poschenreider, 1990; Cumming and Taylor, 1990; Dickinson et al., 1991; Ernst et al., 1992; Farago, 1981; Fernandes and Henriques, 1991; Foy et al., 1978; Hagemeyer, 1993; Hamer, 1986; Kagi and Schaffer, 1988; Karin, 1985; Leita et al., 1993; Mahaffey et al., 1975; Meyer and Heath 1988; Prasad, 1995a; Rauser, 1990b, 1995; Reddy and Prasad, 1990; Robinson et al., 1993; Ross, 1994; Roy et al., 1988; Steffens, 1990; Tamaddon and Hogland, 1993; Taylor and Stadt, 1990; Tomsett and Thurman, 1988). High concentration of trace metals is one of the most important environmental stress factors and the extent to which a plant can survive is determined by its sensitivity to metal toxicity.

The global trace metal pollution is increasing in the environment due to mining, industry, road traffic, and anthropogenic activity. Aten et al. (1986) reported trace metals in rainwater of Geneva and New York during 1983. Deterioration of forests in Europe and North America has been attributed to trace metal toxicity (Godbold et al., 1988; Kahle, 1993). Acid rain increases soil acidity and releases normally

207

sparingly soluble aluminum and trace metals into the soil solution (Campbell et al., 1988; Lazerte, 1986; Nelson and Campbell, 1991).

Trace metal pollution is gaining in importance day by day due to its obvious impact on human health through the food chain. Several vegetable, fruit, and cereal crops are reported to accumulate trace metals (Pieczonka and Rosopulo, 1985; Mejuto-Marti et al., 1988; Shane et al., 1988; Rivai et al., 1990a, b). Mercury is bioaccumulated through food chains in aquatic ecosystems. High concentration of mercury was found in predatory fishes from man-made lakes and in areas that formerly received mercury effluents; such concentrations pose risks for human health, especially among fish consumers (Lodenius, 1990).

Unlike animals, plants are sedentary and are vulnerable to varying concentrations of metals. Plants, therefore, must adapt themselves to the prevailing conditions for their survival, resulting in acquisition of a wide range of metal-tolerance mechanisms both between species and among genotypes within a species. Trace metals interact with plants and display different functional aspects (Table 8-1) and the organisms adjust themselves to toxic environs by evolving adaptive strategies.

TRACE METAL AVAILABILITY TO PLANTS

Trace metal pollution is increasing in the environment due to mining, industrialization, and anthropogenic activity (Groten and Vanbladeren, 1994; Jackson and Alloway, 1991; Merrington and Alloway, 1994; Ross, 1994; Tamaddon and Hogland, 1993). Trace metals are abundantly used in surface coatings, pigment formulation, manufacture of batteries, stabilization of polyvinylchloride (plastics), manufacture of automobiles, military applications, and the fertilizer and pesticide industries (Fig. 8-1) (Ross, 1994). Cadmium has applications where high stability and resistance to heat, cold, and light are required. Trace metals released into the environment ultimately tend to concentrate in soils and sediments, a reservoir pool for plants. Thus the availability of trace metals to rooted plants is regulated by pH, Eh (redox potential), and other physicochemical parameters. Certain phosphate fertilizers have been found to contain high levels of cadmium and other trace metals (Campbell et al., 1988; He and Singh, 1993, 1994; Roberts et al., 1994). In the United States the production and use of cadmium has been decreasing with time as less hazardous substitutes have become available (Fig. 8-2).

Forms of trace metals in soils can be defined on the basis of selective extractions, for example, water soluble, exchangeable, reducible, diethyltriamine pentaacetic acid (DTPA) extractable, and organically bound (insoluble) forms (He and Singh, 1994). Changes in redox potential can affect the partitioning of trace metals among these forms. It was demonstrated that an increase in redox potential (Eh) from -150 to 500 mV led to a decrease in exchangeable trace metal and an increase in the reducible form (Khalid et al., 1981). Thus availability of trace metals to plants will be regulated by the pH and Eh (Ahumada and Schalscha, 1993; Hooda and Alloway, 1994). Chloride levels would also be expected to affect trace metal availability. Soil sodium chloride had an antagonistic effect on trace metal toxicity.

(cont'd on page 216)

TABLE 8-1 Trace Metal Plant (Flowering Plants) Interactions

Taxon	Report of the Investigation	Reference
Acer saccharum	Effect of Cd on growth and water movement	Lameroux and Chaney (1977)
Agrostis sp.	Heavy-metal-tolerant populations and other grasses	Jowett (1958)
	Occurrence of metallothioenein in roots tolerant to Cu	Rauser and Curvetto (1980)
A. capillaris	Relationship between tolerance to heavy metals	Humphreys and Nicholls (1984)
	Differential tolerance of three cultivars to Cd, Cu, Pb, Ni, and Zn	Symenoides et al. (1985)
A. gigantea	Cd-thiolate protein	Rauser et al. (1983)
	Quantification of metallothionein in small root samples exposed to Cd	Rauser (1984a)
	Isolation and partial purification of Cd-binding protein from roots	Rauser (1984b)
	The amount of Cd associated with Cd-binding protein in roots	Rauser (1986)
	HPLC characterization of Cd-binding protein	Rauser et al. (1986)
A. tenuis	Pb- and Zn-tolerant populations	Bradshaw (1952)
	Combined tolerance to Cu, Zn, and Pb	Karataglis (1982)
	The subcellular distribution of Zn and Cu in the roots of metal-tolerant clones	Turner (1970)
	ATP-ase from the roots: effect of pH, Mg, Zn, and Cu	Veltrup (1982)
Avena sativa	Oxidative damage by excess of Cu	Luna et al. (1994)
	Characterization of Cu-binding protein from roots grown on excess Cu	Tukendorf and Baszynski (1985)
Betula	Zn tolerance	Brown and Wilkins (1985)
Brassica sp.	Characterization of Cd-binding complexes from leaves	Wagner (1984)
	Inducible Cd-binding complexes	Wagner and Trotter (1982)
	Cd bioavailability in soils treated with sewage sludges	Jackson and Alloway (1991)
B. juncea	PC production in the Se-tolerant plant	Speiser et al. (1992)
B. napus	Heavy-metal-tolerant transgenic plant	Mishra and Gedamu (1989)
Calochortus sp.	Heavy metal accumulator	Fiedler (1985)
Daucus	Cd sulfide crystallites in Cd-binding proteins	Reese and Winge (1990)
Datura innoxia	Precursor-product relationship of poly (y-glutamylcysteinyl) glycine biosynthesis	Berger et al. (1989)
	Poly(y-glutamylcysteinyl) glycine synthesis and binding with Cd	Delhaize et al. (1989a)
	Effect of Cd on gene expression in Cd-tolerant and Cd-sensitive cells	Delhaize et al. (1989b)
	Selection, isolation, and characterization of Cd-tolerant suspension culture	Jackson et al. (1984)
	Biosynthesis of Poly(y-glutamyl-cysteinyl) glycines in Cd-tolerant cells	Robinson et al. (1988)

TABLE 8-1 *(Continued)*

Taxon	Report of the Investigation	Reference
Deschampsia cespitosa	Metal cotolerance	Cox and Hutchinson (1979)
	Multiple and cotolerance to metals: Adaptation, preadaptation, and "cost"	Cox and Hutchinson (1981)
		Schultz and Hutchinson (1988)
	Role of metallothionein-like proteins in Cu tolerance	
Dictyostelium	Ubiquitin gene expression induced by Cd	Taubenberger et al. (1988)
Eichhornia crassipes	Cd in naturally occurring plant	Cooley and Martin (1979)
	Heavy metal distribution in three biotypes	Cooley et al. (1979)
	Uptake of As, Cd, Pb, and Hg from polluted water	Chigbo et al. (1982)
	Occurrence of two Cd-binding components in roots when cultivated in a Cd-rich medium	Fujita (1985)
	Purification and characterization of the Cd-binding complex as reported above	Fujita and Kawanishi (1986)
	Metal specificites on introduction and binding affinities of heavy-metal-binding complexes in root	Fujita and Nakano (1988)
	Effects of heavy metals	Kay et al. (1984)
	Heavy metal removal from polluted water	Muramoto and Oki (1983)
	Uptake of Cu	Sutton and Blackburn (1971)
	Sorption of heavy metals from metal containing solutions	Tatsuyama et al. (1979)
Fagus sylvatica	Effects of Cd and Pb on growth and mineral nutrition	Breckle and Kahle (1992)
	Variations of trace metals concentration in wood	Hagemeyer et al. (1992)
	Development of xylem and shoot growth of young seedlings	Hagemeyer et al. (1993)
Glycine max	Differing sensitivity of photosynthesis and transpiration to Pb contamination	Bazzaz et al. (1974b)
	Cd-binding components	Casterline and Barnett (1982)
	Cd distribution and chemical fate	Cataldo et al. (1981)
	Excess copper induces a cytosolic Cu, Zn super oxide dismutase in root	Chongpraditnum et al. (1992)
	The inhibition of metabolic processes by Cd and Pb	Huang et al. (1974)
	Isolation of a gene for a metallothionein-like protein	Kawashima et al. (1991)
	Genotypic differences in Al tolerance	Koltz and Horst (1988)
	Cd influence on respiration and activities of several enzymes	Lee et al. (1976)
Helianthus annus	Effect of Cd on photosynthesis and transpiration of excised leaves	Bazzaz et al (1975)
	Inhibition of photosynthesis by Pb	Bazzaz et al (1975)

Taxon	Report of the Investigation	Reference
Holcus lanatus	Environment-induced Cd tolerance	Baker (1984)
	Induction and loss of metal tolerance	Baker et al. (1986)
	Tolerance to Pb, Zn, and Cd factorial combination	Coughtrey and Martin (1978)
	Cd-tolerance and toxicity, oxygen radical process, and molecular damage in Cd-tolerant and Cd-sensitive clones	Hendry et al. (1992)
	Interactive effects of Cd, Pb, and Zn on root growth of two metal-tolerant genotypes	Symeonidis and Karataglis (1992)
Hordeum vulgare	Critical levels of 20 potentially toxic elements	Davis and Beckett (1978)
	Comparative analysis of element composition of root and leaves of seedlings grown in the presence of Cd, Mo, Ni, and Zn	Brune and Dietz (1995)
Lactuca sativa	Efficiency of H^+-ATPase activity on Cd uptake by four cultivars	Costa and Morel (1994)
	Cd bioavailability in soils treated with sewage sludge	Jackson and Alloway (1991)
	Cd uptake by lettuce grown on Cd-polluted soil	Kuo and Huang (1993)
Lupinus albus	Cd excretion, a possible mechanism of Cd tolerance	Costa and Morel (1993)
Lycopersicon esculentum	Partial characterization of Cd-binding protein from the roots	Bartolf et al. (1980)
	Photosynthetic activities of Cd-treated tomato plants	Baszynski et al. (1980)
	Selection and characterization of Cd-tolerant cells	Huang et al. (1987)
	Effect of glutathione on phytochelatin synthesis	Mendum et al. (1990)
	In vivo measurement of Cd transport and accumulation in stems	Petit and Van de Geijn (1978)
	Cd sulfide crystallites in Cd-binding protein	Reese and Winge (1990); Reese et al. (1992)
	Phytochelatin synthesis and glutathione levels in response to heavy metals	Scheller et al. (1987)
	Accumulation of nonprotein metal binding polypeptides in Cd-resistant cells	Steffens et al. (1986)
	Inducible Cd-binding protein	Wagner and Trotter (1982)
Mimulus guttatus	Cu tolerance in Californian populations	Allen and Sheppard (1971)
	Cu tolerance governing genes	Macnair (1977)
	Cd-binding polypeptides properties	Reese and Wagner (1987a)
	Isolation of Cu-binding complex	Robinson and Thurman (1986)
	Cu phytochelatin	Salt et al. (1989)

TABLE 8-1 *(Continued)*

Taxon	Report of the Investigation	Reference
Nicotiana tabacum	Cd-enhanced gene expression in suspension culture cells	Hirt et al. (1989)
	Cd tolerance in cell culture and tolerance to temperature stress	Huang and Goldsbrough (1988)
	Relationship between Cd, Zn, Cd peptide, and organic acids in suspension cells	Kortz et al. (1989)
	Inheritance and expression of the mouse MT gene	Maiti et al. (1989)
	Heavy-metal-tolerance transgenic plant	Mishra and Gedamu (1989)
	Effects of buthionine-sulfoximine on Cd-binding peptides treated with Cd, Zn, and Cu	Reese and Wagner (1987a)
	Properties of Cd-binding peptides	Reese and Wagner (1987b)
	Subcellular localization of Cd and Cd-BP, implications of transport function for Cd-binding protein	Vögeli-Lange and Wagner (1990)
Oryza sativa	Isolation of Cd-binding protein from Cd-treated plant	Kaneta et al. (1983)
	Chemical form of Cd and other trace metals	Kaneta et al. (1987)
	Limiting steps on photosynthesis in Cu-treated plants	Lidon and Henriques (1991)
	Effect of Cd, Ni on lipid composition, Mg-ATPase activity, and fluidity of plasma membrane	Ros et al. (1990)
	Cd content in rice and its daily intake in various nations	Rivai et al. (1990a)
	Cd content in rice and rice field soils in China, Indonesia, and Japan with reference to soil type and daily intake from rice	Rivai et al. (1990b)
	Cd-induced peroxidase activity and isozymes	Reddy and Prasad (1992)
	Cd and Ni effects on mineral nutrition and interaction with ABA and GA	Rubio et al. (1994)
Petunia sp.	Inhibition of hsp 70 mRNA processing during $CdCl_2$	Winter et al. (1988)
Phaseolus vulgaris	Cd-induced decrease of water stress resistance: effect of Cd on endogenous ABA	Barcelo et al. (1986a, b)
	Cd-induced structural and ultrastructural disorders in bush bean	Barcelo et al. (1988a, b)
	Patterns of root respiration associated with the induction of aluminum tolerance	Cummings et al. (1992)
	Ethylene biosynthesis and Cd toxicity	Fuhrer (1982)
	Absorption and translocation of Cd	Hardiman and Jacoby (1984)
	Influence of Cd on water relations, stomatal resistance, and ABA	Poschenreider et al. (1989)

Taxon	Report of the Investigation	Reference
	Distribution of Cd and induced Cd-binding protein in roots, stems, and leaves	Leita et al. (1991)
	Carbohydrate levels and photoassimilate export from leaves exposed to excess Co, Ni, and Zn	Samarakoon and Rauser (1979)
	Inhibition of photosynthesis by treatment with toxic concentration of Zn: effect on RUBPC/RUBPO	Van Assche and Clijsters (1986a)
	Effect of Zn on electron transport and phosphorylation	Van Assche and Clijsters (1986b)
Pisum sativum	Activities of Cu-containing proteins in-Cu depleted leaves	Ayala and Sandman (1988)
	Effect of Cr on activities of hydrolytic enzymes in germinating pea seeds	Dua and Sawhney (1991)
	Isolation and partial characterization of Cd-binding protein	Grunhage et al. (1985)
	Influence of Se on uptake and toxicity of Cu and Cd	Landberg and Greger (1994)
	Cu-binding protein and Cu tolerance	Palma et al. (1990)
	Regulation of glutathione synthesis by Cd	Ruegsegger et al. (1990)
Raphanus sativa	Cd induced changes in the composition and structure of light harvesting chlorophyll a/b protein complex	Krupa (1988)
	Trace metal effect on growth	Salim et al. (1993)
Rauvolfia serpentina	Induction of heavy metal sequestering phytochelatin by Cd	Grill and Zenk (1985)
	Phytochelatins protect plant enzymes from heavy metal toxicity	Kneer and Zenk (1992)
Silene cucubalus	Cu-induced damage to the permeability barrier in roots	De Vos et al. (1989)
	Effect of Cu on fatty-acid-tolerant and -sensitive taxa	De Vos et al. (1993)
	Metallothionein role in Cu tolerance	Lolkema et al. (1984)
S. maritima	Ecophysiological aspects of Zn tolerance	Baker (1978)
S. vulgaris	Cu and Co tolerance in closely related taxa within the genus	Baker et al. (1983)
	Genetic control of Cu tolerance	Schat and Bookum (1992)
	Phytochelatin induction and differential metal tolerance	Schat and Kalff (1992)
	Cd tolerance and cotolerance	Verkleij and Prast (1989)
	Phytochelatin role in Cd tolerance	Verkleij et al. (1990)
Sorghum sp.	Arsenate-inducible heat-shock proteins	Howarth (1990)
Spinacia oleraceae	Photosynthetic apparatus of leaves	Baszynski et al. (1982)
	Characteristics of the photosynthetic apparatus of Cu-nontolerant plant exposed to excess Cu	Baszynski et al. (1988)

TABLE 8-1 *(Continued)*

Taxon	Report of the Investigation	Reference
	Short-term dark incubation with sulfate, chloride, and selenate on the glutathione content of leaf discs	De Kok and Kuiper (1986)
	Effect of Zn and Cd on photosynthesis	Hampp et al. (1976)
	Toxic effect of Cu on PSII	Hsu and Lee (1988)
	Response to excess of Cu and Cd	Tukendorf (1993)
	Characteristics of Cu-binding proteins in chloroplasts of Cu-tolerant plants to excess Cu	Tukendorf (1989)
	Cu-binding protein tolerant to excess Cu	Tukendorf et al. (1984)
Tamarix aphylla	Excretion of Cd ions	Hagemeyer and Waisel (1988)
	Uptake of Cd and Fe by excised roots	Hagemeyer and Waisel (1989b)
Thlaspi ochro- leucum	Ecophysiological and ultrastructural effects of Cu	Ouzounidou et al. (1992)
Triticum aestivum	Induction of microsomal membrane proteins in roots of an Al-resistant cultivars under conditions of Al stress	Basu et al. (1994)
	Residual Cd in soil and accumulation in grain	Brams and Anthony (1988)
	Differential aluminum tolerance of high-yielding, early-maturing Canadian wheat	Briggs et al. (1989)
	Effect of Cu on active and possible Rb influx in roots	Jensen and Aoalsteinson (1989)
	Influence of Se on uptake and toxicity of Cu and Cd	Landberg and Gregor (1994)
	Chemical form of Cd and other heavy metals	Kaneta et al. (1987)
	Mn effects of photosynthesis and chlorophyll	Macfie and Taylor (1992)
	Mn effects on production of organic acids in Mn-tolerant and Mn-sensitive cultivars	Macfie et al. (1994)
	Mn tolerance	Moroni et al. (1991a)
	Chlorophyll and leaf growth as a measure of Mn tolerance	Moroni et al. (1991b)
	Heat shock and protection against metal toxicity	Orzeck and Burke (1988)
	Citrate reverses the inhibition of root growth caused by Al	Ownby and Popham (1989)
	Crop rotation and tillage practices on Cd accumulation in grain	Oliver et al. (1993)
	Distribution of Cd, Cu, and Zn in fruit	Pieczonka and Rosopulo (1985)
	Mechanisms of Al tolerance—role in N nutrition and differential pH induced by winter and spring cultivars	Taylor (1985, 1988a, b); Taylor and Foy (1985a–d)

Taxon	Report of the Investigation	Reference
	Al tolerance is independent of rhizosphere pH	Taylor (1988a, b)
	Differential uptake and toxicity of ionic and chelated Cu cultivars	Taylor and Foy (1985a, b)
	Kinetics of Al uptake by excised roots of Al-tolerant and Al-sensitive cultivars	Zhang and Taylor (1989, 1990)
	Interactive effects of Cd, Cu, Mn, Ni, and Zn on root growth in solution culture	Taylor and Stadt (1990)
	Effects of biological inhibitors on kinetics of Al uptake by excised roots and purified cell wall material of Al-tolerant and Al-sensitive cultivars	Zhang and Taylor (1991)
Typha latifolia	Cu and Ni tolerance in clones from contaminated and uncontaminated environments	Taylor and Crowder (1984)
V. mungo	Effect of Cd, Cu, and Zn on the growth	Kalyanaraman and Sivagurunathan (1993)
V. sinensis	Death of epidermal cells by Cu and Hg	Meyer and Heath (1988)
Wolffia globosa	Indicator of heavy metal	Garg and Chandra (1994)
Zea mays	Differing sensitivity of photosynthesis and transpiration to Pb contamination	Bazzaz et al., 1974a
	Effect of pH on Cd distribution in maize inbred lines	Florijn and Van Beuischem (1993a, b)
	PC concentration and binding state of Cd in roots of maize genotypes differing in shoot and root Cd partitioning	Florijn et al. (1993a)
	Cd distribution in maize inbred lines, and evaluation of structural and physiological characteristics	Florijn et al. (1993b)
	Zn and Cd in corn plants	Jones et al. (1988)
	X-ray microanalytical study and distribution of Cd in roots	Kahn et al. (1984)
	Compartmentation of Cd, Cu, Pb, and Zn in seedlings and metallothionein induction	Leblova et al. (1986)
	Localization of Pb	Malone et al. (1974)
	Mobilization of Cd and other trace metals from soil by root exudates	Mench and Martin (1991)
	Metal binding properties of high-molecular-weight soluble exudates	Mench et al. (1987)
	Measurements of Pb, Cu, and Cd bindings with mucilage exudates from roots	Morel et al. (1986)
	Regulation of assimilatory sulfate reduction by Cd	Nussbaum et al. (1988)
	Quantification of metallothionein in small root samples exposed to Cd	Rauser (1984b)
	Cd binding proteins in roots	Rauser and Glover (1984)

TABLE 8-1 *(Continued)*

Taxon	Report of the Investigation	Reference
	Changes in seedling glutathione exposed to Cd	Rauser (1986, 1990a)
	Cysteine, y-glutamylcysteine and glutathione levels: Distribution and translocation in normal and Cd-exposed plants	Rauser et al. (1991)
	Changes in glutathione and phytochelatins in roots exposed to Cd	Tukendorf and Rauser (1990)

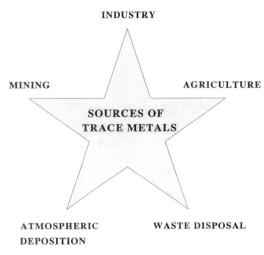

Figure 8-1. *Availability of trace metals to plants. Sources include mining (smelting, river dredging, mine spoils and tailings, metal industries, etc.), industry (plastics, textiles, microelectronics, wood preservatives, refineries, etc.), atmospheric deposition (urban refuse disposal pyrometallurgical industries, automobile exhausts, fossil fuel combustion, etc.), agriculture (fertilizers, pesticides, etc.), and waste disposal (sewage sludge, leachate from land fill, etc.).*

Seedlings of *Sesamum indicum* (an important oil seed crop), when treated with combinations of 1 mM Cd + Pb, Cd + Cu (1–1) and NaCl (2 and 10 S/m), antagonized the metal toxicity with regard to biomass accumulation and nitrate assimilation (Bharati and Singh, 1994). The soil organic matter of different vegetational cover preferentially binds to selected trace metal ions (Krosshavn et al., 1993).

In aquatic plants, trace metals are taken up not only through the roots but also by the shoot system (Ornes and Sajwan, 1993). Trace metal interactions at the shoot or root surface will be affected by the H^+ ion (pH) and calcium concentrations; in both cases, the availability of trace metals appears to be reduced by competition with high levels of calcium or low pH. In addition, the availability of trace metal will also be sensitive to the presence of dissolved organic matter (DOM) in the soil/

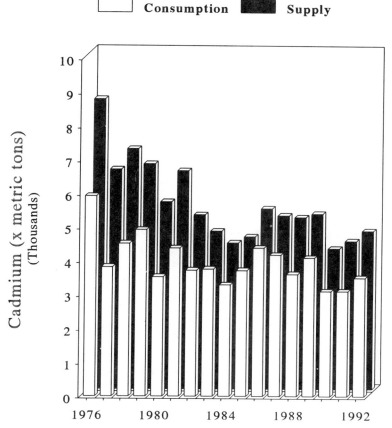

Figure 8-2. *Cadmium supply and apparent consumption in the United States. Supply includes previous year's stock, production, and imports for consumption. Consumption excludes exports, year-end stocks.* (Source: Minerals Year Books, 1976 to 1992, prepared by staff of the Bureau of Mines, US Department of the Interior.)

sediment/interstitial water or in the ambient water surrounding the shoots. Although Cd is sparingly bound to DOM compared to Cu, Pb, Hg, Fe, or Al, nevertheless its bioavailability will tend to decrease in the presence of DOM (Campbell et al., 1988; Hirsch and Banin, 1990; Khalid et al., 1981).

Mining is one of the important sources of trace metals. Thus plants growing in a mining refuse area have adapted to a hostile environment by evolving tolerance, predominantly by complexing the toxic metal ions and by synthesizing phytochelatins (Grill et al., 1988b; Kubota et al., 1988). The flux of toxic trace metals would depend upon the exploration activity of metalliferous rocks and on edaphic and meteorological conditions (Merrington and Alloway, 1994).

Extractable contents of trace metals in agricultural soils have been reported.

Sewage sludge disposal is one of the main sources of trace metal bioavailability and mobility to vegetables and crop plants (Aller and Deban, 1989; Hooda and Alloway, 1994; Jackson and Alloway, 1991; Jeng and Singh 1993; Lamy et al., 1993). Trace metal accumulation in food grains and vegetables is of concern for human health (Groten and Vanbladeren, 1994; Mahaffey et al., 1975; Sanchez-Camazano et al., 1994). The capacity of aquatics to take up trace metals from the water, producing an internal concentration severalfold greater than their surroundings, has been shown for many species (Ornes and Sajwan, 1993; St-Cyr and Campbell, 1994a,b; Yurukova and Kochev, 1994). Trace metal threat to aquatic organisms depends not only on the concentration of toxic metals in the surrounding medium, but also on its uptake and accumulation by the organisms. Cadmium circulation in Finnish forest biocoenosis, involving aquatic plants, terrestrial plants, mushrooms, and fauna, was vividly explained (Nuorteva, 1990).

TRACE METAL UPTAKE AND TRANSPORT WITHIN PLANT

Kinetics of cadmium uptake were investigated in intact soybean plants (Cataldo et al., 1983). The pathway of cadmium movement within the root to the stele is less clear. Cuttler and Rain (1974) assumed symplastic transport of cadmium across the root cortex. Uptake and distribution of cadmium in maize inbred lines was investigated by Florijn and Van Beusichem (1993a, b). Foliar uptake of cadmium was reported in pea *(Pisum sativum)* and sugar beet *(Beta vulgaris)* (Greger et al., 1994). In *Tamarix aphylla,* the uptake of cadmium and iron by excised roots was reported by Hagemeyer and Waisel (1989b). Several physicochemical factors affect the uptake of metals by plant organs, of which pH is the most important, particularly from solutions (Hatch et al., 1988; Jarvis et al., 1976). Lettuce cultivars accumulated cadmium when grown on cadmium-polluted soil (Kuo and Huang, 1993).

Trace metal distribution in plant tissues and trace metal behavior in xylem exudates and other tissues are indications of a change in chemical form resulting from complexation with plant-produced ligands, which are involved in transport and metabolism. In soybean, about 98% of the accumulated cadmium is strongly retained by roots, and only 2% is transported to the shoot system (Cataldo et al., 1983). In vascular plants, the cells that transfer cadmium from the roots to the shoots are located in the vascular bundles. Thus trace metal movement is regulated by vascular tissues (Kuppelwieser and Feller, 1991).

The carrier proteins present in the membranes of the vascular tissues regulate the uptake of trace metals (Leita et al., 1991). The pathway of cadmium movement within the root to the stele is less clear than that for water, but is assumed to occur as symplastic transport across the root cortex (Kahle, 1993). Passage of cadmium through the stelar system is quite similar to that of solute movement through porous adsorptive media (Hardiman and Jacoby, 1984). Translocation of cadmium to the tops would thus be related not only to water flow but also to cadmium uptake by the root system (Hagemeyer and Waisel, 1988). In oat *(Avena sativa)* roots, cadmium

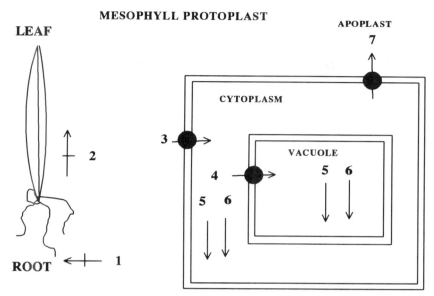

Figure 8-3. *Schematic representation of the potential mechanisms involved in trace-metal tolerance. (1) Exclusion by the root. (2) Retention by root. (3) Low permeability of the plasma membrane. (4) Rapid channeling into vacuole. (5) Precipitation. (6) Binding or complexation. (7) Extrusion from cytoplasm into the apoplast. Processes 1–4 and 6 have been reported in barley seedlings. (Courtesy of Prof. Dr. Karl-Josef Dietz, Universität Würzburg, Germany. Reproduced with the permission of Blackwell Science Ltd., Oxford, UK.)*

transport from cytosol to vacuole across the tonoplast is demonstrated through Cd^{2+}/H^+ antiport activity (Salt and Wagner, 1993).

Compartmentation, complexation, and transport are the primary and basic mechanisms involved in zinc tolerance in barley *(Hordeum vulgare)* (Fig. 8-3) (Brune et al., 1994). In wheat, transport of Rb and Sr to the ear in mature excised shoots was affected by temperature and stem length and also perhaps by the ligands in the xylem (Kuppelwieser and Feller, 1991). Long-distance transport and local accumulation of ^{115}Cd was measured in tomato plants *(Lycopersicon esculentum)*.

PLANTS AS TRACE METAL INDICATORS AND SCAVENGERS

Several plants have been used as indicators of metal contamination. Brooks et al. (1977) used herbarium specimens of certain indicator plants to detect nickeliferous rocks. In south Finnish forest areas some indicator plants for Al, Fe, Zn, Cd, and Hg with different degrees of damage have been identified (Kanerva et al., 1988). A number of indicator plants have been identified in unpolluted Finnish Lapland to monitor the levels of Al, Fe, Zn, Cd, and Hg (Sarkela and Nuorteva, 1987).

Water hyacinth *(Eichhornia crassipes)* attracted the attention of scientists for its ability to accumulate trace metals (Kay and Haller, 1986; Kay et al., 1984). Trace metals have been reported in plant tissues of water hyacinth (Cooley et al., 1979). Metal-binding complexes have been identified in water hyacinth root tissue (Fujita, 1985; Fujita and Nakano, 1988). Cadmium-binding complexes from the root tissue have been purified and characterized (Fujita and Kawanishi, 1986). It is being used as biofilter to remove toxic trace metals from waste and polluted water (Muramoto and Oki, 1983; Sutton and Blackburn, 1971; Tatsuyama et al., 1979).

TRACE METAL TOXICITY TO PLANTS

Growth

Toxic trace metals (cadmium, lead) have altered the growth and mineral nutrition of beech *(Fagus sylvatica)* (Breckle and Kahle, 1992). The growth of silver maple seedlings was affected by cadmium (Lameroux and Chaney, 1977). Cadmium and nickel reduced the growth, net photosynthesis, and carbohydrate distribution in rice plants (Moya et al., 1993). Aluminum-inhibited root growth in wheat *(Triticum aestivum)* was reversed by citrate. In barley, too, trace metals inhibited root growth, and thus such growth can be considered as a parameter to measure tolerance to toxic trace metals (Brune and Dietz, 1995; Ernst et al., 1992; Taylor, 1989a, b, 1991).

Quantification of the inhibitory effects of metal ions on root elongation when supplied in simple metal salt solutions is the most commonly used method for monitoring growth (Godbold and Knetter, 1991). Advantages of this method are that it is: simple (plants are allowed to root in metal amended and control solutions), rapid (frequently only a few days growth in treatment solutions is required), and easy to perform. Indices of tolerance derived from ratios between data for treatment and control solutions are calculated and used to characterize individuals in populations (Symeonidis and Karataglis, 1992).

The tolerance of *Agrostis tenuis* Sibth. populations growing in mining areas was investigated by exposing tillers to different concentrations of metals in nutrient solutions. An index of tolerance was calculated by comparing root length with specimens grown in solutions without trace metals (Karataglis, 1982). Higher plants grown on cadmium-containing substrate show disturbed water balance. Under long-term exposure to cadmium, almost all physiological processes will be affected and the recognition of primary effects would be almost impossible.

Species of *Agrostis* (Poaceae) have different mechanisms of tolerance to trace metals. Relationships between different species and metal complexation processes in roots and plant tissues have been characterized (Bradshaw, 1952; Humphreys and Nicholls, 1984; Jowett, 1958; Karataglis, 1982; Rauser, 1984a, b, 1986, 1993; Rauser and Curvetto, 1980; Rauser et al., 1983, 1986; Symenoides et al., 1985; Turner, 1970; Veltrup, 1982). Toxic trace metals have changed the ultrastructure of plants. Cadmium induced structural and ultrastructural changes in the vascular system of *Phaseolus vulgaris* (Barcelo et al., 1988a, b). *Thlapsi ochroleucum* (Cruciferae) suffered changes in its ultrastructure due to exposure of excess copper (Ouzounidou et al., 1992).

Photosynthesis

Photosynthesis is invariably affected in plants exposed to excess trace metals (Clijsters and Van Assche, 1985; Rascio et al., 1993; Stiborova et al., 1986a, b). Cadmium had direct and indirect effects on photosynthesis in sugar beet *(Beta vulgaris)* (Greger and Orgen, 1991). Photosynthetic apparatus of copper-sensitive spinach exposed to excess copper was investigated by Baszyński et al., (1982, 1988). Further, the copper deficiency on the photosynthetic apparatus of higher plants was also demonstrated (Baszyński et al., 1978). It has been shown that trace metals interfere with the functioning of the photosynthetic apparatus in higher plants (Baszyński, 1986; Baszyński et al., 1980; Baszyński and Tukendorf, 1984).

Photosynthesis was inhibited in corn and sunflower by lead (Bazzaz et al., 1975). Cadmium reduced photosynthesis and transpiration in corn and sunflower (Bazzaz et al., 1974b). However, corn and soybean exhibited differing degrees of sensitivity for photosynthesis and transpiration to excess lead (Bazzaz et al., 1974a, b). Cadmium changed the light reactions of chloroplasts (Bazzaz and Govindjee, 1974). In clover and lucerne, cadmium and lead induced several changes in gas exchange and water relations (Becerril et al., 1989). Wheat leaves of different insertion levels have different degrees of sensitivity to cadmium and nickel as a function of photosynthesis and water relations (Bishnoi et al., 1993a, b). In *Zea mays* cadmium altered the photosynthesis and enzymes of photosynthetic sulfate and nitrate assimilation pathways (Ferretti et al., 1993). Zinc and cadmium inhibited photosynthetic CO_2 fixation and Hill activity of isolated spinach chloroplasts (Hampp et al., 1976). Zinc inhibited the electron transport of photosynthesis in isolated barley chloroplasts (Tripathy and Mohanthy, 1980).

In *Phaseolus vulgaris,* zinc affected the ribulose-1,5-bisphosphate carboxylase/oxygenase (Van Assche and Clijsters, 1986a). Copper exerts toxic effects on photosystem II (PSII) of spinach chloroplasts (Hsu and Lee, 1988; Li and Miles, 1975). Varying concentrations of copper had limiting effects on photosynthesis of rice plants (Lidon and Henriques, 1991). In radish cotyledons, cadmium induced changes in the compositioin and structure of the light-harvesting chlorophyll *a/b* protein complex 2 (Krupa 1988; Krupa et al., 1987). Cadmium altered photosynthesis, transpiration, and dark respiration of excised silver maple leaves (Lameroux and Chaney, 1978). Weigel (1984, 1985), demonstrated the inhibition of photosynthetic reactions of isolated intact chloroplast and photosynthetic reactions of mesophyll protoplasts by cadmium.

Lead inhibited PSII in isolated chloroplasts (Miles et al., 1972; Wong and Givindjee, 1976). Cadmium influenced the net photosynthesis in rice (Misra et al., 1989; Moya et al., 1993). Cadmium and nickel affected the photosynthesis and enzymes of the photosynthetic carbon-reduction cycle in pigeon pea *(Cajanus cajan)* (Sheoran et al., 1990a, b).

Fluorescence spectroscopy has been employed recently to study the toxic functions of trace metals at extremely low concentrations of environmentally realistic levels. Chloroplast photochemical reactions inhibited in wheat seedlings with low concentrations of cadmium were investigated by analyzing the electron transport activities and changes in fluorescence yield (Atal et al., 1991). Similarly, whole leaf fluorescence was used as a technique for measuring tolerance of plants to trace metals (Homer et al., 1980). In *Phaseolus vulgaris,* photosynthetic functions were studied by fluorescence analysis (Krupa et al., 1993). Free radical generation and molecular damage were studied in cadmium-tolerant and cadmium-sensitive clones of *Holcus lanatus* (Hendry et al., 1992).

Cadmium is an effective inhibitor of plant metabolism, particularly of photosynthetic process in higher plants (Table 8-2). The linear relationship between net photosynthesis and inhibition of transpiration suggests that the response to cadmium might be due to the closure of stomata (Bazzaz et al., 1974a, b; Huang et al., 1974). Cadmium inhibits net photosynthesis by increasing both stomatal and mesophyll resistance to CO_2 uptake (Lameroux and Chaney, 1978).

In higher plants, net photosynthesis is reduced by trace metal accumulation in the leaves. Photosynthetic CO_2 fixation capacity of type A chloroplasts was inhibited by 50% after 5 minutes incubation in reaction mixtures containing different concentrations of various trace metals. Bazzaz and Govindjee (1974) showed a 50% inhibition of PSII electron flow after a 10 minutes dark incubation of chloroplasts from *Zea mays* with 0.2 μM cadmium nitrate. In *Zea mays* simultaneous inhibition of photosynthesis and transpiration by nickel suggested that the metal's primary effect should be on stomatal function. Cadmium primarily affected the photosynthetic pigments, before photosynthetic function. Cadmium and lead decreased total chlorophyll and chlorophyll *a/b* ratio in higher plants. Generally carotenoids were less affected by trace metals, resulting in a lower chlorophyll/carotenoid ratio in higher plants. Both permanent stomatal closure and increased ethylene production may be responsible for senescence induction by cadmium (Fuhrer, 1982).

Cadmium inhibited PSII, probably of the water splitting system at the level of manganoprotein (Van Duijvendijk-Matteoli and Desmet, 1975). Cadmium at low concentrations also acts as a photophosphorylation inhibitor in spinach chloroplasts (Lucero et al., 1976). According to Weigel (1985), cadmium affects photosynthesis by inhibition of different reaction steps of the Calvin cycle and not by interaction with photochemical reactions in the thylakoid membrane.

The main site of trace metal action was at the level of the water splitting system at the oxidizing side of PSII (Arndt, 1974). Tripathy and Mohanty (1980) reported the site of inhibition closer to the PSII reaction center and an additional site of inhibition at the plastoquinone level. Bivalent cations play a major role in the activation of rubisco (ribulose, 1,5-bisphosphate carboxylase/oxygenase) and in the equilibrium between CO_2 and O_2 binding by this protein (Lorimer, 1981). Inhibition of net photosynthetic CO_2 fixation rate and increase of the CO_2 compensation point were observed in intact *Phaseolus vulgaris* after zinc treatment (Van Assche et al., 1988). The membrane-bound photosynthetic reactions in isolated mesophyll protoplasts were not impaired by cadmium concentrations, which drastically inhibited CO_2 fixation (Weigel, 1985).

Noncyclic photophosphorylation was very sensitive to trace metals in vitro. The toxic concentration was 20 μM for lead and 5 μM for mercury. Copper and cadmium inhibited both cyclic and noncyclic photophosphorylation at 1.7 μM and

TABLE 8-2 Cadmium Effect on Photosynthetic Characteristics, Gas Exchange Functions and Chloroplasts of Selected Angiosperms[a]

Taxon	Cd Dose	Exposure Duration	Photosynthetic Function	Reference
Acer saccharinum	0.045, 0.09, and 0.18 mM	16, 40, 64 h	PS, TR, and R	Lameroux and Chaney (1978)
Beta vulgaris	5 and 20 μM	4 weeks	PS, Chl	Greger and Orgen (1991)
Cajanus cajan	5, 10, 15, and 20 mM	45–105 d	PS	Sheoran et al. (1990a)
	5 and 10 mM	2 and 6 d	PS and enzymes of PCR cycle	Sheoran et al. (1990b)
Glycine max	30 and 100 μM	12, 24, and 48 h	CP	Ghoshroy and Nadakavukaren (1990)
Hordeum vulgare	1, 10, 100, and 1000 μM	7 d	PS, Chl	Stiborova et al. (1986a, b, 1987)
Medicago sativa	5, 10 g m^{-3}	4 weeks	PS, ET, WR, and GE	Becerril et al. (1989)
Oryza sativa	3 g m^{-3}	15 d	PS	Misra et al. (1989)
	0.01 and 0.1 mM	5 and 10 d	PS	Moya et al. (1993)
Phaseolus vulgaris	1, 2.5, 5, 10, and 20 g m^{-3}	20 d	PS, CAR	Barcelo et al. (1986a)
	1, 10, 20, and 50 μM	7 d	Chl	Krupa et al. (1992)
	10, 20, and 50 μM	10 h dark, 7 h light	Chl	Padmaja et al. (1990)
Raphanus sativa	0.2 mM	48 h	PS	Krupa (1988)
	1, 10, 100, and 200 μM	6, 12, 24, and 48 h	PS	Krupa et al. (1987)
Tamarix aphylla	9, 45, and 90 μM	9 d	TR	Hagemeyer and Waisel (1989a)
Trifolium pratense	5 and 10 g m^{-3}	4 weeks	PS, ET, WR, and GE	Becerril et al. (1989)
Triticum aestivum	30, 60, 90, and 120 μM	24 h	PS	Atal et al. (1991)
Zea mays	10, 100, and 250 μM	14 d	PS, Chl	Ferretti et al. (1993)
	10, 100, and 250	11 d	Chl and PS	Rascio et al. (1993)
	25, 50, and 100 g m^{-3}	2 h	Chl	Singh (1993)
	10 μM	30 min	PS	Stiborova and Leblova (1985)
	1, 10, 100, and 1000 μM	7 d	Chl, PS	Stiborova et al. (1987)

[a]CAR = carotenoid, GE = gas exchange, CP = chloroplasts, Chl = chlorophyll, ET = electron transport, PS = photosynthesis, R = respiration, TR = transpiration, WR = water relations.

45 μM, respectively. Long-term exposure of whole plant to cadmium may affect chlorophyll with consequences for both the chloroplast development in young leaves and the inhibition of photosynthesis in mature leaves (Stobart et al., 1985). In cadmium-treated plants, the decrease in leaf area was not only due to the reduced cell size, but also due to decreased intercellular spaces. Most of the chloroplasts were intact or only slightly damaged, and some cells developed swollen chloroplasts with several plastoglobuli (Barcelo et al., 1988a, b). Cadmium reduced net photosynthesis, chlorophyll, accessory pigments, and PSII, but PSI was not affected. In isolated chloroplasts, lead inhibited PSII (Miles et al., 1972). The fine structure of chloroplast in cadmium-treated plants was degenerated, similar to senescence response, marked by the occurrence of large plastoglobules and disorganization of the lamellar structure, mainly granal stacks (Baszyński et al., 1980). Cadmium altered the content of the specific phosphatidyl glycerol fatty acid $trans\Delta^3$ hexadecenoic acid, widely accepted as a component responsible for the oligomerization of chlorophyll protein complex (Krupa et al., 1987).

Uncoupling of photophosphorylation does not occur in the presence of cadmium, as dissipation of the proton gradient across the thylakoid membrane was not observed (Weigel, 1985). In bush bean (*Phaseolus vulgaris* cv. Contender) plants, chlorosis was not noticed in the primary leaves, but electron microscopic studies revealed disorders in the disposition and stacking of thylakoid membrane, with chloroplasts of trifoliate leaves being more affected than those of primary leaves (Barcelo et al., 1986a, b). Poschenrieder et al. (1989) reported a significant increase of bush bean leaf abscissic acid (ABA) after 120 hours exposure to cadmium. Cadmium was found to inhibit ABA accumulation during drying of excised leaves. The authors also concluded that the direct effects of cadmium in leaves might account for both decreased cell expansion growth and increased stomatal resistance. The decreased cell wall extensibility may be a cause of reduced cell expansion.

Cataldo et al. (1981) investigated the cadmium distribution in plant tissues and reported that the behavior of cadmium in xylem exudates and tissues was suggestive of a change in chemical form resulting from complexation with plant-produced ligands involved in transport and metabolism. In soybean seedlings cadmium interfered with respiration and activities of several enzymes connected with respiration (Lee et al., 1976).

Stomatal Function

Stomatal movements provide the leaf with an opportunity to change the partial pressure of CO_2 and the rate of transpiration. Increased stomatal resistance in cadmium-treated plants has been reported (Bazzaz et al., 1974a, b; Kirkham, 1979; Pearson and Kirkham, 1981). Interference with stomatal function is considered to be a primary mode of action of cadmium and several other metals (Becerril et al., 1989).

The inhibition of stomatal opening in plants exposed to cadmium may depend on both cadmium concentration and exposure time, and also on the degree of toxicity suffered by the plants (Barcelo et al., 1986a, b). A direct effect of cadmium on the ion and water movement in the guard cells has yet to be proven, but cannot be excluded, particularly in relation to K^+ and Ca^{2+} (Barcelo et al., 1986a, b). During absorption, cadmium is strongly retained by roots with only 2% of the accumulated cadmium being transported to leaves, and as much as 8% being transported to seeds during seed filling (Cataldo et al., 1983).

Enzymes

Enzymes of trace-metal-tolerant and trace-metal-sensitive populations of *Silene cucubalus* and their interactions with some other trace metals in vitro and in vivo have been investigated (Mathys, 1975). Cd, Ni, and Cr are reported to alter the activities of hydrolytic enzymes responsible for mobilization of food reserves in germinating pea seeds *(Pisum sativum)*. (Bishnoi et al., 1993a, b; Dua and Sawhney, 1991). Cadmium influenced the respiration rate and activities of several enzymes in soybean seedlings. (Lee et al., 1976). Copper is known to interfere with oxidative enzymes in oat *(Avena sativa)* leaves (Luna et al., 1994).

Zinc inhibits RUBP carboxylase activity without affecting its oxygenase activity (Van Assche and Clijsters, 1986a, b). Total soil phytotoxicity is reflected by the increase in activity of enzymes such as malic enzyme, glucose-6-dehydrogenase, and peroxidase in the leaf (Van Assche and Clijsters, 1987).

Cadmium exerts its toxicity through membrane damage and inactivation of enzymes, possibly through reaction with—SH—groups of proteins (Fuhrer, 1982; Mathys, 1975; Obata and Umebayashi, 1993). SH-inactivation was suggested to explain the small inhibitory effects of lead and cadmium on the activity of the chloroplast enzyme RUBP carboxylase and phosphoribulokinase in vitro. Full inhibition of these enzyme activities was only observed in the presence of millimolar metal concentrations. Trace metals inhibits ATPase activity and also interfere with membrane integrity (Ros et al., 1990).

Sheoran et al. (1990a, b) investigated the effect of Cd^{+2} and Ni^{+2} on photosynthesis and enzymes of the photosynthetic carbon reduction cycle (PCR) in pigeon pea *(Cajanus cajan)* and concluded that Cd^{+2} and Ni^{+2} reduced photosynthesis indirectly through decrease in chlorophyll content and by affecting stomatal conductance and electron transport system more than the enzymes of the PCR cycle such as fructose-1,6-biphosphatase, 3-PGA kinase, rubisco, NADP-gly-P-dehydrogenase, and NAD-gly-P-dehydrogenase.

Peroxidase induction was observed in *Glycine max* (Lee et al., 1976) and *Phaseolus vulgaris* (Van Assche et al., 1988). Cadmium induced peroxidase activity in roots and leaves of *Oryza sativa;* roots showed 10 to 20 fold higher activity than leaves (Reddy and Prasad, 1992). Localization of cadmium in vacuoles, synthesis of phytochelatins (PCs), and binding to trace metals occur mainly in roots (Grill et al., 1989a, b). Peroxidase induction is a general response of higher plants to uptake of toxic amounts of metals such as Cd, Cu, Pb, Zn, and Ni. An increase in leaf peroxidase activity of some wild dicots was ascribed to the lead content of automobile exhaust. In addition to peroxidase, the activities of other enzymes, namely, malic enzyme, glucose-6-P-dehydrogenase, isocitrate dehydrogenase, and glutamate dehydrogenase also increased in *Phaseolus vulgaris* when treated with zinc and cadmium (Van Assche et al., 1988).

Toxicity of trace metals is manifested in several ways. Membrane function disruption is one of the significant toxicological symptom of trace metals. In tolerant and sensitive *Silene cucubalus* this was demonstrated with copper, which acts on fatty acid composition and peroxidation of lipids in the roots altering the uptake of water and solutes (De Vos et al., 1993). In rice, cadmium and nickel act on the lipid composition, Mg-ATPase activity, and fluidity of plasma membranes. Membrane damage due to lipid peroxidation caused by metals is mediated by activated oxygen radicals (hydrogen peroxide, hydroxyl, and superoxide radicals) and could be quenched by the induction of specific enzymes like peroxidase, superoxide dismutase, and catalase (De Vos and Schat, 1991). Increase in peroxidase is generally considered as an indication of plant aging (Hazell and Murray, 1982). Based on the increase in activity of peroxidase and hydrolytic enzymes (ribonuclease, deoxyribonuclease, and acid phosphatase) in *Glycine max* treated with cadmium, Lee et al. (1976) proposed that cadmium causes an anticipated senescence response.

Starch and sugar concentration in leaves of *Betula, Picea,* and *Pinus* in two industrial areas of Sweden polluted by Pb, Cd, Zn, and Cu were found to be significantly higher than in unpolluted areas, indicating an inhibition of hydrolysis of starch and sucrose by the reported pollutants (Behnke et al., 1991). It is also reported that chromium significantly depressed activities of hydrolytic enzymes, namely, ribonuclease, acid phosphatase, phytase, invertase, α-amylase, and β-amylase, in germinating pea seeds (Dua and Sawhney, 1991).

Water Stress

Plant water relations as affected by trace metals was critically reviewed (Barcelo and Poschenreider, 1990). Cadmium induced decrease of water-stress resistance in bush bean plants *Phaseolus vulgaris* cv. Contender by affecting endogenous ABA, water potential, relative water content, and cell wall elasticity (Barcelo et al., 1986a, b; Poschenrieder et al., 1989). Clover and lucerne changed water relations due to excess of cadmium and lead (Becerril et al., 1989). Cadmium-induced changes in water economy in beans was attributed to the involvement of ethylene formation (Fuhrer et al., 1981). Cadmium and nicel adversely affected water relations in wheat leaves of different insertion levels (Bishnoi et al., 1993a, b).

Cadmium generally decreased the water-stress tolerance of plants, causing turgor loss at higher relative water content and leaf water potential than in nontreated control plants. Cadmium increased the bulk elastic modulus and therefore decreased the cell wall elasticity. Low cell wall elasticity seems to be an important cause of the low water-stress tolerance in cadmium-toxic plants (Becerril et al., 1989). The treated plants showed decreased transpiration rate and increased stomatal resistance

(Bazzaz et al., 1974a, b; Kirkham 1979). Xylem obstruction by cadmium-induced cell wall degradation products has been suggested as a cause of reduced water transport (Fuhrer et al., 1981; Lameroux and Chaney 1977, 1978). Wilting in cadmium-treated plants is due to the decreased water transport and probably to reduced water absorption (principally because of reduced root growth).

Both cadmium and lead adversely affected water status and gas exchange in lucerne *(Medicago sativa)* and clover *(Trifolium pratense)* (Becerril et al., 1989). However, the water potentials measured and the decrease of water-use efficiency calculated after lead treatment indicated that, in spite of a strong stomatal closure, transpiration was still remarkable. Treatment with similar concentrations of lead and cadmium in leaves revealed that lead produced a severe water stress, whereas cadmium caused only a slight water stress.

Low concentrations of cadmium in certain experiments increased transpiration. The transpiration rate dropped, probably as a result of root damage due to cadmium toxicity. The transpiration behavior indicates that the effect of water vapor pressure on the degree of stomatal opening is small (Hagemeyer and Waisel, 1989a). Under certain conditions some trace metals reduce transpiration of plants by direct interference with stomatal regulation. Light had a strong effect on transpiration rates of the cadmium-treated plants (Hagemeyer and Waisel, 1989a, b). Cadmium influenced the transpiration on plants more via its effect on the water flow through the roots than via its effects on stomatal aperture (Hatch et al., 1988).

Afflux of Cations

Inactivation of enzymes involved in the selective permeability of membranes, such as ATPases, by copper was demonstrated by Veltrup (1982). It is also reported that the permeability of membranes depends on the degree of lipid peroxidation (Dhindsa et al., 1982). The lipid-peroxidation process is believed to be initiated by free radicals, such as the hydroxyl radical generated in the Haber–Weiss reaction and peroxyl and alkoxyl radicals produced during the decomposition of organic hydroperoxides (De Vos et al., 1989). Copper-induced lipid peroxidation in *Silene cucubalus* suggested that direct free radical formation leading to lipid peroxidation might play an important role in the mechanism of copper-induced damage to the permeability barrier in roots of higher plants in vivo (De Vos et al., 1989).

MECHANISMS OF TRACE-METAL TOLERANCE IN PLANTS

Varying levels of adaptations to toxic concentrations of trace metals are ubiquitously found in all organisms. Mechanisms of copper tolerance have been schematically explained by Ernst et al. (1992) (Fig. 8-4). In general, the possible mechanisms of metal tolerance in plants are: (1) metal binding to cell wall, (2) reduced transport across the cell membrane, (3) active afflux, (4) compartmentalization, and (5) chelation (Tomsett and Thurman, 1988). "Chelation" is the most prevalent mechanism and vast information is available on this subject.

Cell wall

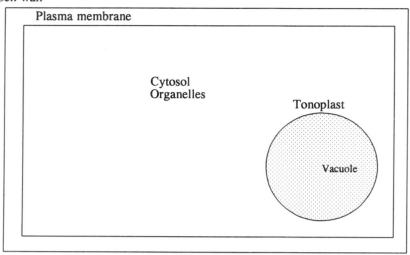

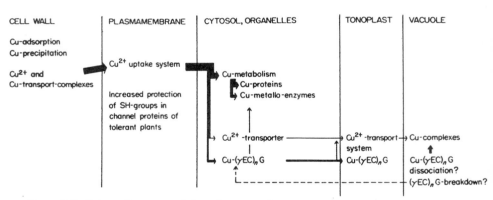

Figure 8-4. *Schematic representation of copper-tolerance mechanisms. (Courtesy of Prof. Dr. W. H. O. Ernst, Faculty of Biology, Vrije Universiteit, Amsterdam, The Nertherlands. Reproduced with the permission of Blackwell Science Ltd., Oxford, UK.)*

Trace-Metal Accumulation and Excretion

Cadmium-bearing granules were found in the cytoplasm and vacuoles of differentiating and mature cells and in nuclei of undifferentiated cells, but were absent from cell walls and epidermal cells (Rauser and Ackerley, 1987). Nickeliferous rocks were mapped by analysis of herbarium specimens of indicator plants (Brooks et al., 1977).

Accumulation is an important strategy in response to trace metals (Baker, 1981; Baker et al., 1985; Brams and Anthony, 1988; Fielder, 1985; Jaffre et al., 1976). There is some relationship between tolerance and accumulation characteristics in higher plants (Kuboi et al., 1987). Cereals are reported to accumulate trace metals

(Malone et al., 1974; Pieczonka and Rosopulo, 1985; Piotrowska and Dudka, 1994; Rivai et al., 1990a, b). Toxic trace-metal accumulation in crop plants and its consequences for human health are of great concern (Wagner, 1993). Plants suffer different degrees of damage due to trace-metal accumulation, and thus serve as indicators (Kanerva et al., 1988).

Excretion is one of the important mechanisms of trace-metal tolerance. *Lupinus albus* and *Tamarix aphylla* are known to excrete excess cadmium (Costa and Morel, 1993; Hagemeyer and Waisel, 1988).

Plants could achieve tolerance to trace metals by two strategies (Baker, 1987): (1) avoidance, by which a plant is protected externally from the influence of the stress, or (2) tolerance, by which a plant survives the effect of internal stress. Thus tolerance is conferred by specific physiological mechanisms that collectively enable it to function normally even in the presence of high concentrations of potentially toxic elements. Mine spoils and other soils contaminated with Cd, Cu, Zn, and Pb show natural colonization by species that have strategies of avoidance or tolerance of metal toxicity (Baker and Proctor, 1990). Evolutionary processes mediating the selection of tolerant individuals and ecotypic differentiation of adapted population on metalliferous soils are: (1) natural selection of tolerant genotypes, (2) constitutional tolerance, and (3) inducible tolerance. Plants adapted to metal-contaminated ecosystems could be used as indicators for exploration of metals (Brooks et al., 1977; Clark and Clark, 1981; Sarkela and Nuorteva, 1987).

Metal-Binding Complexes (MBC) (= Phytochelatins)

Phytochelatins (PCs) or polyglutamylcysteinyl glycines play an important role in metal tolerance (Ernst 1990; Florijn et al., 1993a, b; Grill et al., 1985; Jackson et al., 1987; Rauser, 1995; Salt et al., 1989; Verkleij et al., 1990). Plants growing in a mining refuse area have invariably synthesized the metal-sequestering peptides, that is, the PCs (Grill et al., 1988b). PCs, the trace-metal-binding peptides of plants, are synthesised from glutathione by a specific glutamyl cysteine dipeptidyl *trans*-peptidase (phytochelatin synthase) (Grill et al., 1989a, b). PCs, a class of trace-metal-binding peptides from plants, are considered to be functionally analogous to metallothioneins (Grill et al., 1989a, b). Fabales synthesize homophytochelatins to sequester trace metals using homoglutathione as precursor (Grill et al., 1986a). In Poaceae trace-metal-induced peptides are of hydroxymethyl phytochelatins [(glutamyl cysteine)*n*-serine] (Klapheck et al., 1994). Labile sulfide and sulfite are essential ingredients in phytochelatin complexes (Eanetta and Steffens, 1989; Spieser et al., 1992). A gel electrophoresis assay was developed for detecting phytochelatins (Abrahamson et al., 1992). In *Datura innoxia,* precursor-product relationship of polyglutamyl cysteinyl glycine biosynthesis was experimentally established and implicated in metal tolerance (Berger et al., 1989; Delhaize et al., 1989a, b). Induction of trace-metal-sequestering PCs by trace metals in cell cultures was demonstrated (Grill et al., 1985, 1988a). PCs protect plant enzymes from trace-metal poisoning, and metal-requiring apoenzymes have been reactivated by PCs (Kneer and Zenk, 1992; Thumann et al., 1981).

PC synthesis and glutathione levels in response to trace metals in tomato cells and maize roots established precursor-product relationship (Mendum et al., 1990; Rauser et al., 1991; Scheller et al., 1987; Tukendorf and Rauser, 1990). Some doubts have been expressed as to whether PCs are involved in differential metal tolerance, or whether they merely reflect metal imposed strain (Schat and Kalff, 1992). Evidence against a role for PCs was presented by De Knecht et al. (1992) in *Silene Vulgaris*.

The second international meeting on metallothionein (MT) and lower molecular weight metal-binding proteins, held in 1985, suggested a revised nomenclature for MTs (Fowler et al., 1987). Polypeptides can be termed as "metallothioneins" when they share several of the following characteristics of equine renal metallothionein: low molecular weight, high metal content, high cysteine (Cys) content with the absence of aromatic amino acids and histidine, an abundance of Cys-X-Cys sequence, where X is an amino acid other than Cys, spectroscopic features characteristic of metal thiolates, and metal thiolate clusters. The following classes of metal thiolate polypeptides were considered:

Class 1: Polypeptides with locations of cysteine closely related to those in equine renal metallothionein (as Cys-Cys and Cys-X-Cys clusters, where X is an amino acid other than Cys)

Class 2: Polypeptides with locations of cysteine only distantly related to those in equine renal metallothionein (these cannot be easily aligned with those of equine renal MT)

Class 3: Atypical, nontranslationally synthesized metal thiolate polypeptides (i.e., not synthesized on an mRNA template)

Classes 1 and 2, but not class 3, MTs have been isolated from a wide variety of animals and certain fungi; however, more than 200 species of plants tested so far, together with a few fungi, are known to have class 3 MTs (Gekeler et al., 1989; Theisen and Blincoe, 1988; Tomsett and Thurman, 1988).

PCs were synthesized enzymically by PC synthase (γ-glutamyl cysteine dipeptidyl transferase), a 25 kDa protein that removes a γ-glutamyl cysteine from one molecule of GSH and couples it to another GSH) in higher plants (Gekeler et al., 1989; Rauser 1995).

Löffler et al. (1989) purified the PC synthase to homogeneity in cell cultures of *Silene cucubalus* (Caryophyllaceae), *Podophyllum peltatum* (Berberidaceae), *Eschscholtzia californica* (Papaveraceae), *Beta vulgaris* (Chenopodiaceae), and *Equisetum giganteum* (Pteridophyte, Equisetaceae). This enzyme catalyzes the formation of metal-chelating peptides (PCs) from glutathione in the presence of trace metal ions. Incubation of PC under standard conditions, as described by Löffler et al. (1989), in the absence of trace metal ions does not lead to the formation of the PC peptides. However, addition of cadmium to the incubation mixture instantaneously reactivated the enzymes.

PCs are considered to have an important role in the cellular metal homeostasis (Steffens, 1990). The enzymatically inactive metal-requiring apoforms of diamine oxidase and of carbonic anhydrase were reactivated by copper and zinc PC complexes, respectively (Thumann et al., 1991). When the metal chelator diethyl dithiocarbamate was added to the purified diamine oxidase from pea *(Pisum sativum)* seedlings, it removed copper ions from the enzyme. However, addition of various concentrations of $CuSO_4$ to the apoform of the enzyme resulted in maximal restoration of its activity to 80% (Fig. 8-5). PCs (γ-GluCys)$_n$Gly, that is, [(γ-EC)$_n$ G)] are also implicated in trace-metal detoxification (Jackson et al., 1987; Kneer et al., 1992).

PC synthesis is induced not only by trace metals but also by a variety of multiatomic anions such as SeO_4^{-2}, SeO_3^{-2}, and AsO_4^{-3}. PCs have been shown to bind cadmium and copper directly (Reese et al., 1988) and are believed to bind lead and mercury by competition with cadmium (Abrahamson et al., 1992). It is presumed that trace metals cause cell death in plants by interfering with the activity of enzymes, and possibly with structural proteins, through metal sensitive groups such as -SH or histidyl groups, leading to inactivation of catalytically active or structural

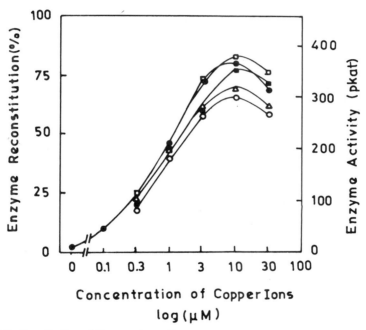

Figure 8-5. *Reactivation of the metal requiring apoenzyme of diamine oxidase from* Pisum sativum *by free copper ions and copper-phytochelatin complexes. The copper ions supplied at varying levels as free ions (●——●), complexes to di(γ-glutamyl-cysteinyl)glycine (PC2, ■——■), tetra(γ-glutamyl cysteinyl)glycine (PC4, △——△), penta(γ-gltamyl cysteinyl)glycine (PC5, ○——○), or to the copper-phytochelatin complex as isolated from plant tissue (PCn, □——□), which consist predominantly of a mixture of PC2 and PC3. The level of reconstitution reflects maximal observed reactivation during an incubation period of 30 minutes. (Courtesy of Prof. Dr M. H. Zenk, Lehrstuhl für Pharmazeutische Biologie der Universität München, Germany; Reproduced with the permission of the Elsevier Science Publishers).*

proteins (Kneer and Zenk, 1992; Van Assche and Clijsters, 1990). Therefore, plants, as an adaptive strategy, might have developed some physiological and biochemical mechanisms to tolerate metal toxicity. PCs reduce cytoplasmic toxicity to certain concentration of trace metals by complexation or binding. Now it is believed that PC-metal complexes are less toxic to cellular plant metabolism than free metal ions. The best indirect evidence for such an assumption is available from the experiments of Steffens et al. (1986), who demonstrated that in tomato cells selected for cadmium tolerance, PCs are accumulated at considerably higher level than in normal sensitive cells.

Direct evidence toward the role of PCs in protecting plant enzymes was reported by Kneer and Zenk (1992) using suspension cell cultures of *Rauvolfia serpentina* and treating them with 100 μM CdCl$_2$ containing 100 μCi ^{109}CdCl$_2$. They reported that a series of metal-sensitive plant enzymes, such as alcohol dehydrogenase, glyceraldehyde-3-phosphate dehydrogenase, nitrate reductase, ribulose-1,5-bisphosphate carboxylase, and urease, tolerate cadmium in the form of a PC complex in amounts from ten- to a thousandfold greater than are tolerated if the cadmium is a free metal ion. PCs reactivated metal-poisoned nitrate reductase in vitro up to a thousandfold better than chelators such as glutathione and citrate, thereby proving the extraordinary sequestering potential of the PCs (Kneer and Zenk, 1992).

Acid labile S^{-2} and sulfite in PC-metal complexes of tomato cells were isolated (Eanetta and Steffens, 1989; Steffens et al., 1986). Verkleij et al. (1990) reported increased levels of S^{-2} in PC-Cd complexes from metal-resistant isolates of *Silene vulgaris*. A high-molecular-weight PC-Cd complex from tomato (Reese et al., 1992) and *Brassica juncea* (Speiser et al., 1992) was reported.

It was found that copper-sensitive and copper-tolerant populations of *S. vulagris* both produce PCs when exposed to copper (Schat and Kalff, 1992). However, the total nonprotein -SH content of the roots responding to the external Cu concentration was population dependent. Further, the authors also found that the dose response curves for nonprotein -SH accumulation and root growth were very consistent. When populations of *S. vulagris* are exposed to their own no-effect concentration, or to 50%-effect concentration of copper for root growth, tolerant and nontolerant plants synthesized equal PC contents in the root apex, which is the primary copper target. Therefore, it was concluded that PCs are not decisively involved in differential copper tolerance in *S. vulgaris* (Schat and Kalff, 1992). Variations in PC-SH production seem to be a mere consequence of variations in tolerance. Differential copper tolerance in the above plant does not appear to rely on differential PC production. Induction and loss of cadmium tolerance was also reported in *Holcus lanatus* and several other grasses (Baker et al., 1986).

Stabilization of MBCs by Sulfide Ions

Occurrence of free sulfide in PCs of various plants and algae has been reported (Gekeler et al., 1988). PCs in higher plant cell cultures had concentrations of sulfide ranging from about 1% in, for example, *Rauvolfia serpentina*, (Grill et al., 1985) to 30% in *Glycine max* (Grill et al., 1986a, b). In metal-tolerant varieties of *Silene cucubalus,* the cadmium-binding protein contains a higher concentration of labile sulfide and binds to a greater amount of cadmium than does the cadmium-binding protein induced in sensitive varieties (Verkleij et al., 1990). However, in a recent study using extended X-ray absorption fine structure measurements, it has been shown that the immediate environment of metal in the native $Cd-PC_n$ complexes consists of four sulfur atoms; they contain discrete $Cd(Cys)_4$ units, and sulfide is of no significance as a ligand (Strasdeit et al., 1991). Rates of Cd–PC complex formation were higher in tolerant *Datura innoxia* cells than in sensitive ones (Delhaize et al., 1989a, b). These findings show that, although PCs produced in sensitive and tolerant cells are alike, yet they form complexes differently.

Cadmium-sulfide crystallites in $Cd-(\gamma EC)_n G$ peptide complexes from hydroponically grown tomato plants (*Lycopersicon esculentum* P. Mill. cv Golden Boy) have been characterized (Reese et al., 1992). The elution profile of the cadmium-peptide extract of the metal-treated plants showed two peaks when subjected to anion-exchange chromatography (the first peak eluting at an ionic strength of approximately 550 mosmol and the second near 650 mosmol). The sulfide–cadmium ratios differed considerably between the first and second peaks. The first had a maximum S–Cd ratio of 0.41, and for the second peak, $S/Cd = 0.13$. The distribution coefficients (K_{av}) of Sephadex G-50 elution profiles (rechromatographed for separation) of the first peak are approximately 0.53, 0.59, and 0.66; those of the second peak are near 0.58, 0.63, and 0.71. Quantification of sulfide in the eluted fractions clearly demonstrated the correlation between S–Cd ratios and the UV transition near 285 nm. The peptide complex ($K_{av} = 0.53$) upon acidification to pH 1.5 volatilized the sulfide. Absorption spectroscopy of the sample after reneutralization showed that the transitions at 260 and 285 nm did not appear but there was the appearance of a transition near 250 nm. Loss of the two transitions as a result of acidification was directly correlated with loss of sulfide, which was observed by the characteristic odor of H_2S given off during the treatment and by the loss of activity of the reneutralized complexes in direct sulfide assay. Addition of 1 mol equivalent of sulfide to the cadmium-peptide complex in vitro resulted in a red shift of the transition to near 315 nm (Reese et al., 1992).

Glutathione

Glutathione is the precursor for the synthesis of metal-binding complexes. Cadmium is reported to change glutathione and PCs in roots of maize seedlings (Rauser, 1990a; Tukendorf and Rauser, 1990) and *Pisum sativum* (Ruegsegger et al., 1990). Sulfate, chloride, and selenate during short-term dark incubation influence the glutathione content of spinach leaf discs (De Kok and Kuiper, 1986). Regulation of thiol contents and alteration of thiol pools in maize exposed to cadmium could be regulated by intermediates and effectors of glutathione synthesis (Farago and Brunold 1994; Meuwly and Rauser, 1992). Direct evidence of glutathione involvement in PC synthesis was demonstrated in tomato cells (Mendum et al., 1990; Scheller et al., 1987). Recently a new homologue of glutathione, glutamyl cysteinyl glutamic acid, was discovered in maize seedlings exposed to cadmium (Meuwly et al.,

1993). Glutathione depleted due to copper induced PC, causing oxidative stress in *Silene cucubalus* (De Vos et al., 1992).

Stress Proteins

Cadmium induced 70 kDa protein in roots of maize (*Zea mays* L.), which was precipitated by hsp70 antibodies. In vitro phosphorylation assay shows that the hsc70 (heat shock proteins that are induced by factors other than heat were termed as "hsp cognates"—hsc) is a phosphoprotein. Phosphoamino acid analysis of immunoprecipitated hsc70 by paper chromatography showed that serine and tyrosine are phosphorylated in control seedlings. Only serine is phosphorylated in cadmium-treated seedlings, not tyrosine, in spite of increase in hsc70 amount (Reddy and Prasad, 1993). This could possibly be due to inhibition of a tyrosine kinase. This change in hsp70 cognate phosphorylation might be playing the role of a chaperon. Possibly the hsc70 in maize might be acting to limit and rescue the damage to proteins caused by environmental stress like other chaperons (Reddy and Prasad, 1993). Similar results were also noticed in cadmium-treated rice seedlings (Reddy and Prasad, 1995).

Cadmium induced a number of stress proteins ranging in molecular weight from 10 to 70 kDa. In *Oryza sativa,* cadmium-induced proteins are 70, 42, 40, 26, 23, 15, and 11 kDa protiens. The 15 kDa protein is speculated as a *SalT* gene product. Reddy and Prasad (unpublished data), using *SalT* cDNA probe, detected certain changes in related RNA in control and cadmium-treated rice. *SalT* cDNA probe hybridized to RNA from control and cadmium-treated rice, and there was an increase in *SalT* expression in coleoptiles and roots. This indicates that cadmium induces *salT* gene expression due to osmotic or dehydration stress. Lameroux and Chaney (1977, 1978) and Poschenrieder et al. (1989) reported the cadmium induced dehydration in cells by retardation of water uptake. Cadmium-induced protein phosphorylation changes, as reported by Reddy and Prasad (1993, 1995), might be the signal transduction mechanism by which transcription may be modulated in response to cadmium stress and other environmental stresses.

The barley and maize seedlings exhibited retardation in shoot and root growth after exposure to copper, cadmium, and lead. The zinc ions practically did not influence these characteristics. The total protein content of barley and maize roots declined with an increase in trace-metal concentration (Stiborova et al., 1986a, b). Total glutathione content was decreased upon exposure to cadmium, more so in roots than in shoots. Robinson and Jackson (1986) suggested that glutathione is somehow involved in the biosynthesis of PCs. Rauser and Curvetto (1980) have isolated copper-binding complexes from *Agrostis gigantea.*

MT Genes and Engineering of Metal-Tolerant Plants

Two major groups of complexes were isolated from higher plants, including in vitro cultures: (1) 8–14 kDa complexes similar to those of MTs in some aspects, (2) 1.5–4 kDa complexes, the PCs = $(\gamma\text{-EC})_n G$ (Rauser, 1990b; Reddy and Prasad, 1990; Steffens, 1990). The first are thought to be aggregates of the second, and contained a number of amino acids with glutamic acid, cysteine, and glycine as the major constituents. In earlier reports where MT-like proteins were purified, some percentage of metal ions also bound to low molecular weight peptides, probably PCs. In *Datura innoxia* 80% of the cadmium was bound to low-molecular-weight (γ-EC)$_n$G and about 15% to MT-like proteins (Robinson et al., 1987). Amino acid composition of this MT-like protein is similar in some respects to that of mammalian and microbial MTs, but contains more glutamine/glutamate than these. In maize about 72% of the cadmium bound to low-molecular weight peptides, possibly (γ-EC)$_n$G, and about 18% bound to MT-like protein of 11.35 kDa (Leblova et al., 1986). For copper, about 60% was bound to MT-like protein and only about 25% to low-molecular weight peptides, that is, (γ-EC)$_n$G (Leblova et al., 1986). This may indicate co-occurrence of both MT and (γ-EC)$_n$G in plants.

MT genes in plants have since been reported (De Miranda et al., 1990; Evans et al., 1990; Tommey et al., 1991). A pea gene (designated $PsMT_A$) showed homology to class I and II MT genes from different organisms (Evans et al., 1990). Macnair (1977) reported major gene for copper tolerance in *Mimulus guttatus*. A cDNA sequence from *M. guttatus* with a single open reading frame encoding a putative 72 amino acid polypeptide sequence when compared with protein database, 19 of 23 matches found were MTs, the top 8 matches being from *Neurospora crassa,* sea urchin, chinese hamster, and *Drosophila* (De Miranda et al., 1990). This strong similarity with class I and II MTs is due to two domains of $^{14}/_{15}$ amino acids, each of which contain six cysteine residues arranged exclusively as Cys-X-Cys clusters.

Kawashima et al. (1991) deduced the amino acid sequence of soybean MT-like protein and compared it with MT genes of rat, *N. crassa, M. guttatus,* and *Pisum sativum*. The N-terminal domain exhibited significant homology to *N. crassa* MT ($^9/_{13}$, 70%) and the b-domain of rate MT ($^9/_{14}$, 64%). However, by contrast the middle domain contains several aromatic and hydrophobic amino acids, and it showed little homology to other MTs. No cysteine was detected in the middle domain. The deduced amino acid sequence of the soybean protein is 51% and 59% homologous to corresponding sequences from *M. guttatus* and pea, respectively. These proteins from *M. guttatus, P. sativum,* and soybean show close semblance to each other in the following three respects:

1. The number of amino acid residues range from 72 *(M. guttatus)* to 79 (soybean).
2. the N- and C- terminals of these three proteins contain cysteine clusters and show a high degree of homology in terms of amino acid sequence.
3. The middle domains of all the three proteins are relatively rich in hydrophobic amino acid residues and contain no cysteine (Kawashima et al., 1991).

An MT gene from *M. guttatus* was also isolated from copper-tolerant clones and repressed by copper (De Miranda et al., 1990). Pea roots also synthesized (γ-EC)$_n$G upon exposure to cadmium salts and $PsMT_A$ transcript was abundant in roots that have not been exposed to elevated concentration of trace metals (De Miranda et al., 1990). Further, they suggested that the (γ-EC)$_n$G may detoxify cadmium and possibly excess of other metals in pea roots, and that the putative $PsMT_A$ product may have a role in essential trace-metal metabolism. However, Grill et al. (1988b) reported $(\gamma\text{-EC})_n G$ induction in cell cultures in standard media and proposed the role in homeostasis of essential metals, the microelements.

Chinese hamster MT-II is expressed and functional in *Brassica campestris* leaf tissue when the tissue is infected with a cloned CMV pCa-BB1 containing cDNA of chinese hamster MT-II gene (Lefevbre et al., 1987). Trace-metal-tolerant transgenic *Brassica napus* and *Nicotiana tabacum* plants were obtained by infecting them with *Agrobacterium tumefacies* containing human MT-II processed gene on a disarmed *Ti* plasmid (Mishra and Gedamu, 1989). Mouse MT gene expression was studied in transgenic tobacco (Maiti et al., 1989). However, in such transgenic plants synthesis of metal-inducible proteins, including PCs, and their role in metal tolerance warrant detailed investigation.

CONCLUSIONS

Many of the laboratory investigations dealing with trace metals using vascular plants involved a wide range of metal levels and duration. Such bioassays have initiated cellular defence mechanisms and resulted in identification of various toxicity and detoxification mechanisms at organ/organism, cellular, and molecular level. The promising areas for future research include the following.

Synthesis of MBC (metal binding complexes) involving glutathione or its homologue is the crux of the metal-tolerance mechanism. Alteration in the functions of the enzymes of glutamyl cycle under metal stress is of special interest (Grill *et al.*, 1989a, b; Mendum et al., 1990; Meuwly et al., 1993; Ruegsegger et al., 1990; Tukendorf and Rauser, 1990). Mucilage secreted from meristems and other plant parts is reported to act as ligand for toxic metals like Al, Pb, Cd, and Cu (Horst et al., 1982; Morel et al., 1986). Trace metals have been removed from polluted soil with grass silage juice (Leidman et al., 1994). Bark of trees has been used for passive monitoring of airborne trace metals (Walenhorst et al., 1993). Chemically modified bark of *Pinus pinaster* had certain ligands that adsorb trace metal ions and thus help to develop biofilters for industrial application to clean effluents (Vázquez et al., 1994).

Cadmium induced protein phosphorylation changes in rice *(Oryza sativa)* and corn *(Zea mays)* (Reddy and Prasad, 1993, 1995). Phosphorylation is considered to be a signal transduction mechanism and is reported to involve transient increase in calcium concentration and changes in calcium/calmodulin-dependent phosphorylation. Thus some of the cadmium-inducible protein phosphorylation changes reported for rice and corn may be mediated by calcium/calmodulin, as was observed in other organisms (Behra, 1993; Karez et al., 1990).

A genetic approach, as opposed to physiological/biochemical investigations, may provide understanding of the mechanisms of metal tolerance. Some studies have been conducted in this direction on mechanism of metal tolerance by selecting metal-sensitive and metal-tolerant strains (Howden and Cobett, 1992; Howden et al., 1995a, b). Genetics of trace-metal tolerance and genetic improvement of metal hypersensitive genotypes of agricultural, horticultural, and silvicultural plants may emerge as a challenging subject (Schat and Bookum, 1992; Schat et al., 1993). Transgenic production of trace-metal excluders might emerge as a priority area (Mishra and Gedamu, 1989).

Modeling of metal uptake, soil-plant root interactions, the influence of crop rotation, tillage practices, and trace-metal interactions with humic substances in decomposing litter might elucidate important aspects of trace metal bioaccumulation. Phytotoxicity of trace metals, uptake by plant roots, and interactive functions of the trace metals are complex processes and detailed modeling is needed (Conzonno et al., 1994; Munch, 1993; Oliver et al., 1993; Rao and Mathur, 1994; Taylor et al., 1991, 1992).

Given that trace metals are known as free radical generators, use of chemicals as protectants for preventing membrane damage might have some application for enhancing the crop productivity (De Vos and Schat, 1991; Hendry et al., 1992; Meharg, 1993, 1994; Ros et al., 1990; Singh, 1993).

In vitro (cell culture) investigations are relevant not only to understanding metal tolerance but also to gain knowledge about enzymological aspects and metal ion-homeostasis. Cellular and molecular basis of thermoprotection of trace metals and heat-shock protein induced trace metal tolerance need critical investigation (Fig. 8-6) (Howarth 1990; Neumann et al., 1994; Orzech and Burke, 1988; Repka, 1993).

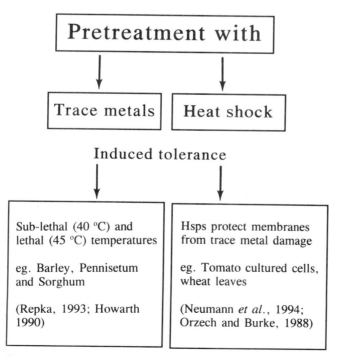

Figure 8-6. *Pretreatment with heavy metals induced tolerance to high temperature in crop plant seedlings. Heat shock induced heavy metal tolerance to cell cultures of tomato and wheat.*

ACKNOWLEDGMENTS

Trace metal research in the author's laboratory is being funded by CSIR [38(901)/ 95 EMR-II], New Delhi.

REFERENCES

Abrahamson, S. L., D. M. Speiser, and D. W. Ow (1992), *Anal. Biochem.*, *20*, 239–243.

Ahumada I. T. and E. B. Schalscha (1993), *Agrochimica.*, *37*, 281–289.

Allen, W. R. and P. M. Sheppard (1971), *Proc. Roy. Soc. London.*, *177B*, 177–196.

Aller, A. J. and L. Deban (1989), *Sci. Total Environ.*, *79*, 253–270.

Arndt, U. (1974), *Environ. Pollut.*, *6*, 181–194.

Atal, N., P. P. Saradhi, and P. Mohanty (1991), *Plant Cell Physiol.*, *32*, 943–951.

Aten, C. F., J. B. Bourke, and J. C. Walton (1986), *Bull. Environ. Contam. Toxicol.*, *36*, 918–923.

Ayala, M. B. and G. Sandmann (1988), *Physiol. Plant.*, *72*, 801–806.

Baker, A. J. M. (1978), *New Phytol.*, *80*, 635–642.

Baker, A. J. M. (1981), *J. Plant Nutrit.*, *3*, 643–654.

Baker, A. J. M. (1984), *Chemosphere*, *13*, 585–598.

Baker, A. J. M. (1987), *New Phytol.*, *106*, 93–111.

Baker, A. J. M. and J. Proctor (1990), *Plant Syst. Evol.*, *173*, 91–108.

Baker, A. J. M., R. P. Brooks, A. J. Pease, and F. Malaisse (1983), *Plant Soil, 73*, 377–385.

Baker, A. J. M., R. P. Brooks, and W. J. Kersten (1985), *Taxon, 34*, 89–95.

Baker, A. J. M., C. J. Grant, M. H. Martin, S. C. Shaw, and J. Whitebrook (1986), *New Phytol.*, *102*, 575–587.

Barcelo, J. and C. Poschenreider (1990), *J. Plant Nutrit.*, *13*, 1–37.

Barcelo, J., C. Cabot, and C. Poschenrieder (1986a), *J. Plant Physiol.*, *125*, 27–34.

Barcelo, J., C. Poschenrieder, I. Andreu, and B. Gunse (1986b), *J. Plant Physiol.*, *125*, 17–25.

Barcelo, J., M. D. Vázquez, and C. Poschenreider (1988a), *Bot. Acta.*, *101*, 254–261.

Barcelo, J., M. D. Vázquez, and C. Poschenreider (1988b), *New Phytol.*, *108*, 37–49.

Bartolf, M., E. Brennen, and C. A. Price (1980), *Plant Physiol.*, *66*, 438–441.

Basu, A., B. Basu, and G. J. Taylor (1994), *Plant Physiol.*, *104*, 1007–1013.

Baszyński, T. (1986), *Acta Soc. Bot. Polish.*, *55*, 291–304.

Baszyński, T. and A. Tukendorf (1984), *Folia Soc. Sci. Lubliensis.*, *26*, 31–39.

Baszyński, T., M. Ruszkowska, M. Krol, A. Tukendorf, and D. Wolinska (1978), *Z. Pflanzenphysiol.*, *89*, 207–216.

Baszyński, T., L. Wajda, M. Krol, D. Wolinska, Z. Krupa, and A. Tukendorf (1980), *Physiol. Plant.*, *48*, 365–370.

Baszyński, T., M. Król, Z. Krupa, M. Ruszkowska, U. Wojieska, and D. Wolińska (1982), *Z. Pflanzenphysiol.*, *108*, 385–395.

Baszyński, T., A. Tukendorf, M. Ruszkowska, E. Skórzyńska, and W. Maksymiec (1988), *J. Plant Physiol., 132*, 708–713.

Bazzaz, M. B. and Govindjee (1974), *Environ. Lett., 6*, 1–12.

Bazzaz, F. A., G. L. Rolfe, and R. W. Carlson (1974a), *Plant Physiol., 32*, 373–376.

Bazzaz, F. A., G. L. Rolfe, and R. W. Carlson (1974b), *J. Environ. Qual., 3*, 156–158.

Bazzaz, F. A., R. W. Carlson, and G. L. Rolfe (1975), *Physiol. Plant, 34*, 326–329.

Becerril, J. M., A. Munoz-Rueda, P. Aparicio-Tejo, and C. Gonzalez-Murua (1988), *Plant Physiol. Biochem., 26*, 357–363.

Becerril, J. M., C. González-Murua, A. Munoz-Rueda, and M. R. De Felipe (1989), *Plant Physiol. Biochem., 27*, 913–918.

Behnke, H. D., K. Esser, K. Subiltzki, M. Runge, and H. Ziegler (Eds.) (1991), *Progress in botany, structural botany, physiology, genetics, taxonomy and geobotany 52*, 395–398, Springer-Verlag, Berlin, Heidelberg, Germany.

Behra, R. (1993), *Arch. Environ. Contam. Toxicol., 24*, 21–27.

Berger, J. M., P. J. Jackson, N. J. Robinson, L. D. Lujan, and E. Delhaize (1989), *Plant Cell Rep., 7*, 632–635.

Bharti, N. and R. P. Singh (1994), *Phytochemistry, 35*, 1157–1161.

Bishnoi, N. R., I. S. Sheoran, and R. Singh (1993a), *Biol. Plant., 35*, 583–589.

Bishnoi, N. R., I. S. Sheoran, and R. Singh (1993b), *Photosynthetica, 28*, 473–479.

Bradshaw, A. D. (1952), *Nature, 169*, 1098.

Brams, E. and W. Anthony (1988), *Plant Soil, 109*, 3–8.

Breckle, S.-W. and H. Kahle (1992), *Vegetatio, 101*, 43–53.

Briggs, K. G., G. J. Taylor, I. Sturges, and J. Hoddinott (1989), *Can. J. Plant Sci., 69*, 61–69.

Brooks, R. R., J. Lee, R. D. Reeves, and T. Jaffre (1977), *J. Geochem. Expl., 7*, 49–57.

Brown, M. T. and D. A. Wilkins (1985), *New Phytol., 99*, 101–106.

Brune, A. and K. J. Dietz (1995), *J. Plant Nutrit., 18*, 853–868.

Brune, A., W. Urbach, and K. J. Dietz (1994), *Plant Cell Environ., 17*, 153–162.

Brune, A., W. Urbach, and K. J. Dietz (1995), *New Phytol., 129*, 403–409.

Campbell, P. G. C., A. G. Lewis, P. M. Chapman, A. A. Crowder, W. K. Fletcher, B. Imber, S. N. Luoma, P. M. Stokes, and M. Winfrey, (1988), *Biologically available metals in sediments*. National Research Council, Canada. No. 27694. Assoc. Comm. Sci. Criteria for Environ. Qual., p. 298.

Casterline, J. L., Jr., and N. M. Barnett (1982), *Plant Physiol., 69*, 1004–1007.

Cataldo, C. D., T. R. Garland, and R. E. Wildung (1981), *Plant Physiol., 68*, 835–839.

Cataldo, C. D., T. R. Garland, and R. E. Wildung (1983), *Plant Physiol., 73*, 844–848.

Chigbo, F. E., W. S. Ralph, and L. S. Fred (1982), *Environ. Pollut., Ser. A, 27*, 31–36.

Chongpraditnum, P., S. Morti, and M. Chino (1992), *Plant Cell Physiol., 33*, 239–244.

Clark, R. K. and S. C. Clark (1981), *New Phytol., 87*, 799–815.

Clijsters, H. and F. Van Assche (1985), *Photosynth. Res., 7*, 41–40.

Conzonno, V., C. Dirisio, M. Tudino, R. E. Ballsells, and A. F. Cirelli (1994), *Fres. Environ. Bull., 3*, 1–5.

Cooley, T. M. and D. F. Martin (1979), *Chemosphere, 2*, 75–78.

Cooley, T. M., D. F. Martin, W. C. Durden, and B. D. Perkins, Jr. (1979), *Water Res.*, *13*, 343–348.

Costa, G. and J. L. Morel (1993), *J. Plant Nutrit.*, *16*, 1921–1929.

Costa, G. and J. L. Morel (1994), *J. Plant Nutrit.*, *17*, 627–637.

Coughtrey, P. J. and M. H. Martin (1978), *New Phytol.*, *81*, 147–154.

Cox, R. M. and T. C. Hutchinson (1979), *Nature, 279*, 231–233.

Cox, R. M. and T. C. Hutchinson (1981), *J. Plant Nutrit.*, *3*, 731–741.

Cumming, J. R. and G. J. Taylor (1990), In: *Stress responses in plants: Adaptation and acclimation mechanisms,* (R. C. Alscher and J. R. Cumming, Eds.), Wiley-Liss, Inc., New York, pp. 329–356.

Cumming, J. R., A. B. Cumming, and G. J. Taylor (1992), *J. Exp. Bot.*, *43 (253)*, 1075–1081.

Cuttler, J. M. and D. W. Rain (1974), *Plant Physiol.*, *54*, 67–71.

Davis, R. D. and P. H. T. Beckett (1978), *Plant Soil, 68*, 835–839.

De Knecht, J. A., P. L. M. Koevoets, J. A. C. Verkleij, and W. H. O. Ernst (1992), *New Phytol.*, *122*, 681–688.

De Kok, L. J. and P. J. C. Kuiper (1986), *Physiol. Plant.*, *68*, 477–482.

Delhaize, E., N. J. Robinson, and P. J. Jackson (1989a), *Plant. Mol. Biol.*, *12*, 487–497.

Delhaize, E., P. J. Jackson, L. D. Lujan, and N. J. Robinson (1989b), *Plant Physiol.*, *89*, 700–706.

De Miranda, J. R., M. A. Thomas, D. A. Thurman, and A. B. Tomsett (1990), *FEBS Lett.*, *260*, 277–280.

De Vos, C. H. R. and H. Schat (1991), In *Ecological responses to environmental stress* (J. Rozema and J. A. C. Verkleij, Eds.), Kluwer Academic, Dordrecht, Netherlands, pp. 22–30.

De Vos, C. H. R., H. R. Schat, R. Vooijs, and W. H. O. Ernst (1989), *J. Plant Physiol.*, *135*, 164–169.

De Vos, C. H. E., M. J. Vonk, R. Vooijs, and H. Schat (1992), *Plant Physiol.*, *98*, 853–858.

De Vos, C. H. R., W. M. T. Bookum, R. Vooijs, H. Schat, and L. J. De Kok (1993), *Plant Physiol Biochem.*, *31*, 151–158.

Dhindsa, R. S., P. P. Dhindsa, and T. A. Thorpe (1982), *J. Exp. Bot.*, *32*, 93–101.

Dickinson, N. M., A. P. Turner, and N. W. Lepp (1991), *Water, Air Soil Pollut.*, *57 & 58*, 627–633.

Dua, A. and S. K. Sawhney (1991), *Environ. Exp. Bot.*, *31*, 133–139.

Eanetta, N. T. and J. C. Steffens (1989), *Plant Physiol.*, *89*, 76.

Ernst, W. H. O. (1990), *Plant Cell Physiol.*, *13*, 913–921.

Ernst, W. H. O., J. A. C. Verkleij, and H. Schat (1992), *Acta Bot. Neerl.*, *41(3)*, 229–249.

Evans I. M., L. N. Gatehouse, J. A. Gatehouse, N. J. Robinson, and R. R. D. Coy (1990), *FEBS Lett.*, *62*, 29–31.

Farago, M. E. (1981), *Co-ordinat. Chem. Rev.*, *36*, 155–182.

Farago, S. and C. Brunold (1994), *J. Plant Physiol.*, *144*, 433–437.

Fernandes, J. C. and F. S. Henriques (1991), *Bot. Rev.*, *57*, 246–273.

Ferretti, M., R. Ghisi, L. Merlo, F. D. Vecchia, and C. Passera (1993), *Photosynthetica, 29,* 49–54.

Fiedler, P. L. (1985), *Am. J. Bot., 72,* 1712–1718.

Florijn, P. J. and M. L. Van Beusichem (1993a), *Plant Soil, 150,* 25–32.

Florijn, P. J. and M. L. Van Beusichem (1993b), *Plant Soil, 153,* 79–84.

Florijn, P. J., J. A. De Knecht, and M. L. Van Beusichem (1993a), *J. Plant Physiol., 142,* 537–542.

Florijn, P. J., J. A. Nelemans, and M. L. Van Beusichem (1993b), *Plant Soil, 154,* 103–109.

Fowler, B. A., C. E. Hilderrand, Y. Kojima, and M. Webb (1987), *Experentia, 52,* 19–22.

Foy, C. D., R. L. Chaney and M. C. White (1978), *Annu. Rev. Plant Physiol., 29,* 435–444.

Fuhrer, J. (1982), *Plant Physiol., 70,* 162–167.

Fuhrer, J., G. T. Geballe, and C. Fries (1981), *Plant Physiol., 67,* 55.

Fujita, M. (1985), *Plant Cell Physiol., 26,* 295–300.

Fujita, M. and T. Kawanishi (1986), *Plant Cell Physiol., 27,* 1317–1325.

Fujita, M. and K. Nakano (1988), *Agric. Biochem., 52,* 2335–2336.

Garg, P. and P. Chandra (1994), *Environ. Monitor. Assess., 29,* 89–95.

Gekeler, W., W. E. Grill, E. L. Winnacker, and M. H. Zenk (1988), *Archiv. Microbiol., 150,* 197–202.

Gekeler, W., W. E. Grill, E. L. Winnacker, and M. H. Zenk (1989), *Z. Naturforsch., 44C,* 361–369.

Ghoshroy, S. and M. J. Nadakavukaren (1990), *Environ. Exp. Bot., 30,* 187–192.

Godbold, D. L. and C. Knetter (1991), *J. Plant Physiol., 138,* 231–253.

Godbold, D. L., E. Fritz, and A. Huttermann (1988), *Proc. Natl. Acad. Sci. USA, 85,* 3888–3892.

Greger, M. and E. Orgen (1991), *Physiol. Plant, 83,* 128–138.

Greger, M., M. Johansson, A. Stihl, and K. Hamza (1994), *Physiol. Planta, 88,* 563–570.

Grill, E. and M. H. Zenk (1985), *Naturwissenschaften, 72,* 432–433.

Grill, E., E. L. Winnacker, and M. H. Zenk (1985), *Science, 230,* 674–676.

Grill, E., W. Gekeler, E. L. Winnacker, and M. H. Zenk (1986a), *FEBS Lett., 205,* 47–50.

Grill, E., E. L. Winnacker, and M. H. Zenk (1986b), *FEBS Lett., 197,* 115–120.

Grill, E., J. Thumann, E. L. Winnacker, and M. H. Zenk (1988a), *Plant Cell Rep., 7,* 375–378.

Grill, E., E. L. Winnacker, and M. H. Zenk (1988b), *Experentia, 44,* 539–540.

Grill, E., S. Loffler, E. L. Winnacker, and M. H. Zenk (1989a), *Proc. Natl. Acad. Sci. USA, 86,* 6838–6842.

Grill, E., E. L. Winnacker, and M. H. Zenk (1989b), *Proc. Natl. Acad. Sci. USA, 84,* 439–443.

Groten, J. P. and P. J. Vanbladeren (1994), *Trends Food Sci. Technol., 5,* 50–55.

Grunhage, L., D. I. Weigel, and H. J. Jager (1985), *J. Plant Physiol., 119,* 327–334.

Hagemeyer, J. (1993), In *Plants as biomonitors—Indicators for heavy metals in the terrestrial environment* (B. Markert, Ed.), VCH Weinheim, New York, pp. 541–556.

Hagemeyer, J. and Y. Waisel (1988), *Physiol. Plant, 73,* 541–546.

Hagemeyer, J. and Y. Waisel (1989a), *Physiol. Plant, 75,* 280–284.

Hagemeyer, J. and Y. Waisel (1989b), *Physiol. Plant, 77,* 247–253.

Hagemeyer, J., A. Lülfsmann, M. Perk, and S.-W. Breckle (1992), *Vegetatio., 101,* 55–63.

Hagemeyer, J., D. Lohrmann, and S.-W. Breckle (1993), *Water, Air Soil Pollut., 69,* 351–361.

Hamer, D. H. (1986), *Annu. Rev. Biochem., 55,* 913–951.

Hampp, R., K. Beulich, and H. Zeigler (1976), *Z. Pflanzenphysiol., 77,* 336–344.

Hardiman, R. T. and B. Jacoby (1984), *Physiol. Plant, 61,* 670–674.

Hatch, D. J., L. H. P. Jones, and R. G. Burau (1988), *Plant Soil Bot., 105,* 121–126.

Hazell, P. and D. R. Murray (1982), *Z. Pflanzen., 108,* 87–92.

He, Q. B. and B. R. Singh (1993), *Acta Agric. Scandina. Sect B., Soil Plant Sci., 43,* 142–150.

He, Q. B. and B. R. Singh (1994), *Water, Air Soil Pollut., 74,* 251–265.

Hendry, G. A. F., A. J. M. Baker, and C. F. Ewart (1992), *Acta Bot. Neerl., 41,* 271–181.

Hirsch, D. and A. Banin (1990), *J. Environ. Qual., 19,* 366–372.

Hirt, H., G. Casari, and A. Barta, (1989), *Plant, 179,* 414–420.

Homer, J. R., R. Cotton, and E. H. Evans (1980), *Oecologia, 45,* 88–89.

Hooda, P. S. and B. J. Alloway (1994), *Water, Air Soil Pollut., 74,* 235–250.

Horst, W. J., A. Wagner, and H. Marschner (1982), *Z. Pflanzenphysiol., 105,* 435–444.

Howarth, C. J. (1990), *J. Exp. Bot., 41,* 877–883.

Howden, R. and C. S. Cobett (1992), *Plant Physiol., 99,* 100–107.

Howden, R., P. B. Goldsbrough, C. A. Anderson, and C. S. Cobett (1995a), *Plant Physiol., 107,* 1059–1066.

Howden, R., C. A. Anderson, P. B. Goldsbrough, and C. S. Cobett (1995b), *Plant Physiol., 107,* 1067–1073.

Hsu, B, and J. Lee (1988), *Plant Physiol., 87,* 116–119.

Huang, B. and P. B. Goldsbrough (1988), *Plant Cell Rep., 7,* 119–122.

Huang, C. Y., F. A. Bazzazz, and L. N. Vanderhoeff (1974), *Plant Physiol., 54,* 122–124.

Huang, B., E. Hatch, and P. B. Goldsbrough (1987), *Plant Sci., 52,* 211–221.

Humphreys, M. O. and M. K. Nicholls (1984), *New Phytol., 95,* 177–190.

Jackson, P. A. and B. J. Alloway (1991), *Plant Soil, 132,* 179–186.

Jackson, P. J., E. J. Roth, P. R. McClure, and C. M. Naranjo (1984), *Plant Physiol., 75,* 914–918.

Jackson, P. J., C. J. Unkefer, J. A. Doolen, K. Watt, and N. J. Robinson (1987), *Proc. Natl. Acad. Sci. USA, 84,* 6619–6623.

Jaffre, T., R. R. Brooks, J. Lee, and R. D. Reeves (1976), *Science, 193,* 579–580.

Jarvis, S. C., L. P. H. Jones, and H. J. Hoeppe (1976), *Plant Soil, 44,* 179–191.

Jeng, A. S. and B. R. Singh (1993), *Soil Sci., 156,* 240–250.

Jensen, P. and S. Aoalsteinsson (1989), *Physiol. Plant, 75,* 195–200.

Jones, R., K. A. Prohaska, and M. S. E. Burgess (1988), *Water, Air Soil Pollut., 37,* 355–363.

Jowett, D. (1958), *Nature, 182,* 816–817.

Kagi, J. H. R. and A. Schaffer (1988), *Biochem.*, *27*, 8509–8515.

Kahle, H. (1993), *Environ. Exp. Bot.*, *33*, 99–119.

Kahn, D. H., J. G. Duckett, B. Frankland, and J. B. Kirkham (1984), *Z. Pflanzen Physiol.*, *115*, 19–28.

Kalyanaraman, S. B. and P. Sivagurunathan (1993), *J. Plant Nutrit.*, *16*, 2029–2042.

Kanerva, T., O. Sarin, and P. Nuorteva (1988), *Ann. Bot. Fenn.*, *25*, 275–279.

Kaneta, M., H. Hikichi, S. Endo, and N. Sugiyama (1983), *Agric. Biol. Chem.*, *47*, 417–418.

Kaneta, M., H. Hikichi, S. Endo, and N. Sugiyama (1987), *Environ. Health Persp.*, *65*, 33–37.

Karataglis, S. S. (1982), *Oikos, 38*, 234–241.

Karez, C. S., D. Allem, D. Renzis, M. Gnassiabarelli, M. Romeo, and S. Puiseuxdao (1990), *Plant Cell Environ.*, *13*, 483–487.

Karin, M. (1985), *Cell, 41*, 9–10.

Kawashima, I., Y. Inokuchi, M. Chino, M. Kimura, and N. Shimizu (1991), *Plant Cell Physiol.*, *32*, 913–916.

Kay, S. H. and W. T. Haller (1986), *Bull. Environ. Contam. Toxicol.*, *37*, 239–245.

Kay, S. H., W. T. Haller, and L. A. Garrard (1984), *Aquat. Toxicol.*, *5*, 117–128.

Khalid, R. A., R. P. Gambrell, and W. H. Patrick, Jr. (1981), *J. Environ. Qual.*, *10*, 523–528.

Kirkham, H. B. (1979), *J. Environ. Qual.*, *6*, 201–205.

Klapheck, S., W. Fliegner, and I. Zimmer (1994), *Plant Physiol.*, *104*, 1325–1332.

Kneer, R. and M. H. Zenk (1992), *Phytochem.*, *31*, 2663–2667.

Kneer, R., T. M. Kutchan, A. Hochberger, and M. H. Zenk (1992), *Archiv. Microbiol.*, *157*, 305–310.

Koltz, F. and W. J. Horst (1988), *J. Plant Physiol.*, *132*, 701–707.

Kortz, R. M., B. P. Evangelou, and G. J. Wagner (1989), *Plant Physiol.*, *91*, 780–787.

Krosshavn, M., E. Steinnes, and P. Varskog (1993), *Water, Air Soil Pollut.*, *71*, 185–193.

Krupa, Z. (1988), *Physiol. Plant, 73*, 518–524.

Krupa, Z., E. Skórzyńska, W. Maksymiec, and T. Baszyński (1987), *Photosynthetica, 21*, 156–164.

Krupa, Z., G. Oqüist, and N. P. A. Huner (1992), *Acta Physiol Plant*, 14, 71–76.

Krupa, Z., G. Oqüist, and N. P. A. Huner (1993), *Physiol Plant, 88*, 626–630.

Kuboi, T., A. Noguchi, and J. Yazaki (1987), *Plant Soil, 104*, 275–280.

Kubota, K., H. Nishizone, S. Suzuki, and F. Ishii (1988), *Plant Cell Physiol.*, *29*, 1029–1033.

Kuo, T. and Y. Huang (1993), *J. Agric. Assoc., China, 161*, 27–32.

Kuppelwieser, H. and U. Feller (1991), *Plant Soil, 132*, 281–299.

Lameroux, R. J. and W. R. Chaney (1977), *J. Environ. Qual.*, *6*, 201–205.

Lameroux, R. J. and W. R. Chaney (1978), *Physiol. Plant, 43*, 231–236.

Lamy, I., S. Bourgeois, and A. Bermond (1993), *J. Environ. Qual.*, *22*, 731–737.

Landberg, T. and M. Greger (1994), *Physiol. Plant, 90*, 637–644.

Lazerte, B. (1986), *Water, Air Soil Pollut.*, *31*, 569–576.

Leblova, S., A. Mucha, and E. Spirhanzova (1986), *Biologia, 41*, 777–785.

Lee, K. C., B. A. Cunnigham, G. M. Paulson, G. H. Liang, and R. B. Moore (1976), *Physiol. Plant, 35*, 4–6.

Lefevbre, D. D., B. L. Miki, and J. F. Laliberte (1987), *Biotechnol., 5*, 1053–1056.

Leidman, P., K. Fischer, D. Bieniek, F. Nusslein, and A. Kettrup (1994), *Chemosphere, 28*, 383–390.

Leita, L., M. Contin, and A. Maggioni (1991), *Plant Sci., 77*, 139–147.

Leita, L., M. Denobili, C. Mondini, and M. T. B. Garcia (1993), *J. Plant Nutrit., 16*, 2001–2012.

Li, E. H. and C. D. Miles (1975), *Plant Sci. Lett., 5*, 33–40.

Lidon, F. C. and F. S. Henriques (1991), *J. Plant Physiol., 138*, 115–118.

Lodenius, M. (1990), *Environmental mobilization of mercury and cadmium*. Dept. of Environ. Conser., Univ. Helsinki, Finland, 32 pp.

Löffler, S., A. Hochberger, E. Grill, E. L. Winnacker, and M. H. Zenk (1989), *FEBS Lett., 258*, 42–46.

Lolkema, P. C., M. H. Donker, A. J. Schouten, and W. H. O. Ernst (1984), *Planta, 162*, 174–179.

Lorimer, G. H. (1981), *Annu. Rev. Plant Physiol., 32*, 349–388.

Lucero, H. A., C. S. Andreo, and R. M. Vallejo (1976), *Plant Sci. Lett., 6*, 309–313.

Luna, C. M., C. A. Gonzalez, and V. S. Trippi (1994), *Plant Cell Physiol., 35*, 11–15.

Macfie, S. M. and G. J. Taylor (1992), *Physiol. Plant, 85*, 467–475.

Macfie, S. M., E. A. Cossins, and G. J. Taylor (1994), *J. Plant Physiol., 143*, 135–144.

Macnair, M. R. (1977), *Nature, 268*, 428–430.

Mahaffey, K. R., P. E. Corneliussen, C. F. Jelinek, and J. A. Fiorino (1975), *Environ. Health Persp., 12*, 63–69.

Maiti, I. B., G. J. Wagner, R. Yeargan, and A. G. Hunt (1989), *Physiol. Plant, 91*, 1020–2024.

Malone, C., D. E. Koeppe, and R. J. Miller (1974), *Plant Physiol., 53*, 388–394.

Mathys, W. (1975), *Physiol. Plant, 33*, 161–165.

Meharg, A. A. (1993), *Physiol. Plant, 88*, 191–198.

Meharg, A. A. (1994), *Plant Cell Environ., 17*, 989–993.

Mejuto-Marti, M., M. Bollain-Rodriguez, C. Herrero-Latorre, and F. Bermejo-Martinez (1988), *J. Agric. Food Chem., 36*, 293–295.

Mench, M. and E. Martin (1991), *Plant Soil, 132*, 187–196.

Mench, M., J. J. Morel, and A. Guckert (1987), *Biol. Fertil. Soils, 3*, 165–169.

Mendum, M. L., S. C. Gupta, and P. B. Goldsbrough (1990), *Plant Physiol., 93*, 484–488.

Merrington, G. and B. J. Alloway (1994), *Water, Air Soil Pollut., 73*, 333–344.

Meuwly, P. and W. E. Rauser (1992), *Plant Physiol., 99*, 8–15.

Meuwly, P., P. Thibault, and W. E. Rauser (1993), *FEBS Lett., 336*, 472–476.

Meyer, A. L. F. and M. C. Heath (1988), *Can. J. Bot., 66*, 613–623.

Miles, C. D., J. R. Brandle, D. J. Daniel, O. Chu-Der, P. D. Schnore, and D. J. Uhlik (1972), *Plant Physiol., 49*, 820–825.

Mishra, S. and L. Gedamu (1989), *Theor. Appl. Genet., 78*, 161–168.

Misra, A. K., V. Sarkunan, S. K. Miradas, S. K. Nayak, and P. K. Nayar, (1989), *Curr. Sci., 58,* 1398–1400.

Morel, J. L., M. Mench, and A. Guckert (1986), *Biol. Fertile Soils, 2,* 29–34.

Moroni, J. S., K. G. Briggs, and G. J. Taylor (1991a), *Euphytica., 56,* 107–120.

Moroni, J. S., K. G. Briggs, and G. J. Taylor (1991b), *Plant Soil, 136,* 1–9.

Moya, J. L., R. Ros, and I. Picazo (1993), *Photosynth. Res., 6,* 75–80.

Munch, D. (1993), *Sci. Total. Environ., 138,* 47–55.

Muramoto, S. and Y. Oki (1983), *Bull. Environ. Contam. Toxicol., 30,* 170–177.

Nelson, W. O. and P. G. C. Campbell (1991), *Environ. Pollut., 71,* 91–130.

Neumann, D., O. Lichtenberg, D. Gunther, K. Tschiersch, and L. Nover (1994), *Planta, 194,* 360–367.

Nuorteva, P. (1990), *Metal distribution patterns and forest decline seeking Achille's heels for metals in Finnish forest biocoenoses,* Dept. Environ. Conser., Helsinki Univ., Finland, Pub. No. 11, pp. 1–77.

Nussbaum, S., D. Schmutz, and C. Brunold (1988), *Plant Physiol., 88,* 1407–1410.

Obata, H. and M. Umebayashi (1993), *Plant Soil, 155 & 156,* 533–536.

Oliver, D. P., J. E. Schultz, K. G. Tiller, and R. H. Merry (1993), *Aust. J. Agric. Res., 44,* 1221–1234.

Ornes, W. H. and K. S. Sajwan (1993), *Water, Air Soil Pollut., 69,* 291–300.

Orzech, K. A. and J. J. Burke (1988), *Plant Cell Environ., 11,* 711–714.

Ouzounidou, G., E. P. Eleftheriou, and S. Karataglis (1992), *Can. J. Bot., 70,* 947–957.

Ownby, J. D. and H. P. Popham (1989), *J. Plant Physiol., 135,* 588–591.

Padmaja, K., D. D. K. Prasad, and A. R. K. Prasad (1990), *Photosynthetica, 24,* 399–405.

Palma, J. M., J. Yanez, M. Gomez, and L. A. del Rio, (1990), *Planta, 81,* 487–495.

Pearson, C. H. and K. B. Kirkham (1981), *J. Plant Nutrit., 3,* 309–318.

Petit, C. M. and S. C. Van de Geijn (1978), *Planta, 138,* 137–143.

Pieczonka, K. and A. Rosopulo (1985), *Fresenius Analy. Chem., 322,* 697–699.

Piotrowska, M. and S. Dudka (1994), *Water, Air Soil, 73,* 179–88.

Poschenrieder, C., B. Gunse, and J. Barcelo (1989), *Plant Physiol., 90,* 1365–1371.

Prasad, M. N. V. (1995a), *Environ. Exp. Bot., 35,* 525–545.

Prasad, M. N. V. (1995b), *Photosynthetica, 31,* 635–640.

Rao, S. and S. Mathur (1994), *J. Irrig. Drain. Eng., 120,* 89–96.

Rascio, N., F. Dallavecchia, M. Ferretti, L. Merlo, and R. Ghisi (1993), *Arch. Environ. Contam. Toxicol., 25,* 244–249.

Rauser, W. E. (1984a), *Plant Physiol., 74,* 1025–1029.

Rauser, W. E. (1984b), *J. Plant Physiol., 116,* 253–260.

Rauser, W. E. (1986), *Plant Sci., 43,* 85–91.

Rauser, W. E. (1990a), *Plant Sci., 70,* 155–166.

Rauser, W. E. (1990b), *Annu. Rev. Biochem., 59,* 61–86.

Rauser, W. E. (1995), *Plant Physiol., 109,* 1141–1149.

Rauser, W. E. and C. A. Ackerley (1987), *Can. J. Bot., 65,* 643–646.

Rauser, W. E. and N. R. Curvetto (1980), *Nature, 287,* 563–564.

Rauser, W. E. and J. Glover (1984), *Can. J. Bot., 62,* 1645–1650.

Rauser, W. E., H. Hartmann, and U. Weser (1983), *FEBS Lett., 164,* 102–104.

Rauser, W. E., P. E. Hunziker, and J. H. R. Kagi (1986), *Plant Sci., 45,* 105–109.

Rauser, W. E., R. Schupp, and H. Rennenberg (1991), *Plant Physiol., 97,* 112–122.

Reddy, G. N. and M. N. V. Prasad (1990), *Environ. Exp. Bot., 30,* 252–264.

Reddy, G. N. and M. N. V. Prasad (1992), *Biochem. Archiv., 8,* 101–106.

Reddy, G. N. and M. N. V. Prasad (1993), *Biochem. Archiv., 9,* 25–32.

Reddy, G. N. and M. N. V. Prasad (1995), *J. Plant Physiol., 145,* 67–70.

Reese, R. N. and G. J. Wagner (1987a), *Biochem. J., 241,* 641–647.

Reese, R. N. and G. J. Wagner (1987b), *Plant Physiol., 84,* 574–577.

Reese, R. N. and D. R. Winge (1990), *Plant Physiol., 89,* 723.

Reese, R. N., C. A. White, and D. R. Winge (1992), *Plant Physiol., 89,* 723.

Reese, R. N., R. K. Mehera, E. B. Tarbet, and D. R. Winge (1988), *J. Biol. Chem.,* 263, 4186–4192.

Repka, V. (1993), *Biol. Plant., 35,* 617–627.

Rivai, I. F., H. Koyama, and S. Suzuki (1990a), *Bull. Environ. Contam. Toxicol., 44,* 910–916.

Rivai, I. F., H. Koyama, and S. Suzuki (1990b), *Jap. J. Health Human Ecol., 56,* 168–177.

Roberts, A. H. C., R. D. Longhurst, and M. W. Brown (1994), *New Zealand J. Agric. Res., 37,* 119–129.

Robinson, N. J. and P. J. Jackson (1986), *Physiol. Plant, 67,* 499–506.

Robinson, N. J. and D. A. Thurman (1986), *Planta, 169,* 192–197.

Robinson, N. J., K. Barton, C. M. Naranjo, L. O. Sillerud, J. Trewhella, K. Watt, and P. J. Jackson (1987), *Experentia, 52,* 323–327.

Robinson, N. J., R. L. Ratliff, P. J. Anderson, E. Dehlaize, J. M. Berger, and P. J. Jackson (1988), *Plant Sci., 56,* 197–204.

Robinson, N. J., M. Tommeya, C. Kuske, and P. J. Jackson (1993), *Biochem. J., 295,* 1–10.

Ros, R., D. T. Cooke, R. S. Burden, and C. S. James (1990), *J. Exp. Bot., 41 (225),* 457–462.

Ross, S. M. (Ed.) (1994), *Toxic metals in soil plant systems.* Wiley, Chichester, UK, 469 pp.

Roy, A. K., A. Sharma, and G. Talukader (1988), *Bot. Rev., 54,* 145–178.

Rubio, M. I., I. Escrig, C. Martinezcortinam, F. J. Lopezbenetm, and A. Sanz (1994), *Plant Growth Reg., 14,* 151–157.

Ruegsegger, A., D. Schmutz, and C. Brunold (1990), *Plant Physiol., 93,* 1579–1584.

Salim, R., M. M. Alsubu, and A. Atallah, (1993), *Environ. Internat., 19,* 393–404.

Salt, D. E. and G. J. Wagner (1993), *J. Biol. Chem., 268,* 12297–12302.

Salt, D. E., D. A. Thurman, A. B. Tomsett, and A. K. Sewell (1989), *Proc. Roy. Soc. London Ser. B., 236,* 79–89.

Samarakoon, A. B. and W. E. Rauser (1979), *Plant Physiol., 63,* 1165–1169.

Sanchez-Camazano, M., M. J. Sanchezmartn, and L. F. Lorenzo, (1994), *Sci. Total Environ., 147,* 163–168.

Sarkela, M. and P. Nuorteva, (1987), *Ann. Bot. Fennici.*, *24*, 301–305.

Schat, H. and W. M. T. Bookum (1992), *Heredity, 68*, 219–229.

Schat, H. and M. M. A. Kalff (1992), *Plant Physiol.*, *99*, 1475–1480.

Schat, H., E. Kuiper, W. M. T. Bookum, and R. Vooijs (1993), *Heredity, 70*, 142–147.

Scheller, H. V., B. Huang, E. Hatch, and P. B. Goldsbrough (1987), *Plant Physiol.*, *85*, 1031–1035.

Schultz, C. L. and T. C. Hutchinson (1988), *New Phytol.*, *110*, 163–171.

Shane, B. S., C. B. Littman, L. A. Essick, W. A. Gutenmann, G. J. Dess, and J. Lisk (1988), *J. Agric. Food Chem.*, *36*, 328–333.

Shoeran, I. S., N. Aggarwal, and R. Singh (1990a), *Plant Sci.*, *129*, 243–249.

Shoeran, I. S., H. R. Singal, and R. Singh (1990b), *Photosynth. Res.*, *23*, 345–351.

Singh, V. P. (1993), *J. Plant Growth Reg.*, *12*, 1–3.

Speiser, J. L., S. L. Abrahamson, G. Banuelos, and D. W. Ow (1992), *Plant Physiol.*, *99*, 817–821.

St-Cyr, L. and P. G. C. Campbell (1994a), *Can. J. Plant Sci.*, *72*, 429–439.

St-Cyr, L., P. G. C. Campbell, and K. Guertin (1994b), *Hydrobiologia.*, *291*, 141–156.

Steffens, J. C. (1990), *Annu. Rev. Plant Physiol. Plant Mol. Biol.*, *41*, 553–575.

Steffens, J. C., D. F. Hunt, and B. G. Williams (1986), *J. Biol. Chem.*, *261*, 13879–13882.

Stiborova, M. and S. Leblova (1985), *Photosynthetica, 19*, 500–503.

Stiborova, M., M. Doubravova, and S. Leblova (1986a), *Biochem. Physiol. Pflanzen.*, *181*, 373–379.

Stiborova, M., M. Doubravova, A. Brezninova, and A. Friedrich (1986b), *Photosynthetica, 20*, 418–425.

Stiborova, M., M. Doubravova, and A. Brezninova (1987), *Biol. Plant.*, *29*, 453–467.

Stobart, A. K., W. T. Griffiths, I. Ameen-Bukhari, and R. P. Sherwood (1985), *Physiol. Plant, 63*, 293–298.

Strasdeit, H., A. K. Duhme, R. Kneer, M. H. Zenk, C. Hermes, and H. F. Nolting (1991), *J. Chem. Soc. Comm.*, *16*, 1129–1130.

Sutton, D. L. and R. D. Blackburn (1971), *Hyacinth Contr. J.*, *9*, 18–20.

Symeonidis, L. and S. Karataglis (1992), *BioMetals, 5*, 173–178.

Symenoides, L., T. McNeilly, and A. D. Bradshaw (1985), *New Phytol.*, *101*, 309–315.

Tamaddon, F. and W. Hogland (1993), *Waste Manag. Res.*, *11*, 287–295.

Tatsuyama, K., H. Egawa, H. Yamamoto, and M. Nakamura (1979), *Weed Res., Jap.*, *24*, 260–263.

Taubenberger, A. M., J. Hahman, A. Noegel, and G. Gerisch (1988), *J. Cell. Sci.*, *90*, 51–58.

Taylor, G. J. (1985), *Can. J. Bot.*, *63*, 2181–2186.

Taylor, G. J. (1988a), *Commun. Soil Sic. Plant Anal.*, *19*, 1217–1227.

Taylor, G. J. (1988b), *Can. J. Bot.*, *66*, 694–699.

Taylor, G. J. (1988c), The physiology of aluminum phytotoxicity, In Aluminum and its role in biology. *Metal ions in biological systems* (H. Sigel and A. Sigel, Eds.), *Vol.* 24, pp. 123–163. Marcel Dekker, Inc. New York.

Taylor, G. J. (1988d), The physiology of aluminum tolerance, In Aluminum and its role in biology, *Metal ions in biological systems* (H. Sigel and A. Sigel, Eds), *Vol.* 24, pp. 165–198. Marcel Dekker, Inc. New York.

Taylor, G. J. (1989a), *Plant Physiol. Biochem., 27,* 605–611.

Taylor, G. J. (1989b), Aluminum toxicity and tolerance in plants, In *Acid precipitation, Vol. 2, Biological and ecological effects* (D. C. Adriano and A. H. Johnson, Eds.), pp. 329–361. Springer-Verlag Beriln Heidelberg

Taylor, G. J. (1991), *Current topics in plant biochemistry and physiology, Vol. 10,* pp. 57–93. Univ. Missouri-Columbia, USA.

Taylor, G. J. and A. A. Crowder (1983), *Can. J. Bot., 61,* 63–73.

Taylor, G. J. and A. A. Crowder (1984), *Can. J. Bot., 62,* 1304–1308.

Taylor, G. J. and C. D. Foy (1985a), *Can. J. Bot., 63,* 1271–75.

Taylor, G. J. and C. D. Foy (1985b), *Am. J. Bot., 72,* 695–701.

Taylor, G. J. and C. D. Foy (1985c), *Am. J. Bot., 72,* 702–706.

Taylor, G. J. and C. D. Foy (1985d), *J. Plant Nutrit., 8,* 811–824.

Taylor, G. J. and K. J. Stadt (1990), Interactive effects of cadmium, copper, manganese, nickel and zinc on root growth of wheat *(Triticum aestivum)* in solution culture, In *Plant nutrition—Physiology and applications* (M. L. van Beusichem, Ed.), Kluwer Academic Publishers, Dordrecht, Netherlands, 317–322.

Taylor, G. J., K. J. Stadt, and M. R. Dale (1991), *Can. J. Bot., 69,* 359–367.

Taylor, G. J, K. J. Stadt, and M. R. Dale (1992), *Environ. Exp. Bot., 32,* 281–293.

Theisen, M. O. and C. Blincoe (1988), *Biol. Trace Element Res., 16,* 239–251.

Thumann, J., E. Grill, E. L. Winnacker, and M. H. Zenk (1991), *FEBS Lett., 284,* 64–69.

Tommey, A. M., J. Shi, W. P. Lindsay, P. E. Urwin, and N. J. Robinson (1991), *FEBS Lett., 292,* 48–52.

Tomsett, A. B. and D. A. Thurman (1988), *Plant Cell Environ., 11,* 388–394.

Tripathy, B. C. and P. Mohanthy (1980), *Plant Physiol., 66,* 1174.

Tukendorf, A. (1989), *J. Plant Physiol., 135,* 280–284.

Tukendorf, A. (1993), *Photosynthetica, 28,* 573–575.

Tukendorf, A. and T. Baszyński (1985), *J. Plant Physiol., 120,* 57–63.

Tukendorf, A. and W. E. Rauser (1990), *Plant Sci., 70,* 155–166.

Tukendorf, A., S. Lyszcz and B. Baszyński (1984), *J. Plant Physiol., 115,* 351–360.

Turner, R. G. (1970), *New Phytol., 69,* 725–731.

Van Assche, F. and H. Clijsters (1986a), *J. Plant Physiol., 125,* 355–360.

Van Assche, F. and H. Clijsters (1986b), *Physiol. Plant., 66,* 717–721.

Van Assche, F. and H. Clijsters (1987), *Med. Fac. Landbouww. Rijkksuniv. Gent., 52,* 1819–1824.

Van Assche, F. and H. Clijsters (1990), *Plant Cell Environ., 13,* 195–206.

Van Assche, F., C. Cardinales, and H. Clijsters (1988), *Environ. Pollut., 52,* 103–115.

Van Duijvendijk-Matteoli, M. A. and G. M. Desmet (1975), *Biochem. Biophys. Acta., 408,* 164–169.

Vázquez, G., G. Antorrena, J. Gonzlez, and M. D. Doval (1994), *Bioresource Technol., 48,* 251–255.

Veltrup, W. (1982), *Z. Pflanzenphysiol.*, *108*, 457–469.

Verkleij, J. A. C. and J. E. Prast (1989), *New Phytol.*, *111*, 637–645.

Verkleij, J. A. C., P. Koevoets, J. V. Riet, R. Bank, Y. Nijdam, and W. H. O. Ernst (1990), *Plant Cell Environ.*, *13*, 913–921.

Vögeli-Lange, R. and G. J. Wagner (1990), *Plant Physiol.*, *92*, 1086–1093.

Wagner G. J. (1993), *Adv. Agron.*, *51*, 173–212.

Wagner, G. J. (1984), *Plant Physiol.*, *76*, 797–805.

Wagner, G. J. and M. M. Trotter (1982), *Plant Physiol.*, *69*, 804–809.

Walenhorst, A., J. Hagemeyer, and S.-W. Breckle (1993), Passive monitoring of airborne pollutants, particularly trace metals, with tree bark. In *Plants as biomonitors—Indicators for heavy metals in the terrestrial environment* (B. Markert, Ed.), VCH Weinheim, New York, pp. 523–540.

Weigel, H. J. (1984), *J. Plant Physiol.*, *119*, 179–189.

Weigel, H. J. (1985), *Physiol. Plant*, *63*, 192–200.

Winter, J., D. R. Wright, N. Dick, C. Gasser, R. Farley, and D. Shah (1988), *Mol. Gen. Genet.*, *211*, 315–319.

Wong, D. and Govindjee (1976), *Photosynthetica*, *10*, 241–254.

Youssef, R. A. and M. Chino (1991), *Water, Air Soil Pollut.*, *57 & 58*, 249–258.

Yurukova, L. and K. Kochev (1994), *Bull. Environ. Contam. Toxicol.*, *52*, 627–632.

Zhang, G. and G. J. Taylor (1989), *Plant Physiol.*, *91*, 1094–1099.

Zhang, G. and G. J. Taylor (1990), *Plant Physiol.*, *94*, 577–584.

Zhang, G. and G. J. Taylor (1991), *J. Plant Physiol.*, *138*, 533–539.

Part II

Anthropogenic, Biotic Factors

9

Allelochemicals

S. Rama Devi, F. Pellissier, and M. N. V. Prasad

INTRODUCTION

In the process of evolution, higher plants have acquired a number of biosynthetic pathways through which a variety of secondary metabolites are synthesized and accumulated. A wide array of these compounds are released into the environment in appreciable quantities via root exudation and as leachates during litter decomposition and are known to play a major role in allelopathy as allelochemicals.

The term "allelopathy," coined by Molisch (1937), generally refers to the direct or indirect detrimental effect of one plant (including microorganisms) on the germination, growth, or development of other plants through the production of chemicals that escape into the environment. It differs from competition wherein plants compete for a common resource (Fig. 9-1). Allelopathic interactions between plants have been implicated in the patterning of vegetation and weed growth in agricultural systems (Aldrich, 1987; Rice, 1987) and in inhibition of growth of several crops (Liu and Lovette, 1993).

In recent years, allelopathy in agriculture has received considerable attention for two reasons:

1. Allelochemicals decrease the agricultural and silvicultural yields (Chou, 1982; Chou et al., 1981; Moitra et al., 1994).
2. Allelochemicals can be beneficial as natural pesticides.

Thus the science of allelopathy is gaining momentum.

The indiscriminate application of synthetic pesticides has resulted in increasing resistance in organisms, causing severe environmental pollution and health hazards. Use of allelochemicals from plants for this purpose would be environmentally

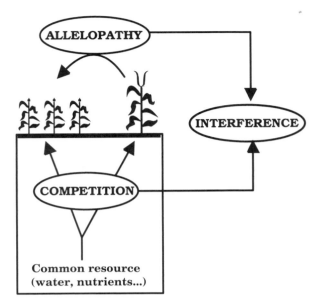

Figure 9-1. *Two major types of interferences among plants that can influence crop yields: Allelopathy (mediated via the action of allelochemicals, and important in sustainable agriculture) and competition among plants for a common resource).*

friendly, due to the fact that natural chemicals are renewable and easily degradable. For example, legume leaf mulches of *Flemingia macrophylla, Gliricidia sepium,* and *Leucaena leucocephala* are being used to control weeds (i.e., as herbicides) (Budelman, 1988a). *Inula* exhibited antifungal activity on *Helminthosporium sativum* and *Fusarium oxysporum* (Qasem et al., 1995). *Amaranthus retroflexus, Chenopodium murale,* and *Lepidium draba* had allelopathic effects on vegetable crops and wheat (Qusem, 1994, 1995). The experiments of Mondoza and Ilag (1980) suggest the possibility of using mimosine (prominent constituent in *Leucaena leucocephala*) as a biocide against fungal pathogens. This approach plays a key role in sustainable management of renewable natural resources, particularly in the agricultural, agroindustrial, and silvicultural production sectors.

ABBREVIATIONS USED IN TABLES

BEN   Benzoic acid
CAF   Caffeic acid
CIN   Cinnamic acid
CHL   Chlorogenic acid
COU   Coumaric acid
DHBA   Dihydroxy benzoic acid
EAL   Ellagic acid

FER	Ferulic acid
GAL	Gallic acid
GEN	Gentisic acid
HBA	Hydroxy benzoic acid
HQU	Hydro quinone
MYR	Myrestic acid
PAL	Palmitic acid
PCAT	Protocatechuic acid
***p*-COU**	*p*-Coumaric acid
***p*-HBA**	*p*-Hydroxy benzoic acid
PHC	*p*-Hydroxy cinnamic acid
PHL	Phlorizidine
***p*-HPA**	*p*-Hydroxy phenylacetic acid
PZ	Phlorizin
SAL	Salicylic acid
SYR	Syringic acid
VAN	Vanillic acid

CHEMISTRY OF ALLELOCHEMICALS

Chemical Nature

Plants synthesize a variety of compounds and several of them have been implicated in allelopathy. Most of these are secondary metabolites and are identified from soil (Whitehead, 1964) and plants (Glass, 1974; Putnam and Tang, 1986). These chemicals range from simple gases, aliphatic compounds, to complex multiringed aromatic acids including acetic and butyric acids, long chain fatty acids, quinones, simple phenols, phenolic acids derived from cinnamic and benzoic acids, coumarins, flavonoids, several hydrolyzable and condensed tannins, terpenoids, alkaloids, and various nitrogenous compounds (Harborne, 1980; Mandava, 1985; Rice, 1987). The allelochemicals isolated from various plants have been listed (Table 9-1). Although a wide range of chemical compounds possessing allelochemic action have been isolated, specific chemicals involved in allelopathy remain obscure. Generally allelochemicals are grouped under the following categories.

Aliphatic Compounds   Though these compounds do not comprise a large portion of the allelochemicals identified, a few of them, such as methanol, butanol, oxalic, formic, butyric, and lactic acids, are known to inhibit the germination of seeds and growth of plants (Rice, 1984).

Unsaturated Lactones   Most of the well-known antibiotics, such as patulin and penicillic acid, are simple lactones and are inhibitory to several microorganisms

TABLE 9-1 Phytotoxic Phenolics in Aqueous Leaf and Soil Extracts

Plant	Source of Phenolic	Phenolics[a]												
		CAF	FER	GAL	p-HBA	HQU	COU	p-COU	PCAT	PHC	p-HPA	PZ	SYR	VAN
Acacia auriculiformis[b]	Leaf	?	+	+	–	?	?	+	+	?	+	?	?	?
Adenostoma	Leaf	?	+	–	+	+	–	+	–	?	?	+	+	+
	Soil	?	+	–	+	–	–	+	+	?	?	–	+	+
Arctostaphylos	Leaf	?	+	+	+	+	–	–	+	?	?	–	–	+
	Soil	?	+	–	+	–	+	+	–	?	–	?	+	+
Cryptomeria japonica	Leaf	?	+	?	–	?	?	+	?	?	–	?	+	+
	Soil	?	+	?	–	?	?	+	?	?	?	?	+	+
Cunninghamia lanceolata	Leaf	?	+	+	+	?	?	–	+	?	?	?	+	+
	Soil	?	+	+	+	?	?	–	+	?	?	?	+	+
Eucalyptus camadulensis[b]	Leaf	?	?	+	+	?	?	+	+	?	+	?	?	?
Leucaena leucocephala (cv K-8)	Leaf	–	–	+	+	?	?	+	+	?	–	?	?	?
	Soil	–	–	+	–	?	?	?	+	?	–	?	?	+
Phyllostachys edulis	Leaf	?	+	?	+	?	?	+	?	+	+	?	+	+
	Soil	?	+	?	+	?	?	+	?	?	+	?	+	+

[a] + = present, – = absent, ? = below detectable level.
[b] Unpublished data.

Source: Adapted from Devi and Prasad (1990). See also the references therein.

and higher plants (Mandava, 1985; Rice, 1984). However, their role in allelopathy is not well understood.

Fatty Acids and Lipids Several fatty acids from both terrestrial and aquatic plants are known to exhibit toxicity to plants growing in their vicinity. A well-known unsaturated fatty acid chlorellin has been isolated from an alga *Chlorella* (Rice, 1984). Several fatty acids, namely, nonanoic, decanoic, lauric, myristic, palmitic, steric, and linoleic acids, from *Chlamydomonas* are inhibitory to the growth of another alga *Haematococcus* (Rice, 1984). Similar compounds isolated from *Polygonum* inhibited the growth of *Azotobacter* and *Rhizobium* (Ibrahim et al., 1983).

Terpenoids These comprise a large portion of allelochemicals that have been isolated from plants, particularly those growing in arid and semiarid regions. These include α-pinene, camphor, cineole, and dipentene, are produced by *Salvia* spp. (Lovett, 1986), *Amaranthus* (Bradow and Connick, 1988a), *Eucalyptus* (del Moral and Muller, 1970), *Artemisia* (Kil et al., 1992), and *Pinus* (Nehlin et al., 1994), and are well known for their allelopathic potential to crops and weeds. Several sesquiterpenes with allelochemic activity have been isolated from *Melilotus* (Macias et al., 1994).

Cyanogenic Glucosides Various cyanogenic glucosides, such as dhurrin and amygdalin, are widely recognized for their allelopathic potential to various crops (Mandava, 1985; Rice, 1984). Most of the Brassicaceae members are known to produce a large amount of these glycosides, which on hydrolysis produce isothiocyanates with equal biological potential (Choesin and Boerner, 1991).

Aromatic Compounds These comprise the largest portion of allelochemicals, and include phenols, phenolic acids, cinnamic acid derivatives, quinones, coumarins, flavonoids, and tannins.

Simple Phenols Among the simple phenols, hydroquinones and arbutin have been isolated from the leachates of *Arctostaphylos* (Rice, 1984) and are known to inhibit the growth of several plants.

Benzoic Acid and Derivatives Benzoic acid and its derivatives, such as hydroxy benzoic acid and vanillic acid, are the most common acids involved in allelopathy. The compounds have been isolated from cucumber (Yu and Matsui, 1994), *Avena* (Perez and Ormeno-Nunez, 1991a), *Rorippa sylvestris* (Yamane et al., 1992), and *Sorghum* (Nicollier et al., 1985). These compounds are also identified from soils (Mandava, 1985).

Cinnamic Acid and Derivatives Most of these compounds are derived from the shikimic acid pathway and are widespread in plants. They have been identified from *Parthenium* (Kohli et al., 1985), cucumber (Yu and Matsui, 1994), sunflower (Macias et al., 1993), and guayule (Scholman et al., 1991). Other cinnamic acid derivatives, such as chlorogenic, caffeic, p-coumaric, and ferulic acids, are widely distributed among plant kingdom (Bushra et al., 1987; Eussen and Niemann, 1981; Mallik and Tesfai, 1988; Miller et al., 1991; Nicollier et al., 1985; Perez and Ormeno-Nunez, 1991b) and are inhibitory to a variety of crops and weeds. The toxic effects of these compounds are pronounced due to long persistence in the soil (Rice, 1984), and several cinnamic acid derivatives have been identified as germination inhibitors (Sathiyamoorthy, 1990).

Quinones and Derivatives   Several of the quinones and their derivatives are derived from the shikimic acid. Juglone and related naphthoquinones have been isolated from walnut (Thiboldeaux et al., 1994). Naphthoquinone and α-naphthol are isolated from sunflower (Rice, 1984). Most of these derivatives are well-known antibiotics such as helminthosporin, tetracyclin, and aureomycin (Whittaker and Feeny, 1971).

Coumarins   Coumarins, the lactones of hydroxy cinnamic acid, are known to occur in several plants. Some of the coumarins, such as methyl esculin, have been identified from *Ruta* (Aliotta et al., 1994), *Avena* (Porwal and Gupta, 1987), and *Imperata* (Eussen and Niemann, 1981). Among others scopolin, scopoletin, and furanocoumarins are well known for their potential to inhibit growth of several crop plants (Rice, 1984).

Flavonoids   A large variety of flavonoids, such as phlorizin from apple residues (Rice, 1984) and its breakdown products such as glycosides of kaempeferol, quercetin, and myrcetin, are well-known allelochemicals (Harborne, 1980; Rice 1984).

Tannins   Tannins, which include both hydrolyzable and condensed varieties, are well known for their inhibitory effects due to their capacity to bind to the proteins. The common hydrolyzable tannins, such as gallic, ellagic, trigallic, tetragallic, and chebulic acids, are widely distributed in the plant kingdom (Harborne, 1980; Swain 1977). Most of these are also identified from the forest soils in concentrations required to inhibit nitrification (Harborne 1980; Ponder, 1987). The condensed tannins, which arise by the oxidative polymerization of the catechins, are well known for their potential to inhibit nitrifying bacteria in forest soils (Ponder, 1987) and to slow the rate of decomposition of organic matter in soil (Fisher, 1987) which is very important in mineral cycling.

Alkaloids   Though several alkaloids have been isolated from a variety of plants, only few of them are identified as having allelopathic potential. Some of the alkaloids, such as cocaine, physostigmine, caffeine, quinine, strychine, and cinchonin, are well-known seed germination inhibitors (Rice, 1984).

Biosynthesis

A great majority of allelochemicals are secondary metabolites derived either from acetate or shikimic acid (Fig. 9-2) (Rice, 1984). Among several groups of allelochemicals, terpenoids, steroids, flavonoids, water soluble organic acids, straight

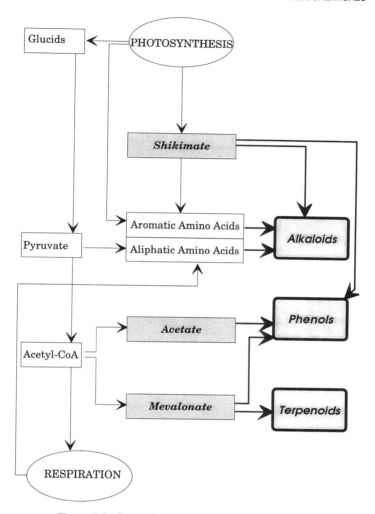

Figure 9-2. *Biosynthetic pathways of allelochemicals.*

chain alcohols, aliphatic aldehydes, ketones, simple unsaturated fatty acids, long chain fatty acids, polyacetylenes, naphthoquinones, anthraquinones, complex quinones, and phloroglucinol are derived from acetate. The allelochemicals derived from shikimic acid include simple phenols, benzoic acid and its derivatives, cinnamic acid and its derivatives, coumarins, sulfides and glycosides, alkaloids and cyanohydrins, some of the quinone derivatives, and hydrolyzable and condensed tannins. The biosynthetic pathways of several phenolic compounds, which include most of the benzoic acid and cinnamic acid derivatives, have been studied to a greater extent, and a number of enzymes involved in the biosynthetic pathways have been purified and characterized and are known to be influenced by both abiotic and biotic factors (Bolwell et al., 1986; Liang et al., 1989). Approximately 2% of the photosynthetic carbon has been diverted to the biosynthesis of several flavonoid and phenolic compounds. The concentrations of these compounds in the tissues vary according to the rate of their biosynthesis, storage, and degradation, and are also affected by the internal balance of plant growth regulators (Harborne, 1980).

Availability in Plants and Soils

Several factors influence the level of allelochemicals in plants and soil. Most of the flavonoids are exuded by roots at concentrations as high as 25 mg g^{-1} fresh weight, and the concentration of flavonoids usually ranges from 1 nM g^{-1} to 7.5 mM g^{-1} fresh weight (Rao, 1990). However, the concentration differs with the plant. It has been found that 100–120 kg ha^{-1} of phenolic allelochemicals are added to the soils every year (Guenzi et al., 1967). The range of concentration of some of the allelochemicals, namely, *p*-hydroxy benzoic, vanillic, *p*-coumaric, ferulic, and caffeic acids, are estimated to be 0.2, 0.9, 1.1, 3.4, and 0.9 mg g^{-1} soil, respectively (Siqueira et al., 1991).

MODE OF RELEASE OF ALLELOCHEMICALS

The mode of release of allelochemicals from donor plant into the environment comprises an important part of allelopathy. A variety of allelochemicals are synthesized and stored in different plant cells, either freely or in bound form (Gershenzon,

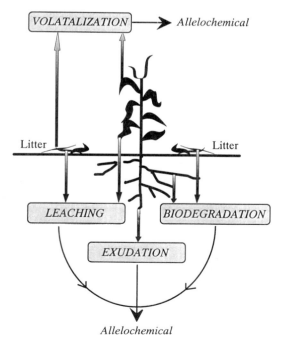

Figure 9-3. *Major routes of entry of allelochemicals into the environment.*

1993), and are released into the environment in response to various abiotic and biotic stresses (Rice, 1984). The mode of release of a particular allelochemical depends on its chemical nature (Berg and Staff, 1980; Mandava, 1985).

Four major routes of entry of allelochemicals into the environment have been recognized, based on their chemistry (Fig. 9-3): (1) volatilization, (2) leaching, (3) root exudation, and (4) decomposition of plant residue (Mandava 1985; Narwal; 1994, Rice; 1984).

Volatilization

The release of allelochemicals by volatilization is often confined to plants producing terpenoids and monoterpenes (Fischer et al., 1994). The allelopathic potential of the volatiles is well documented (Table 9-2). The common genera that release volatiles include *Artemisia, Salvia, Parthenium,* and *Eucalyptus* (Rice, 1984). *Eucalyptus* has been known to contain a variety of terpenoids that are toxic to germination and seedling growth of numerous crop plants (del Moral and Muller, 1970). In addition, a large number of Brassicaceae members *(Brassica juncea, B. nigra, B. rapa, B. oleracea)* are known to release volatiles that inhibit the germination and growth of lettuce and wheat (Oleszek, 1987). Apart from their toxic effects on plants, volatile allelochemicals have been shown to have insecticidal (Nehlin et al., 1994) and feeding deterrent potential (McCall et al., 1994). Furthermore, the toxicity of vola-

TABLE 9-2 Allelopathic Potential of Volatiles

Plant Name	Inhibitory Effect on Target Plant	Chemical Nature of Volatile	Reference
Amaranthus palmeri	Germination of tomato, onion, carrot	2-Octanone, 2-nana-none, 2-heptanone	Bradow and Connick (1987, 1988a, b)
Artemisia princeps var. *orientalis*	Autotoxic and inhibitory to callus development of lettuce, *Eclipta prostrata*	nd	Kil et al. (1992)
Brassica juncea, B. napus, B. rapa	Germination of lettuce and wheat	nd[a]	Oleszek (1987)
Eucalyptus globulus	Germination and growth of crop plants	Variety of terpenes	del Moral and Muller (1970)
Gossypium spp.	Feeding deterrent to herbivores	Hexyl acetate, 2-methyl butyrate	McCall et al. (1994)
Heliotropium europeum	Stimulated growth of buckwheat, radish	nd	Grechkanev and Rodionov (1971)
Pinus sylvestris	Toxic to carrot psyllid	α-pinene, camphene, β-pinene, 3-carene, terpinene, terpino-lene	Nehlin et al. (1994)
Salvia reflexa	Seed germination and seedling growth	Monoterpenes, α-pinene, β-pinene, cineole	Lovett (1986)

[a] nd = not determined.

tiles is prolonged, due to their adsorption to soil particles; they remain inhibitory to plants for several months. In the desert and mediterranean ecosystems, the release of allelochemicals through volatilization is frequently observed, due to the prevalence of high temperatures, and influences the patterning of vegetation.

Leaching

Leaching is the removal of substances from plants by the action of rain, snow, fog, dew, or mist. It offers a greater portion of allelochemicals to the environment (Kumari and Kohli, 1987). The degree of leachability, however, depends on the type of plant tissue, age of the plant, and the amount and nature of precipitation (Molina et al., 1991). The major allelochemicals released via leaching include a variety of organic and inorganic substances such as phenolic compounds, terpenoids, and alkaloids (Rice, 1984), and most of these chemicals have been well characterized for their toxic effects on the surrounding plants and microorganisms under field and laboratory conditions (Macias et al., 1994). A number of seed and leaf leachates have been identified that are toxic to various crop and noncrop plants (Table 9-3). The amount of rainfall determines the leachability of a particular compound and its toxicity either to the leached plant (autotoxic) and/or to an adjacent plant (allotoxic).

Root Exudation

In several instances, the suppression in crop yields has been attributed to toxins released by roots of various crop and weed species adjacent to crop plants (Alsaadawi et al., 1985). Several of the compounds released via roots are known to reduce

TABLE 9-3 Allelopathic Potential of Leachates

Plant Name	Inhibitory Effect on Target Plant	Chemical Nature	Reference
Abutilon ohata	Growth of competing seedlings and seed affecting fungi	Delphinidin, cyanidin, quercetin, myrcetin, catechin, epicatechin	Paszkowski and Kremer (1988)
Artemisia	Growth of barley, lettuce, chrysanthemum	nd	Kil and Yun (1992)
Brassica napus	Germination of soybean	Allylisothiocyanate	Choesin and Boerner (1991)
Brassica rapa (L.) Metzger var. *botrytis* L. s.-var. *cymosa* Lm.	Growth of barley, rye radish	nd[a]	Jimenez-Osornio and Gliessmann (1987)
Calmintha ashei	Germination and growth of *Rudberkia hirta* and *Leptochloa dubia*	(+) Evodone and des acetyl calaminthone	Weidenhamer et al. (1994)
Camelina	Seedling growth of lin	Benzylamine	Lovett (1987)
Datura stramonium	Growth of wheat and soybean	Scopolamine, hyoscyamine	Lovett et al. (1981)
Eucalyptus globulus	Growth of crop plants	nd	Molina et al. (1991)

[a] nd = not determined.

seed germination, root and shoot growth, nutrient uptake, and nodulation (Table 9-4) (Pandya et al., 1984; Weston, 1986; Yu and Matsui, 1994). Though root exudates comprise only 2–12% of the total photosynthates of the plant, they contribute a great deal toward allelopathy, as most of the well-known allelochemicals are root exudates. It has been observed that factors such as plant age, nutrition, light, and moisture would influence root exudation both quantitatively and qualitatively (Einhellig, 1987).

TABLE 9-4 Allelopathic Potential of Some Root Exudates

Plant Name	Inhibitory Effect on Target Plant	Chemical Nature	Reference
Asparagus officinalis	Autotoxic	nd	Young (1984)
Avena spp.	Shoot and root growth, ear length of wheat	Scopoletin, VAN	Schumacher et al. (1982, 1983); Porwal and Gupta (1987)
	Growth of wheat	Scopoletin, BEN, COU, VAN	Perez and Ormeno-Nunez (1991a)
Bidens pilosa	Leaf area, growth, dry matter of maize, sorghum, kidney bean, lettuce	nd	Stevens and Tang (1985, 1987)
Celosia argentea L.	Nodulation in pigeon pea, mothbean	nd	Pandya et al. (1984)
Chenopodium murale	Shoot and ear head length, dry matter of wheat	nd	Porwal and Gupta (1987)
Chinese cabbage	Growth of mustard and autotoxic	nd	Akram and Hussain (1987)
Cucumis sativus	Growth of *Lactuca sativa*	BEN, CHL, MYR, palmitic acid	Yu and Matsui (1994)
Lycopersicon esculentum	Growth of lettuce and brinjal	nd	Kim and Kil (1987)
Medicago sativa	Growth of soybean, maize, barley, radish	nd	Tsuzuki and Kawagoe (1984)
	Cell suspension culture of cabbage, tomato	Canavanine	Miersch et al. (1992)
Parthenium hysterophorus	Growth and nodulation of kidney bean	nd	Kanchan and Jayachandra (1979)
Quack grass	Root growth, dry matter, nodulation, N_2 fixation	nd[a]	Lovett and Jokinen (1984); Weston (1986)
Rorippa sylvestris	Lettuce seedling growth	Hirsutin and pyro catechol, HBA, VAN	Yamane et al. (1992)
Triticum aestivum	Growth of wild oats	Hydroxamic acid	Perez and Ormeno-Nunez (1991a)

[a] nd = not determined.

Decomposition of Plant Residue

The decomposition of plant residues adds a large quantity of allelochemicals to the rhizosphere (Goel, 1987). The inhibitory potential of several compounds released during decomposition have been listed (Table 9-5). Factors that influence this process include the nature of plant residue, soil type, and the conditions of decomposi-

TABLE 9-5   Allelopathic Potential of Decomposing Residues

Plant Name	Inhibitory Effect on Target Plant	Chemical Nature	Reference
Agropyron repense	Seedling growth of alfalfa, maize, soybean	5-Hydroxy indole, 3-acetic acid	Touchette et al. (1988); Hagin (1989)
Artemisia princeps	Growth, dry weight and caloric content of *Lactuca, Plantago, Chrysanthemum,* and *Achyranthus*	nd	Yun and Kil (1992)
Chenopodium album and *C. murale*	Germination and growth of wheat, rye, maize, chickpea, soybean, mustard; nutrient uptake of maize, soybean, and tomato	nd	Bhowmik and Doll (1984); Goel (1987); Mallik and Tesfai (1988); Qasem and Hill (1989)
Cyperus rotundus L.	Yield of tomato, rice, cabbage, cucumber, carrot, soybean, cotton	Polyphenols and sesquiterpenes	Zimdahl (1980); Chivinge (1985)
Cyperus esculentus L.	Germination and growth of sugar beet, peas, lettuce, tomato, maize, soybean, tobacco	FER, HBA, SYR, VAN	Mallik and Tesfai (1988); Singh et al. (1988); Lolas (1986)
Imperata cylindrica	Growth of maize, rice, sorghum, tomato	Scopolin, scopoletin, BEN, CHL, COU, GEN, VAN	Eussen and Niemann (1981)
Parthenium hysterophorus	Germination of rape seed	Parthenin, coronopilin, CAF, *p*-COU	Kohli et al. (1985); Jarvis et al. (1985)
Setaria viridis L.	Growth of soybean, maize, sorghum	nd[a]	Bhowmik and Doll (1982); Mallik and Tesfai (1988)
Sorghum halepense L.	Germination and growth of sunflower, tomato, radish	CHL, COU, HBA, VAN	Nicollier et al. (1985)
Xanthium spp.	Germination and growth of wheat, maize, tobacco, pear millet, chickpea, rape seed, lettuce	BEN, CAF, CHL, COU, EA	Bushra et al. (1987); Ambasht and Rai (1981); Cutler (1985)

[a] nd = not determined.

tion (Mason-Sedum et al., 1986). The decomposing plant materials are never equally distributed throughout the soil, and therefore as the roots grow through the soil, at some points they may come in contact with decomposing residue and be affected by allelochemicals. The compounds released into the soil are subjected to transformation by soil microflora and produce even more biologically active products than the parent compounds (Blum and Shafer, 1988). Investigations using aqueous plant extracts have shown that water-soluble inhibitors present in crop plants can be quickly released during decomposition.

The toxicity arising from plant residues provides some of the challenging problems and opportunities for agronomists and weed scientists. Toxins from the decomposing crop residues can affect young crop plants sown between mature plants, that is, in relay cropping. Likewise the stubble of the preceding crop in multiple sequential cropping can affect the next crop. In areas where stubble mulch farming is practiced for soil and water conservation, toxins from the stubbles have proved toxic to certain crops in rotation. These residues influence not only the crop emergence, growth, and productivity, but also influence similar aspects of weed growth.

REGULATION OF ALLELOCHEMICAL PRODUCTION AND RELEASE

Abiotic Regulation

The synthesis and the release of allelochemicals into the environment are influenced by a variety of abiotic factors (Fig. 9-4), factors which thus determine the level and

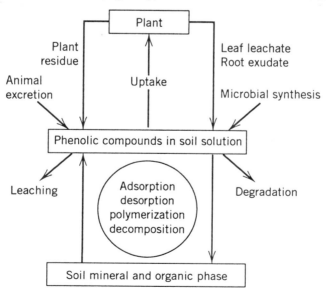

Figure 9-4. *Major processes regulating phenolics in plant-soil system. Reprinted with permission from Siqueira, J. O., M. G. Nair, R. Hammerschmidt, and G. R. Safir (1991).* Critical Reviews in Plant Science, *10 (1), p. 67. Copyright CRC Press, Boca Raton, Florida.*

toxicity of the allelochemicals in the environment (Berg, 1986). A great deal of research has been done on the effects of various factors on the production (quality and quantity), release, and persistence of allelochemicals in the environment.

Light   Light plays an important role in the synthesis of allelochemicals, rather than in their release into the environment. In addition to synthesis, light is also known to accelerate the release of some of the allelochemicals such as acetophenone, furanocoumarins, and furanoquinones (Downum, 1992). A similar case was observed for toxicity of polyacetylenes and thiophenes from Asteraceae toward herbivorous insects (Chapagne et al., 1986). The light-mediated synthesis, however, depends on the quality and quantity of light perceived by the plant and the photoperiod.

Exposure of tobacco and sunflower to ionizing radiation increased the amount of various phenolic compounds (Koeppe et al., 1970a). Plants exposed to X-ray-irradiation exhibited a dose-dependent increase in scopoletin levels and, after an initial temporary increase, a decrease in total chlorogenic acid, even with the highest dose tested (Koeppe et al., 1970a).

Much work has been done on the influence of UV light on the synthesis of many allelochemicals. In several instances, a substantial rise in a variety of flavonoids and phenolic compounds has been noticed in higher plants, as well as an increase in several microorganisms. These changes were thought to have a role in defense mechanisms. However, their synthesis in response to UV light varied with the part of the plant, the order of accumulation being young leaves > old leaves > stems > roots (Koeppe et al., 1976).

An increase in synthesis of ferulic and *p*-coumaric acids in potato tuber discs was observed upon exposure to red light when compared to far-red light. Short exposure to red light is also known to increase the alkaloid content, whereas plants exposed to far-red light showed increased levels of chlorogenic acid and other soluble phenolics (Tso et al., 1970).

Increased synthesis of chlorogenic acid has been observed in response to short exposure to visible light compared to dark (Zucker, 1963). In addition to the intensity of light, day-length has also been observed to influence the synthesis. Long days are shown to increase the concentration of phenolic acids and terpenes in plants, regardless of the day-length required for flowering (Burbott and Loomis, 1967; Taylor, 1965; Zucker, 1969). Taylor (1965) observed increased synthesis of chlorogenic and isochlorogenic acids, flavonoid aglycones, and quercetin glycosides in response to long days in *Xanthium pennsylvanicum*. Burbott and Loomis (1967) noticed greater increase in the synthesis of monoterpenes on long days in *Mentha piperata*. Day-length is also known to influence the level of enzymes involved in the synthesis of various allelochemicals. Prolonged day-length increased several cinnamic acid derivatives in *Xanthium* (Taylor, 1965).

Temperature   In general, high temperatures are known to enhance the allelochemical effect (Einhellig and Ecrich, 1984). In *Sorghum*, the threshold concentration of ferulic acid for inhibition was 0–0.2 mM at 37°C while the concentration was 0–0.4 mM at 29°C (Einhellig, 1987), which indicates the importance of temperature in determining the relative sensitivity of a plant to a particular allelochemical. Glass (1976) demonstrated an enhancement of allelopathic effect at low temperature similar to that found at high temperature. Steinseik et al. (1982) observed leachates of wheat causing more inhibition of germination and growth of *Ipomea hederacea* at 35°C than at 30°C or 25°C. Bhowmik and Doll (1983) noticed an alteration of allelochemical effect of *Amaranthus retroflex* on *Setaria glauca* connected with changes in temperature as well as in photosynthetic flux density. Blaschke (1977) observed the influence of temperature and humidity on the production of ethylene from the decomposing litter of *Picea abies*.

Temperature influences the release of allelochemicals rather than their synthesis. Increased temperatures favor the exudation of allelochemicals by roots. Chilling increases the concentration of total chlorogenic acid in leaves and stems but decreases it in roots. Chou (1986) observed a pronounced toxicity of *Leucaena* on the growth of understory vegetation at high temperatures and considered this to be an adaptive mechanism to prevent the growth of other weeds competing for water.

Soil Adsorption Soil adsorption is of considerable importance for volatiles to exert their action, as most of the volatiles released into the environment get adsorbed to soil particles and persist for longer durations (Einhellig, 1987). In the soil the allelochemicals are decomposed by soil microorganisms depending on their specificities. Ferulic, cinnamic, and coumaric acids are important in the synthesis of lignin and the same compounds are released during lignin degradation and persist in the soil by getting adsorbed to soil particles. It is evident that combinations of allelopathic compounds often exert additive or synergistic action against the growth of plants and microorganisms (Rice, 1984).

Soil Texture The soil texture also influences the allelochemical toxicity. Most of the volatile allelochemicals adsorbed to the colloidal particles of the soil remain in an active form for prolonged periods and migrate to sites of action in plants (Muller and del Moral, 1966). Seedlings grown in sandy soil were more affected, compared to loam, probably due to the greater retention of allelochemicals in the loam reducing the availability of allelochemicals to act on the plant (del Moral and Muller, 1970). The stability of a particular compound in the soil depends on its chemical nature in addition to various environmental factors. The combination of allelochemicals often exerts additive or synergistic actions against the growth of plants and microorganisms. The synergistic action is of considerable significance in natural and agronomic communities.

Soil Aeration The toxicity of phenolic compounds present in the soil is influenced mostly by soil aeration and tillage practices. The soil oxygen determines the type of products formed, their accumulation, and persistence. Under aerobic conditions the decomposition of organic compounds by microbes is faster, and anaerobic conditions favor the production of methane, hydrogen sulfide, ethylene, acetic acid, lactic acid, butyric acid, and various phenolics that have been shown to have toxic effects under laboratory studies (Chou and Chiou, 1979; Chou et al., 1977). The oxygen availability not only determines the type of compounds formed during decomposition of plant material, but also affects the activity, concentration, and persistence in soil (Parr and Renszer, 1962).

Mineral Deficiencies Inhibition in the uptake of mineral nutrients by plants under allelopathic conditions has often been observed (Blum et al., 1992; Booker et al., 1992; Glass, 1973, 1974). However, several investigations have demonstrated the decrease of allelochemical effect with increase in nutrient level (Glass, 1976; Stowe and Osborne, 1980). Amelioration of allelochemical-mediated inhibition by increased supplies of nitrogen, phosphorus, and potassium has been observed in *Solidago, Festuca,* and *Helianthus* (Fisher et al., 1978; Hall et al., 1983). In contrast, Bhowmik and Doll (1984) demonstrated the inability of increased levels of nitrogen and phosphorus to reduce the allelopathic potential of five annual weeds. This is further evident from the studies of Buchholtz (1971), who showed failure of increased fertilization to overcome the allelopathic effect of quack grass on corn or to reduce the autotoxicity of *Trifolium*. Thus it is noteworthy that, though increased fertilization can overcome some allelopathic effects, operation of allelopathic mechanisms in nature cannot be overlooked, and in fact these studies demonstrate the importance of interaction of two stress factors in nature. Moreover, a large number of minerals are known to affect the biosynthesis of allelochemicals. An increase in synthesis of scopolin and caffeic and chlorogenic acids has been observed in plants grown in boron-deficient soil (Dear and Arnoff, 1965; Watanabe et al., 1961). Similar increases in the levels of scopolin and chlorogenic acid were also observed in plants grown in calcium- and magnesium-deficient soils (Armstrong et al., 1971; Loche and Chouteau, 1963). Increase in total chlorogenic and isochlorogenic acids in roots, stems, and leaves of nitrogen-deficient sunflower was observed (del Moral, 1972). The studies on the synthesis of the allelochemicals in nitrogen-deficient soils is of great importance in allelopathic studies because there are large areas of land deficient in nitrogen, and this is extremely important in connection with allelopathic mechanisms operating in revegetation of infertile old fields. Koeppe et al. (1976) observed an increase in the isomers of chlorogenic acid in sunflowers grown in phosphorus-deficient soils. Chouteau and Loche (1965) reported decreased concentrations of chlorogenic acid and increased levels of scopolin in tobacco leaves when plants were grown in potassium-deficient soils. Plants grown in sulfur-deficient soils showed increased chlorogenic acid concentration (Lehman and Rice, 1972). Moreover, larger amounts of phenolic compounds were leached from intact roots, dried roots, and aerial parts of phosphate-deficient plants than from phosphate-sufficient ones. These leachates contained scopolin, but none of the isomers of chlorogenic acid or caffeic acid. Thus the pronounced increase in concentrations of inhibitors in plants resulting from phosphorus deficiency are probably very important in the allelopathic mechanisms operating during old field succession.

Soil pH The pH of the soil also influences the release of allelochemicals via modifying the microbe-mediated decomposition of litter. Low pH of the soil is known to decrease the decomposition of the litter and decrease the release of allelochemicals (Berg, 1986; Berg and Agren, 1984; Berg and Staff, 1980).

Moisture Interference of allelochemicals with plant water balance is well known. Reductions in the water potentials are observed with ferulic acid and coumaric acid in sunflower, *Abutilon,* and *Kochia* (Colton and Einhellig, 1980; Einhellig and Schon, 1982). In several instances the allelopathic effects are increased at low water potentials (Choesin and Boerner, 1991). Sodium chloride-imposed water stress and drought stress resulted in substantial increase in concentration of total chlorogenic and isochlorogenic acids in roots, stems, and leaves. Compared to their individual effects, various combinations of stress are often found to have pronounced effects on the synthesis of allelochemicals. Combinations of water stress and exposure to supplemental UV light increased the concentration of total chlorogenic and isochlorogenic acids more than either factor alone with normal nitrogen (Einhellig, 1987).

Biotic Regulation of Allelochemicals Production and Release

Plants In addition to various abiotic factors regulating the level of allelochemicals in the environment, plants have developed a variety of mechanisms that are involved in the detoxification of allelopathic compounds (Shimabukuro, 1985; Shimabukuro et al., 1982). These include phase I reactions like oxidation, reduction, and/or hydrolysis. Oxidation reactions involve the peroxidases, monoxygenases, and other oxygenase enzymes. These reactions are the most common in plants.

Phase II reactions include conjugation reactions that involve the conjugation of allelochemicals with an endogenous substrate to form a new compound. The importance of conjugation reactions lies in the greater water solubility of the conjugates. The structural modification of the allelochemical during the conjugation reduces the toxicity and the higher molecular weights of the conjugates impose a limitation for its free movement. The most important conjugation mechanisms in plants include glucosylation, which involves the conjugation of allelochemicals with glucose mediated by an enzyme glucosyl transferase. A large number of glucosyl transferases, depending on the allelochemical being conjugated, have been identified and purified (Sun and Hrazdina, 1991; Yalpani et al., 1992).

Phase III reactions include further conjugation of phase II conjugates to other endogenous substrates, which results in tertiary conjugates. This mechanism is the most important process in the ultimate reduction of the toxicity of the allelochemicals to considerable extent. In general these conjugation reactions are the ultimate processes involved in the detoxification of the allelochemicals; for the most part they occur at sites of continuing metabolic activity in all organisms. The final toxicity of the compounds, however, is determined by the extent to which an organism is able to reduce the accumulation of allelochemicals from site of action.

Age of the Plant Organs   Koeppe et al. (1976) showed increased concentration of scopolin and chlorogenic acid in tobacco leaves with increase in the age of the plant. Total isochlorogenic acid decreased with age of leaves from apex (Koeppe et al., 1970b). In stems, chlorogenic as well as isochlorogenic acids decreased with increase in age of the plants. In another study, Wardle et al. (1993) observed more inhibitory nature of thistle at early bolting stage and at the stage when bolting plants are dying. Plants during flowering phase are found to be more allelopathic than during vegetative stage (Ahmed and Wardle, 1994; Wardle et al., 1993)

Pathogens and Predators   There is convincing evidence that infection increases the synthesis of various secondary metabolites, among which many are allelopathic in nature, and offers resistance to at least some plants against pathogens. Woodhead (1981) demonstrated increased concentration of phenolics in *Sorghum* infected with downey mildew or rust.

Much of the research on the factors regulating the synthesis or release often deal with phenolics and alkaloids, and literature is rather scanty on the regulation of other allelochemicals. Further research is needed to unravel the involvement of various factors governing the synthesis and release of allelochemicals. These studies indicate that in addition to having their own individual effects, all these environmental stresses tend to enhance the allelopathic potential under natural conditions. Thus the impact of allelochemicals on plant growth should be assessed with regard to both the levels of allelochemicals and physical and chemical factors of the environment. Agronomists should manage allelopathy indirectly by acting on related factors such as nutrient level, soil moisture, and precipitation.

Microbes   Soil microorganisms play a major role in determining the concentration as well as the potential of allelochemicals. Several microbes and fungi are known to use a variety of allelochemicals as their carbon source and thus reduce their concentration in the soil. The utilization of several phenolic compounds by denitrifying bacteria has been observed (Hoagland and Williams, 1985). The most important process that largely decreases the level of allelochemicals in the soil is their transformation by soil bacteria (Hoagland and Williams, 1985; Siqueira et al., 1991). In a few instances, microbial products exhibit either equal or greater biological activity than the parent compound. Such transformations are observed for benzoxazilinone and its derivatives of rye (Siqueira et al., 1991) and for oxidation of hydrojuglone to juglone (Rice, 1984). It has been observed that certain crops stimulate phenolic degrading organisms in the soil (Siqueira et al., 1991). Such changes in soil microbial populations is of great significance in detoxification of phenolic compounds. The detoxification of allelochemicals by microbes also involves conjugation reactions similar to those observed for higher plants (Siqueira et al., 1991)

MODE OF ACTION OF ALLELOCHEMICALS ON PLANT PHYSIOLOGICAL PROCESSES

Though much evidence has been documented on growth inhibition due to allelochemicals (Table 9-6), the physiological and molecular mechanisms leading to such inhibition have not been critically investigated. The most probable reason for the

TABLE 9-6   Inhibitory Effect of Allelochemicals on Seed Germination and Seedling Growth

Source Plant	Target Plant	Chemical Nature	Reference
Avena Triticum	Wild oats	nd[a]	Perez and Ormeno-Nunez (1991b)
Chrysoma pauciflosculosa	*Leptochloa, Schizachyrium,* radish, lettuce	Polyacetylenes	Menelaou et al. (1992)
Cinchona	*Catharanthus* and *Ocimum*	Quinoline alkaloids	Aerts et al. (1991)
Cucumis	Lettuce	BEN, *p*-HBA, 2,5, myristic DHBA, PAL, and steric acid	Yu and Matsui (1994)
Festuca	Alfalfa and Italian rye grass	nd	Smith and Martin (1994)
Ipomea batatus	*Cyperus esculentus*	nd	Harrison and Peterson (1991)
Medicago sativa	Cucumber	nd	Ells and McSay (1991)
	Autotoxic	nd	Hegde and Miller (1990, 1992)
Mikania micrantha	*Asystatia, Chrysopogon*	nd	Ismail and Mah (1993)
Parthenium argentatum	Broccoli, cotton, cantaloupe, lettuce, cauliflower	*p*-Anisic acid, BEN, CIN	Schloman et al. (1991)
Pilocarpus	Lettuce	Bergapten, donatin, goudotianus, imperatorin	Macias et al. (1993)
Pluchea lanceolata	*Brassica juncea,* asparagus bean	Formononetin 7 glucoside	Inderjit and Dakshini (1992a, b)
Polygonum	Lettuce	Anthraquinone, Physcion, emodin	Inoue et al. (1992)
Sorghum bicolor	*Abutilon, Datura, Amaranthus, Digitaria, Echinocloa*	nd	Einhellig and Souza (1992)

TABLE 9-6 *(Continued)*

Source Plant	Target Plant	Chemical Nature	Reference
Areimisia princeps	*Achyranthus japonica*	nd	Yun and Kil (1992)
Empetrum her-maphroditum	*Populus tremula*	nd	Zackrisson and Nils-son (1992)
Giant foxtail, yel-low foxtail, bahia grass, barnyard grass	Alfalfa, rye grass	nd	Martin and Smith (1994)
Kalmia	*Picea mariana*	COU, FER, GEN, VAN, SYR	Zhu and Mallik (1994)
Melilotus messa-nenis	*Lepidium sativum,* lettuce	Lupane triterpenes, lupeol, betulin	Macias et al. (1994)
Rorippa sylvestris	Lettuce	p-HBA, SAL, SYR, VAN	Yamane et al. (1992)
Ruta graveolens	Radish	5-methoxy pso-ralon, 4-hydroxy coumarin	Aliotta et al. (1994)

[a] nd = not determined.

lack of understanding of the mechanism of action is that most of the allelopathic effects under natural conditions are either synergistic or additive, and it is difficult to study the mode of action of a particular allelochemical separately. Though few attempts have been made to elucidate the mode of action, no definite mechanism of action for any particular chemical has been proposed. The threshold concentration of the compound required to affect plant processes varied with the type of the compound, plant species, and the concentration (Rice, 1987). The effect of allelochemicals on some of the plant physiological processes have been dealt with below.

Growth

Allelochemical effect on growth of various crop and weed species has been well documented (Table 9-6). Many of the allelochemicals have been found to reduce the growth of root and shoot, reduce leaf expansion, and cause wilting of plants in a variety of crop and vegetable plants (Blum and Dalton, 1985; Blum et al., 1984; Einhellig et al., 1985; Holappa and Blum, 1991; Klein and Blum, 1990; Waters and Blum, 1987). Cell division and elongation are essential phases of growth and development. Thus alterations in these processes would have a direct effect on growth of plants. A decrease in cell wall elongation of *Oryza sativa* coleoptiles due to ferulic acid has been correlated to decrease in the growth of rice seedlings (Kamisaka et al., 1990; Tan et al., 1991, 1992). The decrease in cell elongation was further noticed in cucumber root cells treated with umbelliferone (Jankay and Muller, 1976), while volatiles from *Salvia* (cineole and camphor) reduced both cell division and cell elongation of cucumber hypocotyls (Muller, 1965). The decrease in cell wall elongation is thought to be due to the accumulation of more lignin, making the cell walls more rigid (Tan et al., 1991, 1992). In addition to elongation, a few of the allelochemicals, namely parasorbic acid, coumarin, and scopoletin, inhibit root mitosis (Avers and Goodwin, 1956). Though these studies reveal the interference of allelochemicals with cell division and elongation, no conclusive evidence has been documented to show the relative contribution of these processes to allelochemical-mediated growth inhibition.

Interaction with Hormones

It has long been observed that many of the allelochemicals interact with plant growth regulators, either synergistically or additively, while exerting their action. A variety of allelochemicals have been shown to alter the levels of endogenous indole acetic acid by modifying the activity of indole acetic acid oxidase, an enzyme that acts on indole acetic acid. Altered indole acetic acid level was observed in ferulic acid-treated seedlings (Tayal and Sharma, 1985) and synergistic interaction between indole acetic acid and phenols in rooting of oat and pea seedlings has also been observed (Tomaszewski and Thimann, 1966). The decrease in root length in presence of indole acetic acid and ferulic acid indicates an antagonistic nature of ferulic acid to indole acetic acid (Tayal and Sharma, 1985). Similar interactive effect of phenol and auxin in regulating the rooting of mangrove was reported (Kathiresan et al., 1990). Depending on their action on indole acetic acid, most of the phenolic compounds are divided into two groups (Lee et al., 1980): either those that promote indole acetic acid destruction (*p*-coumaric, hydroxy benzoic, vanillic, syringic, and phloretic acids), or those that inhibit indole acetic acid destruction (chlorogenic, caffeic, ferulic, and protocatechuic acids). Inhibition of auxin-mediated elongation of rice hypocotyl cells by ferulic acid has been observed (Ishii and Saka, 1992; Kamisaka et al., 1990). Furthermore, many of the phenolic allelochemicals are known to interact with gibberellic acid, either by binding to the molecule or by inhibiting the gibberellic acid-mediated process(es). Rasmussen and Einhellig (1979) observed an inhibition of gibberellic acid-stimulated germination of *Sorghum* by ferulic acid together with *p*-coumaric acid and vanillic acid. A variety of tannins inhibited the gibberellic acid-induced growth, accompanied by reduction in amylase and acid phosphatase synthesis in endosperm half seeds of barley (Corcoran et al., 1972; Jacobsen and Corcoran, 1977). Similar inhibition of hydrolytic enzymes has been observed in maize seeds treated with ferulic acid (Devi and Prasad, 1992). Moreover, the regulation of growth by an allelochemical is dependent on its affinity to abscisic acid. Ray and Laloraya (1983) observed interaction of phenolic compounds with gibberellic acid and abscisic acid in addition to their action on processes mediated by them. The growth inhibition of cucumber seedlings due to ferulic acid and other phenolic compounds has been correlated to increased abscisic acid levels (Holappa and Blum, 1991; Li et al., 1993).

Though these observations revealed the interactive nature of allelochemicals with plant growth regulators, no conclusive evidence has been demonstrated relating such interactions to growth regulation. Thus these findings suggest that allelochemicals to some extent interact with growth regulators. Further research is needed to elucidate the mechanisms involved in such interactions.

Enzymes

Many of the allelochemicals are known to modify either the synthesis and or activity of various enzymes in vitro and in vivo. The activity of glucose 6-phosphate dehydrogenase was inhibited with ferulic acid and other phenolic compounds (Hoover et al., 1977). Often most of the allelochemicals are shown to exert dual effects in regulating the enzyme activities, that is, low concentrations tend to activate while high concentrations inhibit the enzyme activity. Increase in activity of some of the oxidative enzymes (peroxidase, catalase, indole acetic acid oxidase) has been observed in ferulic acid-treated maize seedlings (Devi and Prasad, 1996a), together with considerable rise in phenylpropanoid enzymes such as phenylalanine ammonialyase and cinnamylalcohol dehydrogenase, thus activating oxidative and phenylpropanoid metabolisms. Further, under most of the allelopathic situations, a decrease in the activities of nitrogen fixing enzymes such as nitrate reductase has been observed (Dhir et al., 1992; Jain and Srivastava, 1981). In maize, Devi and Prasad (1992) reported the inhibitory nature of ferulic acid to hydrolytic enzymes, that is, amylase, maltase, invertase, protease, and acid phosphatase of maize seeds, which are involved in the mobilization of reserve food material; the authors ascribed these changes to growth inhibition.

Respiration

Allelochemicals have been shown to inhibit or alter respiration depending on the age of the plant. Respiratory metabolism showed a decrease or an increase, depending on the chemical nature and concentration of the allelochemicals involved. Inhibition of O_2 uptake by isolated mitochondria was caused by by volatile monoterpenes of *Salvia,* while no effect on root respiration has been observed (Muller et al., 1969).

In addition to whole plants, many of the allelochemicals have been tested for their effect on respiration using isolated mitochondria. Juglone inhibited respiration by inhibiting the coupled intermediates of oxidative phosphorylation and slowing the electron flow to oxygen. In contrast, allelochemicals such as coumarin, various aldehydes, cinnamic acid, and benzoic acid are shown to stimulate oxygen uptake by yeast cells at 1 mM concentration (Van Sumere et al., 1971). However, no correlation has been drawn between growth and decrease in respiration when tested with phenolics. Ravanel (1986) observed the capacity of flavones and flavonols to uncouple the oxidative phosphorylation of isolated mitochondria of potato and mung bean. However, the potential to inhibit phosphorylation depends on the nature of substituted groups on the ring (Koeppe and Miller, 1974; Ravanel et al., 1982). In addition to these, cinnamic and benzoic acid derivatives are known to inhibit the oxidation of succinate and malate (Moreland and Novitzky, 1987; Ravanel and Tissut, 1986). Many of the quinones are shown to uncouple the oxidative phosphorylation (Moreland and Novitzky, 1987). A number of chalcones are known to inhibit ATP synthesis in mitochondria by uncoupling the electron transfer and phosphorylation (Ravanel et al., 1982; Tereda, 1980). Some of the flavonoids are also known to inhibit respiration by inhibiting the mitochondrial ATPases and interfering with energization of the membrane (Lang and Racker, 1974; Moreland and Novitzky, 1987). Several of the phenolic compounds are also known to reduce the respiration of mycorrhizal fungi (Boufalis and Pellissier, 1994; Pellissier, 1993a).

Though these studies claim the reduced respiratory metabolism is due to allelochemicals, no insight into the changes at enzyme level has been gained. Studies with specific site determining compounds resulted in chemical reaction with the allelochemicals due to their same redox potentials, making the studies more complicated.

Photosynthesis

Most of the allelopathic situations lead to a decrease in the dry matter production accompanied by a decrease in the photosynthetic ability of the plants. Patterson (1981) reported the potential of ferulic, cinnamic, *p*-coumaric, vanillic, and gallic acids to decrease net photosynthesis. Similar reduction in net CO_2 assimilation of three grasses has been observed (Lodhi and Nickell, 1973). Devi and Prasad (1996b) reported a decrease of net CO_2 assimilation of maize with concomitant reduction in stomatal conductance in ferulic acid-treated seedlings. Further, Blum and Rebbeck (1989) observed considerable reduction in the proportion of carbon allocated to roots, in addition to reduction in the photosynthetic rates and chlorophyll content. Inhibitory nature of coumarins and cinnamic and benzoic acids to *Lemna* photosynthesis has been reported (Einhellig et al., 1985). In addition to their action at whole plant level, many of the allelochemicals are also shown to inhibit CO_2-dependent evolution of O_2 and photophosphorylation of isolated thylakoids (Arntzen et al., 1974; Einhellig et al., 1970; Moreland and Novitzky, 1987). This reduction has been ascribed to an inhibition in the electron transport rates and a decrease in the efficiency of membrane ATPases of chloroplast (Moreland and Novitzky, 1987). Mersie and Singh (1993) observed an inhibition in the photosynthetic efficiency of the chloroplasts of velvet leaf by a number of phenolic compounds.

In most of the situations, decrease in photosynthesis has been correlated to decrease in chlorophyll content and stomatal conductance (Chander et al., 1988; Choudhary 1990; Patterson, 1981; Todorov et al., 1992). The experiments dealing with isolated chloroplasts revealed the action of allelochemicals on electron transport and photophosphorylation. Sharma and Singh (1987) observed a dose-dependent curtailment in the hill reaction of rice with salicylic, tannic, and benzoic acids. Similar reduction in the electron transport rates of chloroplasts of maize seedlings treated with ferulic acid has been reported (Devi and Prasad, 1996b). Another well-known flavonol, kaempeferol, is known to inhibit photophosphorylation of pea thylakoids by acting as an energy transfer inhibitor (Arntzen et al., 1974). Allelochemicals such as dihydrochalcone glucoside, phlorizin, and several phenolic compounds are known to interfere with structural or functional organization of the chloroplast membrane (Arntzen et al., 1974; Moreland and Novitzky, 1987; Muzafarov et al., 1988).

The studies of allelochemical action on photosynthesis, either at whole plant or chloroplast level, revealed the potential of these compounds to decrease the metabolism. In fact, some of the allelochemicals, namely the substituted phenols, mimic the action of phenolic herbicides in inhibiting photosynthesis, and the possibility of close relation of some of the allelochemicals to potential herbicides cannot be ruled out.

Stomatal Conductance

Many of the allelochemicals are known to induce stomatal closure and affect the stomatal conductance depending on concentration. Studies with ferulic, p-coumaric, and cinnamic acids showed their ability to reverse the abscisic acid-mediated closure of stomata by increasing the K^+ uptake of guard cells when tested at low concentrations (Laloraya et al., 1986; Manthe et al., 1992; Purohit et al., 1991, 1994). However, at high concentrations, the same phenolics are known to induce stomatal closure (Purohit et al., 1991) by decreasing the turgidity of guard cells, as many of the phenolics are well known for their potential to decrease water potentials (Blum and Rebbeck, 1989; Klein and Blum, 1990). Tannic acid is known to induce stomatal closure at 1 mM concentration (Einhellig, 1971). The water extracts of velvet leaf, *Kochia,* and *Helianthus* are known to induce stomatal closure in *Sorghum* and soybean when introduced into the nutrient solution (Colton and Einhellig, 1980; Einhellig and Schon; 1982, Einhellig et al., 1985). In addition to stomatal movement, several phenolic compounds are also known to interfere with stomatal conductance (Balke, 1985; Blum and Dalton, 1985; Einhellig and Kuan, 1971; Einhellig et al., 1985; Patterson, 1981). Scopoletin and chlorogenic acid at 0.5 and 1 mM concentrations decreased the stomatal conductance of tobacco and sunflower in a manner similar to their effect on photosynthesis (Einhellig and Kuan, 1971). Ferulic acid at low concentrations (0.25 and 0.5 mM) increased the photosynthesis and stomatal conductance of maize, while high concentrations (1 and 3 mM) inhibited the process (Devi and Prasad, 1996b).

Water Relations

The uptake of water, which is essential to maintain turgor in the cell, is shown to be adversely affected by a number of allelochemicals. A considerable decrease in leaf water potentials and water utilization of soybean, cucumber, *Sorghum,* and tomato (Blum and Dalton, 1985; Blum et al., 1985; Patterson 1981) has been observed with ferulic, p-coumaric, caffeic, and gallic acids. In contrast, Blum et al. (1985) and Klein and Blum (1990) did not observe any change in the water potentials of bean plants, indicating differential sensitivity of plants to allelochemicals. Further, a decrease in stomatal conductance, transpiration, and leaf expansion have been ascribed to reduction in the water potentials (Einhellig et al., 1985), due to reduction in the osmotic potential and turgor pressure. Thus the impact of allelochemicals on the tested plants might have been mediated through alterations in the plant water balance.

Ion Uptake

The uptake of water and nutrients from soil are inhibited under most of the allelo-pathic situations. At several instances, a decrease in the mineral content or a decrease in the capacity of the plants to absorb minerals from the soil or absorbing solution has been observed. Bergmark et al. (1992) showed an inhibition of the uptake of nitrate, ammonium, and potassium ions by maize roots when grown in the presence of ferulic acid. Booker et al. (1992) observed an inhibition in the uptake of potassium. Inhibition of the uptake of phosphate by barley, *Glycine max,* and maize roots (Devi, 1994; Glass, 1975; McClure et al., 1978), and potassium absorption by roots of *Hordeum vulgare* (Glass, 1974) and *Avena sativa* (Harper and Balke, 1981) with exogenous ferulic acid has been observed. Lyu and Blum (1990) observed an inhibition in the capacity of roots to absorb potassium and phosphorus. Salicylic acid, even at low concentrations, has been shown to inhibit potassium absorption (Klein and Blum, 1990), and low concentrations of ferulic acid (100 mM) are known to inhibit the uptake of rubidium by paul's scarlet rose (Danks et al., 1975). The inhibition was observed to be more pronounced at low pH (Devi, 1994; Harper and Balke, 1981). In addition to an inhibition in the uptake capacity of the roots, often a decrease in the mineral accumulation has been observed under most of the allelopathic situations. A decrease in potassium and magnesium levels in *Sorghum* has been observed with ferulic acid (Glass, 1976). Increased nitrogen and decreased phosphate levels in pig weed have been observed when grown in soils rich in chlorogenic acid (Hall et al., 1983). In addition to crop plants, such mineral imbalances are also observed for forest plants. Leachates of *Festuca* decreased the phosphate levels and increased the potassium levels of *Liquidambar styraciflua* (Walters and Gilmore, 1976). Similar effect with mulches of *Solidago* and aster on sugar maple has been reported (Fisher et al., 1978). Though the inhibitory potential of allelochemicals to accumulate minerals is evident from these studies, the level of toxicity appears to depend on the test plant, concentration of the test compound, and various biotic and abiotic factors. In addition to their direct action on plants, allelochemicals also exert indirect influence on mineral absorption by inhibiting mycorrhizal fungi in the soil, indicating the complexity of the system to evaluate the mode of action of allelochemicals (Perry and Choquette, 1987).

The potential of the allelochemicals depends on their lipid partition coefficient, that is, their capacity to be soluble in membrane lipids. Depending on the partition coefficients, Glass (1975) categorized the potential of benzoic and cinnamic acid derivatives to inhibit phosphate absorption. The higher the lipid partition coefficient, the more inhibitory the allelochemical is to nutrient absorption. Glass and Dunlop (1974) demonstrated the ability of most of the phenolic compounds to depolarize the membranes and correlated the rate of depolarization and inhibition of ion uptake. The inhibition was found to be noncompetitive (Glass, 1973, 1975; McClure et al., 1978). Thus the inhibition of ion uptake is of considerable significance under field conditions, particularly under acidic conditions, where allelochemical effects are more pronounced.

Membrane Permeability

Many of the allelochemicals, particularly many of the phenolic derivatives, flavonols, and flavonoids are shown to alter membrane properties (Moreland and Novitzky, 1987). In fact, reduction in many of the physiological processes such as ion uptake, respiration, and photosynthesis are thought to be due to alteration in the structural and functional integrity of the membranes. In several instances, depolarization of the membranes has been observed with allelochemicals. Benzoic acid and cinnamic acid derivatives are known to depolarize the membrane, and the extent of disruption depends on the capacity to be soluble in the membrane lipids (Glass, 1973). In addition to depolarization, many of the phenolic allelochemicals interfere with membrane ATPases, which provide the necessary energy for their function (Moreland and Novitzky, 1987). Salicylic acid is known to decrease ATP content of oat roots (Balke, 1985). Several phenolic compounds, quinones, terpenoids, and flavonoids are known to uncouple the oxidative phosphorylation (Moreland and Novitzky, 1987). Thus the interference of allelochemicals with ATP synthesis or its availability to various transport processes would alter the membrane functions and alter the plant metabolism.

Apart from depolarization of the membranes by allelochemicals, an increased permeability of the membranes to both cations and anions, together with disruption of the ion gradient and efflux of some of the ions, has been observed (Balke, 1985). Murphy et al. (1993) observed efflux of potassium ions when the roots are treated with salicylic acid; they ascribed this to a decrease in the membrane potential or structural alterations in the membrane due to greater capacity of salicylic acid to solubilize in the membrane. Similar reduction in the membrane permeability due to altered membrane lipids has been observed (Hayeshi et al., 1992). Though the studies could not lead to a decisive conclusion on the primary action of allelochemicals, it is certain that allelochemicals do interfere with membrane functions while exerting their action.

The studies on the allelochemical effect on various plant physiological processes and possible modes of action suggest that these compounds do not act on a single definite process but have multiple sites on which to exert their action. In plants all these processes are interdependent, and any change in one process would certainly have an effect on other processes. In such cases, the relative sensitivity of a particular process is most important when proposing the mode of action of a particular allelochemical. Research using combinations of various allelochemicals simulating a particular natural condition is of greater importance to evaluate the toxicity of allelochemicals and to hypothesize their modes of action.

ALLELOPATHY IN TERRESTRIAL ECOSYSTEMS

Agrosystems

Vast evidence is available on the allelopathic interactions in agricultural ecosystems (Narwal, 1994; Rice, 1984). In addition to competition, a large part of yield reduction of many crops has been ascribed to allelopathy. Allelopathic interferences are more pronounced where stubble mulching is practiced, whether in monoculture, crop rotation, or mixed cropping systems (Guenzi et al., 1967). The major sources of allelochemicals in agricultural systems is root exudates, leaching from plant parts, and decomposing residue (Narwal, 1994). Broadly, four types of interactions are noticed in agroecosystems, namely: (1) weeds versus crops, (2) crops versus weeds, (3) crops versus crops, and (4) weeds versus weeds (Rice, 1984).

Weeds Versus Crops   Under field conditions weed infestation is a major problem, reducing the crop yield. Allelopathic potential of weeds (exudates, leachates, and residues) has been listed (Table 9-7). In most of the situations, toxic effects of allelochemicals are exerted by a direct inhibition of the crop plant's nitrogen fixation or by an indirect effect on the growth of nitrogen-fixing microorganisms such as *Rhizobium* and *Azotobacter* (Alsaadawi and Rice, 1982; Mallik and Tesfai, 1988; Wardle et al., 1994; Weston and Putnam, 1985). Allelochemicals inhibit the nutrient uptake in cereals and legumes (Bansal and Kalia, 1984; Bhowmik and Doll, 1984; Buchholtz, 1971). Owing to their greater persistence in soil, weed residues offer a major source of allelochemicals in addition to leachates and root exudates. Weed toxicity is more pronounced in monoculture than in crop rotation and mixed cropping. Thus proper selection of crop species for different agricultural practices should be followed to minimize the crop loss. Apart from their inhibitory effects, allelochemicals of some weeds such as *Cyperus, Setaria, Agropyron* are known to increase the growth and yield of maize, *Sorghum,* soybean, wheat, and rye (Bhowmik and Doll, 1982; Chivinge, 1985; Hagin, 1989).

Crops Versus Weeds   Compared to weeds, only few crops have been evaluated for their potential to inhibit weeds in various cropping systems. In most cases, allelopathic effects are exerted either by growing crops or by crop residues. The major sources of allelochemicals are root exudates and leachates of growing crops (Einhellig and Souza, 1992). Allelopathic potential of some crop plants on weeds has been listed (Table 9-8). Growing crops such as barley (Putnam and De Frank, 1983), *Sorghum* (Weston et al., 1989), and soybean (James et al., 1988) are known to suppress weeds by their phytotoxic exudates (Einhellig and Souza, 1992). Crop residues from alfalfa (Abdul Rahman and Habib, 1989) and sunflower (Leather, 1983a) have been shown to be toxic to weeds, decreasing the seed germination. For a few crops such as wheat (Shilling et al., 1985), corn (Anaya et al., 1987), and soybean (Parkinson et al., 1987), both growing plants as well as their crop residues have equal potential to inhibit weeds.

A few crop plants such as wheat, pea (Purvis et al., 1985a, b), and sunflower (Leather, 1983a, b) are known to increase weed germination and growth, suggesting the differential sensitivity of weeds to crop allelochemicals. Wild species of crop plants have been shown to have a greater potential to inhibit weeds. Thus the genetic traits that regulate the allelochemical production can be exploited to breed new varieties with natural pesticide or weedicide properties.

TABLE 9-7 Allelopathic Potential of Weeds on Crops

Donor Crop	Target Crop	Inhibitory Effect	Chemical Nature	Reference
Abutilon theophrasti	Soybean, corn	Crop yield, seedling growth	nd	Holm et al. (1979)
Agropyron repens	Corn	Crop yield	nd	Bhowmik and Doll (1982)
Avena fatua	*Triticum aestivum*	Seedling growth	Scopoletin, *p*-HBA, VAN, COU	Perez and Ormeno-Nunez (1991b)
Carduus nutans	*Trifolium repense*	Nitrogen fixation	nd[a]	Wardle et al. (1994)
Cyperus esculentus	Sugarbeet, peas, soybean wheat, corn	Germination growth	*p*-HBA, VAN, SYR, FER, *p*-COU	Drost and Doll (1980); Mallik and Tesfai (1988)
Parthenium hysterophorus	Wheat, barley, sorghum, maize, groundnut, radish, pigeon pea, rape seed	Germination, seedling growth	Parthenin, CIN, cocornopilin, *p*-COU, alkaloids	Ambika and Jaya-chandra, (1980); Kohli et al. (1985); Srivastava et al. (1985)
Pluchea lanceolata	*Brassica juncea, Vigna unguiculata*	Seedling growth, germination, nodulation	Formononetin 7-0 glucoside	Inderjit and Dakshini (1991, 1992a)
Rorippa sylvestris	*Lactuca sativa*	Seedling growth	SAL, *p*-HBA, VAN, pyrocatechol, hirsutin, syringin, pyrocatechol, hirsutin, SYR	Yamane et al. (1992)

[a] nd = not determined.

280

TABLE 9-8 Allelopathic Potential of Crops on Weeds

Donor Plant	Target Plant	Inhibitory Effect	Chemical Nature	Reference
Ipomea batatus	*Cyperus esculentus*	Seedling growth	nd[a]	Harrison and Peterson (1991)
Sorghum bicolor	*Abutilon, Datura, Amaranthus retroflex, Setaria viridis*	Seedling growth	Sorgoleone	Einhellig Souza (1992)
Triticum aestivum, Secale cereale	*Avena fatua*	Seedling growth	Hydroxamic acid	Perez and Ormeno-Nunez (1991a)
Zea mays L.	*Chenopodium album*	Seedling growth	nd	Dzyubenko and Petrenko (1971)

[a] nd = not determined.

Crops Versus Crops   Allelopathic interactions among crop plants are of considerable importance when intercropping or crop-rotation are practiced. The effects can be beneficial or harmful, depending upon the crop species grown in a particular agricultural system (Table 9-9).

Monoculture   Under monoculture practices often autotoxicity of several crops has been noticed, due to the accumulation of phytotoxins from the decomposing residue of the previous crop. Autotoxic effects are well observed in rice (Chou et al., 1981), sugarcane (Wu et al., 1976), *Sorghum* (Burgos-Leon et al., 1980), and alfalfa (Miller et al., 1991).

Crop Rotation   Allelopathy has been implicated in yield decline in crops when grown after *Sorghum,* corn (Wozniak et al., 1981), mung bean (Ventura et al., 1984), and sunflower (Schon and Einhellig, 1982). The allelopathic effect of pearl millet on seedling growth of wheat, barley, lentil, chickpea, soybean, rape, and mustard was observed (Gupta *et al.,* 1987; Narwal, 1994). In most of the crop rotations, stimulatory as well as inhibitory effects have been observed, depending on the rotating crops. Soybean and corn rotation, for example, has been exploited by farmers for a long time and stimulatory effect of allelochemicals has been mostly observed in such rotation (Narwal, 1994). Similar increase in corn yields in rotation with tobacco has also been observed (Rizvi and Rizvi, 1987). In contrast, inhibition of growth and yield of rice when grown in rotation with tobacco has been observed (Rizvi and Rizvi, 1987). Similar yield reduction of *Sorghum* has been observed when grown in rotation with sunflower (Schon and Einhellig, 1982).

Intercropping   Allelopathic interactions play a greater role in determining the growth and yield of component crops in intercropping and mixed cropping systems. In these agricultural practices, legume association is often known to increase the yield of cereals by increasing the nutrient availability (Lastuvka, 1970), improving ion exchange (Rakhteenko et al., 1973), and providing partial weed control due to allelochemicals (Jimenez-Osornio et al., 1983). Allelochemicals of root exudates determine the growth of component crops. Moreover, all the plants in an agricultural system are not equally sensitive to allelochemicals. Therefore, ideal crop rotations and proper selection of crops for mixed cropping system will avoid the negative effects of allelochemicals and the beneficial crop combinations can be recommended for various agronomic practices.

Weeds Versus Weeds   Allelopathic effects of weeds on other weeds have not been extensively studied. Inhibitory effects of some weeds on other weeds has been shown however (Table 9-10). The allelopathic effect of weeds on other weeds may be due to competition for water and nutrients.

   The studies reveal the potential role of allelopathy in agroecosystems, particularly in the management of weeds in crop fields. Crop allelochemicals can be exploited for weed control by using phytotoxic crop residues as mulches and cover

TABLE 9-9 Allelopathic Potential of Crops on Other Crops

Donor Plant	Target Plant	Inhibitory Effect	Chemical Nature	Reference
Cumumis sativus	*Lactuca sativa*	Seedling growth	2-hydroxy benzothiazole, p-thiocyanato phenols p-HBA, CIN, PAL, MYR	Yu and Matsui (1994)
Glycine max	Wheat	Germination, growth, yield	nd	Huber and Abney (1986)
Helianthus annus	*Lactuca sativa*	Germination and growth	CIN, CAF, COU	Macias et al. (1993)
Medicago sativa	Barley, autotoxic, and grain yield	Growth, lint	FER, p-COU, SYR, CIN, BEN, VAN	Hairston et al. (1987); Hicks et al. (1989); Lodhi et al. (1987)
Parthenium argentatum	Cauliflower, cotton, guayule, lettuce, pepper, tomato	Seed germination	Anisic, BEN, GAL, p-HBA, CAF, VAN, CIN, COU	Scholoman et al. (1991)
Saccharum officinarum	Autotoxic	Yield	p-HBA, p-COU	Wu et al. (1976)
Triticum vulgare	Cotton, soybean	Yield, number of nodules, N_2 fixation	na[a]	Rice et al. (1960)

[a]nd = not determined.

TABLE 9-10 Allelopathic Potential of Weeds on Weeds

Donor Plant	Target Plant	Inhibitory Effect	Chemical Nature	Reference
Asparagus officinalis	Autotoxic	Yield and growth	CAF	Miller et al. (1991)
Festuca arundinacea	Birds foot trefoil	Growth	nd	Luu et al. (1989)
Mikania micrantha	Asystasia intrusa, Chrysopo-gan aciculatus, Paspalum conjugatum	Germination and seedling growth	nd[a]	Ismail and Mah (1993)
Polygonum sachalinense	Amaranthus viridis, Phleum pratense, crab grass	Seedling growth	Anthraquinone, emodin, physcion	Inoue et al. (1992)

[a]nd = not determined.

crops, by growing allelopathic crops in crop rotation, crop mixtures, or intercropping, by using trap crops to control parasite weeds, and finally and most importantly, by breeding crops for greater suppression effect on weeds. The proper selection of crops for crop rotation and intercropping can avoid the negative effects of crops and can reduce the use of herbicides and weedicides.

In pasture/grassland ecosystems some of the herbaceous weeds are known to occupy significant portion of pasture area present and have the potential to compete with other forage species by exerting the allelopathic effect (Wardle et al., 1994). Ahmed and Wardle (1994) demonstrated the ability of *Senecio* to inhibit seed germination and seedling growth of associated pasture forage species by release of allelochemicals. Different forage species in association with ragweed showed differential sensitivity to allelochemicals. In general, legume forage crops exhibited more sensitivity to allelochemicals than grass forage crops (Wardle et al., 1993). Thus it is of agronomic importance to study such interactions among forage crops and their responses to allelochemicals from various weed species and even other forage crops (Purvis, 1990).

Forestry Systems

Allelopathic interactions have been identified in all types of forests but to a different degree. Most of the time, such interactions have been shown to result in failure of forest trees to regenerate naturally, (Fisher, 1987; Horsley, 1987; Leibundgut, 1976), decrease in tree biomass production (Walters and Gilmore, 1976), or change in the ability of trees to associate with nitrogen-fixing bacteria (Jobidon and Thibault, 1981) and mycorrhizal fungi (Perry and Choquette, 1987). Rare are the reports of positive effects of allelopathy in forests ecosystems; hence it is of considerable importance in selection of tree species for various forestry programs.

Few studies have been done on allelopathy in tropical forests (Table 9-11). In such climate, leaching is the main source for the release of allelochemicals (Ferreira et al., 1992; Konar and Kushari, 1989), although relatively high temperatures allow volatilization (Oster et al., 1990). One of the more interesting observations concerning allelochemicals in tropical forests is that a positive effect is often observed, probably due to the high amount of rain throughout the year, which reduces the concentration of allelochemicals in soils and promotes seed germination or growth of target species.

Forests of desertic and semidesertic regions are very scarce, and it is well known that dynamics of vegetation do not exist in such regions (Barry, 1991). Few studies have been done in these regions with three dominant genera, *Eucalyptus* sp., *Acacia* sp., and *Cupressus* sp., which produce volatile allelochemicals (Friedman, 1987) and one can suppose that allelopathy is a component of competition for water (<200 mm yr^{-1}) between trees and "understory" species.

A lot of volatile aromatic allelochemicals are released in mediterranean forests (Table 9-11). Producer-plants are trees (*Quercus* sp., *Pinus* sp.) or phryganic plants (*Artemisia* sp., *Salvia* sp., *Origanum* sp.). In such ecosystems, subject to frequent fires (Thanos and Marcou, 1993), allelopathy is involved in vegetation dynamics.

TABLE 9-11 Allelopathy in Forest Ecosystems

Donor Plant	Target Plant	Inhibitory Effect	Chemical Nature	Reference
		Tropical Forests		
Eperua falcata	*Lepidium sativum*	Germination, seedling growth	Abscisic acid	Oster et al. (1990)
Quercus eugeniaefolia	*Cucumis* spp.	Seedling growth	nd[a]	Gliessman (1978)
		Desertic and Semidesertic Forests		
Eucalyptus microtheca	Understory spp.	Germination, seedling growth	α-Pinene, camphene, cineole	Al-Mousawi and Al-Naib (1976)
		Mediterranean Forests		
Acacia delabata	Understory spp.	Germination, seedling growth	Phenolic acids	Reigosa et al. (1984)
		Temperate Forests		
Athyrium filix-femina	*Picea abies*	Germination, seedling growth	Phenols	Pellissier (1993b)
Abies alba	*Understory spp.*	Germination, seedling growth	nd	Drapier (1983)
		Boreal Forests		
Empetrum herma-phroditum	*Pinus sylvestris*	Termination, seedling growth	Phenols	Nilsson (1992)

[a]nd = not determined.

Muller et al. (1968) showed that, allelochemicals decreased quantitatively in burned areas of a californian phryganic ecosystem.

Allelochemicals produced in temperate forests are mainly released by leaching or exudation from trees and understory species. A lot of studies have shown their effect (mostly negative) on tree seed germination, seedling growth, mycorrhizal infection, and biological processes involved in natural regeneration (Table 9-11). In such ecosystems, soil acts as a buffer of allelochemicals. Though adsorption to mineral colloids and detoxification by microorganisms reduces the biological effects of allelochemicals, water balance, low pH, and temperature tend to increase their effect. This means that allelopathy in temperate forest is most often a constraint that foresters aim to reduce.

Boreal forests, from the point of view of allelopathy, can be considered as a mixture of mediterranean (mediation by spontaneous fires) and temperate (soil-buffered allelopathy) forests. Modes of release and effects of allelochemicals are very similar to those described in temperate forests, but postfire regeneration can also occur.

Agroforestry System

Several nitrogen-fixing perennials are being used in tropical agroforestry systems. *Leucaena leucocephala* is one prominent example and is an important multipurpose tree used for forage, fuel, paper, pulp, timber, and for soil improvement (Budelman and Vandapol, 1992; Mittal et al., 1992). The only problem with *Leucaena* is that its foliage contains potential allelochemicals (Chou and Kuo, 1986; Prasad and Subhashini, 1994; Reddy and Prasad, 1989) (Table 9-12).

Chou and Kuo (1986) reported a number of allelochemicals, namely gallic, protocatechuic, *p*-hydroxy phenylacetic, *p*-hydroxy benzoic, vanillic, caffeic, syringic, cinnamic, and ferulic acids [determined by paper and thin layer chromatography,

TABLE 9-12 Quantitative Composition of Phytotoxic Phenolics Present in Leaves of *Leucaena leucocephala* Variety K28

Phytotoxic Phenolics	Amount ($\times 10^{-4}$ μmol)	
	Young	Mature
Gallic acid	4.215	2.153
Protocatechuic acid	12.05	1.816
p-Hydroxybenzoic acid	2.439	1.551
p-Hydroxy phenylacetic acid	nd	2.685
Vanillic acid	0.467	0.492
Caffeic acid	3.772	2.637
p-Hydroxycinnamic acid	9.528	4.015
Ferulic acid	1.323	0.343
Total	33.746	15.692

Source: From Chou and Kuo (1986), with the permission of the authors and of Plenum Publishing Corp., USA.

UV-visible spectrophotometry, and high performance liquid chromatography (HPLC)] (Fig. 9-5) from *Leucaena* leaf and soil extracts of *Leucaena* high density plantation (40,000 ha^{-1}) at Kaoshu, in Taiwan. *Leucaena* leaf mulch exhibited allelopathic effect on weeds in a grassland ecosystem. In high-density *L. leucoceph- ala* plantations of Taiwan, it was observed that 80% of the ground was bare on the *Leucaena* floor, when compared to the grassland control area (Chou, 1986) (Figs. 9-6 and 9-7). Using HPLC, several allelochemicals, namely gallic, *p*-hydroxy ben- zoic, ferulic, and protocatechuic acids, have been identified from the leaf and soil extracts of a *L.leucocephala* high-density plantation (20,000 ha^{-1}) in the Hydera- bad, Deccan region of India (Reddy and Prasad, 1989). The above allelochemicals, including mimosine from foliage and seeds, are released by leaching and decompo- sition of *L. leucocephala* residues and also as root exudates.

Leaf mulches of *Leucaena* suppressed the growth of crop plants, weeds, and some forest plant species. Mimosine inhibited growth of crop plants, namely, *Phaseolus mungo, Oryza sativa,* and forest plant species *Acacia confusa, Alnus formosana, Causurina glauca,* and so on (Chou and Kuo, 1986; Prasad and Subhas-

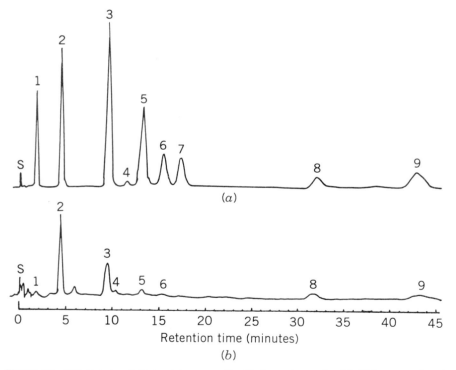

Figure 9-5. *The spectra of high-performance liquid chromatography. (a) Of known authentic compounds. (b) Of* Leucaena *K28 leaf extract. "S" indicates the solvent peak of the starting point. The numbers indicated on the peaks indicate: (1) gallic acid, (2) protocatechuic acid, (3) p-hydroxy benzoic acid, (4) p-hydroxy phenylacetic acid (5) vanillic acid, (6) caffeic acid, (7) syringic acid, (8) p-hydroxy cinnamic acid, (9) ferulic acid. (With the permission of Chou and Kou, 1986, and Plenum Publishing Corp., USA.)*

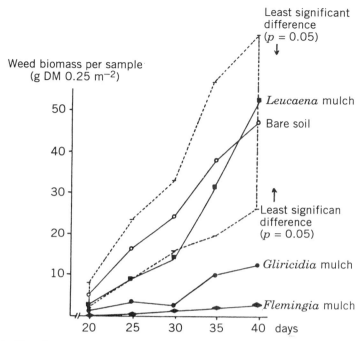

Figure 9-6. *The effect of mulches of* Gliricidia sepium, Flemingia macrophylla, *and* Leucaena leucocephala *on the development of the weed biomass as compared to the weed biomass collected on an unmulched soil. (Note: The area in the figure within the dotted lines represents the weed biomass quantities that do not significantly differ from the weed biomass found on the untreated experimental units.) (Reprinted by permission of Budelman, 1988 and Kluwer Academic Publishers, The Netherlands).*

hini, 1994). Leaf dry matter productivity of *L. leucocephala* in humid tropical environment has been investigated, including studies on characteristics of leaf mulches and nutrient release during decomposition (Table 9-13). Such nutrient release is useful in alley-cropping system, while the effects of leaf mulch are useful for weed control in a cropping environment (Budelman, 1988a, b, c, 1989a, b).

The suppression of the growth of a number of field crops by several tree species, namely *Eucalyptus, Acacia nilotica, Azadirachta, Dalbergia, Zizyphus, Morus alba,* and *Prosopis,* was attributed to allelopathy. However, a great variability in response of crops to given tree species has been observed, and this needs to be considered while selecting the crops and trees for a successful agroforestry system (Narwal, 1994).

Auxiliary roles (viz. capacity for leaf dry matter productivity, nutrient composition, decomposition of leaf mulch, retarded weed emergence, effect of leaf mulches in soil, temperature reduction at different depths, effect of leaf mulch on soil moisture conservation, etc.) of perennials need to be investigated for their allelopathic interactions in different cropping systems (Budelman, 1988a, b, c, 1989a, b; Budelman and Vandapol, 1992; Mittal et al., 1992; Tomar et al., 1992).

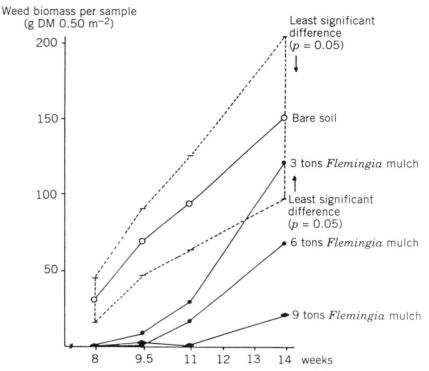

Figure 9-7. *The effect of increasing quantities of mulch of* Flemingia macrophylla *on the development of the weed biomass as compared to the weed biomass collected on an unmulched soil. (Note: The area in the figure within the dotted lines represents the weed biomass quantities that do not significantly differ from the weed biomass found on the untreated experimental units.) (Reprinted by permission of Budelman, 1988 and Kluwer Academic Publishers, The Netherlands).*

TABLE 9-13   Dry Matter Content, Decomposition and Leaf Surface-to-Weight Ratio of the Leaf Mulch Materials Used[a]

Species	Dry Matter Content (%)		Half-Life Value (days)	Leaf Surface to Weight Ratio ($cm^2 \, g^{-1}$ DM)
	2 Year Old Trees	3 Year Old Trees		
Flemingia macrophylla	28.6	27.6	53.4	214.5
Glicricidia sepium	21.3	—	21.9	180.2
Leucaena leucocephala	26.5	—	30.7	193.7

[a] *Note:* The "half-life" value represents the number of days to reach a 50% loss of the mulch material.

Source: Budelman (1988a), with the permission of the author and Kluwer Academic Publishers, The Netherlands.

ALLELOPATHY IN AN AQUATIC ECOSYSTEM

Compared to a terrestrial ecosystem, allelopathy in an aquatic ecosystem is fragmentary and inadequately studied. Though the real mechanisms operating under such conditions have not been investigated to any great extent, a few studies have been made in this direction. The allelopathic mechanisms operating at various levels in an aquatic ecosystem have been critically reviewed (Gopal and Goel, 1993). For many years, the inhibition of phytoplankton growth by macrophytes has been mostly ascribed to competition for nutrients, light, and pH changes. However, with recent advances in chemistry, a number of chemicals with allelopathic potential have been identified from aquatic plants, chemicals that play a greater role in structure and dynamics of aquatic plant communities.

Many of the *Eleocharis* spps. have been shown to inhibit growth of Cyanophyceae and Chlorophyceae members (Van Aller et al., 1985) and aquatic weeds such as *Potamogeton, Hydrilla,* and *Elodea* (Ashton et al., 1985) in laboratory as well as field conditions. *Eleocharis* sps. are also known to decrease nitrogen content of *Lemna* (Sutton and Portier, 1991). Autotoxicity of *Typha* has been demonstrated by McNaughton (1968), who suggested a role in population regulation in these communities. Aqueous extracts of *Azolla* and *Potamogeton* have been shown to inhibit the growth of *Lemna* (Sutton and Portier, 1989). Inability of *Ceratophyllum* spp. to grow together with *Hydrilla* has been observed (Kulshreshtha and Gopal, 1983). Various allelochemicals isolated from *Pistia,* such as linoleic acid and gamma r-linolenic acids, are shown to interfere with growth of *Scenedesmus* and *Selenastrum* spp. Inhibition of shoot weight and reduction in number of tubers of *Hydrilla* has been observed in the presence of *Sagittaria* (Sutton, 1990).

In addition to living plants, various dead and decomposing macrophytes are shown to inhibit the growth of several other species that grow in a particular aquatic system. Carter and Grace (1986) observed the inability of *Polygonum* to grow in the decomposing litter of *Justicia.* The soils dominated by *Typha* are well known for their toxicity due to allelochemicals release, as is the inability of *Typha* to further establish due to autotoxication. In an aquatic ecosystem, littoral zones produce a large amount of allelochemicals. Szczepanska (1971) observed the growth inhibition of *Phragmites* by plant litter of *Eleocharis, Schoenoplectus, Typha,* and *Equisetum.* Decomposing *Taxodium distichum* is shown to inhibit the growth of *Hydrilla* (Dooris et al., 1988) and *Scenedesmus* (Dooris et al., 1982). Suppression of phytoplankton blooms by *Chara* was observed (Crawford, 1979; Horecka, 1991). Inhibitory nature of *Brasenia schreberi* to *Chlorella* and *Anabaena* is reported (Elakovich and Wotten, 1987).

Several phytoplanktonic algae are known to inhibit macrophyte growth. The growth inhibition of *Elodea* and duckweed by *Microcystis* was reported by Kirpenko (1986). Several planktonic algae such as *Scenedesmus, Chlorella, Aphanotheca* spp., *Euglena, Merismopedia,* and *Coelastrum* are inhibitory to *Eicchornia crassipes* (Sharma, 1985).

The chemistry of aquatic allelochemicals reveals a wide range of compounds from both living and decomposing aquatic plants. Among these the most prominent are the long chain fatty acids, which comprise a large portion of allelochemicals from aquatic plants. Chlorellin, an allelochemical produced by *Chlorella* (green alga), is a long chain fatty acid (Rice, 1984). Long chain fatty acids with allelopathic activity are also isolated from *Polygonum* (Van Aller et al., 1985), *Eleocharis* (Stevens and Merrill, 1980), and *Typha* (Aliotta et al., 1990). Most of the Charophyceae members are reported to contain another group of allelochemicals called allomones such as dithiolane and trithiolane (Anthoni and Christopherson, 1982).

Several of the Rhodophyceae members are known to contain a variety of sulfur-containing compounds, which are found to inhibit the growth of *Erythrophyllum* (Izac et al., 1982), *Ceramium* (Iakawa et al., 1973), and *Ceratophyllum* (Wium-Anderson et al., 1982). Similar to terrestrial plants, a number of aquatic plants such as *Zostera, Vallisneria,* and several members of Lemnaceae are known to contain a variety of phenolic compounds, namely, ferulic, vanillic, hydroxy benzoic, caffeic, gallic, protocatechuic, and gentisic acids (Cheng and Riemer, 1989; Quackenbush et al., 1986; Wallace et al., 1969).

Flavonoids are also found to occur largely in several members of Lemnaceae (McClure, 1970). Most of the Nymphaeceae members are abundant in alkaloids and are known to have greater allelopathic potential.

Thus the allelopathy in an aquatic system plays an important role in regulating the organization and dynamics of aquatic plant communities. The allelopathic interactions in an aquatic ecosystem can be exploited for biological control of some of the noxious plants by introducing a plant with greater allelopathic potential or artificially introducing the allelochemical into a body of water infested with weeds.

CONCLUSIONS

As the impact of allelochemicals on the plant growth is increasing, the need to understand the mechanism of action of these chemicals has become inevitable to exploit them in the management of agroecosystems for better yields.

For proper management of weeds and crops, these compounds should be tested in various combinations to find out the threshold concentrations for their activity. Crop varieties need to be screened, or new varieties need to be developed, to exploit the potential of allelochemicals for controlling weeds. Similarly, if crop varieties allelopathic to pathogens are developed, their residues can be used simultaneously for disease control.

In agroforestry, the agricultural crops are grown together with forest plants to increase the yields. However, the major limitation of this system has been the release of allelochemicals during the decomposition of forest plants litter, allelochemicals which are toxic to crop plants. Therefore, it is necessary management to select the complementary tree species and crop plants for better yields in agroforestry.

Allelochemicals in the synthesis of the biopesticides and growth promoters is emerging as another important area. Since most of the chemicals are known to be toxic to a variety of weeds and pests, they can be exploited as natural pesticides, which can replace the synthetics (Miles et al., 1993)

The beneficial effects of allelochemicals can be used as growth promoters. Developments in crop management based on allelopathic studies could not only increase production but reduce expenditures on farm labor, and use of synthetic agrochemicals. It is even possible that allelochemical production can be induced by genetic manipulation into cultivars, to provide an inexpensive, safe, and permanent means of biological pest control.

REFERENCES

Abdul-Rahman, A. A. and S. A. Habib (1989), *J. Chem. Ecol., 15*, 2289–2300.

Aerts, R. J., W. Snoeijer, E. V. D. Meijdent, and R. Verpoorte (1991), *Phytochem., 30*, 2947–2951.

Ahmed, M. and D. A. Wardle (1994), *Plant Soil, 164*, 61–68.

Akram, M. and F. Hussain (1987), *Pak. J. Sci. Ind. Res., 30*, 918–921.

Aldrich, R. J. (1987), in *Allelochemicals: Role in agriculture and forestry* (G. R. Waller, Ed.), Vol. 330, American Chemical Society, Washington, DC, pp. 300–312.

Aliotta, G., M. Della Greca, P. Monacco, G. Pinto, A. Pollio, and L. Previtera (1990), *J. Chem. Ecol., 16*, 2637–2646.

Aliotta, G., G. Cafiero, V. De Feo, and R. Sacchi (1994), *J. Chem. Ecol., 20*, 2761–2775.

Al-Mousawi, A. H. and F. A. G. Al-naib (1976), *Bull. Biol. Res. Cent., 7*, 17–23.

Alsaadawi, I. S. and E. L. Rice (1982), *J. Chem. Ecol., 8*, 993–1009.

Alsaadawi, I. S., J. K. Aluquili, S. M. Alhadithy, and A. J. Alrubea (1985), *J. Chem. Ecol., 11*, 1737–1746.

Ambasht, R. S. and Y. Rai (1981), *Annu. Conf. Indian. Soc. Weed Sci.* (abstracts), Indian Society of Weed Science, Hyderabad, India, p. 2.

Ambika, S. R. and Jayachandra (1980), *Curr. Sci., 49*, 874–875.

Anaya, A. L., L. Ramos, R. Cruz, J. G. Hernandez, and V. Nava (1987), *J. Chem. Ecol., 13*, 2083–2101.

Anthoni, U. and A. Christopherson (1982), *Tetrahedron, 38*, 2425–2427.

Armstrong, G. M., L. M. Rohrbaugh, E. L. Rice, and S. H. Wender (1971), *Proc. Pak. Acad. Sci., 51*, 41–43.

Arntzen, C., S. V. Falkanthal, and S. Bobick (1974), *Plant Physiol., 53*, 304–306.

Ashton, F. M., J. M. Ditomasco, and L. W. J. Anderson (1985), *J. Aquati. Plant Manage., 22*, 52–56.

Avers, C. J. and R. H. Goodwin (1956), *Am. J. Bot., 43*, 613–620.

Balke, N. E. (1985), in *The chemistry of allelopathy* (A. C. Thompson, Ed.), Vol. 268, American Chemical Society, Washington, DC, pp. 161–178.

Bansal, G. L. and A. K. Kalia (1984), *J. Nucl. Agric. Biol., 13*, 89–90.

Barry, J. P. (1991), *Rev. Geographie Alpine, 1*, 55–70.

Berg, B. (1986), *For. Ecol. Manage., 15*, 195–213.

Berg, B. and G. Agren (1984), *Can. J. Bot.*, *62*, 2880–2888.

Berg, B. and H. Staff (1980), *Ecol. Bull.*, *32*, 373–390.

Bergmark, C. L., W. A. Jackson, R. J. Volk, and U. Blum (1992), *Plant Physiol.*, *98*, 639–645.

Bhowmik, P. C. and J. D. Doll (1982), *Agron. J.*, *74*, 601–606.

Bhowmik, P. C. and J. D. Doll (1983), *J. Chem. Ecol.*, *9*, 1263–1280.

Bhowmik, P. C. and J. D. Doll (1984), *Agron. J.*, *76*, 383–388.

Blaschke, H. (1977), *Karst. Flora, Bd.*, *166*, 203–209.

Blum, U. and B. R. Dalton (1985), *J. Chem. Ecol.*, *11*, 279–301.

Blum, U. and J. Rebbeck (1989), *J. Chem. Ecol.*, *15*, 917–928.

Blum, U. and R. Shafer (1988), *Soil Biol. Biochem.*, *20*, 793–800.

Blum, U., B. R. Dalton, and J. O. Rawlings (1984), *J. Chem. Ecol.*, *10*, 169–191.

Blum, U., B. R. Dalton, and J. R. Shann (1985), *J. Chem. Ecol.*, *11*, 619–641.

Blum, U., T. M. Gerig, A. D. Worsham, L. D. Holappa, and L. D. King (1992), *J. Chem. Ecol.*, *14*, 2191–2222.

Bolwell, G. P., C. L. Cramer, C. J. Lamb, W. Schuch, and R. A. Dixon (1986), *Planta*, *169*, 97–107.

Booker, F., U. Blum, and D. L. Discus (1992), *J. Exp. Bot.*, *43*, 649–655.

Boufalis, T. and F. Pellissier (1994), *J. Chem. Ecol.*, *20*, 2283–2289.

Bradow, J. M. and W. J. Connick, Jr. (1987), *J. Chem. Ecol.*, *13*, 185–202.

Bradow, J. M. and W. J. Connick, Jr. (1988a), *J. Chem. Ecol.*, *14*, 1617–1632.

Bradow, J. M. and W. J. Connick, Jr. (1988b), *J. Chem. Ecol.*, *14*, 1633–1648.

Buchholtz, K. P. (1971), in *Biochemical interactions among plants* (Environ. Physiol. Subcomm. US Natl. Comm. for IBP, Eds.), National Academy of Science, Washington, DC, pp. 86–89.

Budelman, A. (1988a), *Agrofor. Syst.*, *6*, 137–145.

Budelman, A. (1988b), *Agrofor. Syst.*, *7*, 33–45.

Budelman, A. (1988c), *Agrofor. Syst.*, *7*, 47–62.

Budelman, A. (1989a), *Agrofor. Syst.*, *8*, 39–51.

Budelman, A. (1989b), *Agrofor. Syst.*, *8*, 53–66.

Budelman, A. and F. Vandapol (1992). *Agrofor. Syst.*, *19*, 187–206.

Burbott, A. J. and W. D. Loomis (1967), *Plant Physiol.*, *42*, 20–28.

Burgos-Leon, W., F. Gaury, R. Nicou, J. L. Chopart, and Y. Dommergues (1980), *Agron. Trop.*, *35*, 319–334.

Bushra, I., H. Farrukh, and B. Forhat (1987), *Pak. J. Sci. Ind. Res.*, *30*, 550–553.

Carter, M. F. and J. C. Grace (1986), *Aquat. Bot.*, *23*, 341–349.

Chander, P., D. S. Bhatia, and C. P. Malik (1988), *Ind. J. Plant Physiol.*, *31*, 217–221.

Chapagne, D. E., J. T. Arnason, B. J. R. Philogene, P. Morand, and J. Lam (1986), *J. Chem. Ecol.*, *12*, 835–858.

Cheng, T. S. and D. N. Riemer (1989), *J. Aquat. Plant Manage.*, *27*, 84–89.

Chivinge, O. A. (1985), *Zimbabwe. Agric. J.*, *82*, 151–152.

Choesin, D. N. and R. E. Boerner (1991), *Am. J. Bot.*, *78*, 1083–1090.

Chou, C. H. (1982), in *Proceedings of seminar on allelochemics and pheramones*, Taipei, R. O. C., pp. 27–64.

Chou, C. H. (1986), in *The science of allelopathy* (A. R. Putnam and C. S. Tang, Eds.), Wiley, New York, pp. 57–74.

Chou, C. H. and S. J. Chiou (1979), *J. Chem. Ecol.*, *5*, 839–859.

Chou, C. H. and Y. L. Kuo (1986), *J. Chem. Ecol.*, *12*, 1431–1448.

Chou, C. H., Y. C. Ching, and H. H. Cheng (1981), *J. Chem. Ecol.*, *7*, 741–752.

Chou, C. H., T. J. Lin, and C. I. Kao (1977), *Bot. Bull. Acad. Sin.*, *18*, 45–60.

Choudhary, B. L. (1990), *Comp. Physiol. Ecol.*, *15*, 87–90.

Chouteau, J. and J. Loche (1965), *C.R. Acad. Sci.*, *260*, 4568–4588.

Colton, C. E. and F. A. Einhellig (1980), *Am. J. Bot.*, *67*, 1407–1413.

Corcoran, M. R., T. A. Geissman, and B. O. Phinney (1972), *Plant Physiol.*, *49*, 323–330.

Crawford, S. A. (1979), *Hydrobiol.*, *62*, 17–31.

Cutler, H. G. (1985), in *Bioregulators for pest control* (P. A. Hedin, Ed.), Vol. 276, American Chemical Society, Washington, DC, pp. 455–468.

Danks, M. C., J. S. Fletcher, and E. L. Rice (1975), *Am. J. Bot.*, *62*, 749–755.

Dear, J. and S. Arnoff (1965), *Plant Physiol.*, *40*, 458–459.

del Moral, R. (1972), *Oecologia, 9*, 289–300.

del Moral, R. and C. H. Muller (1970), *Am. Midl. Nat.*, *83*, 254–282.

Devi, S. R. (1994), Ph.D. Thesis, University of Hyderabad, Hyderabad, India.

Devi, S. R. and M. N. V. Prasad (1990), *Biochem. Arch.*, *6*, 75–82.

Devi, S. R. and M. N. V. Prasad (1992), *J. Chem. Ecol.*, *18*, 1981–1990.

Devi, S. R. and M. N. V. Prasad (1996a), *Biol. Plant, 38*, 387–395.

Devi, S. R. and M. N. V. Prasad (1996b), Photosynthetica, 32, 117–127.

Dhir, K. K., L. Rao, K. J. Singh, and K. S. Clark (1992), *Biol. Plant, 34*, 409–413.

Dooris, P. M. and D. F. Martin (1988), *J. Aquat. Plant Manage.*, *26*, 72–73.

Dooris, P. M., W. S. Silver, and D. F. Martin (1982), *J. Environ. Sci. Health, 17*, 639–646.

Downum, K. R. (1992), *New Phytol.*, *122*, 401–420.

Drapier, J. (1983), These. Doct. 3 cycle, Univ. Nancy I et INRA Champenoux, France, 109 pp.

Drost, D. C. and J. D. Doll (1980), *Weed Sci.*, *28*, 229–233.

Dzyubenko, N. N. and N. I. Petrenko (1971), in *Physiological-biochemical basis of plant interactions in phytocenosis* (A. M. Gnodzinsky, Ed.), Vol. 2, Naukova, Dumka, Kiev, Ukraine, pp. 60–66.

Einhellig, F. A. (1971), *Proc. S.D. Acad. Sci.*, *50*, 205–209.

Einhellig, F. A. (1987), in *Allelochemicals: Role in agriculture and forestry* (G. R. Waller, Ed.), Vol. 330, American Chemical Society, Washington, DC, pp. 343–357.

Einhellig, F. A. and P. C. Ecrich (1984), *J. Chem. Ecol.*, *10*, 161–170.

Einhellig, F. A. and L. Kuan (1971), *Bull. Torrey. Bot. Club, 98*, 155–162.

Einhellig, F. A. and M. K. Schon (1982), *Can. J. Bot.*, *60*, 2923–2930.

Einhellig, F. A. and I. F. Souza (1992), *J. Chem. Ecol.*, *18*, 1–11.

Einhellig, F. A., E. L. Rice, P. G. Risser, and S. H. Wender (1970), *Bull. Torrey. Bot. Club, 97*, 22–33.

Einhellig, F. A., M. Stille Muth, and M. K. Schon (1985), in *The chemistry of allelopathy* (A. C. Thompson, Ed.), Vol. 268, American Chemical Society, Washington, DC, pp. 170–195.

Elakovich, S. D. and J. W. Wooten (1987), *J. Chem. Ecol., 13,* 1935–1940.

Ells, J. E. and A. E. McSay (1991), *Hort. Sci., 26,* 368–370.

Eussen, J. H. H. and G. J. Niemann (1981), *Z. Pflanzen Physiol., 102,* 263–268.

Ferreira, A. G., M. E. A. Aquila, U. S. Jacobi, and V. Rizvi (1992), in *Allelopathy: Basic and applied aspects* (S. J. H. Rizvi and V. Rizvi, Eds.), Chapman and Hall, New York, pp. 243–250.

Fischer, N. H., G. B. Williamson, J. D. Weidenhamer, and D. R. Richardson (1994), *J. Chem. Ecol., 20,* 1355–1380.

Fisher, R. F. (1987), in *Allelochemicals: Role in agriculture and forestry* (G. R. Waller, Ed.), Vol. 330, American Chemical Society, Washington, DC, pp. 176–184.

Fisher, R. F., R. A. Woods, and M. R. Glavicic (1978), *Can. J. For. Res., 8,* 1–9.

Friedman, J. (1987), in *Allelopathy: Role in agriculture and forestry* (G. R. Waller, Ed.), Vol. 330, American Chemical Society, Washington, DC, pp. 53–68.

Gershenzon, J. (1993), in *Insect-plant interactions* (E. A. Bernays, Ed.), Vol. 5, CRC Press, Boca Raton, FL, pp. 105–173.

Glass, A. D. M. (1973), *Plant Physiol., 51,* 1037–1041.

Glass, A. D. M. (1974), *J. Exp. Bot., 25,* 1104–1113.

Glass, A. D. M. (1975), *Phytochemistry, 14,* 2127–2130.

Glass, A. D. M. (1976), *Can. J. Bot., 54,* 2440–2444.

Glass, A. D. M. and J. Dunlop (1974), *Plant Physiol., 54,* 855–858.

Gliessman, S. R. (1978), *Trop. Ecol., 19,* 200–208.

Goel, U. (1987), *Acta Bot. Ind., 15,* 129–130.

Gopal, B. and U. Goel (1993), *Bot. Rev., 59,* 155–210.

Grechkanev, O. M. and V. I. Radionov (1971), in *Physiological-biochemical basis of plant interactions in phytocenosis* (A. M. Grodzinsky, Ed.), Vol. 2, Naukova, Dumka, Kiev, Ukraine, pp. 88–100.

Guenzi, W. D., T. M. McCalla, and F. A. Norstadt (1967), *Agron. J., 59,* 163–165.

Gupta, K., S. S. Narwal, and D. S. Wagle (1987), *Haryana Agric. Univ. J. Res., 17,* 208–215.

Hagin, R. D. (1989), *J. Agric. Food. Chem., 37,* 1143–1149.

Hairston, J. E., J. O. Stanford, D. F. Pope, and D. A. Hormeck (1987), *Agron. J., 79,* 281–286.

Hall, A. B., U. Blum, and R. G. Fites (1983), *J. Chem. Ecol., 9,* 1213–1222.

Harborne, J. B. (1980), in *Secondary plant products* (E. A. Bells and B. V. Charlwood, Eds.), Springer Verlag, Berlin, Heidelberg, Germany.

Harper, J. R. and N. Balke (1981), *Plant Physiol., 68,* 1349–1353.

Harrison, H. F. and J. K. Peterson (1991), *Weed Sci., 39,* 308–312.

Hayeshi, T., S. Todoriki, and A. Nagao (1992), *Env. Exp. Bot., 32,* 265–271.

Hegde, R. S. and D. A. Miller (1990), *Crop Sci., 30,* 1255–1259.

Hegde, R. S. and D. A. Miller (1992), *Agron. J., 84,* 940–946.

Hicks, S. K., C. W. Wendt, J. R. Gannaway, and R. B. Baker (1989), *Crop Sci., 29,* 1057–1061.

Hoagland, R. E. and R. D. Williams (1985), in *The chemistry of allelopathy* (A. C. Thompson, Ed.), Vol. 268, American Chemical Society, Washington, DC, pp. 301–326.

Holappa, L. D. and U. Blum (1991), *J. Chem. Ecol., 17,* 865–886.

Holm, L., J. V. Pancho, J. P. Herberger, and D. L. Plucknett (1979), *A geographical atlas of world weeds,* Wiley, New York.

Hoover, J. D., S. H. Wender, and E. D. Smith (1977), *Phytochemistry, 16,* 199–201.

Horecka, M. (1991), *Arch. Protistenkd., 139,* 275–278.

Horsley, S. B. (1987), in *Allelochemicals: Role in agriculture and forestry* (G. R. Waller, Ed.), Vol. 330, American Chemical society, Washington, DC, pp. 205–212.

Huber, D. M. and T. S. Abney (1986), *J. Agron. Crop Sci., 157,* 73–78.

Iakawa, M., V. M. Thomas, L. J. Buckley, and J. J. Uebel (1973), *J. Phycol., 9,* 302–305.

Ibrahim, S., I. S. Alsaadawi, E. L. Rice, and K. B. Tommy (1983), *J. Chem. Ecol., 6,* 761–774.

Inderjit and K. M. M. Dakshini (1991), *J. Chem. Ecol., 17,* 343–352.

Inderjit and K. M. M. Dakshini (1992a), *J. Chem. Ecol., 18,* 713–718.

Inderjit and K. M. M. Dakshini (1992b), *Am. J. Bot., 79,* 977–981.

Inoue, M., H. Nishimura, H. H. Li, and J. Mizutani (1992), *J. Chem. Ecol., 18,* 1833–1840.

Ishii, T. and H. Saka (1992), *Plant Cell Physiol., 33,* 321–324.

Ismail, B. S. and L. S. Mah (1993), *Plant Soil, 157,* 107–113.

Izac, R., D. Stierle, and J. Sims (1982), *Phytochemistry, 21,* 229.

Jacobsen, A. and M. R. Corcoran (1977), *Plant Physiol., 59,* 129–133.

Jain, A. and H. S. Srivastava (1981), *Physiol. Plant, 51,* 339–342.

James, K. L., P. A. Banks, and K. J. Karnok (1988), *Weed Technol., 2,* 404–409.

Jankay, P. and W. H. Muller (1976), *Am. J. Bot., 63,* 126–132.

Jarvis, B. B., N. B. Pena, M. M. Rao, N. S. Comezoglu, T. F. Comezoglu, and B. N. Mandava (1985), in *The chemistry of allelochemicals* (A. C. Thompson, Ed.), Vol. 268, American Chemical Society, Washington, DC, pp. 149–160.

Jimenez-Osornio, J. J. and S. R. Gliessman (1987), in *Allelochemicals. Role in agriculture and forestry* (G. R. Waller, Ed.), Vol. 330, American Chemical Society, Washington, DC, pp. 262–274.

Jimenez-Osornio, J. J., K. Schultz, A. L. Anaya, J. Nieto de Pascual, and O. J. Espejo (1983), *J. Chem. Ecol., 9,* 1011–1025.

Jobidon, R. and J. R. Thibault (1981), *Bull. Torrey Bot. Club, 108,* 413–418.

Kamisaka, S., S. Takeda, K. Takahasi, and K. Shibata (1990), *Physiol. Plant, 78,* 1–7.

Kanchan, S. D. and Jayachandra (1979), *Plant Soil, 53,* 27–35.

Kathiresan, K., G. A. Ravishankar, and L. V. Venkataraman (1990), *Curr. Sci., 59,* 430–431.

Kil, B. S. and K. W. Yun (1992), *J. Chem. Ecol., 18,* 39–51.

Kil, B. S., K. W. Yun, and S. Y. Lee (1992), *J. Chem. Ecol., 18,* 1455–1462.

Kim, Y. S. and B. S. Kil (1987), *Korean J. Bot., 30,* 59–68.

Kirpenko, N. I. (1986), *Gidrobiol. Zhurn.*, *22*, 48–50.

Klein, K. and U. Blum (1990), *J. Chem. Ecol.*, *16*, 455–463.

Koeppe, D. E. and R. J. Miller (1974), *Plant Physiol.*, *54*, 374–378.

Koeppe, D. E., L. M. Rohrbaugh, E. L. Rice, and S. H. Wender (1970a), *Radia. Bot.*, *10*, 261–265.

Koeppe, D. E., L. M. Rohrbaugh, E. L. Rice, and S. H. Wender (1970b), *Phytochemistry*, *9*, 297–301.

Koeppe, D. E., L. M. Southwick, and J. E. Bittell (1976), *Can. J. Bot.*, *5*, 593–599.

Kohli, R. K., A. Kumari, and D. B. Saxena (1985), *Acta Univ. Agric. (Brno.) Fac. Agron.*, *33*, 253–264.

Konar, J. and D. P. Kushari (1989), *Bull. Torrey Bot. Club*, *116*, 339–343.

Kulshreshtha, M. and B. Gopal (1983), *Aquat. Bot.*, *16*, 207–209.

Kumari, A. and R. K. Kohli (1987), *Weed Sci.*, *35*, 629–632.

Laloraya, M. M., C. Nozzolillo, S. Purohit, and S. Lisa (1986), *Plant Physiol.*, *81*, 253–258.

Lang, D. R. and E. Racker (1974), *Biochem. Biophys. Acta.*, *333*, 180–186.

Lastuvka, Z. (1970), in *Physiological-biochemical basis of plant interactions in phytocenosis* (A. M. Grodzinsky, Ed.), Vol. 1, Narkova, Dumka, Kiev, Ukraine, pp. 37–40.

Leather, G. R. (1983a), *J. Chem. Ecol.*, *9*, 983–989.

Leather, G. R. (1983b), *Weed Sci.*, *31*, 37–42.

Lee, T. T., A. N. Stafford, J. J. Jevnikar, and A. Stoessl (1980), *Phytochemistry*, *19*, 2277–2280.

Lehman, R. H. and E. L. Rice (1972), *Am. Midl. Nat.*, *87*, 71–80.

Leibundgut, C. (1976), *J. For. Sci.*, *9*, 621–635.

Li, H. H., M. Inoue, H. Nishimura, J. Mizutani, and E. Tsuzuki (1993), *J. Chem. Ecol.*, *19*, 1775–1787.

Liang, X., M. Dron, C. Cramer, R. A. Dixon, and C. J. Lamb (1989), *J. Biol. Chem.*, *264*, 14486–14498.

Liu, D. L. and J. V. E. Lovette (1993), *J. Chem. Ecol.*, *19*, 2231–2244.

Loche, J. and J. Chouteau (1963), *C.R. Hebd. Seances Acad. Agric. Fr.*, *49*, 1017–1026.

Lodhi, M. A. K. and G. L. Nickell (1973), *Bull. Torrey Bot. Club*, *100*, 159–165.

Lodhi, M. A. K., R. Bilal, and K. K. A. Milik (1987), *J. Chem. Ecol.*, *13*, 1881–1882.

Lolas, P. C. (1986), *Weed Res.*, *26*, 1–8.

Lovett, J. V. (1986), in *Science of allelopathy* (A. R. Putnam and C. S. Tang, Eds.), Wiley, New York, pp. 75–100.

Lovett, J. V. (1987), in *Allelochemicals, Role in agriculture and forestry* (G. R. Waller, Ed), Vol. 330, American Chemical Society, Washington, DC, pp. 156–175.

Lovett, J. V. and K. Jokinen (1984), *J. Agric. Sci.*, *56*, 1–7.

Lovett, J. V., J. Levitt, A. M. Duffield, and W. G. Smith (1981), *Weed Res.*, *21*, 165–170.

Luu, K. T., A. G. Matches, C. J. Nelson, E. J. Peters, and G. B. Garner (1989), *Crop Sci.*, *29*, 407–412.

Lyu, D. G. and U. Blum (1990), *J. Chem. Ecol.*, *16*, 2429–2439.

Macias, F. A., R. M. Verela, A. Torres, and J. G. Mollinillo (1993), *Phytochemistry, 34,* 669–674.

Macias, F. A., A. M. Simonet, and M. D. Esteban (1994), *Phytochemistry, 36,* 1369–1379.

Mallik, M. A. B. and K. Tesfai (1988), *Plant Soil, 11,* 177–182.

Mandava, N. B. (1985), in *The chemistry of allelopathy* (A. C. Thompson, Ed.), Vol. 268, American Chemical Society, Washington, DC, pp. 33–54.

Manthe, B., M. Schultz, and H. Schnabl (1992), *J. Chem. Ecol., 18,* 1525–1539.

Martin, L. D. and A. E. Smith (1994), *Crop Protec., 13,* 388–392.

Mason-Sedum, W., R. S. Jessop, and J. V. Lovett (1986), *Plant Soil., 93,* 3–16.

McCall, P. J., T. C. J. Turlings, J. Loughrin, A. T. Proveaux, and J. H. Tumilnson (1994), *J. Chem. Ecol., 20,* 3039–3050.

McClure, J. W. (1970), in *Phytochemical phylogeny* (J. B. Harborne, Ed.), Academic Press, New York, pp. 233–265.

McClure, P. R., H. D. Groos, and W. A. Jackson (1978), *Can. J. Bot., 56,* 764–767.

McNaughton, S. J. (1968), *Ecology, 49,* 367–369.

Menelaou, M. A., M. Foroozesh, G. B. Williamson, F. R. Fronczek, H. D. Fischer, and N. H. Fischer (1992), *Phytochemistry, 31,* 3769–3771.

Mersie, W. and M. Singh (1993), *J. Chem. Ecol., 19,* 1293–1301.

Miersch, J., C. Juhlke, G. Sternkopf, and G. J. Krauss (1992), *J. Chem. Ecol., 18,* 2117–2128.

Miles, D. H., V. Chittawang, P. A. Hedin, and U. Kokpol (1993), *Phytochemistry, 32,* 1427–1430.

Miller, H. G., M. Ikawa, and L. C. Pierce (1991), *Hort. Sci., 26,* 1525–1527.

Mittal, S. P., S. S. Grewal, Y. Agnihotri, and A. D. Sud (1992), *Agrofor. Syst., 19,* 207–216.

Moitra, R., D. C. Ghosh, S. K. Majumdar, and S. Sarkar (1994), *Ind. J. Agric. Sci., 64,* 156–159.

Molina, A., M. J. Reigosa, and A. Carballeira (1991), *J. Chem. Ecol., 17,* 147–160.

Molisch, H. (1937), *Allelopathie,* Gustav Fischer, Jena, Germany.

Mondoza, E. M. T. and L. L. Ilag (1980), *Leucaena News Lett., 1,* 23–25.

Moreland, D. F. and W. P. Novitzky (1987), in *Allelochemicals, Role in agriculture and forestry* (G. R. Waller, Ed.), Vol. 330, American Chemical Society, Washington, DC, pp. 247–261.

Muller, C. H. (1965), *Bull. Torrey Bot. Club, 92,* 38–45.

Muller, C. H. and R. del Moral (1966), *Bull. Torrey Bot. Club, 93,* 130–137.

Muller, W. H., P. Lorber, and B. Haley (1968), *Bull. Torrey Bot. Club, 95,* 415–422.

Muller, W. H., P. Lorber, B. Heley, and K. Johnson (1969), *Bull. Torrey Bot. Club, 96,* 89–96.

Murphy, T. M., I. Raskin, and A. J. Enyedi (1993), *Env. Exp. Bot., 33,* 267–272.

Muzafarov, E. A., A. F. Cojocaru, G. N. Nazarova, and N. L. Cojocaru (1988), *Stud. Biophys., 7,* 151–160.

Narwal, S. S. (1994), *Allelopathy in crop production,* Scientific Publishers, Jodhpur, India, pp. 19–161.

Nehlin, G., I. Valterova, and A. K. Borg-Karlson (1994), *J. Chem. Ecol.*, *20*, 771–783.

Nicollier, G. F., D. F. Pope, and A. C. Thompson (1985), in *The chemistry of allelochemicals* (A. C. Thompson, Ed.), Vol. 268, American Chemical Society, Washington, DC, pp. 208–218.

Nilsson, M. C. (1992), *Dissertation in forest vegetation ecology*, Swedish Univ. Agricul. Sci., Umea, 35 pp.

Oleszek, W. (1987), *Plant Soil, 102*, 271–274.

Oster, U., S. Martin, and R. Wolfhart (1990), *J. Naturforsch. Sect. Biosci.*, *45*, 835–844.

Pandya, S. M., V. R. Dave, and K. G. Vyas (1984), *Sci. Cult.*, *50*, 161–162.

Parkinson, V., Y. Efron, L. Bello, and R. Dashiell (1987), *Plant Prot. Bull. FAO, 35*, 51–54.

Parr, J. F. and H. W. Renszer (1962), *Soil Sci. Soc. Am. Proc.*, *26*, 552–556.

Paszkowski, W. L. and R. J. Kremer (1988), *J. Chem. Ecol.*, *14*, 1573–1582.

Patterson, D. T. (1981), *Weed Sci.*, *29*, 53–59.

Pellissier, F. (1993a), *J. Chem. Ecol.*, *19*, 2105–2114.

Pellissier, F. (1993b), *Acta Oecol.*, *14*, 211–218.

Perez, F. J. and J. Ormeno-Nunez (1991a), *Phytochemistry, 30*, 2199–2202.

Perez, F. J. and J. Ormeno-Nunez (1991b), *J. Chem. Ecol.*, *17*, 1037–1043.

Perry, D. A. and C. Choquette (1987), in *Allelochemicals: Role in agriculture and forestry* (G. R. Waller, Ed.), Vol. 330, American Chemical Society, Washington, DC, pp. 185–194.

Ponder, F., Jr. (1987), in *Allelochemicals: Role in agriculture and forestry* (G. R. Waller, Ed.), Vol. 330, American Chemicals Society, Washington, DC, pp. 195–204.

Porwal, M. K. and O. P. Gupta (1987), *Trop. Agric.*, *4*, 276–279.

Prasad, M. N. V. and P. Subhashini (1994), *J. Chem. Ecol.*, *20*, 1689–1696.

Purohit, S., M. M. Laloraya, and S. Bharti (1991), *Physiol. Plant, 81*, 79–82.

Purohit, S., M. M. Laloraya, S. Bharti, and C. Nozzolollo (1994), *J. Exp. Bot.*, *43*, 103–110.

Purvis, C. E. (1990), *Plant Prot. Q.*, *5*, 55–59.

Purvis, C. E., R. S. Jessop, and J. V. Lovett (1985a), *Weed Res.*, *25*, 415–421.

Purvis, C. E., R. S. Jessop, and J. V. Lovett (1985b), *Proc. Aust. Agron. Conf.* (Abstract), Aust. Soc. Agron. Univ., Tasmania, Hossrt, 375 pp.

Putnam, A. R. and J. De Frank (1983), *J. Chem. Ecol.*, *8*, 1001–1010.

Putnam, A. R. and C. S. Tang (1986), *The science of allelopathy*, Wiley, New York.

Qasem, J. R. (1994), *Allelopathy J.*, *1*, 29–40.

Qasem, J. R. (1995), *Weed Res.*, *35*, 41–49.

Qasem, J. R. and T. A. Hill (1989), *Weed Res.*, *29*, 349–356.

Qasem, J. R. and T. A. Hill (1994), *Weed Res.*, *34*, 109–118.

Qasem, J. R., A. S. Al-Abed, and H. A. Abu-Blan (1995), *Phytopath. Mediterranea, 34*, 7–14.

Quackenbush, R. C., D. Bunn, and W. Lingren (1986), *Aquat. Bot.*, *24*, 83–84.

Rakhteenko, I. N., I. A. Kaurov, and I. F. Minko (1973), in *Physiological biochemical basis of plant interactions in phytocenosis* (A. M. Grokzinsky, Ed.), Vol. 4, Naukova, Dumka, Kiev, Ukraine, pp. 16–19.

Rasmussen, J. A. and F. A. Einhellig (1979), *Plant Sci. Lett., 14,* 69–74.

Rao, A. S. (1990), *Bot. Rev., 56,* 1–84.

Ravanel, P. (1986), *Phytochemistry, 25,* 1015–1020.

Ravanel, P. and M. Tissut (1986), *Phytochemistry, 25,* 577–583.

Ravanel, P., M. Tissut, and R. Douce (1982), *Phytochemistry, 21,* 2845–2850.

Ray, S. D. and M. M. Laloraya (1983), *Can. J. Bot., 62,* 2047–2052.

Reddy, G. N. and M. N. V. Prasad (1989), in *Proc. Bioenergy Soc. 5th Conven. & Symp.* (R. N. Sharma, O. P. Vimal, H. L. Sharma, and K. S. Rao, Eds.), Vol. 5, BESI, New Delhi, pp. 463–466.

Reigosa, M. J., J. F. Casal, and A. Carballeira (1984), *Stud. Oecol., 5,* 135–150

Rice, E. L. (1984), *Allelopathy,* Academic Press, London.

Rice, E. L. (1987), in *Allelochemicals, Role in agriculture and forestry* (G. R. Waller, Ed.), Vol. 330, American Chemical Society, Washington, DC, pp. 8–22.

Rice, E. L., W. T. Penfound, and L. M. Rohrbaugh (1960), *Ecology, 41,* 224–228.

Rizvi, S. J. H. and V. Rizvi (1987), in *Allelochemicals, Role in agriculture and forestry* (G. R. Waller, Ed.), Vol. 330, American Chemical Society, Washington, DC, pp. 69–75.

Sathiyamoorthy, P. (1990), *J. Plant Physiol., 136,* 120–121.

Scholman, Jr, W. W., A. S. Hilton, and J. J. McGrady (1991), *Bioresour. Technol., 35,* 191–196.

Schon, M. K. and F. A. Einhellig (1982), *Bot. Gaz., 143,* 505–510.

Schumacher, W. J., D. C. Thill, and G. A. Lee (1982), *Proc. North Am. Symp. Allelopathy,* Urbana, IL, Abstract No. 14.

Schumacher, W. J., D. C. Thill, and G. A. Lee (1983), *J. Chem. Ecol., 9,* 1235–1245.

Sharma, K. P. (1985), *Aquat. Bot., 22,* 71–78.

Sharma, R. and G. Singh (1987), *Ann. Bot., 60,* 189–190.

Shilling, D. G., R. A. Liebl, and A. D. Worsham (1985), in *The chemistry of allelopathy* (A. C. Thompson, Ed.), Vol. 268, American Chemical Society Symp. Series, Washington, DC, pp. 243–271.

Shimabukro, R. H. (1985), in *Weed physiology* (S. O. Duke, Ed.), Vol. 2, CRC Press, Boca Raton, FL, pp. 215–240.

Shimabukro, R. H., G. L. Lamourex, and D. S. Frear (1982), in *Biodegradation of pesticides,* (F. Matsumara and C. R. K. Murara, Eds.), Plenum Press, New York, pp. 21–26.

Singh, C. M., N. N. Angiras, and S. D. Singh (1988), *Ind. J. Weed. Sci., 20,* 63–66.

Siqueira, J. O., M. G. Nair, R. Hammerschmidt, and G. R. Safir (1991). *Cri. Rev. Plant Sci., 10,* 63–121.

Smith, A. E. and L. D. Martin (1994), *Agron. J., 86,* 243–246.

Srivastava, J. N., J. P. Shukla, and R. C. Srivastava (1985), *Acta Bot. Ind., 13,* 194–197.

Steinseik, J. W., L. R. Oliver, and F. C. Collins (1982), *Weed Sci., 30,* 495–497.

Stevens, K. L. and G. B. Merrill (1980), *J. Agric. Food Chem., 28,* 644–646.

Stevens, G. A., Jr. and C. S. Tang (1985), *J. Chem. Ecol., 11,* 1411–1425.

Stevens, G. A., Jr. and C. S. Tang (1987), *J. Trop. Ecol., 3,* 91–94.

Stowe, L. G. and A. Osborne (1980), *Can. J. Bot., 58,* 1149–1153.

Sun, Y. and G. Hrazdina (1991), *Plant Physiol.*, *95*, 556–563.

Sutton, D. L. (1990), *J. Aquat. Plant Manage.*, *28*, 20–22.

Sutton, D. L. and Portier, K. M. (1989), *J. Aquat. Plant Manage.*, *27*, 90–95.

Sutton, D. L. and Portier, K. M. (1991), *J. Aquat. Plant Manage.*, *29*, 6–11.

Swain, T. (1977), *Annu. Rev. Plant Physiol.*, *28*, 479–501.

Szczepanska, W. (1971), *Ekol. Pol.*, *25*, 437–445.

Tan, K. S., T. Hoson, Y. Masuda, and S. Kamisaka (1991), *Physiol. Plant, 83*, 397–403.

Tan, K. S., T. Hoson, Y. Masuda, and S. Kamisaka (1992), *J. Plant Physiol.*, *140*, 460–465.

Tayal, M. S. and S. M. Sharma (1985), *Ind. J. Plant Physiol.*, *28*, 271–276.

Taylor, A. O. (1965), *Plant Physiol.*, *40*, 273–280.

Tereda, H. (1980), *Biochim. Biophys. Acta, 639*, 225–242.

Thanos, C. A. and S. Marcou (1993), *Proc. Int. Symp. Pinus brutia Ten.*, Ministry of Forestry, Ankara, Turkey, pp. 176–183.

Thiboldeaux, R. L., R. L. Lindroth, and J. W. Tracy (1994), *J. Chem. Ecol.*, *20*, 1631–1641.

Todorov, D., V. Alexieva, D. Karanov, and V. Velikova (1992), *J. Plant Growth Reg.*, *11*, 233–238.

Tomar, V. P. S., P. Narain, and K. S. Dadhwal (1992), *Agrofor. Syst.*, *19*, 241–252.

Tomaszewski, M. and K. V. Thimann (1966), *Plant Physiol.*, *41*, 1443–1454.

Touchette, R., G. D. Leroux, and J. M. Deschones (1988), *Can. J. Plant Sci.*, *68*, 785–792.

Tso, T. C., M. J. Kasperbauer, and T. P. Sorokin (1970), *Plant Physiol.*, *45*, 330–333.

Tsuzuki, E. and H. Kawagoe (1984), *Bull. Fac. Agric. Univ. Miyazaki.*, *31*, 189–195.

Van Aller, R. T., G. F. Pessony, V. A. Rogers, E. J. Watkins, and H. G. Leggett (1985), in *The chemistry of allelopathy* (A. C. Thompson, Ed.), Vol. 268, American Chemical Society, Washington, DC, pp. 387–400.

Van Sumere, C. F., J. Cottenie, J. De Greef, and J. Kint (1971), *Rec. Adv. Phytochem.*, *4*, 165–221.

Ventura, W., I. Watanabe, H. Komada, M. Nishio, A. D. Cruz, and M. Castillo (1984), *IRRI Research paper series*, No. 99, Int. Rice Res. Inst., Los Banos, Phillipines.

Wallace, J. W., T. J. Mabry, and R. E. Alston (1969), *Phytochemistry, 8*, 93–99.

Walters, D. T. and A. R. Gilmore (1976), *J. Chem. Ecol.*, *2*, 469–479.

Wardle, D. A., K. S. Nicholson, and A. Rahman (1993), *Weed Res.*, *33*, 69–78.

Wardle, D. A., K. S. Nicholson, M. Ahmed, and A. Rahman (1994), *Plant Soil, 163*, 287–297.

Watanabe, R., W. J. McIlirath, J. Skok, W. Chorney, and S. H. Wender (1961), *Arch. Biochem. Biophys.*, *94*, 241–243.

Waters, E. R. and U. Blum (1987), *Am. J. Bot.*, *74*, 1635–1645.

Weidenhamer, J. D., M. Menelaou, F. A. Macias, N. H. Fischer, D. R. Richardson, and G. B. Williamson (1994), *J. Chem. Ecol.*, *20*, 3345–3359.

Weston, L. A. (1986), *Diss. Abst. Int. B, 47*, 1366.

Weston, L. A. and A. R. Putnam (1985), *Crop Sci.*, *25*, 561–565.

Weston, L. A., R. Harmon, and S. Muller (1989), *J. Chem. Ecol.*, *15*, 1855–1865.

Whitehead, D. C. (1964), *Nature, 202,* 417–418.

Whittaker, R. H. and P. P. Feeny (1971), *Science, 171,* 757–758.

Wium-Anderson, S., U. Anthoni, C. Christopherson, and G. Houen (1982), *Oikos, 39,* 187–190.

Woodhead, S. (1981), *J. Chem. Ecol., 7,* 1035–1047.

Wozniak, K. L., W. M. Sullivan, A. D. Flowerday, and R. P. Walden (1981), *Proc. 73rd Annu. Meet. Am. Soc. Agron., 116* (Abstract).

Wu, W. M. H., C. L. Liu, and C. C. Chao (1976), *J. Chin. Agric. Chem. Soc., 96,* 16–37.

Yalpani, N., N. E. Balke, and M. Schulz (1992), *Plant Physiol., 100,* 114–119.

Yamane, A., H. Nishimura, and J. Mizutani (1992), *J. Chem. Ecol., 18,* 683–691.

Young, C. C. (1984), *Plant Soil, 82,* 247–254.

Yu, J. Q. and Y. Matsui (1994), *J. Chem. Ecol., 20,* 21–31.

Yun, K. W. and B. S. Kil (1992), *J. Chem. Ecol., 18,* 1933–1940.

Zackrisson, O. and M. C. Nilsson (1992), *Can. J For. Res., 22,* 1310–1319.

Zhu, H. and A. U. Mallik (1994), *J. Chem. Ecol., 20,* 407–421.

Zimdahl, R. L. (1980), Int. Plant Prot. Centre, Pregon State Univ. Coravallis, OR.

Zucker, M. (1963), *Plant Physiol., 38,* 575–580.

Zucker, M. (1969), *Plant Physiol., 44,* 912–922.

10

Herbicides

Gérard Merlin

PRESENTATION

The term "herbicides" includes chemicals that enhance the growth and the yield of the crop. In temperate countries herbicides are the major type of pesticides used. Thus in 1992 in the West European market, herbicides represented 43% of the total amount, and 85% of pesticides used in the United States in 1987 are herbicides (USDA, 1987). In tropical and subtropical areas a much larger percentage of the pesticides used are insecticides.

A few products dominate the pesticide market; in the United States in 1985 six compounds accounted for nearly 70% of the 440 000 tonnes of synthetic herbicides used: alachlor, atrazine, butylate, trifluralin, metolachlor, and 2,4-D.

Historical Aspects

In 1896 a French farmer applying Bordeaux mixture (copper sulfate, lime, and water) against vine mildew noticed that it caused the leaves of yellow charlock in the vicinity of his vines to turn black; this was probably the origin of the idea of selective herbicides. A few years later it was discovered that by spraying a solution of iron sulfate on a mixture of cereals and weeds, only the weeds were killed (Martin and Woodcock, 1983). The 1930s represent the beginning of the modern area of synthetic organic herbicides, and the large-scale pesticide industry only dates from the end of World War II. In 1943 herbicidal activity of the phenoxyacetic acids were discovered, and two compounds, 2-methyl-1-4-chloro and 2,4-dichloro phenoxyacetic acids (2,4-D and 2,4,5-T), are still in use. In fact, they are the most widely used pesticides in Great Britain (Cremlyn, 1991). The bipyridilium herbicides diquat and paraquat were introduced in 1958. These are very quick-acting herbicides, which are absorbed by plants and translocated, causing dessication of the foliage, but they are strongly adsorbed by clay particles in soil and therefore deactivated.

All over the world, many of the pesticides discovered in the 1950s and 1960s are still extensively used, causing weed resistance in many cases. To solve the growing problems presented by resistant species, there is an urgent need to introduce more selective herbicides with different modes of action. A new standard of pesticidal potency was set by the introduction of sulfonylurea herbicides in the 1970s. They are active in rates of grams rather kilograms per hectare and can be incorporated in integrated pest management programs that include cultural and biological control measures and behavior-modifying chemicals. These programs enable the use of smaller amounts of pesticides and reduce environmental pollution and also slow the emergence of resistant strains.

Herbicides and the Environment

It usually takes a period of about ten years from the date of the initial discovery of a new herbicide to the marketing stage. Different chemicals are tested against weeds. If a chemical shows promise, structural analogues are synthesized and tested to establish the structure-activity relationship leading to the best herbicide. Toxicological and ecotoxicological properties have to be evaluated early in the development of the herbicide, before or when experiments are initiated on a small scale in greenhouses. Most developed countries in the world have comparable requirements. If toxicological studies are satisfactory, then large-scale field trials are initiated. To be successful, commercial herbicides must control weeds, display crop safety, and have low toxicity toward other organisms. Sulfonylurea herbicides fulfill these requirements (Saari and Mauvais, 1996). Nevertheless, to reduce long-term effects on the environment, a decrease in the use of herbicides has been requested by authorities in many countries of the OECD (Bernson and Ekstrom, 1991). Atrazine is a major herbicide that has been heavily used throughout the world, and there is an increasing concern regarding its use, since several reports have shown contamination of surface waters and groundwaters with the herbicide and phytotoxic degradation products (Nagy et al., 1995).

Economic Aspects

The chemical cost of five effective herbicides for the control of rubber vine *(Cryptospegia grandiflora)* in Australia is $476–1863 ha^{-1} (Vitelli et al., 1994). Cotton yield is increased by herbicide treatments; it have been reported that *Sorghum halepense* competition can reduce cotton yield by more 80% (McWhorter, 1989). In cereal crops such as wheat and barley, herbicide application can substantially reduce grain yield, and similar reductions may occur in grass seed crops (Lodge and McMillan, 1994).

CHEMICAL CLASSIFICATION OF HERBICIDES

725 agrochemicals are commonly used, including 241 herbicides (Tomlin, 1994). All these herbicides have various modes of action and physicochemical properties, they can be metabolized in different ways, and some weed species are resistant to their action.

In this chapter, a general overview is proposed; for more details about certain topics, the reader should refer to specific publications.

The 241 commonly used herbicides belong to 30 different chemical families. Mineral herbicides are no longer in common use; organic synthesis herbicides have replaced them. Each family of compounds shares an identical basic structure and has generally the same mode of toxic action (Table 10-1). Compounds belonging to the same family differ from each other by the presence of specific functions or chemical groups. The quantitative activity structure relationship (QASR) is used to determine the most effective derivative(s) in a chemical family. Most of these organic herbicides own one or several benzenic or phenolic nuclei (Fig. 10-1). Many herbicides originate from natural molecules; thus the aryloxyacids simulate the plant hormones such as auxin, which stimulate or regulate the growth (Table 10-1).

The most important herbicides in current use were commercialized in the 1960s and 1970s. Most recent herbicides belong to families of sulfonylureas, cyclohexanediones, pyridones (and derivatives), or, again, the quinolincarboxylic acids. The most commonly used herbicides are sulfonylureas ($\sim 10\%$), ureas ($\sim 9\%$), thiocarbamates ($\sim 6\%$), and triazines ($\sim 7\%$).

ABSORPTION AND TRANSPORT IN PLANT

Herbicides may be divided into two main types: systemic and nonsystemic (contact) herbicides.

Penetration, transport, and therefore translocation within the plant vascular system, are limited for contact herbicides. The earlier herbicides belong to this type, and their major disadvantages are that they are susceptible to the effects of wind, rain, or sunlight, causing low efficiency and ecological problems.

The more recent herbicides are systemic, and they can penetrate the plant cuticle and move through the plant vascular system.

Formulation of the herbicide must bring the active ingredient into a convenient, easy-to-use form for application, storage, and transport in a safe and effective manner.

The translocation of chemicals within plants may be considered in three stages (Jacob and Newmann, 1987):

- *Entry into the Free Space Within the Tissues:* Chemicals pass into free space in contact by diffusion with the external environment after penetrating the leaf cuticle.

TABLE 10-1 Primary Mode of Action of Herbicide Chemical Families

Chemical Families of Herbicides	Some Commercial Products	Primary Modes of Action
Arylalanines	Benzoylprop, flamprop-methyl[a]	Fatty acids synthesis inhibitor (inhibit cell growth)
Arenecarboxylic acids (benzoic acids)	Dicamba,[a] chloramben	Auxin type (auxinlike growth regulator)
Aldehyde	Acrolein[a]	React with sulfydryl groups of enzymes
Amides (benzamides)	Propyzamide,[a] isoxaben	Cell division inhibitors, photosynthesis inhibitors
Anilides	Propanil	Cell division inhibitors, photosynthesis inhibitors
Aryloxyphenoicpropionic acids	Fluazifop®,[a] diclofop methyl®	Fatty acids synthesis inhibitors
Aryloxyalkanoic acids	2,4-D,[a] dichlorprop, MCPA	Auxin-type herbicides, growth inhibitors
Aryloxyalkanamids	Napropamide	Cell division inhibitors
Arylacetic acids	Benazolin	Auxin-type, growth regulators
Benzofuranylalkanesulfonates (benzofurans)	Ethofumesate, benfuresate®[a]	Meristematic growth inhibitors
Bipyridiliums	Diquat,[a] paraquat	Photosynthetic electron transport acceptors
Benzonitriles	Dichlorobenil[a]	Cellulose biosynthesis inhibitors
Benzothiadiazinones	Bentazone	Inhibit CO_2 fixation
Chloroacetanilide	Alachlor,[a] propachlor	Proteins and nucleic acids synthesis inhibitors
Carbamates	Phenmedipham,[a] CIPC, asulam	Cell division and cell expansion inhibitors
Cineoles	Cinmethylin®[a]	Meristematic growth inhibitors
Cyclohexanediones, oximes	Cycloxydim®, sethoxidim®[a]	Mitosis inhibitor, bleaching herbicides
Dinitroanilines	Trifluralin,[a] oryzalin, pendimetham	Cell division inhibitors (mitosis inhibitors)
Dinitrophenols	DNOC, dinoterb	(Photo)oxidative phosphorylations uncouplers
Diphenyl ethers	Bifenox, formesafen,[a] oxifluorfen®	Protoporphyrinogen oxidase inhibitors
Imidazolinones	Imizamethabenz®[a]	Branched-chain amino acids synthesis inhibitors
Nitriles (hydroxybenzonitriles)	Bromoxynil[a]	(Photo)oxidative phosphorylations uncouplers
Organochlorines	Dalapon[a]	Modification of protein structure
Oxadiazolinones	Oxadiazon	Protoporphyrinogen oxidase inhibitors
Organophosphorus	Anilofos, glyphosate[a]	Aromatic amino acids synthesis inhibitors
Pyridine carboxilic acids	Pichloram	Auxine type inhibitors
Pyridones, pyrrolidones	Fluridone,[a] flurochlorydrone®	Carotenoid biosynthesis inhibitors
Pyridazinones	Norflurazon	Phytoene desaturase inhibitors
Quinolincarboxylic acids	Quinclorac®	Auxin type inhibitors
Sulfonylureas	Metsulfuron-methyl®,[a] bensulfuron	Branched-chain amino acids synthesis inhibitors
Thiocarbamates	Molinate, EPTC,[a] prosulfocarb	Fatty acids (+giberellic acid) synthesis inhibitors
Triazines	Atrazine,[a] terbutryn, simazine	Photosynthesis electron transport inhibitors
Triazinones	Metribuzin	Photosynthesis electron transport inhibitors
Ureas	Chlortoluron, isoproturon, DCMU	Photosynthesis electron transport inhibitors

[a] Illustrated in Fig. 10-1.
® Commercialized after 1980s.

308

- *Movement in the Xylem by Free Diffusion:* This process is driven by a pumping mechanism in the roots and the evaporation of water vapor from the leaf surface.
- *Movement in the Phloem:* This process requires metabolic energy, as chemicals are distributed to the growing tissues of the plant. As a rule, compounds possessing carboxyl, hydroxyl, or sulfamoxyl groups are phloem mobile. Phloem-mobility of herbicides is dependent on the lipophilicity acting as the leading factor, modulating the polarity of the compound with a pKa range of 3 to 6.5 (Brudenell et al., 1995). Ploem-mobile herbicides are often weak acids or own dissociating acidic groups. Nonphloem-mobile aryloxyphenoxy alkanoic acids are hydrolyzed by the plant metabolism to phloem-mobile aryloxyphenoxy alkanoic acids. Glyphosate, as an example of zwitterion at physiological pH, has been shown to be phloem-mobile, all sulfonylureas are known to be phloem-mobile, but urea herbicides are reported to have little or no movement in the phloem (Brudenell et al., 1995).

To act as a systemic herbicide by foliar application the compound must have a lipophilic-hydrophilic balance.

Herbicide Uptake

Uptake of the herbicide into the plant can occur at the soil-root or leaf-air interfaces. Herbicides must pass through the nonliving portions (apoplast) of the plant and enter the living parts (symplast) to reach their active site and to have an effect. Movement of a herbicide across the nonliving tissues is complex and depends on various factors, such as characteristics of the herbicide, species and age of the plant, and so on (Devine et al., 1993). Herbicide absorption occurs through roots by passive diffusion through epidermis and cortex. That absorption occurs primarily at the apical end where the Casparian strip (suberized layer) is least developed.

Translocation

After absorption, herbicides often are transported to the site of activity by translocation between nonliving tissues—cell walls and xylem (apoplast)—and living tissues—cell plasmalemma and phloem (symplast). This translocation of herbicides can either be done over relatively short distances, as for paraquat where activity in living cells is near the point of entry, or it can be achieved by the vascular system over longer distances, as is the case for glyphosate, which is transported in the phloem (Dekker and Duke, 1995).

Weakly acidic herbicides are quickly absorbed but not readily translocated to distal parts of the treated weed following foliar application. Limited transport appears to be due to their inhibitory effects on phloem transport (Chao et al, 1994; Devine et al., 1990).

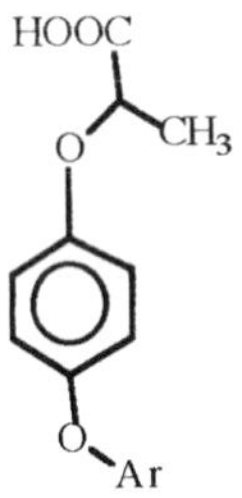

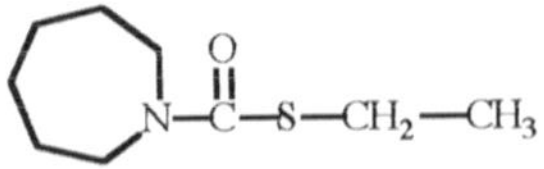

Thiocarbamate: molinate

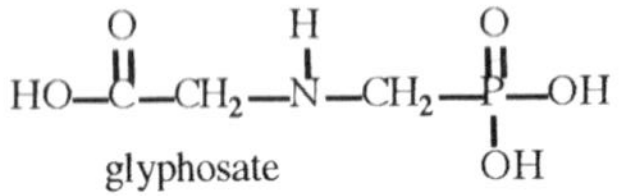

Triazine: atrazine

CH_3
CH_3
CH
NH
N
N
Cl
N
H_3C-H_2C-HN

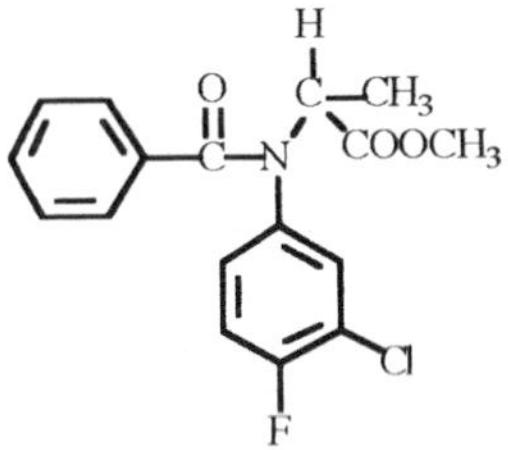

Polycyclic alkaloic acid: (R)-Flamprop-methyl

CH_3
C
$COOCH_3$
H
N
Cl
F
O

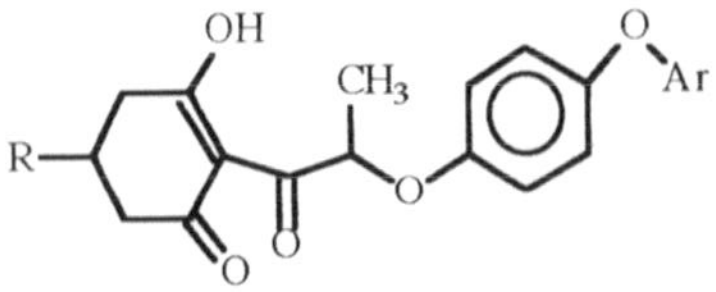

novel cyclic triketones

$O-Ar$
CH_3
OH
O
R

glyphosate

O
H
O
$HO-C-CH_2-N-CH_2-P-OH$
OH

Thiocarbamate: SUTAN

CH_3
$H_3C-HC-H_2C$
$N-C-CH_2-CH_3$
$H_3C-HC-H_2C$
CH_3
O

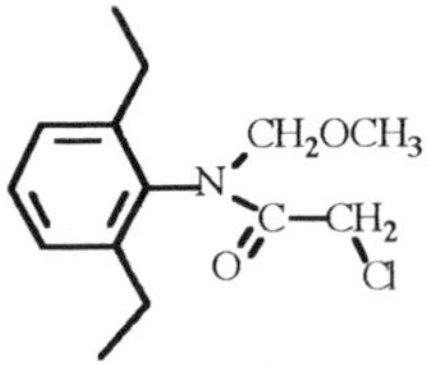

2-chloroacetamide: propachlor

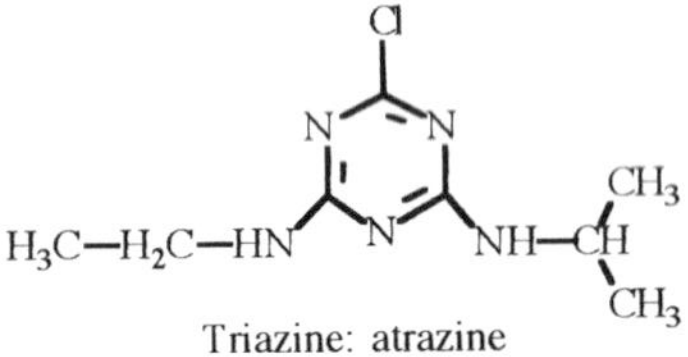

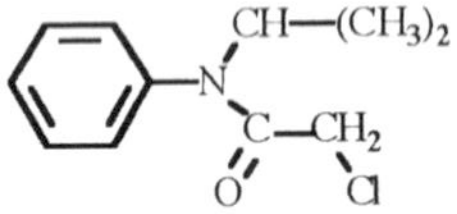

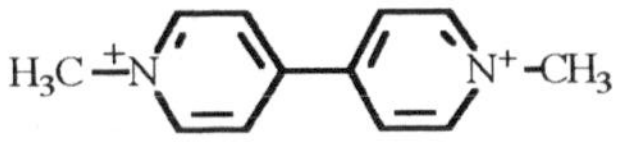

$CH-(CH_3)_2$
CH_2
C
Cl
N
O

$H_3C-H_2C-H_2C$ $CH_2-CH_2-CH_3$
N
O_2N NO_2
CF_3

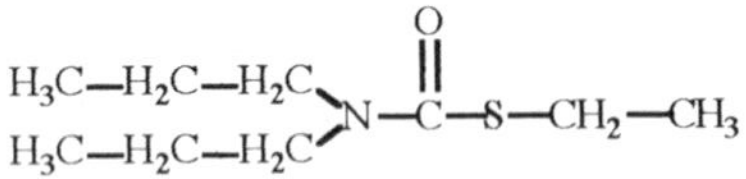

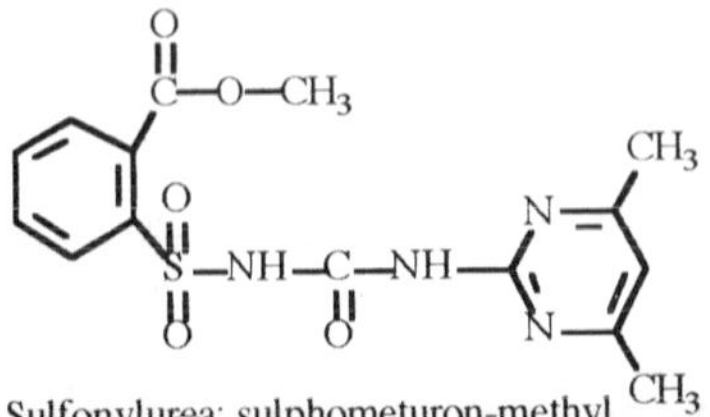

Sulfonylurea: sulphometuron-methyl

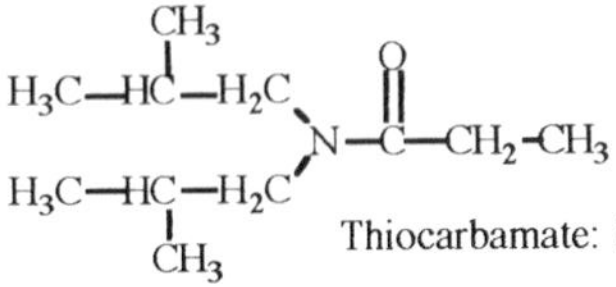

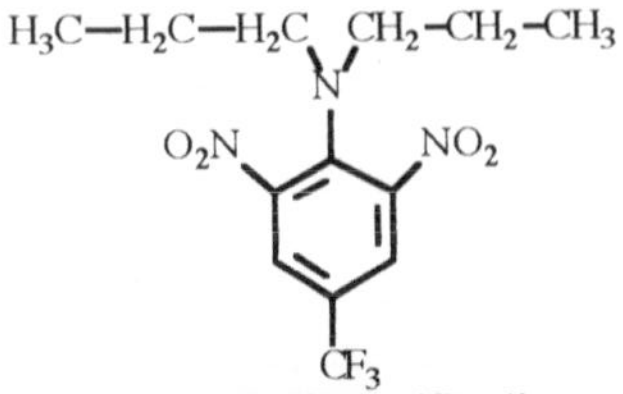

CH_3
CH_3
N
N
$O-CH_3$
O
$S-NH-C-NH$
O O

aryloxyphenoxypropionate
Ar=5-CF$_3$-pyridyl for fluazifop

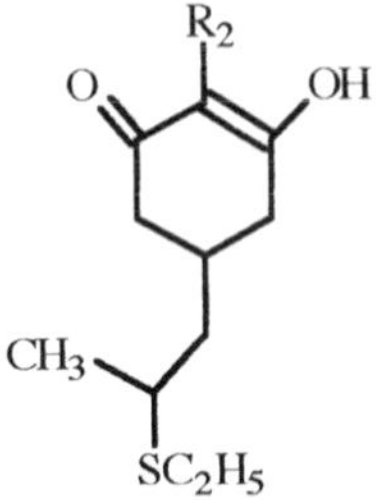

$HOOC$
CH_3
CH
O
$O-Ar$

cyclohexanedione
R2=(C$_3$H$_7$)C=NOC$_2$H$_5$ for sethoxidim

R_2
OH
O
CH_3
SC_2H_5

Thiocarbamate: EPTC

O
$H_3C-H_2C-H_2C$
$N-C-S-CH_2-CH_3$
$H_3C-H_2C-H_2C$

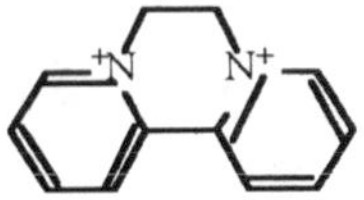

2-chloroacetamide: Alachlor

CH_2OCH_3
$C-CH_2$
N Cl
O

diquat

N^+
N^+ 2 Br$^-$

Bipyridylium

paraquat

H_3C-N^+
N^+-CH_3 2 Cl$^-$

310

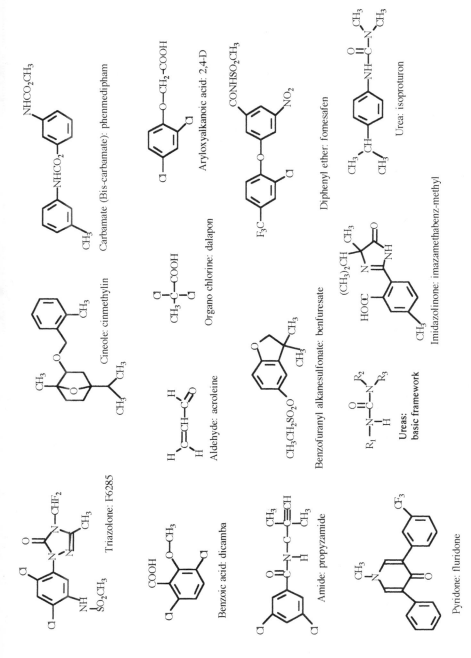

Figure 10-1. Structures of organic herbicides, illustrating some agronomically important chemical families.

Compartmentation

Herbicides can be stocked in several plant locations; some lipophilic herbicides may become immobilized by partitioning into lipid-rich area such as glands or oil bodies (Stegink and Vaughn, 1988). Temporary sequestration in vacuoles before metabolization is also possible (Schmitt and Sandermann, 1982). In the apoplastic part of the leaf (cuticle, cell walls, vacuoles), the distribution of the xenobiotic products is probably directly controlled by the physicochemical properties of the herbicide. In the symplasmatic areas, it is now clear that lipophilic compounds passively diffuse and concentrate in membrane lipids, as for example: phenmedipham and isoproturon (Haddad et al., 1992; Ravanel et al., 1990).

Accumulation of atrazine in tolerant species can be attributed to conversion and retention of the more hydrophilic hydroxyatrazine (Darmstadt et al., 1984).

Some surfactants or other spray additives can increase a herbicide's phytotoxicity. Sodium bisulfate ($NaHSO_4$) at 28 mM increased the efficacy of imazamethabenz on wild oats because foliar absorption and acropetal translocation were increased due to higher solubility of the herbicide (Liu et al., 1995).

MODES OF ACTION

The mode of action may be defined as the sum of anatomical, morphological, physiological, biochemical, and molecular responses, and constitutes the overall phytotoxic action of a herbicide. This process will account for both primary and secondary effects in response to the herbicide (Devine and Shimabukuro, 1994).

The primary mechanism of action may be biochemical or biophysical process or lesion sufficient or not sufficient to cause complete phytotoxicity and secondary effects. Chlorosis and necrosis are not specific symptoms and are the resulting consequences of primary effects. Life is characterized by energetic processes (catabolism), synthesis of materials (anabolism), exchanges with the environment, information and autoreplication (genetic processes), and so on. To kill nondesirable plants in crop cultures, herbicides must act on these life processes. These processes are essentially located in the basic unit of life, which is the cell. The cell is separated from its environment by a lipidic membrane that controls the exchange of materials between the cytoplasm (colloidal suspension of substances) and the external environment and preserves homeostasy (a state of equilibrium).

Bioenergetic Effects

The energy needed for cellular activities in plants comes essentially from two processes: photosynthesis and respiration.

Photosynthesis   Green plants are able to utilize light energy to synthesize carbohydrates from carbon dioxide and water. The following equation summarizes enzymatic reactions:

$$nCO_2 + nH_2O \longrightarrow [CHO]_n + nO_2 \qquad (10\text{-}1)$$

If n is equal to 6, the equation gives glucose as the carbohydrate and photosynthesis is the reverse of respiration.

PSII Perturbations Many modern herbicides effectively block photosynthetic electron transport at the level of the photosystem II (PSII) acceptor site (the D-1 protein) (Table 10-1). The PSII is a multiple-protein complex located in the grana or appressed lamellae; it consists of at least 23 different subunits (Andersson and Styring, 1991). The primary effect is the inhibition of photosynthetic O_2 evolution, reduction of NADP and photophosphorylation; secondary effects include delayed chlorophyll degradation, decreased chlorophyll a/b ratio, and changes in nitrogen metabolism (Fedtke, 1982). In the presence of light, s-triazines are preferentially attached to a high-affinity binding site, the D-1 protein, and compete with quinone. This binding blocks electron transport on the reducing side of PSII from Q_A to Q_B. The redirection of electrons away from the electron transport chain results in the generation of toxic oxi-radicals and other reactive radical species, causing tissue oxidation. Bromoxynil (a benzonitrile herbicide) is a PSII inhibitor with the same mode of action as triazines (Devine et al., 1993). According to current models of the Q_B binding niche where the D-1 protein has five transmembrane segments (I to V), the Q_B binding niche is localized between helices IV and V (Fig. 10-2), plasto-quinone binding involves hydrogen bonding between the carbonyl oxygens of plas-toquinone and the amide backbone of His 215 and the hydroxyl of Ser 264 (or Phe 265) (Tietjen, et al., 1991). Herbicides that bind in the Q_B niche can be categorized into two groups: the urea and triazine families, which strongly interact with Ser 264, and the phenol family, which interacts with His 215 (Trebst, 1987) (Fig. 10-2). PSII inhibitors block photosynthetic electron transport and prevent the reduction of $NADP^+$. But the herbicidal activity is essentially due to the oxidative stress generated when electron transport is blocked, causing destruction of PSII center and the photooxidation of chlorophyll molecules. The binding of PSII inhibitors inter-feres with the degradation of the D-1 protein by a protease, occurring during normal operation of the PSII reaction center, and the damaged D-1 protein cannot be re-placed (Gronwald, 1994).

Effects of sublethal concentrations of urea herbicides are more complex; for methabenzthiazuron (the most common herbicide in the Mediterranean areas for the control of weed in cereals and bean crops) in *V. faba*, photosynthetic rates are transiently stimulated, and nodulation, N_2 fixation, and seed quality and quantity were also enhanced (Vidal et al., 1992).

PS I Perturbations and Photooxidative Stress Some herbicides have been found to generate active oxygen species, by direct involvement in radical produc-tion or by inhibition of the biosynthetic pathway. The antioxidative defense system that destroys active oxygen species generated during photosynthetic electron trans-port is not sufficient to protect the plant, and photooxidative damages are increased. During photosynthesis the formation of active oxygen species is minimized by vari-

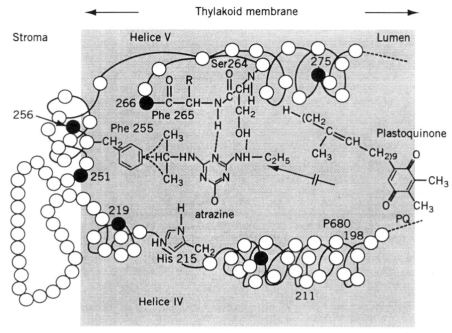

Figure 10-2. Schematic representation of the interaction of atrazine in the Q_{B^-} binding niche of the D1 protein. Hydrogen and hydrophobic bonds are represented by dashed lines. Amino acid residues modified in herbicide-resistant mutants (algae, cyanobacteria, and higher plants) are depicted as darkened circles. Scale is not respected for the Q_{B^-} binding niche. (Adapted from Barber and Andersson, 1992, and Fuerst and Norman, 1991.)

ous regulatory mechanisms present in all plant cells, composed of both nonenzymatic and enzymatic constituents. The nonenzymatic antioxidants are hydrophilic antioxidants—ascorbate and glutathione—and lipophilic antioxidants—α-tocopherol and the carotenoid pigments. Phenolic and flavonoid compounds also scavenge superoxide, singlet oxygen, and hydroxyl radicals. More complex molecules such as phytic acid or phytoferritin form complexes with metals to prevent free-radical production by the iron-catalyzed Haber–Weiss reaction. The enzymatic antioxidative components include superoxide dismutase (SOD), catalase, ascorbate peroxidase (Fig. 10-3), and the enzymes involved in the synthesis and regeneration of the low molecular mass antioxidants (Foyer et al., 1994).

Paraquat and diquat (bipyridyl herbicides) are PSI electron acceptors that mediate the transfers of electrons to O_2, generating toxic oxygen radicals that peroxidate the fatty acid side chains of membrane lipids, leading to loss of membrane integrity (Kunert and Dodge, 1989), and causing a cascade of oxidative reactions resulting in the inactivation of enzymes, lipid peroxidation, protein degradation, DNA strand breaks, and pigment bleaching (Scandalios, 1993). Application of paraquat in two wheat cultivars induced photoinhibition, loss of fresh mass, protein, membrane integrity, and photosynthetic pigments with chlorophyll *b* declining the least and

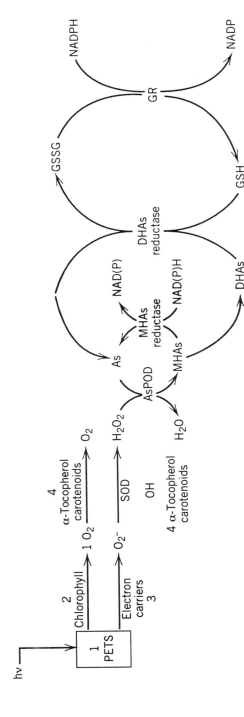

Figure 10-3. Schematic representation of photooxidative mechanisms and associated antioxidative defense system in the chloroplast, including sites of action of herbicides. (1) Herbicides that block photosynthetic electron transport and cause increased excitation energy transfer and then $^1O_2$; (2) Herbicides that induce accumulation of protoporphyrin IX, causing $^1O_2$ generation under light. (3) Herbicides that accept electrons coming from PSI and react with O_2 to produce O_2^- and other toxic species (OH· and H_2O_2). (4) Herbicides that inhibit carotenoids biosynthesis, decreasing antioxidative system.

As: ascorbate oxidase
AsPOD: ascorbate peroxidase
DHAs: dehydroascorbate
GR: glutathione reductase
GSH: reduced glutathione
GSSG: oxidised glutathione
MHAs: monohydroascorbate
PETS: photosynthetic electron transport system
SOD: superoxide dismutase

315

carotenoids declining the most (Kraus et al., 1995). This is the only class of herbicides interacting with PSI that has been commercialized.

PSI is a membrane-bound protein complex that catalyzes oxidation of plastocyanin and reduction of ferredoxin under light conditions. A photon of light is captured by the trap chlorophyll of PSI (P_{700}), which passes an excited electron to the electron acceptor called A0 and then to ferredoxin by several carriers of increasing redox potential (Fig. 10-4). Bipyridyl herbicides (BP), because their redox potential (from -300 to -500 mV) are able to accept electrons coming from iron-sulfur centers or ferredoxin and form a bipyridyl cation radical (Fujii et al., 1990). This cation is unstable and reacts with O_2 to form superoxide:

$$BP^{2+} + e^- \longrightarrow BP\cdot^+ \tag{10-2}$$

$$BP\cdot^+ + O_2 \longrightarrow BP^{2+} + O_2\cdot^- \tag{10-3}$$

Then antioxidative enzyme (superoxide dismutase) can detoxify superoxide by producing hydrogen peroxide and molecular oxygen, which can itself be detoxified by the enzymes of the ascorbate-glutathione cycle following the overall reaction:

$$H_2O_2 + NADPH + H^+ \longrightarrow 2H_2O + NADP^+ \tag{10-4}$$

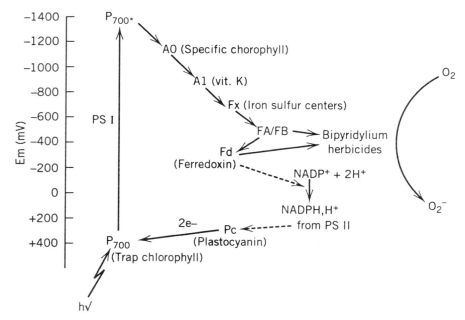

Figure 10-4. *The energetics of electron carriers in PSI and bipyridilium herbicides as electron acceptors.*

Other reactions can occur within the choroplasts involving H_2O_2 such as Fenton's reaction (with trace levels of Fe^{2+}) or various versions of Winterbourn's reactions (Preston, 1994).

All these reactions are linked and the net result is the production of the hydroxyl radical, OH· causing superoxide damages in cells. The radicals attack double bonds in the fatty acid side chains of lipids with the subsequent membrane disruption (Bowyer and Camilleri, 1987; Kunert and Dodge, 1989).

Recently a series of 1,3,4(2H)-isoquinolinetriones have been found to be fast-acting postemergence herbicides, producing symptoms of dessication similar to paraquat. These redox-active compounds are very potent stimulators of the light-dependent consumption of oxygen at PSI, enhancing the generation of superoxide radicals. But the compounds were found to be unstable toward hydrolysis, and this was considered to be a major factor limiting the overall herbicidal effects (Mitchell, 1995). The diphenyl ethers, cyclic imides, and lutidine derivatives cause oxidative stress in plants, acting by inhibition of biosynthetic pathways with accumulation of reactive radical-forming intermediates. These photodynamic herbicides induce abnormal accumulation of photosensitizing tetrapyrolles (e.g., protoporphyrin IX), pigments that are able to cause light-dependent singlet oxygen (Matringe and Scalla, 1988). A 2.2′-bipyridyl treatment and subsequent illumination cause photodynamic damages in plants in toto and inhibition of greening process (grana formation, chlorophyll and carotenoid accumulation, and transformation of prolamellar bodies); the reason may be related to the accumulation of light-harvesting chlorophyll *a/b*-binding proteins in endoplasmic reticulum cisternae, which inhibits their transport to plastids (Mostowska and Siedlecka, 1995). The accumulation of protoporphyrin IX and other chlorophyll precursors, such as magnesium protoporphyrin IX or protochlorophyllide, can be promoted in the dark by 2,2′-dipyridyl in combination with 5-aminolevulinic acid (ALA). Glutamic acid, ALA's natural early precursor, can successfully replace ALA as a component of the photodynamic herbicide. An irradation of treated leaves of wheat and vegetable marrow *(Cucurbita moschata)* by glutamic acid and 2,2′-dipyridyl caused an inhibition of protochlorophyllide reduction (Koleva et al., 1995).

These groups of herbicide can also inhibit protoporphyrinogen oxidase (see the section on biosynthesis below). Photooxidative stress can also be initiated by herbicides that block PSII. In this case, photoelectron transport blocking causes increased transfer of excitation energy from triplet chlorophyll to oxygen, which generates singlet oxygen and superoxide. Another mode of photooxidative stress is inhibition of carotenoid biosynthesis (again, see the section on biosynthesis below), which eliminates important quenchers of the triplet chlorophyll and singlet oxygen (Foyer et al, 1994).

Perturbation of the Membrane Proton Gradient and Respiration   A proton gradient is established at the plasma membrane between inside (negative) and outside (positive) by an electrogenic proton pump requiring energy called H^+ ATPase. The free energy available, or proton motive force ($\Delta \mu H^+$), is the difference between the membrane potential (Em) and the pH difference that exists between the inside and the outside of the membrane. A more negative Em reflects the increased outward transport of H^+ that can be responsible for a decrease in pH (acidification). Any pesticide that perturbs this transmembrane proton gradient may have significant effects on the physiology of plants, because this energy transduction mechanism regulates the active transport of organic and inorganic solutes across the membrane (Fig. 10-5). It also regulates and controls the intracellular pH, and therefore the hormonal control of plant growth, regulation in phloem transport, cell division, and many others functions (Devine and Shimabukuro, 1994). Diclofop and haloxyfop [anyloxyphenoxypropanoate (AOPP) herbicides] under acid forms depolarize the membrane potential by a specific influx of protons (Wright and Shimabukuro, 1987). The characteristic response of susceptible plants is the rapid inhibition of meristematic growth (Shimabukuro, 1990). Dinitrophenol and benzonitrile herbicides (dinoterb, bromoxynil, etc.) are uncouplers of photooxidative phosphorylations and oxidative phosphorylations, respectively, achieved by chloroplasts and mitochondria, blocking ATP synthesis. They increase H^+ membrane permeability, causing disruption of gradient protons (Figs. 10-6 and 10-7) (Moreland and Novitzky, 1987, 1988). The rapid proton influx results in the alkalinization of the outer membrane surface, which decreases the passive uptake of acetate by plant cells contaminated by diclofop, but the rapid inhibition of growth due to depolarization of Em (biophysical mechanism) and the effect on fatty acid synthesis (biochemical mechanism) are distinctly separate events (Wright and Shimabukuro, 1987).

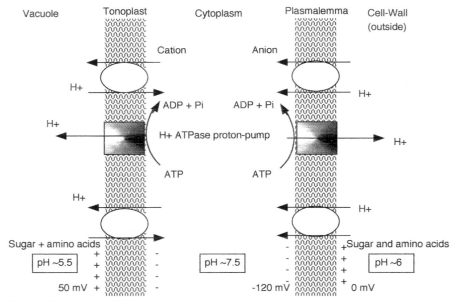

Figure 10-5. *Metabolites transport: chemioosmotic mechanisms involved in H^+ translocation. Active transport is dependent on an electrogenic proton gradient established by the activity of H^+ ATPase proton pump.*

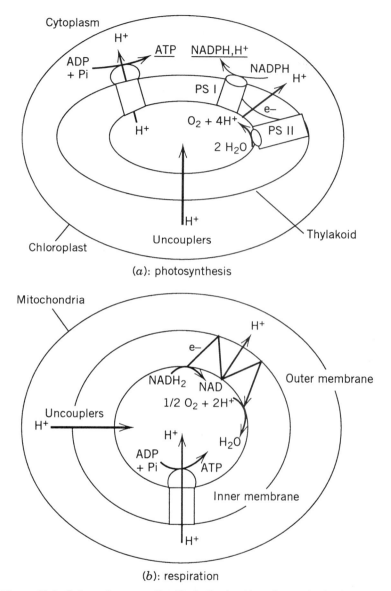

(a): photosynthesis

(b): respiration

Figure 10-6. Schematic uncoupling illustration in chloroplast and mitochondria.

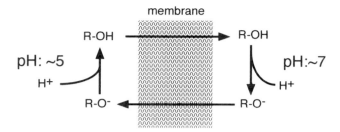

R-OH: phenols - weak acids

Figure 10-7. Schematic mechanism for H⁺ translocation through the membrane by uncouplers.

Biosynthesis (Anabolism)

Inhibition of Fatty Acid Biosynthesis Cyclohexanedione and aryloxyphenoxypropionate groups of herbicides have been shown to act through the inhibition of acetyl-CoA carboxylase (ACCase), which catalyzes the first committed step in fatty acid biosynthesis, acyl lipid biosynthesis (Fig. 10-8) (Rendina and Felts, 1988). These commercial grass-selective herbicides have been used in agriculture since the mid-to-late 1970s.

Lipids biosynthesis is essential for plants. Lipids are involved in the biogenesis and functions of membranes. Fatty acid biosynthesis is located in the chloroplasts and plastids of nongreen tissues and the synthesis of malonyl-CoA catalyzed by ACCase.

ACCase is the key enzyme that regulates the biosynthesis of fatty acids, but many aspects are still not understood (Browse and Somerville, 1991). This is a plastid-localized enzyme exerting strong flux control especially in light-stimulated biosynthesis (Page et al., 1994). This is a biotin enzyme with a prosthetic group involved in transfer of activated CO_2 (Wurtele and Nikolau, 1990).

Inhibition by herbicides such as AOPP and cyclohexanedione (CHD) is reversible, linear, and noncompetitive or nearly competitive, indicating that the herbicides may interact with an acetyl-CoA binding site and perhaps with the release of malonyl-CoA (Rendina et al., 1990). The two herbicide classes overlap with each other and with the CoA substrate only in the small acyl region of the thioester, which implies that the CoA site is largely distinct from the herbicide site, and potent multisite inhbitors can be made by combining features of the substrate and the herbicides. The hydrophobic portion of the aryloxyphenoxypropionate may occupy the same site as the biotin cofactor (Taylor et al., 1994). A new class of cyclic triketones has been prepared by combining the aryloxyphenoxypropionates with the cyclic portion of the cyclohexanedione oximes (combination structures containing features of both herbicide classes). These cyclic triketones have shown the same species selectivity as the parent herbicides; they were bound more than 100 times more tightly by ACCase from monocotyledonous (wheat) than that from dicotyledonous plants (mung bean), and they control weeds without damage to broadleaf species (Rendina et al., 1994).

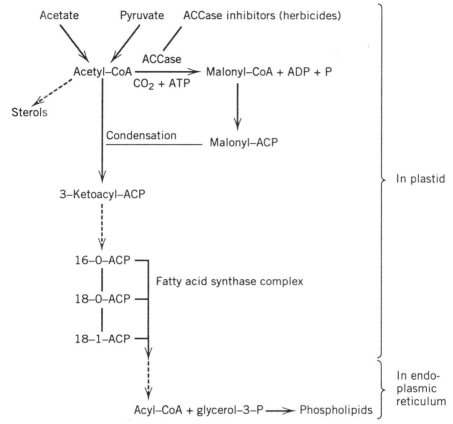

Figure 10-8. *Biosynthetic pathway for fatty acid biosynthesis in plants, including site of action of ACCase inhibitors.*

Acetolactate (Acetohydroxyacid) Synthase Inhibition   Sulfonylurea herbicides inhibit acetolactate synthase (ALS), also known as acetohydroxyacid synthase (AHAS), the first common enzyme of valine, leucine, and isoleucine biosynthesis (Fig. 10-9). This enzyme is essential in the biosynthetic pathways for the branched-chain amino acids valine, leucine, and isoleucine (Ray, 1989). This pathway is not present in animals, and the enzyme is located in the plastids in a multimeric structure. The quaternary structure of ALS is apparently different between prokaryotes and eukaryotes: in yeast and plants only monomers and oligomers of a single subunit are found, the reported molecular weight of monomeric plant ALS varies from 58000 to 72800, and some plants species, such as maize, soybean, canola, or tobacco, express more than one form of ALS. In plants ALS is encoded in the nucleus, synthesized in the cytoplasm, and then transported into chloroplasts via a transit peptid that is cleaved from the mature protein (Saari and Mauvais, 1996). It is nuclear encoded and requires a transit sequence to be imported and processed. Other classes of herbicides inhibited ALS, imidazolinones, triazolopyrimidines,

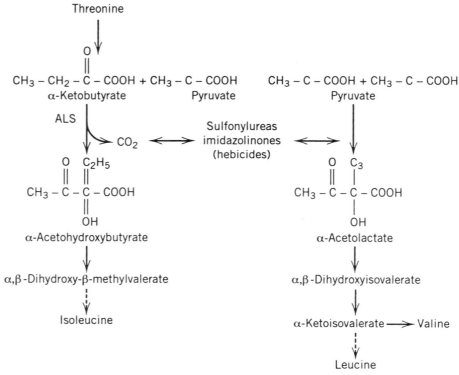

Figure 10-9. *Biosynthetic pathway for branched-chain amino acids, including site of AIS inhibitors.*

and pyrimidyl oxybenzoates. These ALS inhibitors stop growth and then kill plants at only few grams per hectare for certain sulfonylureas.

The ALS inhibitors are also able to affect phloem transport of photosynthates, revealed by an increased accumulation of starch granules in chloroplasts of wild oat treated leaf, but reduced starch levels in the main stem, caused accumulation of the neutral sugar levels of corn leaves, or increased sucrose concentration in pennycress leaf (Bestman et al., 1990; Chao et al, 1994; Shaner and Reider, 1986).

Glutamine Synthetase Inhibition   This enzyme is essential for the assimilation of nitrogen by plants, and its inhibition leads to immediate metabolic dysfunctions. The rapid inhibition of photosynthesis brought about glyoxylate accumulation is the more important effect. When glutamine synthetase is inhibited, the levels of amino acids required for photorespiratory glyoxylate transamination are reduced and glyoxylate accumulates (Dekker and Duke, 1995).

Shikimate Pathway Perturbations   The shikimate pathway corresponds to the major target site of glyphosate. Glyphosate inhibits 5-enolpyruvylshikimate-3-phos-

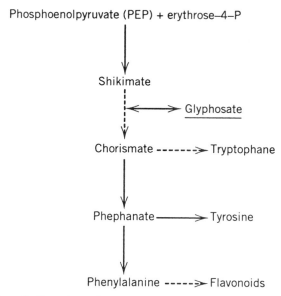

Figure 10-10. *Biosynthetic pathway for aromatic amino acids in plants with site of inhibition of glyphosate.*

phate synthase enzyme and causes aromatic amino acids deficiency, due to a possible accumulation of free shikimate (Fig. 10-10) (Schulz et al., 1990):

$$\text{shikimate-3-P} + \text{PEP} \xrightarrow{\text{5-EPS-3-P synthase}} \text{5-enolpyruvylshikimate-3-P} + \text{PO}_4\text{H}_3 \quad (10\text{-}6)$$

The shikimate pathway including the EPSPS (a nuclear-coded plastid enzyme) is the biosynthetic source of the three aromatic amino acids: phenylalanine, tryptophan, and tyrosine. These amino acids are essential for plant life; they are involved in protein synthesis, auxin biosynthesis, and formation of many other secondary compounds such as phenolic compounds. Thus the blockage of the shikimate pathway can lead to large damages and physiological effects. EPSPS is the primary site of action of glyphosate (N-phosphonomethyl glycine) (Duke, 1988).

Bleaching Activity and Protoporphyrinogen Oxidase Inhibition   The term "bleaching" generally refers to a decrease in the amount of photosynthetic pigments, such as chlorophylls or carotenoids. This pigment deficiency can be caused either by inhibition of biosynthesis of the pigments or by peroxydative destruction of already formed pigments (Kohno et al., 1995).

Impairment of plant pigment biosynthesis is one of the most prominent and attractive objectives of herbicidal mode of action research, because their plant specificity, and a considerable number of bleaching herbicides have been discovered over the past 15 years Babczinski et al., 1995.

Three 1-aryl-3,5-dialkyl-4 substituted pyrazoles inhibited protoporphyrinogen-IX oxidase (proto-IX), causing accumulation of protoporphyrin-IX and light-induced ethane formation (Babczinski et al., 1995).

Protoporphyrinogen oxidase (Protox or PPO) is the last enzyme in the porphyrin pathway that is common to both heme and chlorophyll synthesis pathways (Fig. 10-11). Protoporphyrin-IX (tetrapyrollic pigment) accumulates when Protox is inhibited and this photosensitizing agent generates highly reactive singlet oxygen in the presence of sunlight, causing phytotoxicity (Duke et al., 1994; Matringe et al., 1992). A number of highly active herbicides have been found to be effective inhibitors of Protox (Table 10-1). The new herbicide family of tetrahydroindazoles are another class of inhibitors of protoporphyrinogen oxidase, with a characteristic light-induced dessication, the result of free-radical-initiated lipid peroxidation (Lyga et al., 1994). Potent Protox inhibitors are effective in a nanomolar range, and act as reversible, competitive inhibitors with respect to the substrate, as has been demonstrated by Lineweaver–Burk analysis for different type of compounds (Camadro et al., 1991; Nicolaus et al., 1993, 1995). Recent results suggest that protoporphyrin accumulation is a result of protoporphyrinogen IX accumulation in the herbicide-inhibited chloroplast, followed by export into the cytoplasm and subsequent oxidation to the phototoxic protoporphyrin IX by herbicide-resistant mechanisms (enzymatic oxidations) associated with extraorganellar plant mem-

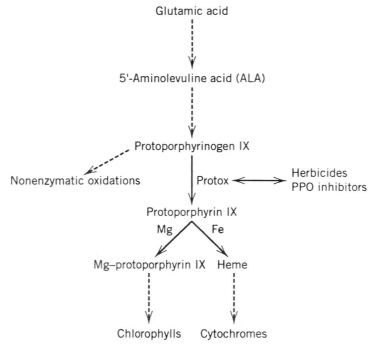

Figure 10-11. *Biosynthetic pathway for chlorophylls and cytochromes, with site of PPO inhibition.*

branes (Jacobs and Jacobs, 1993; Jacobs et al., 1991). These results allow us to explain the extraplastidic accumulation of protoporphyrin IX in plants treated with bleaching herbicides. Diphenyl ethers are bleaching herbicides with toxic effect dependent not only on their ability to cause protoporphyrinogen accumulation by inhibiting protoporphyrinogen oxidase, but also on the rate at which the accumulated protoporphyrinogen is converted to protoporphyrin. In this case protoporphyrinogen may decompose to nonprotoporphyrin (nontoxic) products if it is not rapidly oxidized (Jacobs et al., 1994). Protoporphyrinogen oxidase is supposed to be associated with both mitochondrial and chloroplast membranes in plant cells, but the activity of this enzyme was found not only in crude etioplast and mitochondrial fractions but also in the soluble fraction of tobacco cell lines (Yamato et al., 1994), or in plasma membranes of barley (Lee et al., 1993). The oxidizing activity of these fractions was not inhibited by herbicides that inhibit protoporphyrin oxidase. The soluble protoporphyrinogen-oxidizing enzyme seemed to be a kind of peroxidase assuming the following reactions:

$$\text{protoporphyrinogen IX} + H_2O_2 \xrightarrow{\text{peroxidase}} \text{protoporphyrinogen IX (free radical)} \quad (10\text{-}7)$$

$$\text{protoporphyrinogen IX (free radical)} + O_2 \rightarrow \text{protoporphyrin IX} + O_2^- + H^+ \quad (10\text{-}8)$$

$$O_2 + H^+ \rightarrow \tfrac{1}{2}(H_2O_2 + O_2) \quad (10\text{-}9)$$

This soluble enzyme with herbicide resistance may play an important role in oxidation of protoporphyrinogen IX, which accumulates out of the site of heme and chlorophyll biosynthesis inthe herbicide-treated plants (Yamamoto et al., 1994).

Phytoene Desaturase Inhibition   Phytoene desaturase (PD) is an enzyme involved in carotenoid biosynthesis. Inhibition of this enzyme leads to plants without chlorophyll. Only two PD inhibitors (fluridone and norfluzaron) are actually commercially available (Dekker and Duke, 1995), but recently substituted tetrahydropyrimidones (1,3-diarylperhydropyrimidine-2 ones, "cyclic ureas"), new selective herbicides for preemergence use in cotton and rice, have been shown to function as inhibitors of carotenoid biosynthesis at the PD step. Chlorophyll bleaching was demonstrated to be a secondary photooxidative event caused by the loss of carotenoids as quenching substrates (Babczinski et al., 1995).

Other Modes of Actions

Seed Production Inhibition   Phenoxycarboxylic acids (PCA) such as 2,4-D or 2,4,5-T are the oldest synthetic herbicides, but their molecular site of action is still unknown (Dekker and Duke, 1995). They are called hormone-type herbicides because they mimic in many ways lethal doses of the plant hormone indole acetic acid (Devine et al., 1993). These compounds influence plasma membrane properties by acting at a molecular site in the plasma membrane (Dekker and Duke, 1995). The phenoxyacetic acids have negative effects on flowering, fruit set, and seed set (Andersson, 1994). The seed production in *Chenopodium album* was reduced by up to 70% by MCPA doses as low as one-sixteenth of a normal dose, and exposure to 2,4-D close to flowering seriously reduced the seed production in *Rumex crispus* and other weeds (Andersson, 1994; Pedersen and Rasmussen, 1990). Reductions in seed production by phenoxyacetic acids have been reported by many authors, but this reduction has also been obtained by application of the sulfonylurea herbicides such as chlorsulfuron (Beck, 1990) and metsulfuron (Beck, 1990; Pedersen and Rasmussen, 1990). Low doses (sublethal) of herbicides reduce the seed production substantially; their effect can play an important role in a strategy for controlling certain weed species without injuries to crops, particularly when herbicides are used in combination (Andersson, 1994).

Herbicide residues may affect seedlings during early stages of their development. The persistence data reported for the family of sulfonylurea herbicides indicates a carryover of herbicide residues to a second season, and application of chlorsulfuron or metsulfuron methyl in early stages of seed germination seriously affects the gravity perception center (consisting of the statocystes), and the secretory tissue of the root caps, thus probably disturbing the processes of gravitropism and the protective slime secretion of the roots (Ciamporina and Mistrik, 1993; Fayez et al., 1995).

Mitotic Inhibitors   Mitosis can be affected by herbicides in different ways: indirectly, for example, through the disturbance of energy metabolism, or directly by inhibition of the DNA-synthetic phase. This latter mode of action can be obtained through the effect on several different targets:

- Tubulin self-assembly. Several herbicides appear to have a primary effect on protein involved in cell division. Dinitroanilines can bind to the protein subunit of microtubules (tubulin) to disrupt proper assembly of the microtubules required for cell division and thus meristematic activity in shoots and roots (Devine et al., 1993):

$$\alpha\text{-tubulin} + \beta\text{-tubulin} \xrightarrow{\text{Ca}^{++}} \text{tubulin} <=> \mu\text{-tubules} \qquad (10\text{-}10)$$

- Activity of the microtubule organization center (propham and chlorpropham).
- Cell wall formation inhibition (2,6-dichlorobenzonitrile).
- Coumarin or cellulase biosynthesis inhibition (2,4-D).

Benzamides such as propyzamide prevent spindle formation, thereby interrupting the cell cycle before metaphase, and affecting chromosome migration process (Merlin et al., 1987). Isoxaben inhibits cell wall synthesis by blocking glucose incorporation (Corriot and Scalla, 1991); this mode of action is close to that of dichlobenil, which inhibits cellulosis synthesis.

Interactions Farm chemicals and fertilizer account for approximately 25% of the total operating costs associated with growing flax (Driver and Josephson, 1993). Reducing herbicide rates to lower herbicide costs is one option, but this frequently results in poor weed control, reduced yields, and dissatisfied producers (Wall, 1994). The use of tank mixtures of selected herbicides or application of two or more herbicides sequentially is quite common in modern weed management. The reasons for the use of these mixtures are:

- To increase the spectrum of weed control
- To delay the appearance of weed biotypes resistant to a given herbicide
- To improve crop safety by using minimum doses of herbicide combinations
- To improve weed control under variable weather or soil conditions
- To reduce cost of treatment (Hatzios and Penner, 1985; Zhang et al., 1995).

It has been demonstrated that herbicides may interact before or after entering the plants, and the outcome of the interaction may be synergistic, antagonistic, or additive (Green, 1989; Hatzios and Penner, 1985). Generally, interactions between herbicides were antagonistic more frequently than synergistic; antagonistic interactions occurred more frequently when the target plants were monocot rather than dicot, and in the Asteraceae, Poaceae, or Fabaceae than in the Chenopodiacae or Convolvulacae families (Zhang et al., 1995). The mixture of fluazifop-P and clethodim controls wild oat and green foxtail in flax and is as effective as full rates of either herbicide applied alone; it also represents a 20% reduction in total amount of active ingredient required to control these weeds (Wall, 1994). MCPA reduces fenoxaprop activity in wheat and barley, protecting the crops from herbicide injury (Raymond et al., 1990).

Synergistic interaction between the organophosphate insecticide terbufos and the sulfonylurea herbicide nicosulfuron may result ins ever injury to the crop. Terbufos inhibits the metabolism of nicosulfuron; metabolism of nicosulfuron by corn includes aryl hydroxylation followed by glucose conjugation. Naphthalic anhydride protects corn against nicosulfuron terbufos. Interaction may be by enhancing the activity of the P-450 monooxygenase system (Siminskzy et al., 1995). Ethatmesulfuron interacts synergistically with clopyralid (3,6-dichloro-2-pyridinecarboxylic acid) on common lamb's-quarters *(Chenopodium album)* and redroot pigweed *(Amaranthus retroflexus)*, but under some conditions, the mixture of fluazifop-P with ethametsulfuron suppressed early canola growth and reduced seed yield. Growth and yield responses of canola crops to mixtures of ethametsulfuron with specific grass herbicides depend on the type of graminicide used. No injury results when the ethametsulfuron is mixed with sethoxydim, and yield losses ranging from 59% to 97% result when it is mixed with the following grass herbicides, listed in decreasing order of injury: haloxyfop > fluazifop > fluazifop-P > quizalofop > quizalofop-P (Harker et al., 1995).

Another type of interaction consists of antagonism between the two mixed herbicides under particular conditions. A study conducted with wild oats *(Avena fatua)* indicated that the interactions between the methyl ester of imazamethabenz, a selective postemergence herbicide for wild oat, and MCPA, a growth regulator type herbicide for broadleaf weed control, differed with different formulations of both herbicides and with the growth stages when the herbicides are applied. Antagonism is most likely to occur at an early stage of wild oat development and when the two herbicides are mixed in a same tank, or in the sequence of MCPA followed by imazamethabenz (liquid concentrate).

Influence of Mycorrhizal Colonization

Relationships between vascular plants and arbuscular mycorrhizal fungi have developed over the course of 400 million years of evolution, leading to symbiosis. Few studies have compared the effects of herbicides on mycorrhizal and nonmycorrhizal plants (Hamel et al., 1994). Some herbicides affect arbuscular fungi directly, while others affect the host plants (Garcia-Romera and Ocampo, 1988). Most of the time, herbicides have no effect or negatively affect mycorrhizal fungi; mycorrhizal *Fraxinus americana* have reduced growth in presence of paraquat (Pope and Holt, 1981). Recently, the effects of three commonly used herbicides (paraquat, dichlobenil, and simazine) were examined on apple trees. The response of mycorrhizal plants to herbicides was greater, and the relative elongation rate was more sharply reduced in mycorrhizal than in nonmycorrhizal plants, and it was found that plant mortality was higher among mycorrhizal that nonmycorrhizal apple trees (Hamel et al., 1994).

For certain herbicides, modes of action are unknown (e.g., cyanamide) or unclear (e.g., dalapon).

MECHANISMS OF HERBICIDE RESISTANCE

There were predictions that herbicide resistance would develop in weeds in the 1950s, because pesticide resistance had already appeared in insects and pathogens. The first herbicide-resistant weed, *Senecio vulgaris,* was discovered in 1968 in crops treated with simazine (triazine group) (Gronwald, 1994). It is important to understand the mechanisms of herbicide resistance in plants. Particularly with the availability of many biotechnological tools to incorporate herbicide resistance in crops, the possibility now exists to select herbicides for desirable weed control as well as for desirable environmental and economic qualities at the beginning of the discovery process, and then to incorporate resistance into the crops after these herbicides have been identified (Dekker and Duke, 1995). Resistance can be selected by screening for mutant plants exposed to the herbicide or by introducing a resistance gene into the crop by genetic engineering (Saari and Mauvais, 1995).

Three key factors influence the mechanisms of resistance in plants:

1. Herbicide absorption and translocation
2. Herbicide detoxification and metabolism

3. Sensitivity of the target site(s), since reduced target site sensitivity is frequently a significant factor in herbicide resistance

These mechanisms can be grouped into two categories (Dekker and Duke, 1995):

1. Mechanisms that exclude the herbicide molecule from the active site(s): herbicide uptake, translocation, compartmentation, metabolic detoxification
2. Mechanisms that make the specific site(s) resistant: alteration, mutation, overproduction

Resistance can be due to differential sensitivity of molecular target sites, resulting from alteration or mutation of the different target site protein structures such as s-triazines, ACCase inhibitors, imidazolinones, and so on. The most important mechanism for resistance to the s-triazines herbicides in plants is brought about by alterations to the D-1 protein. Resistance can result from point mutations to the psbA gene at several sites, which lead to amino-acid substitutions. The quinone-binding pocket alterations in the mutant also result in changes in the electron transfer properties in PSII with a reduced rate of transfer (Trebst, 1991). Genes of resistance can immigrate into a weed population from two sources: seed or pollen. *S. ibenica* plants can move over 5 km in about one week (Saari et al, 1994).

Resistance to PSII Inhibitors

Triazine resistance is the most prevalent type of herbicide resistance: 57 resistant biotype weed species were discovered in continuous maize monoculture where triazines had been applied repetitively for several years (LeBaron, 1991). Resistance to PSII inhibiting herbicides is due to modifications at the D-1 protein of the PSII complex (target site), and for a few species resistance is due to enhanced herbicide detoxification. The modifications of the target site consist of modifications of amino acid residues in the Q_B binding site, by point mutation of the psbA gene. A substitution of Gly for Ser 264 decreases atrazine affinity by 1000 (Fuerst et al., 1986).

Maize is highly resistant to chlorotriazines due to a primary mechanism of detoxification involving glutathione conjugation by 3-glutathione-S-tranferase (3-GST) isozymes, with N-dealkylation as a possible secondary mechanism (Shimabukuro, 1985).

Resistance to the Phenyl Urea Herbicides

For chlorotoluron, resistance is not due to differential uptake, translocation, or modified target sites, but to the enhanced ability of resistant plants to metabolize the herbicide, and cytochrome P-450 oxidase enzymes are involved in this mechanism of resistance in blackgrass plants (Jorrin et al., 1990; Kemp et al., 1988). Enhanced metabolism by a broad-spectrum cytochrome P-450 system might explain why resistance to a wide range of different herbicides occurs (James and Kemp, 1995).

Resistance to Acetyl-CoA Carboxylase Inhibitors

A biotype of *Eleusine indica* (C_4 annual grass weed), which had a field history of two or three applications of fluazifop-*p*-butyl per year over 4–5 years, was previously found to show 100-fold resistance to the herbicide under field conditions compared with a susceptible population. The close correlation between the whole plant and ACCase inhibitors suggests that herbicide resistance is conferred by a mutation to the herbicide target site (Leach et al., 1995). A similar mechanism of resistance has been found in a biotype of *Setaria viridis*. The resistance is conferred by an altered form of ACCase that is much less sensitive to a wide range of aryloxy-phenoxypropanoate and cyclohexanedione herbicides (Marles et al., 1993). The resistant *E. indica* biotype is less vigorous than the wild type and produces less biomass, indicating that it may be at a disadvantage when grown with the wild type (Leach et al., 1995).

Resistance to Acetolactate Synthase Inhibitors

Mutations alter herbicide binding to the enzyme; for example, the mutant *Brassica napus* ALS3 gene, contains a point mutation predicting an amino acid substitution (557 Trp to Leu) that confers high levels of resistance (Brandle et al., 1994).

ALS mutations that give resistance to one sulfonylurea by target-site sensitivity changes generally accord some resistance to most other sulfonylurea herbicides (Saari et al., 1994). The nucleotide changes in the ALS gene responsible for resistance have not been elucidated in most of the resistant organisms. Pro mutations (Pro 197 to Ser, Pro 196 to Gln, Pro 196 to Ala) in the ALS gene may be more prevalent than other mutations, resulting in ALS inhibitor resistance in plants. Imidazolinone-resistant weeds are target site mutants with predominantly Ser mutations in the ALS gene (Saari et al., 1994).

Resistance to Bipyridylium Herbicides

The mechanism of resistance to diquat and paraquat was investigated in *Arctotheca calendula* Levyns. These experiments demonstrated that diquat appears at the active site more slowly in the resistant biotype than in the susceptible biotype. The mechanism of resistance is not a result of changes at the active site, decreased herbicide absorption, or decreased translocation, but appears to be due to reduced herbicide penetration to the active site (Preston et al., 1994).

Altogether, resistance to bipyridyl herbicides has been reported for 4 grass species and 12 broadleaf weeds and has become apparent in the field following many exposures (Preston, 1994). In all cases of resistant biotype of different species (except *Hordeum leoprinum*), the paraquat movement into leaf cells to the active site was restricted compared to the susceptible biotype (Preston, 1994).

Resistance to "Bleaching" Herbicides (Protoporphyrinogen Oxidase Inhibitors)

Certain plants have a higher capacity to destroy protoporphyrinogen than to oxidize it, and accumulated protoporphyrinogen would decompose to nonphototoxic products (nonporphyrin) rather than become oxidized to protoporphyrin (Jacobs et al., 1994).

Resistance to Mitotic Inhibitors (Dinitroaniline Herbicides)

Resistance to pendimethalin in blackgrass *(Alopecurus myosuroides),* a major annual grass weed of winter cereal crops, is attributable to an oxidative degradation of the 4-methyl group, similar to that which occurs with resistance to chlorotoluron; the other dinitroaniline herbicides are not metabolized, due to the absence of ring methyl or other groups that are sensitive to oxidative degradation (James and Kemp, 1995). Resistance is also due to an intrinsic change in tubulin protein. In a resistant biotype of goose grass, a β-resistant biotype has been identified. The microtubules of the R biotype are hyperstabilized and therefore less sensitive to destabilization by dinitroanilines (Smeda and Vaughn, 1994).

Resistance by Overproduction and Fitness

Resistance can be conferred by overproduction of the target site, thus diluting the herbicide and allowing enough additional target protein to continue normal functions and growth (Devine et al, 1993). For example, glyphosate resistance is obtained by the overproduction of 5-enolpyruvylshikimate-3-phosphate synthase, a key enzyme in the pathway synthesizing plant aromatic compounds (Dekker and Duke, 1995).

The relative fitness of herbicide-resistant and -susceptible weed biotypes will influence how quickly the resistant population declines once the selection pressure is removed; for example, triazine-resistant weeds produces less biomass and tend to be less fit than their susceptible counterparts (Warwick and Black, 1981). Fitness is a measure of survival and ability to produce viable offspring, and triazine resistance often has negative physiological consequences for the plants. This is not the case for the ALS-resistant plants. For diverse weeds seed germination was realized, and resistant lines germinated faster than susceptible lines (Saari et al., 1994). The psbA mutation (in PSI inhibitor resistant-biotype) reduces plant fitness by causing a reduction in CO_2 fixation, a lowering of quantum yield (which reduces the rate of electron transfer between Q_A and Q_B), and a reduction of biomass production (Gronwald, 1994).

Herbicides are designed to be toxic to plants, but only a few studies have been performed to obtain an understanding of their effects on plant communities and particularly of their effects upon species interactions. An experiment was conducted for assessing multispecies responses to three pesticides (including two herbicides: atrazine and 2,4-D) where pesticides were applied at low concentrations on model plant communities grown in raised beds using soil containing a natural seed bank (Pfleeger and Zobel, 1995). Parameters measured in this study included target plant biomass and flower height, neighborhood percent cover and aboveground biomass by species, and total community aboveground biomass. Community biomass decreased with the tested herbicides at, respectively, 16 and 10.6% of recommended application rate for atrazine and 2,4-D. Atrazine treatment at this concentration caused a radical change in community structure: two dominant species were killed and the community became dominated by three other species. All tested compounds modified relative species abundance, altered dominance, and simplified the treated communities.

The phytotoxicity of herbicides can alter the structure of plant communities; for example, use of 2,4-D in wheat fields (Hume, 1987) reduced the influence of the dominant species, as did use of growth retardants such as mefluidide and paclobutrazol in pastures (Marshall, 1988).

HERBICIDE DETOXIFICATION (METABOLIZATION) IN PLANTS

The biochemical reactions that detoxify herbicides in plants can be regrouped in three phases (Fig. 10-12): activation, conjugation, and compartmentation.

Functional (Activation) Phase

Enzymatic reactions introduce functional groups such as —OH, —COOH, —NH2, and so on, in molecules of herbicides, to increase polarity and to facilitate the action of conjugation enzymes.

Oxidation Because herbicides are relatively stable molecules, activation must occur to involve degradation of the chemical. Oxidation is the major and the primary oxidative mechanism of the metabolism. Oxidative reactions are catalyzed by monooxygenases and include alkyl oxidation, aromatic hydroxylation, epoxidation, N-dealkylation, O-dealkylation, and sulfur oxidation. Aryl (ring phenyl) hydroxylation may be the most common reaction for detoxification (Shimabukuro, 1985).

The rate and extent of metabolism of herbicides determine the susceptibility or tolerance of plant to most herbicides, and the microsomal cytochrome P_{450} monoox-

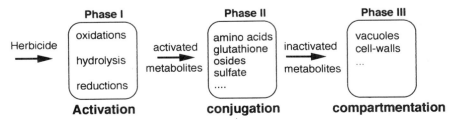

Figure 10-12. *Detoxification mechanisms in plants (schematic view). Amino acids (a-a).*

ygenase system (cyt P_{450} is the major oxidative pathway involved in the metabolism of xenobiotics (Moreland et al., 1995). Cyt P_{450} is an enzyme complex, and its involvement in detoxification of many herbicides is based on the effect of P_{450} inhibitors such as aminobenzotriazol (ABT). Cyt P_{450} is intramembrane heme protein constituted by a constant part, a protoporphyrin IX including binding site with O_2, and a variable part [(molecular mass) M.M, 43 to 60 Kda] binding substrates (Durst, 1991). More than 160 isozymes are now identified and are able to catalyze multiple reactions (Fig. 10-13). To metabolize the herbicide to nonphytotoxic compounds for sulfonylureas, the more common reactions are aryl and aliphatic hydrox-

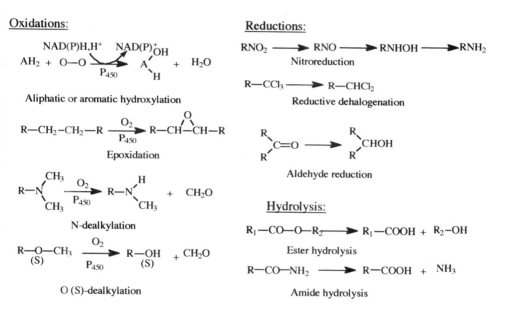

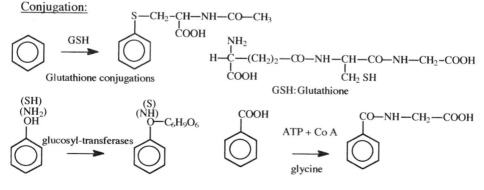

Figure 10-13. *Main reactions for herbicide detoxifications in plants.*

ylation, *O*-dealkylation, and deesterification (Brown et al., 1991). Microsomes isolated from wheat shoots have been shown to metabolize chlorsulfuron, chlortoluron, diclofop, flumetsulam, linuron, and triasulfuron. For dicot, microsomes isolated from excised cotyledons of mung bean *(Vigna radiata)* metabolized three acetamide herbicides (metolachlor, CGA-27704, and alachlor). The three herbicides were *O*-demethylated. The metabolism required a reduced pyridine nucleotide and was affected by several cytochrome P_{450} monooxygenase inhibitors such as CO, tetcyclasis, piperonyl butoxide, 1-aminobenzotriazole, or SKF-525 A.

Cytochrome P_{450} monooxygenases of higher plants mediate the oxidative metabolism of a number of classes of herbicides: phenylureas, sulfonylureas, diclofop, bentazon, metolachlor, and 2,4-D. Pretreatment of seed with various chemicals including herbicide safeners, ethanol, phenobarbital, and so on, results in enhanced metabolism of herbicides. Pretreatment of grain sorghum seed with various safeners resulted in enhanced metabolism of lauric acid, bentazon, and diazinon (Moreland et al., 1993).

Metabolism of aryloxyphenoxypropanoate herbicides is by rapid initial hydrolysis of the ester to the acid. Following deesterification, possibly outside the plasma membrane, diclofop acid (AOPP herbicide) is subject to aryl hydroxylation catalyzed by a cytochrome P_{450} monooxygenase (Zimmerlin and Durst, 1990).

After oxidative metabolism (mediated by Cyt P_{450}), herbicide-resistant weeds exhibit an enhanced capacity to metabolize various types of herbicides such as *s*-triazines, phenylureas, aryloxyphenoxypropionates, cyclohexanediones, imidazolinones, and sulfonylureas by *N*-demethylation and aryl hydroxylation in *Alopecurus myosuroides* R-biotypes or *N*-dealkylation in *Lolium rigidum* (Burnet et al., 1993; Moss, 1990).

Reduction   There is little information about reduction enzymes involved in plant detoxification. These enzymes need reducers, 6 NADPH to reduce $-NO_2$ to $-NH_2$, and also need FAD as cofactor. Aryl nitroreduction is an important reaction in herbicide degradation (Figure 10-13).

Hydrolysis   Herbicide hydrolysis is essentially catalyzed by esterases and amidases (Fig. 10-13). This is a common plant reaction important in detoxification of several herbicides: bromoxynil, cyanazine, propanil, carboxylic acid ester herbicides (2,4-D), and several graminicides such as diclofop methyl, which are hydrolyzed to the active, free acid once in the plant leaf (Dekker and Duke, 1995; Loos, 1975).

Studies on the metabolism of bisethyleneglycolesters of the phenoxyacetic acid 2,4-D and MCPA, respectively, in *Hordeum vulgare* and in *Lycopersicon esculentum* (tomato) have shown that ester hydrolysis was the initial step of metabolism. The resulting free phenoxyacetic acids were first rapidly hydroxylated and then conjugated with monosaccharides, dissacharides, and higher carbohydrates (Barenwald et al., 1993, 1994).

Conjugation Phase

Conjugation in plants is the reaction in which a herbicide metabolite obtained in earlier reactions is joined with an endogenous substrate such as glucose, an amino acid, or glutathione to form a larger compound that is more water-soluble (more polar) and nonphytotoxic, and that is subsequently metabolized.

Glucose conjugation (glucosylation) occurs for herbicides with amino, carboxyl, or hydroxyl functional groups, and more specifically, in the presence of a phenolic, carboxylic, or arylamine moiety (chlorpropham, propanil, diclofop methyl) (Hatzios, 1991). Gucosylation is a common final reaction in the biosynthetic pathway of many plant secondary metabolites, and glucosyltransferases regulate the intracellular concentrations of various endogenous phenolic compounds including cyanogenic glucosides, cytokinin, abscisic acid, and indole-3-acetic acid. These gucosyltransferases also glucosylate xenobiotics and/or xenobiotic metabolites (Gallandt and Balke, 1995). Amino acid conjugation occurs primarily with acidic herbicides through an α-amide bond. For example 2,4-D forms major glutamic and aspartic acid conjugates.

Glutathione conjugation catalyzed by glutathione-S-transferase (GST) is one of the most important types of conjugation because of the wide range of groups of herbicides that can be conjugated: α-chloroacetamides, diphenylethers, thiocarbamates sulfoxides, or 2-chloro-S-triazines (Dekker and Duke, 1995). GST is a cytoplasmic dimeric enzyme that catalyzes a nucleophilic attack at electrophilic sites on some herbicides. Conjugation of acidic herbicides with plant constituents is now a well-established phenomenon; such conjugation usually involves amido or ester linkages (Hatzios, 1991). Conjugates of MCPA in wheat, barley, and oat, and of bromoxynil in barley have been isolated. Residues of MCPA and bromoxynil in postemergence-treated triticale (a new plant produced by crossing wheat and rye) decreased to less than 25 ppb 21 days after application (Cessna, 1994).

Compartmentation Phase

Metabolites are stocked in vacuoles; only a few are incorporated in cell walls (case of aromatic polycyclic herbicides). Some metabolites can form adducts with D.N.A. and are potentially carcinogenic.

For certain herbicides there is no metabolization; this is the case for paraquat and diquat, for which no metabolism has yet been detected for any higher plants (Preston, 1994).

Interactions

Early research has shown that injury caused by the herbicide EPTC (carbamates group) to sugar beets was reduced significantly with bacterial fertilizers such as azobacterin, and accelerated microbial degradation of several other agrochemicals in the root zone of various crops has been observed (Hatzios, 1989).

Metabolism of Toxic Products of Herbicide Action

Toxic O_2 species (superoxide, hydroxyl radicals, and so on,) are produced by the interaction of certain herbicides with O_2 (e.g., bipyridyl herbicides). There are a variety of enzymes present in plants to detoxify these toxic species. Different isoenzymes of superoxide dismutase are localized to the chloroplast, the cytoplasm, and the mitochondrion; ascorbate peroxidase activity can be observed in chloroplast, as can cytosol and glutathione reductase activity, while catalase activity is observed in the cytosol (Preston, 1994).

At minimum and maximum ranges of recommended doses of methabenzthiazuron (urea herbicide), 0.25 and 0.40 g m^{-2} treatment, respectively, for *Vicia faba,* plant tissues detoxified the herbicide and one month after plant emergence, photosynthesis was fully recovered with biomass production even increased at the end of the growth period (Vidal et al., 1995).

CONCLUSION

In 50 years, the utilization of herbicides has become as common an agricultural practice as utilization of fertilizer. The 40 chemical families (and their derivatives) in common use represent a range of varied products adapted to various situations, from total weed control to very selective weed control. The efficiency of these molecules is very great, but factors such as resistant weed species, environmental risks, and high costs necessitate both a reduction of the utilization of herbicides by a best practice, and the application of alternative solutions.

The new chemical herbicide research is currently following three main paths:

1. Systematic screening of all chemical molecules to identify possible herbicidal properties
2. Search for analogs with natural substances, such as vegetable hormone analogs like 2,4-D
3. Development of new molecules based on theoretical knowledge such as the QSAR

Thus in these last 20 years, phenylureas, phenylcarbamates, aryloxyphenoicpropionic acids, and cyclohexanediones have been commercialized, and all these herbicides are more selective for weeds. Many of the recent herbicides belong to the family of the diazinones (thio derivatives). New targets are also investigated, for example, sucrose phosphate synthase is a key enzyme for sucrose biosynthesis in plants and involved in the regulation of carbon partitioning in the leaves (Bruneau et al., 1991).

But other strategies have also been realized, such as strategies to develop herbicide-resistant crop plants by changing the sensitivity of the target enzyme or by introducing a detoxifying enzyme by genetic engineering methods.

Because of herbicide contamination of groundwater, residues in soil and crops, and other negative effects of intensive weed management systems, alternative weed management has been a recent development. Most alternative methods of weed control (such as 40 to 60% reduction in the amount of herbicides used and reduced tillage inputs) are more energy efficient than conventional weed control practices (such as broadcast application of herbicides at recommended rates) (Clements et al., 1995).

Reduction in the amount of herbicides can be achieved in a number of ways. For example, herbicide use may be reduced by banded applications instead of broadcast applications and by restricting applications within a critical period of weed development.

Allelopathy is another possible strategy to control weeds in crops, but much more research on this topic is needed.

REFERENCES

Andersson, L. (1994a), *Swedish J. Agric. Res.*, *24*, 49–56.

Andersson, L. (1994b), *Swedish J. Agric. Res.*, *24*, 95–100.

Andersson, B. and S. Styring (1991), *Curr. Top. Bioenerget.*, *16*, 1–81.

Babczinski, P., G. Sandmann, R. R. Schmidt, K. Shiokawa, and K. Yasui (1995), *Pest. Biochem. Physiol.*, *52*, 33–44.

Barber, J. and B. Andersson (1992), *TIBS, 17*, 61–66.

Barenwald, G., B. Schneider, and H. R. Schutte (1993), *Phytochemistry, 32*, 523–526.

Barenwald, G., B. Schneider, and H. R. Schutte (1994), *J. Plant Physiol.*, *144*, 396–399.

Beck, A. (1990), *Weed Technol.*, *4*, 482–486.

Bernson, V. and G. Ekstrom (1991), *Pesticide Outlook, 2*, 33–36.

Bestman, H. D., M. D. Devine, and W. H. Vanden Born (1990), *Plant Physiol.*, *93*, 1441–1448.

Bowyer, J. R. and P. Camilleri (1987), in *Herbicides* (D. H. Hutson and T. R. Roberts, Eds.), Wiley, New York, pp. 105–145.

Brandle, J. E., M. J. Morrison, J. Hattori, and B. L. Miki (1994), *Crop Sci.*, *34*, 226–229.

Brown, H. M., R. F. Dietrich, W. H. Kenyon, and F. T. Lichtner (1991), In *Brighton Crop Protection Conf.—Weeds,* The British Crop Protection Council, Farnham, UK, pp. 847–856.

Browse, J. and C. Somerville (1991), *Annu. Rev. Physiol. Plant Mol. Biol.*, *42*, 267–506.

Brudenell, A. J. P., D. A. Baker, and B. T. Grayson (1995), *Plant Growth Regul.*, *16*, 215–231.

Bruneau, J. M., A. C. Worrell, B. Cambou, D. Lando, and T. A. Voelker (1991), *Plant Physiol.*, *96*, 473–478.

Burnet, M. W., B. R. Loweys, J. A. Holtum, and S. B. Powles (1993), *Planta, 190*, 182–189.

Camadro, J. M., M. Matringe, R. Scalla, and P. Labbe (1991), *Biochem. J.*, *277*, 17–21.

Cessna, A. J. (1994), *Weed Technol.*, *8*, 586–590.

Chao, J. F., W. A. Quick, A. I. Hsiao, and H. S. Xie (1994), *Weed Sci.*, *42*, 345–352.

Ciamporina, M. and I. Mistrik (1993), *Environ. Exp. Bot., 33,* 11–26.

Clements, D. R., S. F. Weisc, R. Brown, D. P. Stonehouse, D. J. Hume, and C. J. Swanton (1995), *Agric. Ecosyst. Environ., 52,* 119–128.

Corriot, M. F., and R. Scalla (1991), *GFP Acta, 3,* 344–348.

Cremlyn, R. J. (1991), *Agro chemicals: Preparation and mode of action,* Wiley, Chichester, UK, p. 396.

Darmstadt, G. L., N. E. Balke, and T. P. Price (1984), *Pestic. Biochem. Physiol., 21,* 10–21.

Dekker, J. and S. O. Duke (1995), *Adv. Agron., 54,* 69–116.

Devine, M. D. and R. H. Shimabukuro (1994), In *Herbicide resistance in plants* (S. Powles and J. Holtum, Eds.), CRC Press, Boca Raton, FL, pp. 141–170.

Devine, M. D., H. D. Bestman, and W. H. Vanden Born (1990), *Weed Sci., 38,* 1–9.

Devine, M. D., S. O. Duke, and C. Fedtke (1993), *Physiology of herbicide action,* Prentice Hall, Englewood Cliffs, NJ, p. 254.

Driver, H. C. and R. M. Josephson (1993), In *Proc. Manitoba-North Dakota Zero-Tillage Farmers Assoc. Workshop,* Jan. 25–27, Brandon, MB, pp. 163–167.

Duke, S. O. (1988), In *Herbicides: Chemistry, degradation and mode of action* (P. C. Kearney and D. D. Kaufman, Eds.), Vol. III, Dekker, New York, pp. 1–70.

Duke, S. O. H. J. Lee, U. B. Nandihalli, and M. V. Duke (1994), *ACS Symp. Ser., 559,* 191–204.

Durst, F. (1991), In *Les herbicides* (R. Scalla, Ed.), INRA, Paris, pp. 193–236.

Fayez, K. A., I. Gerken, and U. Kristen (1995), In *Structure and function of roots* (F. Baluska, Ed.), Kluwer Academic, Dordrecht, Netherlands, pp. 277–284.

Fedtke, C. (1982), In *Biochemistry and physiology of herbicide action* (C. Fedtke, Ed.), Springer Verlag, Berlin, pp. 19–85.

Foyer, C. H., M. Lelandais, and K. J. Kunert (1994), *Physiol. Plant., 92,* 696–717.

Fuerst, E. P. and M. A. Norman (1991), *Weed Sci., 39,* 458–464.

Fuerst, E. P., C. J. Arntzen, K. Pfister, and D. Penner (1986). *Weed Sci., 34,* 344–353.

Fujii, T., E. Yokoyama, K. Inoue, and H. Sakurai (1990), *Biochem. Biophys. Acta, 1015,* 41–48.

Gallandt, E. R. and N. E. Balke (1995), *Pestic. Sci., 43,* 31–40.

Garcia-Romera, I., and J. A. Ocampo (1988), *Z. Pflanz. Bodenkd., 151,* 225–228.

Green, J. M. (1989), *Weed Technol., 3,* 217–226.

Gronwald, J. W. (1994), In *Herbicide resistance in plants: Biology and biochemistry* (S. Powles and J. Holtum, Eds.), CRC Press, Boca Raton, FL, pp. 27–60.

Haddad, G., A. Ponte-Freitas, P. Ravanel, and M. Tissut (1992), *Plant Physiol. Biochem., 30,* 173–180.

Hamel, C., F. Morin, A. Fortin, R. L. Granger and D. L. Smith (1994), *J. Am. Soc. Hort. Sci., 119(6),* 1255–1260.

Harker, K. N., R. E. Blackshaw, and K. J. Kirkland (1995), *Weed Technol., 9,* 91–98.

Hatzios, K. K. (1989), In *Crop safeners for herbicides* (K. K. Hatzios and R. E. Hoagland, Eds.), Academic Press, San Diego, CA, pp. 3–45.

Hatzios, K. K. (1991), In *Environmental chemistry of herbicides* (R. Grover and A. J. Cessna, Eds.), Vol II, CRC Press, Boca Raton, FL, pp. 141–185.

Hatzios, K. K., and D. Penner (1985), *Rev. Weed Sci., 1,* 1–63.

Hume, H. (1987), *Can. J. Bot., 65,* 2530–2536.

Jacob, F. and S. T. Newmann (1987), In *General principles of the uptake and translocation of fungicides in modern selective fungicides* (H. Lyr, Ed.), Longman, Harlow, Great Britain, pp. 13–22.

Jacobs, J. M. and N. J. Jacobs (1993), *Plant Physiol., 101,* 1181–1187.

Jacobs, J. M., N. J. Jacobs, T. D. Sherman, and S. O. Duke (1991), *Plant Physiol., 97,* 197–206.

Jacobs, J. M., J. M. Wehner, and N. J. Jacobs (1994), *Pest Biochem. Physiol., 50,* 23–30.

James, E. H., and M. S. Kemp (1995), *Pestic. Sci., 43,* 273–277.

Jorrin, J., J. Menendez, E. Romera, A. Taberner, M. Tena, and R. De Prado (1990), *Med. Fac. Landbouww. Univ. Gent., 57/3b,* 1047–1052.

Kemp, M. S., L. N. Newton, and J. C. Caseley (1988), In *European Weed Res. Soc. Symp.,* pp. 121–126.

Kohno, H., C. Ogino, T. Iida, S. Takasuka, Y. Sato, B. Nicolaus, P. Boger, and K. Wakabayashi (1995), *J. Pestic. Sci., 20,* 137–143.

Koleva, A., V. Toneva, and I. Minkov (1995), *Photosynthetica, 31,* 189–196.

Kraus, T. E., B. D. McKersie, and R. Austin-Fletcher (1995), *J. Plant Physiol., 145,* 570–576.

Kunert, K. J., and A. D. Dodge (1989), In *Target sites of herbicide action* (P. Boger and G. Sandmann, Eds.), CRC Press, Boca Raton, FL, pp. 45–63.

Leach, G. E., M. D. Devine, R. C. Kirkwood, and G. Marshall (1995), *Pestic. Biochem. Physiol., 51,* 129–136.

LeBaron, H. M. (1991), In *Herbicide resistance in weeds and crops* (J. C. Caseley, G. W. Cussans, and R. K. Atkin, Eds.), Butterworth-Heinemann, Oxford, UK, pp. 27–43.

Lee, H. J., M. V. Duke, and S. O. Duke (1993), *Plant Physiol., 102,* 881–889.

Liu, S. H., A. Hsiao, W. A. Quick, T. M. Wolf, and J. A. Hume (1995), *Weed Sci., 43,* 40–46.

Lodge, G. M. and M. G. McMillan (1994), *Austral. J. Exp. Agric., 34,* 759–764.

Loos, M. A. (1975), In *Herbicides: Chemistry, degradation and mode of action* (P. C. Kearney and D. D. Kaufman, Eds.), Vol. 2, Dekker, New York, pp. 1–73.

Lyga, J. W., R. M. Patera, M. J. Plummer, P. Halling and D. A. Yuhas (1994), *Pestic. Sci., 42,* 29–36.

Marles, M. A. S., M. D. Devine, and J. C. Hall (1993), *Pestic. Biochem. Physiol., 46,* 7–14.

Marshall, E. J. P. (1988), *J. Appl. Ecol., 25,* 619–630.

Martin, H. and D. Woodcock (1983), *The scientific principles of crop protection,* 7th ed., Arnold, London, p. 285.

Matringe, M. and R. Scalla (1988), *Pestic. Biochem. Physiol., 32,* 164–172.

Matringe, M., J.-M. Camadro, M. A. Block, J. Joyard, R. Scalla, P. Labbe, and R. Douce (1992), *J. Biol. Chem., 267,* 4646–4651.

McWhorter, C. G. (1989), *Rev. Weed Sci., 4,* 85–121.

Merlin, G., F. Nurit, P. Ravanel, J. Bastide, C. Coste, and M. Tissut (1987), *Phytochemistry, 26,* 1567–1571.

Mitchell, G., E. D. Clarke, S. M. Ridley, D. T. Greenhow, K. J. Gillen, and S. K. Vohra (1995), *Pestic. Sci., 44*, 49–58.

Moreland, D. E., and W. P. Novitzky (1987), *Z. Naturforsch., 42c*, 718–726.

Moreland, D. E., and W. P. Novitzky (1988), *Pestic. Biochem. Physiol., 31*, 247–260.

Moreland, D. E., F. T. Corbin, and J. E. McFarland (1993), *Pestic. Biochem. Physiol., 45*, 43–53.

Moreland, D. E., F. T. Corbin, T. J. Fleishmann, and J. E. McFarland (1995), *Pestic. Biochem. Physiol., 52*, 98–108.

Moss, S. R. (1990), *Weed Sci., 38*, 492–496.

Mostowska, A. and M. Siedlecka (1995), *Acta Physiol. Plant., 17*, 21–30.

Nagy, I., F. Compernolle, K. Ghys, J. Vanderleyden, and R. De Mot (1995), *Appl. Environ. Microbiol., 61*, 2056–2060.

Nicolaus, B., G. Sandmann, and P. Boger (1993), *Z. Naturforsch., 48c*, 326–333.

Nicolaus, B., J. N. Johansen, and P. Boger (1995), *Pestic. Biochem. Physiol., 51*, 20–29.

Page, R. A., S. Okada, and J. L. Harwood (1994), *Biochem. Biophys. Acta, 1210*, 369–372.

Pedersen, J. O. and I. A. Rasmussen (1990), *Weeds*, pp. 73–83.

Pfleeger, T. and D. Zobel (1995), *Ecotoxicology, 4*, 15–37.

Pope, P. E. and H. A. Holt (1981), *Can. J. Bot., 59*, 518–521.

Preston, C. (1994), In *Herbicide resistance in plants: Biology and biochemistry* (S. Powles and J. Holtum, Eds.), CRC Press, Boca Raton, FL, pp. 61–82.

Preston, C., S. Balachandran, and B. Powles (1994), *Plant, Cell Environ., 17*, 1113–1123.

Ravanel, P., M. Tissut, F. Nurit, and S. Mona (1990), *Pestic. Biochem. Physiol., 38*, 101–109.

Ray, T. B. (1989), In *Target sites of herbicide action* (P. Boger and G. Sandmann, Eds.), CRC Press, Boca Raton, FL, ch. 6, pp. 152–175.

Raymond, J. A., A. Deschamps, I. Andrew, and W. A. Quick (1990), *Weed Sci., 38*, 62–66.

Rendina, A. R. and J. M. Felts (1988), *Plant Physiol., 86*, 983–993.

Rendina, A. R., A. C. Craig-Kennard, J. D. Beaudoin, and M. K. Breen (1990), *J. Agric. Food Chem., 38*, 1282–1287.

Rendina, A. R., O. Campopiano, E. Marsilii, M. Hixon, H. Chi, and W. S. Taylor (1994), in *8th Int. Congr. Pesticide Chemistry*, in press.

Saari, L. L. and C. J. Mauvais (1996), in *Herbicide resistant crops* (S. O. Duke, Ed.), Lewis Publishers, Chelsea, MI, in press.

Saari, L. L., J. C. Cotterman, and D. C. Thill (1994), In *Herbicide resistance in plants: Biology and biochemistry* (S. Powles and J. Holtum, Eds.), CRC Press, Boca Raton, FL, pp. 83–140.

Scandalios, J. G. (1993), *Plant Physiol., 101*, 7–12.

Schmitt, R. and H. Sandermann (1982), *Z. Naturforsch., 37c*, 772–777.

Schulz, A., T. Munder, H. Hollander-Czytko, and N. Amrhein (1990), *Z. Naturforsch., 45*, 529–534.

Shaner, J. S. and M. L. Reider (1986), *Pestic. Biochem. Physiol., 25*, 248–257.

Shimabukuro, R. H. (1985), in *Weed physiology* (S. O. Duke, Ed.), Vol. II, CRC Press, Boca Raton, FL, pp. 215–240.

Shimabukuro, R. H. (1990), *Plant Growth Regul. Soc. Am. Q., 18,* 37–54.

Siminskzy, B., F. T. Corbin, and Y. Sheldon (1995), *Weed Sci., 43,* 163–168.

Smeda, J. and K. C. Vaughn (1994), In *Herbicide resistance in plants: Biology and biochemistry* (S. Powles and J. Holtum, Eds.), CRC Press, Boca Raton, FL, pp. 215–228.

Stegink, S. J. and K. C. Vaughn (1988), *Pestic. Biochem. Physiol., 31,* 269–275.

Taylor, S. W., M. Hixon, H. Chi, E. Marsilii, and A. R. Rendina (1994), In *8th Int. Congr. Pesticide Chemistry,* in press.

Tietjen, K. G., J. F. Kluth, R. Andree, M. Haug, M. Lindig, K. H. Muller, H. J. Wroblowsky, and A. Trebst (1991), *Pestic. Sci., 31,* 65–72.

Tomlin, C. (1994), *The pesticide manual,* 10th ed., British Crop Protection and the Royal Society of Chemistry, BCPC Publications, Cambridge, UK, p. 1341.

Trebst, A. (1987), *Z. Naturforsch., 42c,* 742–750.

Trebst, A. (1991), *Pestic. Sci., 31,* 65–72.

USDA (1987), in *Agricultural resources: Inputs outlook and situation report,* Washington, DC, Report AR-5 (January), Economic Research Service, US Dept. of Agriculture.

Vidal, D., J. Martinez, C. Bergareche, A. M. Miranda, and E. Simon (1992), *Plant Soil, 144,* 235–245.

Vidal, D., J. Martinez-Guijarro, and E. Simon (1995), *Photosynthetica, 31,* 9–20.

Vitelli, J. S., R. J. Mayer, and P. Jeffrey (1994), *Trop. Grass., 28,* 120–126.

Wall, D. A. (1994), *Weed Technol., 8,* 673–678.

Warwick, S. I. and L. Black (1981), *Can. J. Bot., 59,* 689–693.

Wright, J. P. and R. H. Shimabukuro (1987), *Plant Physiol., 74,* 61–66.

Wurtele, E. S. and B. J. Nikolau (1990), *Arch. Biochem. Biophys., 278,* 179–186.

Yamato, S., M. Katagari, and H. Ohkawa (1994), *Pestic. Biochem. Physiol., 50,* 72–82.

Zhang, J., A. S. Hamill, and S. E. Weaver (1995), *Weed Technol., 9,* 86–90.

Zimmerlin, A. and F. Durst (1990), *Phytochemistry, 29,* 1729–1732.

11

Polyamines

Manchikatla V. Rajam

INTRODUCTION

Plants are constantly challenged by unfavorable fluctuations in their environment, and the accumulation of low molecular weight metabolites such as amino acids (e.g., proline), di- and polyamines (e.g., putrescine), quaternary ammonia compounds (e.g., glycine betaine), and a variety of polyhydroxylated sugar alcohols (polyols) is a virtually universal response to cope with various forms of environmental stresses such as extreme temperatures, salinity, drought, and so on (Galston, 1989; Vernon et al., 1993). Polyamines are among the most widespread of the low molecular weight metabolites that accumulate in stressed plant cells, and it has been suggested that the accumulation of putrescine is one of the biochemical adaptations in higher plants to cope with stress condition (Galston and Kaur-Sawhney, 1990). However, our present state of understanding the significance of polyamine accumulation (particularly the diamine putrescine) for plant stress responses is in its infancy. This article is designed to outline our current knowledge of the effects of various environmental stress factors on the polyamine concentration and metabolism of higher plants. The practical uses of exogenous application of polyamines and chemotherapeutic implications of selective inhibition of microbial polyamine biosynthesis in crop plants is also considered briefly. However, these areas have also been considered in some of the recent excellent reviews (Evans and Malmberg, 1989; Flores, 1990; Flores et al., 1989; Galston and Kaur-Sawhney, 1987; Rajam, 1993; Tiburcio et al., 1993; Walters, 1995).

The naturally occurring polyamines (PAs), spermidine (Spd) and spermine (Spm) and their diamine obligatory precursor, putrescine (Put) are small molecules present ubiquitously in plants (Fig. 11-1). They are polycationic aliphatic nitrogenous compounds. The past two decades have witnessed a growing awareness of the

H$_2$N—(CH$_2$)$_4$—NH$_2$

Putrescine
(Diamine)

H$_2$N—(CH$_2$)$_4$—NH(CH$_2$)$_3$—NH$_2$

Spermidine
(Triamine)

H$_2$N—(CH$_2$)$_3$—NH—(CH$_2$)$_4$—NH—(CH$_2$)$_3$—NH$_2$

Spermine
(Tetraamine)

Figure 11-1. *Structures of common di- and polyamines.*

importance of PAs for the normal functioning of the cells. Several investigations have emphasized the many and varied roles of PAs and associated enzymes in molecular, cellular, and physiological functions such as regulation of nucleic acid synthesis and function, protein synthesis, cell growth and differentiation, embryo development, physical and chemical properties of membranes, hormones, and modulation of enzyme activities (Balasundarum and Tyagi, 1991; Evans and Malmberg, 1989; Flores et al., 1989; Galston, 1983; Galston and Flores, 1991; Galston and Kaur-Sawhney, 1990; Galston and Tiburcio, 1991; Minocha and Minocha, 1995; Rajam 1993; Slocum et al., 1984; Walters, 1995). Many of the biological functions of PAs appear to be attributable to the cationic nature of these molecules, which are highly protonated at physiological pH, and their electrostatic interaction with polyanionic nucleic acids and negatively charged functional groups of membranes and enzymatic or structural proteins in the cell (Slocum et al., 1984; Tiburcio et al., 1993). Significant progress has been made in animal systems toward an understanding of the molecular mechanisms that relate these actions to cellular functions (Balasundarum and Tyagi, 1991). Though much less is known in plants, it has been suggested that most of the cellular functions and interaction of PAs with macromolecular and cellular structures exist as in other well-studied systems (Tiburcio et al., 1993). These compounds have been suggested to be endogenous growth regulatory compounds or intracellular second messengers, mediating the effects of plant hormones (Evans and Malmberg, 1989; Galston, 1983; Smith, 1985).

A major milestone in the development of plant PA research occurred in 1966 when the growth stimulatory action of PAs was demonstrated in dormant tuber explants of *Helianthus tuberosus* by Bagni (1966). Since then numerous investigators have shown the regulatory role of PAs in numerous plant developmental processes, including somatic embryogenesis (Minocha and Minocha, 1995; Sharma and Rajam, 1995), shoot and root formation (Galston and Flores, 1991; Sharma et al., 1995), pollen development (Rajam, 1989), flowering and fruit ripening (Kakkar and Rai, 1993), and senescence (Galston and Kaur-Sawhney, 1987; Tiburcio et al., 1994). The role of PAs in many cellular and molecular processes have also been demonstrated. For instance, PAs have been shown to stimulate phosphorylation of protein kinase (Roux, 1993) and of certain soluble and plasma membrane proteins (Ye et al., 1994), to increase rates of RNA and protein synthesis, and to initiate DNA synthesis (Bueno et al., 1993). Existence of posttranslational modification of PA-bound polypeptides has been reported (Mehta et al., 1991). PAs also polymerize rubisco subunits and bind to some light-harvesting complex proteins (Del Duca et al., 1994; Tiburcio et al., 1994). PAs, especially Put, have been known to accumulate in plants under stress conditions (Galston, 1989). The specific inhibition of fungal PA biosynthesis has also been proved as a new method of protecting higher plants from phytopathogenic fungi (Galston and Weinstein, 1988; Rajam, 1993; Rajam et al. 1985, 1986; Walters, 1995).

However, most of the earlier studies concentrated on only the free PAs (free cations). It is only now that the importance of PAs conjugated to various hydroxycinnamic acids amides (Pfosser, 1993) or bound to specific proteins (Kaur-Sawhney and Applewhite, 1993) is being realized, since they are important in flower development (Kakkar and Rai, 1993) and plant defense responses (Martin-Tanguy, 1985; Rajam, 1993). It has also been noticed that it is not only the presence of "high" but "adequate" PA levels and Put/Spd ratios that are important for somatic embryogenesis (Bajaj and Rajam, 1995; Faure et al., 1991; Sharma and Rajam, 1995) and probably for other physiological processes. The ratio between PAs and other growth regulators may also modulate morphogenetic processes (Altamura et al., 1993).

POLYAMINES

Distribution

The diamine Put, triamine Spd, and tetraamine Spm (Fig. 11-1) are the most common PAs present in all plant cells. In general, plant cells contain fairly large amounts of Spd, considerable amounts of Put, and little Spm. In addition to the above common PAs, certain other PAs also occur and are referred to as uncommon PAs. The uncommon PAs, Nor-Spd, Nor-Spm, pentamine, and hexamine, were first described in thermophilic (Oshima, 1983) and halophilic bacteria (Hamana et al., 1985). Homo-Spd has now been found in many mosses and ferns (Hamana and Matsuzaki, 1985) and in some higher plants (Birecka et al., 1984). Canavalamine and pentamines are also found in legumes (Matsuzaki et al., 1990), while Nor-Spd and Nor-Spm occur in alfalfa following growth under drought conditions (Rodriguez-Garay et al., 1989). Though less understood than the common PAs, the uncommon PAs have been postulated to serve specific protective roles under extreme environments in both bacteria (Oshima, 1983) and higher plants (Flores et al., 1989; Galston and Kaur-Sawhney, 1990).

Different Forms of Polyamines

PAs can be present in three different forms. We've already considered the first form, PAs occurring in free forms (free cations). In their second form, PAs may

TABLE 11-1   **Examples of Some Common
Polyamine Conjugates in Higher Plants**

Hydroxycinnamoylputrescine
Alkylcinnamoylputrescine
Coumaroylputrescine
Feruloylputrescine
Caffeoylputrescine
Caffeoylspermidine
Alkylcinnamoylspermine
Coumaroylagmatine

also be conjugated with a variety of low molecular weight secondary metabolites in plants by the formation of the amide linkage, utilizing esters of coenzyme A for the provision of the activated carboxyl group. PA conjugates with phenolic acids such as caffeic, ferulic, and cinnamic acids are widespread in higher plants (Martin-Tanguy, 1985; Smith et al., 1983) (Table 11-1). Agmatine (Agm) coumaroyl transferase, which catalyzes the biosynthesis of coumaroyl Agm, has been characterized in barley seedlings (Bird and Smith, 1983), while Put hydroxycinnamoyl transferase has been characterized and purified (Negrel et al., 1992) in tobacco cell suspensions.

The third form for PAs is covalently bound to cellular proteins, and this form may account for a significant portion of the metabolic pools. In *Amaranthus* seeds, 100%, 89%, and 60% of Put, Spd, and Spm may be bound, respectively (Flores et al., 1989). Their biosynthesis may be catalyzed by a transglutaminase-like enzyme found in *Helianthus tuberosus* (Serafini-Fracassini et al., 1988) and has been associated with tobacco flowering (Apelbaum et al., 1988), ovary development (Kaur-Sawhney and Applewhite, 1993), and cell division cycle in *H. tuberosus* (Dinella et al., 1992).

Biosynthesis of Di- and Polyamines

Put, which is the obligatory precursor for Spd and Spm formation, is probably formed universally directly by decarboxylation of L-ornithine in a reaction catalyzed by a rate limiting, pyridoxal phosphate-dependent enzyme ornithine decarboxylase (ODC). In plants, however, L-arginine is the major precursor of Put. Arginine is decarboxylated to agmatine by another pyridoxal phosphate-dependent enzyme arginine decarboxylase (ADC) (Fig. 11-2). Agm iminohydrolase catalyzes the conversion of Agm to N-carbomyl Put (NCP), which is converted to Put by NCP amidohydrolase. Alternatively, a single multifunctional enzyme, Put synthase, may convert arginine to Put in *Lathyrus* (Srivenugopal and Adiga, 1981) and cucumber seedlings (Prasad and Adiga, 1987), with the latter also showing ODC activity (Prasad and Adiga, 1986). Put may also be synthesized by decarboxylation of citrulline via NCP (Speranza and Bagni, 1978). An alternative mechanism for Put biosynthesis has been suggested in *L. sativus* by Srivenugopal and Adiga (1981), who reported Put carbomyltransferase activity to be one of four apparently distinct enzyme activities associated with a single, multifunctional enzyme, Put synthase. The

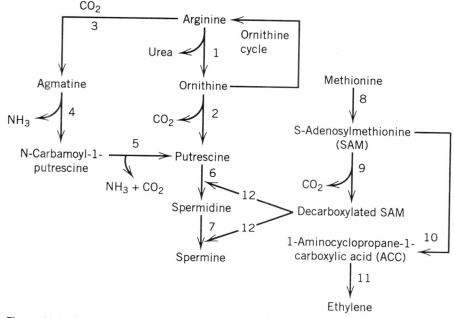

Figure 11-2. *Biosynthesis of di- and polyamines in plants. (1) Arginase. (2) Ornithine decarboxylase. (3) Arginine decarboxylase. (4) Agmatine iminohydrolase. (5) N-Carbamoylputrescine. (6) Spermidine synthase. (7) Spermine synthase. (8) S-Adenosylmethionine (SAM) synthase. (9) SAM decarboxylase. (10) 1-Aminocyclopropane-1-Carboxylic acid (ACC) synthase. (11) ACC oxidase, (12) Propylamino group [$(CH_2)_{3-}NH_2$].*

diamine cadavarine (Cad) is formed from lysine decarboxylation by lysine decarboxylase (LDC) (Schoofs et al., 1983). Arginine and ornithine may also be interconverted through the urea cycle, allowing recovery of nitrogen and carbon from the cycle. Arginine is generally preferred for Put synthesis in higher plants, due to its relative abundance in storage proteins, its involvement in long-distance transport, and its position in the plant economy with high nitrogen to carbon ratio (Altman et al., 1983).

The PAs, Spd and Spm, are synthesized by the addition of an aminopropyl group to one or both primary amine groups of Put by Spd and Spm synthases, respectively. The aminopropyl moiety is derived from decarboxylated S-adenosylmethionine (dcSAM) formed from S-adenosylmethionine (SAM) in a reaction catalyzed by SAM decarboxylase (SAMDC). SAM, in turn, is derived from L-methionine (Fig. 11-2) via SAM synthase. SAM is also involved in ethylene formation. It is converted to 1-aminocyclopropane-1-carboxylic acid (ACC) by SAM synthase, which is then involved in the liberation of ethylene (Fig. 11-2). SAM is also used as a methyl donor in many cellular metabolic reactions (Evans and Malmberg, 1989). An alternative pathway for Spd synthesis has been demonstrated in *L. sativus*, which coexists with the route involving SAMDC and PA synthases. In this case, the aminopropyl moiety is contributed by β-aspartic semialdehyde (Adiga and Prasad, 1985). The biosynthetic pathways for uncommon PAs have been described in thermophilic bacteria (De Rosa et al., 1978) and have been adapted as a working model for the biosynthesis of uncommon PAs in higher plants (Kuehn et al., 1990). ADC and ODC are distributed widely in higher plants and have been purified from various plants (Adiga and Prasad, 1985). SAMDC is also widely distributed in higher plants (Smith, 1985) and has been partially purified and characterized from few plants such as *Lathyrus* and corn (Adiga and Prasad, 1985). In chinese cabbage, SAMDC and Spd synthase have been partially purified and characterized, while Spm synthase has been detected (Yamanoha and Cohen, 1985). Interestingly, ODC-antizyme has been reported to exist in barley seedlings, which may be induced by PAs (Panagiotidis and Kyriakidis, 1985).

Catabolism of Di- and Polyamines

Catabolism of diamines and PAs is brought about by diamine oxidase (DAO) and PA oxidase (PAO), and they serve to regulate intracellular PA levels in plants (Fig. 11-3). DAO is found principally in Leguminaceae though they have also been purified from a few other plants such as *Euphorbia,* rice, barley, and maize (Smith, 1985). DAO is preferentially associated with cell walls and membranes and acts primarily on diamines (Put and Cad), resulting in the formation of pyrroline (in case of Put), NH_3, and H_2O_2; it may also catalyze the breakdown of Spd into aminopropylpyrroline, NH_3, and H_2O_2 (Smith, 1985). On the other hand, PAO has been found principally in Poaceae (Smith, 1985), though it has also been detected in *Medicago sativa* (Bagga et al., 1991). This enzyme may be associated with cell walls. It oxidizes Spd and Spm to give pyrroline and aminopropylpyrroline, respectively, together with diaminopropane (DAP) and H_2O_2 (Adiga and Prasad, 1985;

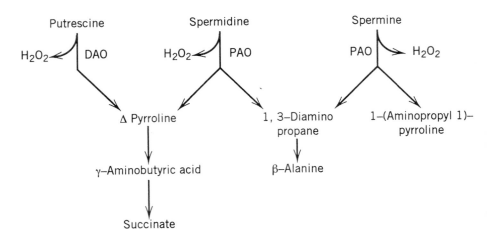

Figure 11-3. *Catabolism of di- and polyamines in plants.*

Smith, 1985). PAO is specific and is usually a monomeric glycoprotein that contains flavin adenosine dinucleotide. The action of DAO or PAO is only the first step in PA degradation and reutilization of carbon and nitrogen of the aliphatic amines (Fig. 11-3). Pyrroline (obtained by the action of DAO on Put or of PAO on Spd) is further oxidized to γ-aminobutyric acid (GABA) by a nicotinamide adenine dinucleotide-dependent dehydrogenase (Flores and Filner, 1985), which is in turn transaminated, and the resulting succinic semialdehyde is further oxidized to succinate, a Krebs cycle intermediate. H_2O_2 produced by DAO and PAO activity can be utilized for the polymerization of lignin and suberin precursors in the cross-linking of extension and polysaccharide-bound phenols. The fates of the NH_3 and DAP produced by DAO and PAO activities, respectively, are, however, poorly understood. In many plant species, conventional PA oxidizing systems are not detectable, and their catabolism may proceed through conjugated intermediates (Santanen and Simola, 1994).

Put also forms the source for the synthesis of nicotine and tropane alkaloids (Slocum et al., 1984). The pyrroline ring of nicotine is produced from Put by oxidation to N-methyl-D'-pyrrolinium salt, catalyzed by N-methyl Put oxidase (Haslam and Young, 1992). ADC may preferentially act as the source of Put for tobacco alkaloids (Tiburcio and Galston, 1986). Put is methylated by Put methyltransferase (PMT), which is then involved in the biosynthesis of tropane alkaloids in several Solanaceae members and members of some other families like Convolvulaceae (Slocum et al., 1984). A new Spm alkaloid, kukoamine B, has been isolated from the root bark of *Lycium chinense,* and its structure has been characterized by Funayama et al. (1995).

Subcellular Compartmentation of Polyamines and Their Metabolic Enzymes

The knowledge of subcellular compartmentation of PAs and their biosynthetic enzymes is scanty. PAs can bind (but not covalently bind) to the cell wall constituents (Goldberg and Pedrizet, 1984), and Spm is generally very scarce (Mariani et al., 1989). Put was found mainly in the cytoplasm, while Spd and Spm were found mainly in the cell wall fraction of carrot cells (Pistocchi and Bagni, 1990). Although the plasma membrane may also represent a binding site, PAs are most likely to accumulate in the cytoplasm (Bagni and Pistocchi, 1990). Early studies suggested that the vacuole, though temporarily, could be a site of Put accumulation as Put can be readily transported across the tonoplast and plasma membrane (Pistocchi et al., 1988). Stress-induced Put accumulation may also occur in the apoplastic regions (DiTomaso et al., 1992). PAs have also been detected in the subcellular organelles like mitochondria (Pistocchi et al., 1990; Torrigiani et al., 1986) and chloroplasts (Torrigiani et al., 1986). In chloroplasts, PAs have been found associated with the thylakoid membranes (Kotzabasis et al., 1993). Spm has also been found in the nucleus (Smith, 1985).

The compartmentation of the metabolic enzymes involved in PA biosynthesis is also known. ADC has been found mainly in the cytoplasm, while ODC may occur in the nucleus (Smith, 1985) in close association with chromatin (Kyriakidis et al., 1983). Both of these enzymes are most probably particulate enzymes and distributed in nuclei, chloroplasts, and mitochondria. SAMDC is found mainly in the soluble cytoplasmic fractions (Torrigiani et al., 1986), though it has also been detected in mitochondria (Walters, 1989). A chloroplastic location has been suggested for LDC (Schoofs et al., 1983) and Spd synthase (Sindhu and Cohen, 1984). DAO and PAO activities have been found to occur mainly in the cell walls (Flores and Filner, 1985). In pea epicotyls, DAO was found to be located in the apoplast (Federico and Angelini, 1986); PAO of oat leaves is apoplastic, while that of barley is symplastic in location (Li and McClure, 1989).

INHIBITORS OF POLYAMINE METABOLISM

The design and synthesis of specific inhibitors of PA metabolism is a logical approach to unravel the physiological significance of PAs. The inhibition of ODC was first reported by Skinner and Johannsen (1972). Since then, a large number of compounds have been described that are effective inhibitors of PA biosynthesis in both prokaryotic and eukaryotic cells (Bey et al., 1987) (Fig. 11-4), and their availability has led to a rapid increase in knowledge of the participation of PAs in the wide variety of living organisms.

The PA inhibitors may be categorized as follows: substrate and product based analogues, and antagonists of cofactors as competitive inhibitors (Bitonti et al., 1987). α-Difluoromethylornithine (DFMO) (Fig. 11-4), one of the first enzyme-activated substrate-based inhibitors to be synthesized (Metcalf et al., 1978), has been extensively used to inhibit ODC enzyme along with monofluoromethylornithine (MFMO) (Kallio et al., 1982). Similarly, the α-fluoromethyl analogues of arginine, difluoromethylarginine (DFMA) (Fig. 11-4) (Kallio et al., 1981), monofluoromethylarginine (MFMA), and agmatine (Bitonti et al., 1987) are potent inhibitors of ADC, while the corresponding analogues of lysine (Poso et al., 1984) and cadeverine are effective inhibitors of LDC (Bitonti et al., 1987). Both MFMO and MFMA were found to be much more effective than DFMO and DFMA (Bey et al., 1987). Methylgloxylbis(guanylhydrazone) (MGBG) (Fig. 11-4) has been found to be an effective inhibitor of SAMDC (Williams-Ashman and Schenone, 1972), and a large number of its analogues have been tried but only with limited success (Pegg and Williams-Ashman, 1987). The synthesis of the α-fluoromethyl derivative of SAMDC has been described (Bey et al., 1977), but unfortunately it is inactive as an inhibitor (Kolb et al., 1982). Bis(cyclohexylammonium) sulfate (CHA), previously known erroneously as dicyclohexylammonium sulfate, has been shown to inhibit the activity of Spd synthase in both eukaryotic (Hibasami et al., 1980) and prokaryotic cells (Bitonti et al., 1982). In addition to inhibitors of PA biosynthesis, inhibitors of PA catabolism are also known and have provided insights into the quantitative significance of interconversion and terminal oxidation of PAs, and into the dynamic aspects of PA metabolism (Seiler, 1987).

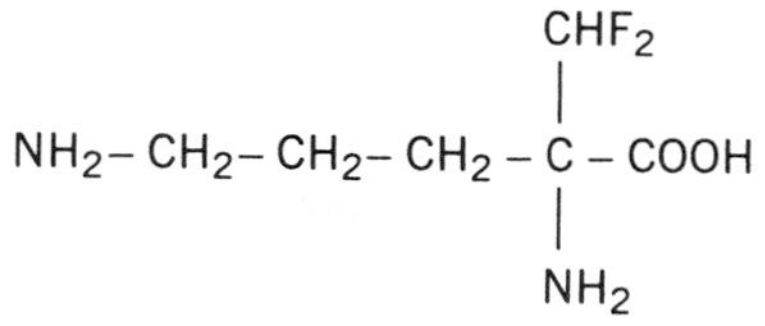

$$CHF_2$$
$$|$$
$$NH_2 - CH_2 - CH_2 - CH_2 - C - COOH$$
$$|$$
$$NH_2$$

Difluoromethylornithine (an inhibitor of ODC)

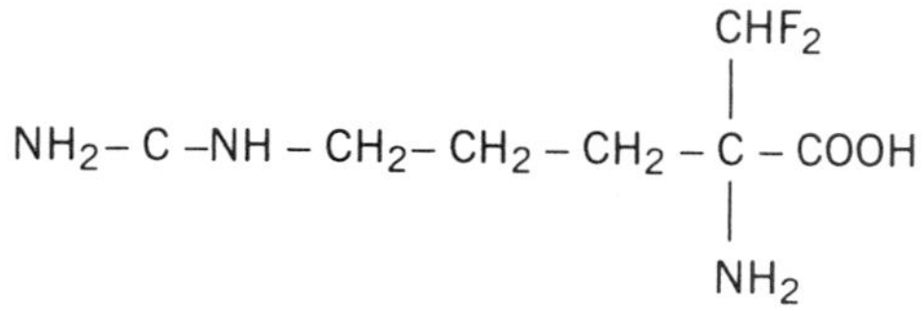

$$CHF_2$$
$$|$$
$$NH_2 - C - NH - CH_2 - CH_2 - CH_2 - C - COOH$$
$$|$$
$$NH_2$$

Difluoromethylarginine (an inhibitor of ADC)

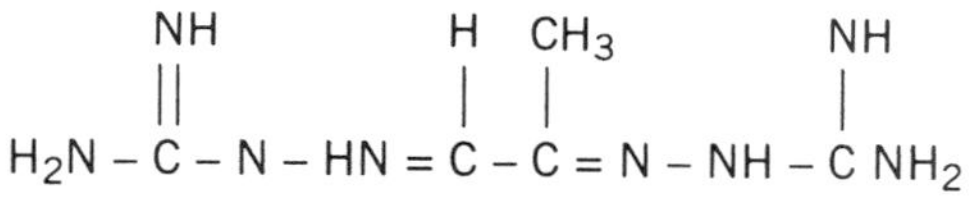

$$NH \quad\quad H \quad CH_3 \quad\quad NH$$
$$|| \quad\quad\quad | \quad | \quad\quad\quad |$$
$$H_2N - C - N - HN = C - C = N - NH - C \, NH_2$$

Methylglyoxal bis (guanylhydrazone)(an inhibitor of SAMDC)

Figure 11-4. *Structures of common inhibitors of PA biosynthesis.*

POLYAMINES AND PLANT STRESS

A variety of reports indicate that PA levels (particularly Put) and the activities of PA biosynthetic enzymes (particularly ADC) change in close correlation with various plant abiotic stresses (Flores, 1990; Flores et al., 1985). In biotic stresses (viral and fungal infections), plant cells accumulate free PAs as well as bound forms conjugated with hydroxycinnamic acids (HCAs) (Martin-Tanguy, 1985). This indicates that PAs are intricately involved in the plant's responses to environmental challenges by changing their growth and developmental patterns (Altman et al., 1982; Evans and Malmberg, 1989; Flores, 1990). However, in most cases, Put accumulation occurred without a corresponding increase in Spd and Spm levels (Young and Galston, 1984). It has also been shown that Put content decreased in osmotically stressed dicot leaves while Spd and Spm contents increased (Tiburcio et al., 1986a). Unfortunately, only free PA levels were measured in most of the studies involving stress responses on PA metabolism. Nevertheless, a rapid increase in bound Put titer has been recorded in tobacco following ozone exposure (Langebartels et al., 1991). It has been suggested that stress-induced changes in PA pattern may be mediated by abscisic acid (ABA) (Aurisano et al., 1993).

Abiotic Stresses

Mineral Deficiency   The first report implicating an important role for PAs in higher plants was a report of an accumulation of Put in barley plants grown under K^+ deficiency (Richards and Coleman, 1952). Since then Put accumulation, particularly in cereals, has been shown to occur in response to varied stresses (Smith, 1973; Young and Galston, 1984). Put was found to accumulate under high salt and deficiencies of K^+, Mg^{2+}, and Ca^{2+} in barley leaves; it declined, however, in phosphorus, sulfur, and nitrogen deficiencies, and on substituting NO_3^- for NH_4^+. K^+ and Mg^{2+} deficiencies were also found to stimulate Put synthesis and accumulation in a number of plant species; no consistent trend could, however, be found between PA levels and mineral deficiency (Basso and Smith, 1974). In wheat seedlings, a stresslike PA pattern could be induced by ABA, which could be reversed by K^+ (Aurisano et al., 1993).

Put accumulation and biosynthesis also increased due to excess NH_4^+ nutrition, and it was suggested that the differential effect of NH_4^+ and NO_3^- on PA content may reflect the role of PAs in buffering cellular pH (Smith, 1983). K^+-deficient peas utilizing NH_4^+ as a nitrogen source accumulated Put to a greater extent than those maintained on NO_3^- (Klein et al., 1979). PA accumulation by tomato plants exposed to interaction stress of ammonium toxicity and potassium deficiency has been reported (Corey and Barker, 1989). A relationship among Put accumulation, ammonium accumulation, ethylene formation, and stress-induced (nutritional stress) symptoms was also reported (Feng and Barker, 1993). In tobacco cell suspension also, greater Put accumulation was found with NH_4^+ than with NO_3^- as the nitrogen source. It was further observed that increase of Put upon depletion of KNO_3 was not due to K^+ deficiency but due to excess NH_4^+ relative to NO_3^-, thus demonstrating that nitrogen source (and level) directly affected PA biosynthesis (Altman and Levin, 1993).

Osmotic Shock   Put has also been found to accumulate in response to osmotic stress, especially in cereals. More massive accumulations of Put were observed in stressed cereal leaf sections (Flores and Galston, 1982, 1984a,b) and protoplasts (Tiburcio et al., 1986a,b) than in intact drought-stressed plants (Turner and Stewart, 1986). In osmotically stressed barley plants, greater Put accumulation occurred in the presence of PO_4^{3-} (Turner and Stewart, 1988). Put accumulation was found to be due not only to activation of the ADC pathway, but also to the inhibition of Spd synthase activity in stressed oat leaves (Tiburcio et al., 1993). In another study, it was shown that the protective effect of PAs on osmotically stressed oat leaves may be due to the PAs' ability to stabilize molecular complexes in thylakoid membranes (Besford et al., 1993). Plants exposed to water stress also showed Put accumulation (Kandpal and Rao, 1985). Osmotic stress induced an increase in Put, Cad, and Spm, while water stress caused a decrease in the levels of these amines, except Cad in intact wheat plants (Foster and Walters, 1991). In excised rice leaves, methyl ester of jasmonic acid mimicked osmotic stress and ABA had effects similar to water stress (Chen and Kao, 1993a). Further, it was shown that arginine and ornithine were preferentially utilized as precursors for Put accumulation during osmotic stress, and for proline accumulation under conditions of water stress (Chen and Kao, 1993a).

Low pH   Put accumulation can also be induced by acid stress (Smith and Sinclair, 1967; Young and Galston, 1983). Mutant tobacco lines with higher endogenous Put levels and ADC activity showed increased tolerance to acid stress (Hiatt and Malmberg, 1988). PA determinations in spruce trees have been suggested to be biochemical indicators of acid rain damage (Konigshofer, 1989). In detached rice leaves, Put levels were found to be regulated by cytosolic pH; the levels were increased by weak acids while weak bases reduced endogenous Put levels (Chen and Kao, 1993b).

Low Oxygen Concentration   A number of Poaceae species were used to study the effect of reduced oxygen availability on PA metabolism and provided evidence of an association between the capacity to accumulate Put and the tolerance to anoxia (Reggiani and Bertani, 1990). Under oxygen deficit, rice coleoptiles elongated and accumulated free, conjugated, and bound Put and, to a lesser extent, conjugated Spd and Spm. In contrast, accumulation of PA conjugates was severely inhibited in seedling roots, which require oxygen to grow. Interestingly, anoxic conditions stimulated increased ADC activity in both coleoptiles and roots (Reggiani et al., 1989). Put accumulation in the coleoptiles was found to be induced by ethylene and mediated by ADC (Lee and Chu, 1992). Recently, purification and synthesis under anaerobic conditions of rice ADC, and synthesis of polyclonal antibodies against ADC has been reported (Reggiani, 1994).

Salinity   After short-term exposure to salt stress, slightly adaptive and nonadaptive tobacco cell lines showed a considerable increase in Spd and Spm and a decrease in Put. After prolonged exposure, however, Put content increased in the adaptive and resistant cell lines at the expense of Spd and Spm (Shevyakova et al., 1985). Put accumulation was noticed in rice in response to salt stress (Prakash et al., 1988). Later, it was shown that the addition of Put can alleviate salt-induced inhibition and early seedling growth (Prakash and Prathapasenan, 1988). Salt stress in germinated rice seedlings also caused an accumulation of Put and concomitant increase in ADC activity, and DFMA has reduced such responses (Basu et al., 1988). Salt-sensitive cultivars accumulated high levels of Put and low levels of Spd and Spm under saline conditions; the reverse was found to be true for salt-tolerant cultivars. These findings led to the suggestion that higher increase in PAs (Spd and Spm) against the low increase in diamine, that is, Put, may be one of the mechanisms of salt tolerance (Krishnamurthy and Bhagwat, 1989). Friedman et al. (1989) have reported that the exposure of mung bean plants to salt stress caused changes in PAs that are coordinated at the organ level; induction of PA biosynthesis enzymes, accompanied by high levels of Put synthesis, was seen in the root system, and Put accumulation was seen in leaves. Friedman et al. (1989) have demonstrated that the glycophytes and halophytes under hypertonic and hypotonic stress, respectively, resulted in an increase in the activity of PA biosynthesis enzymes in roots. In beetroot, PA levels were found to be higher in habituated salt-sensitive callus when compared to normal, salt-tolerant callus (Le Dily et al., 1991). PA levels have been shown to be different in various rice genotypes with respect to NaCl salinity (Basu and Ghosh, 1991). Wheat cultivars differing in salt sensitivity also showed changes in PA metabolism; salt stress decreased Put content, especially in roots, and Spd and Spm titers were increased with a great effect in shoots (Reggiani et al., 1994). These authors have suggested that PA/Put ratio may be used as a good marker in wheat to evaluate different cultivars for salt sensitivity using shoots. The foliar application of Put can ameliorate the effects caused by salt stress in salt-tolerant rice (Krishnamurthy, 1991). In a recent study, it has been shown that NaCl has increased the levels of Spd, but not Spm; in fact it drastically lowered the Put concentration (Lin and Kao, 1995). It has been further shown that the application of precursors of Put synthesis (L-arginine and L-ornithine) as well as Put caused an increase in cellular Put levels, but could not prevent the inhibitory effects of NaCl on rice seedling growth.

Extreme Temperatures Temperature stress can also induce similar stresslike PA patterns. Heat stress induced Put accumulation in rice and wheat along with increased ADC activity (Das et al., 1987). Chilling stress also induced Put accumulation in apple fruits (McDonald and Kushad, 1986), zucchini squash (Kramer and Wang, 1989), and in cereal and forage plants under controlled (Nadeau et al., 1987) and field (Nadeau and Paquin, 1988) conditions, while Spd was the predominant PA that accumulated in cucumber seedlings exposed to chilling stress (Wang, 1987). Put accumulation in zucchini squash in response to chilling appears to be produced from the ODC pathway, rather than from the ADC pathway as in most other stresses (Kramer and Wang, 1990). However, comparison of endogenous PA levels in bean cultivars differing in chilling tolerance revealed that relative changes in Put content rather than absolute levels correlate well with chilling tolerance (Guye et al., 1986). Preconditioning of zucchini squash led to significant increases in Spd and Spm, and reduced chilling injury (Kramer and Wang, 1989). Cold-hardening of wheat also led to increases in the levels of Put and, to a lesser extent, of Spd. It was also found to greatly reduce the plant response following intensive additional stresses of a different nature (Nadeau, 1990). Songstad et al. (1990) reported that PAs can induce chilling tolerance in maize suspension cultures.

Drought It is interesting to note that a new series of uncommon and novel PAs like caldopentamine have been reported in shoot meristems of alfalfa subject to drought stress (Rodriquez-Garay et al., 1989). Such compounds have previously been reported in thermophilic bacteria grown at high temperatures (over 50°C), and they have been implicated in adaptation of thermophilic organisms to extreme temperatures (Oshima, 1983). These observations suggest that plants may be adapted to drought and extreme temperature stress conditions by producing such unusual PAs (Flores et al., 1989).

Atmospheric Pollutants   Plants are exposed to a variety of environmental pollutants, especially heavy metals (e.g., cadmium, zinc, silver, chromium, cobalt, nickel, copper, lead, and mercury), ozone, and radiations. Put concentration was increased significantly in pea root meristem exposed to cadmium (Melnichuk et al., 1983, 1984), in oat seedlings exposed to cadmium (Weinstein et al., 1986), and in collard plants grown on sewage sludge containing a high quantity of cadmium (Hughes et al., 1987). Put concentrations also increased in tomato leaves exposed to high levels of silver (Bertrand et al., 1990), and in leaves from seedlings of pea, tomato, and barley stressed with chromium (Hauschild and Jacobsen, 1990). The ADC activity also followed the pattern of Put accumulation, both of which can be inhibited by DFMA (Weinstein et al., 1986). However, PA biosynthesis in the oat seedling system might not be an effective biomarker for heavy metal stress. Exogenous application of PAs to an ozone-sensitive tobacco cultivar provided protection to the leaves against ozone damage, along with increasing free and conjugated Put and Spd, though the scavenging of oxyradicals could be efficiently done only by PA conjugates (Bors et al., 1989). The increase in free and conjugated Put and a concomitant increase in ADC activity were higher in leaves of an ozone-tolerant line than in those of an ozone-sensitive line (Schraudner et al., 1990). Later, increased Put synthesis was also reported in tobacco (Langebartels et al., 1991) and potato (Reddy et al., 1993) when exposed to ozone.

Put has also been found to accumulate on fumigation with sulfur dioxide (Priebe et al., 1978), herbicide treatment (DiTomaso et al.; 1988, Zheleva et al., 1993), air pollution (Rowland-Bamford et al., 1989), gamma irradiation (Triantaphylides et al., 1990), and UV-B exposure (Predieri et al., 1993). The rapid increase in both free and bound (with trichloroacetic acid-insoluble substances) PAs (Put, Spd, and Spm) was observed in pea leaves subjected to atrazine herbicide, and the exogenously supplied PAs can afford partial protection of herbicide treated plants as they can reduce the effects of herbicide on growth, chlorophyll content, and photosynthetic parameters (Zheleva et al., 1993).

Other Stresses   PAs have also been shown to be involved in some other abiotic stresses. For instance, increased levels of Put have been observed in tobacco under flooding condition, which (flooding) also promoted senescence; however flooding did not affect Spd and Spm titers (Hurng and Kao, 1993). In later studies, Hurng et al. (1994) have shown that Spd content also increased along with Put in flooded leaves of tobacco, and the authors have suggested that PAs are possibly not involved in the regulation of flooding-promoted senescence of tobacco leaves. Interestingly, seasonal fluctuations (variations) can also lead to changes in PA metabolism in different parts of juvenile spruce trees (Konigshofer, 1989) and in needles of Scotch pine *(Pinus sylvestris)* (Sarjala and Savonen, 1994). High levels of Put content were observed in Scotch pine in winter, and the lowest levels in summer, with some variations among trees; Spd content was highest in spring, but fluctuation in Put concentration was more apparent than Spd (Sarjala and Savonen, 1994). Put accumulation can even be seen during injury of fruits (McDonald and Kushad, 1986).

Biotic Stresses

Viruses   Relatively little work has been done on the changes in PA metabolism in plants in response to viral infections. In tobacco (*Nicotiana tabacum* cv. Xanthi n.c.), infection with tobacco mosaic virus (TMV) resulted in increased amounts of feruloyl-Put (FP) and diFP (Martin-Tanguy et al., 1973), and it was confirmed in *N. sylvestris* where highest amounts of diFP were observed at the time of completion of viral particle multiplication. The number of local lesions could be reduced by as much as 90% by exogenous application of coumaroyl-, dicoumaroyl-, or caffeoyl-Put (Martin-Tanguy et al., 1976). Interestingly, a tobacco mutant that does not synthesize HCAs showed hypersensitive reaction to TMV infection, which was lethal (Martin and Martin-Tanguy, 1981).

There was a rapid increase in ODC activity on virus infection of *N. tabacum* (Negrel et al., 1984). The HCAs synthesized during the hypersensitive reaction bound to the cell wall and were found to be good substrates of peroxidases in vitro and to form insoluble polymers in vivo. It has been suggested that they play a role in host-pathogen interaction (Negrel and Smith, 1984). Since their formation is associated with virus multiplication and their presence has been shown to retard or inhibit viral multiplication, HCAs have been implicated in playing a role in virus resistance in plants (Martin-Tanguy, 1985). A decrease in free Put was suggested to be a step between ethylene and morphological alterations in tomato plants infected with citrus exocortis viroid (Belles et al., 1991), and it was suggested that this decrease correlated with an ethylene-mediated decrease in ODC activity, and that it was not due to a decrease in ADC activity or to a breakdown of PA conjugates (Belles et al., 1993).

Fungi   An antifungal factor, *p*-coumaroyl agmatine, was isolated (Stoessl, 1965); its dimers, called "hordatines," were found to be similar in action to streptomycin (Venis, 1969), and their occurrence was suggested to be responsible for the resistance in young barley seedlings to *Helminthosporium sativum* (Stoessl and Unwin, 1978). In K^+-deficient barley leaves infected with powdery mildew, there was an accumulation of Put (Sinclair, 1969), while infection of wheat by stem rust fungus led to the formation of HCA amides (Samborski and Rohringer, 1970). HCA amides also appeared when the carnation cuttings were induced to be resistant using a water-soluble elicitor of *Phytophthora infestans* (Ponchet et al., 1982).

In barley, infection by brown rust fungus led to increases in the levels of Spd, which was highest at the time of fungal sporulation, and of three unknown PAs. Additionally, three more PAs (unidentified) appeared on infection (Greenland and Lewis, 1984). Localization studies later revealed a high activity of ODC within the fungal pustule (Bailey et al., 1987). PA levels were also found to increase in turnip roots on infection with *Plasmodiophora brassicae,* but only in the "clubbed" regions (Walters and Shuttleton, 1985). Infection of barley leaves with the powdery mildew fungus *(Erysiphe graminis* f.sp. *hordei)* also induced increase in the levels of PAs and their biosynthetic enzymes in the first leaves, but not in roots (Walters et al., 1985). Their levels were found to be highest in the regions around the mildew pustules, which remain green and physiologically active while the rest of the leaf senesces, and were suggested to be responsible for the formation of "green islands" (Walters and Wylie, 1986). These elevated free PA levels in the "green islands" were found to result from reductions in conjugated pools, rather than from de novo synthesis (Coghlan and Walters, 1990). Put, Spd, and Spm levels were increased in rusted leaves (inoculated with *Puccinia graminis* f. sp. *tritici)* and these changes were accompanied with small increase in cytosolic and bound forms of ODC (Foster and Walters, 1992).

In three near-isogenic lines of wheat differing in resistance to stem rust fungus *(P. graminis* f. sp. *tritici),* Spm levels remained unchanged in all the lines. Put/Agm contents were unaffected in the resistant line but increased in the moderately resistant and susceptible lines; Spd levels remained unaffected in the resistant line, increased in the moderately resistant line, while decreasing in the susceptible line (Machatschke et al., 1990). It was suggested that while increase in Put may be a general response to stress, increase in Spd levels may be responsible for resistance to the fungus (Machatschke et al., 1990).

Similar results were obtained in another set of near-isogenic lines of wheat differing in their resistance to *P. recondita* (Bharti and Rajam, unpublished results). Cad levels and DAO activity were constitutively higher in a chickpea cultivar (Sultano) resistant to the necrotrophic fungus *Ascochyta rabiei* as compared with the susceptible cultivar (Calia), and this suggested that PA metabolism may have a role in the disease resistance in the plant (Angelini et al., 1993)

Other Pathogens   It has also been shown that PAs and activity of enzymes involved in PA biosynthesis increase in tissues of crown galls and hairy roots of dicotyledonous plants induced by the soil pathogens *Agrobacterium tumefaciens* and *A. rhizogenes,* respectively (Kulpa et al., 1985). Similar observations were reported in club root galls (Walters, 1989). However, the increased PA biosynthesis in these tissues may reflect its importance in cell division of both plant and pathogen rather than any specific role in pathogenesis (Walters, 1989).

SENESCENCE

PAs have been extensively demonstrated to be involved in various senescence processes occurring in plants, and have been referred to as "antisenescence agents" (reviewed by Galston and Kaur-Sawhney, 1987; Kaur-Sawhney and Galston, 1991; Tiburcio et al., 1994). They have been shown to increase protein, RNA, and DNA synthesis in aging oat protoplasts (Altman et al., 1977; Kaur-Sawhney et al., 1980), to reduce RNAase activity and chlorophyll loss (Kaur-Sawhney and Galston, 1979), and to inhibit specific protease activity of senescing oat leaves (Kaur-Sawhney et al., 1982a). However, transfer of leaves to light retards senescence (Kaur-Sawhney and Galston, 1979) along with increased ADC activity and PA titers (Kaur-Sawhney et al., 1982b). DAP, a catabolic product of Spd and Spm, has also been shown to inhibit protease activity, chlorophyll breakdown, and ethylene production in detached oat (Shih et al., 1982) and rice (Cheng et al., 1984) leaves. PAs have also been shown to retard senescence in dormant tubers of *Helianthus tuberosus* (Serafini-Fracassini et al., 1980), apple fruit tissue (Apelbaum et al., 1982), suspension cultures of several plants (Muhitch et al., 1983), aging rice panicles (Sen and Ghosh, 1984), barley leaves (Rodriguez et al., 1987) and embryo (Nielsen, 1990), and detached wheat leaves (Huang et al., 1990).

The reported antisenescence activity of PAs is manifested over the same range of concentrations in which they are known to exist in plant tissues (Slocum et al., 1984), and it increases with the increment of amino groups, that is, Spm > Spd > Put = Cad (Galston and Kaur-Sawhney, 1987).

The process of senescence has been correlated with decreasing PA levels in intact bean and detached raddish leaves (Altman and Bachrach, 1981), oat protoplasts (Kaur-Sawhney et al., 1985), pea ovaries (Carbonell and Navarro, 1989), nodules of *Vigna mungo* (Lahiri et al., 1992), petunia flowers (Botha and Whitebread, 1992), and detached leaves of barley (Coghlan and Walters, 1990) and rice (Chen and Kao, 1993a). In a slow-ripening land race of tomato in which senescence was delayed, Put levels were higher than in normal tomato or in its revertant (Dibble et al., 1988). Both ADC and ODC biosynthetic pathways of PAs seem to be active during senescence of *V. mungo* (Lahiri et al., 1992).

Aging and senescence are, however, not correlated with reduced PA levels as has been noticed in tobacco leaves and potato tubers (Altman and Bachrach, 1981), *Heliotropium* leaves (Birecka et al., 1984), and intact wheat tissues (Peeters et al., 1993). In light- and dark-grown barley seedlings and senescing leaves, PAs were comparable under almost all conditions. Furthermore, exogenous Spd did not retard senescence in dark, and Spd and Spm increased it in light (Srivastava et al., 1983). DFMA alone or in combination with DFMO had no effect during dark-induced senescence in intact leaves of oat, barley, and *Heliotropium* (Birecka et al., 1991). In wheat, no correlation could be found between PA content and onset of flooding-induced senescence of tobacco leaves, though ethylene was shown to be involved (Hurng et al., 1994).

However, in cut carnation flowers, PA levels were increased during senescence (Roberts et al., 1984). Inhibition of PA biosynthesis using DFMA, DFMO, and MGBG promoted ethylene production and senescence, while inhibition of ethylene increased Spm levels and delayed senescence, suggesting that ethylene and PAs may compete for SAM during senescence (Roberts et al., 1984), as had been noticed earlier in aged orange peel discs (Even-Chen et al., 1982). Exogenous Put and Spd resulted in greater ethylene production and reduced bloom longevity (Downs and Lovell, 1986). A later study, however, revealed exogenous PAs were effective in delaying the senescence of carnation buds but were ineffective when applied to open flowers, probably due to higher levels of Put in the latter (Upfold and van Staden, 1991).

The mechanism of inhibition of senescence by exogenous PAs may be related to their possible ability to inhibit ethylene biosynthesis or to stabilize membranes (Evans and Malmberg, 1989; Tiburcio et. al., 1994). PAs have been shown to inhibit ethylene production in mung bean hypocotyls (Suttle, 1981) and peas (Apelbaum et al., 1985), and have been linked with retardation of senescence of oat leaves (Fuhrer et al., 1982). However, Ke and Romani (1988) noted very little interdependence of ethylene biosynthesis and senescence in pear fruit cells. In detached tomato fruit, ACC (a precursor of ethylene) synthesis and accumulation was not due to decreased PA levels (Casas et al., 1990).

There is increasing evidence to suggest that exogenous PAs may inhibit senescence by stabilization of membranes (Tiburcio et al., 1994). PAs have been demonstrated to stabilize plasmalemma (Naik and Srivastava, 1978) and thylakoid membranes (Popovic et al., 1979). This is supported by the fact that addition of Ca^{2+} counteracts the antisenescence properties of PAs (Cheng et al., 1984; Kaur-Sawhney and Galston, 1979; Srivastava et al., 1983). It has been shown, however, that in the absence of Ca^{2+}, PAs destroy rather than maintain membrane stability (DiTomaso et al., 1989). However, Roberts et al. (1986) have cautioned that PAs may reduce membrane fluidity and reflect membrane rigidification rather than show true physiological processes. Another distinctly mechanistic hypothesis is that the senescence signal itself acts by decreasing PAs. There was no decrease of PAs prior to the appearance of early symptoms of senescence in apical pea buds; during senescence, there was a marked decline in PA levels (Smith and Davies, 1985).

Recently, the antisenescence property of PAs has been explained at the level of ADC gene expression (Tiburcio et al., 1994). In the presence of Spm, there is activation of transcription of the ADC gene, and the translation product of ADC mRNA, the inactive 60 kDa precursor protein, accumulates, suggesting that the posttranslational processing (clipping) of this molecule is inhibited with a consequent decrease in the active ADC form. Under these conditions, in spite of the increased levels of ADC mRNA, the inhibition by Spm of ADC clipping leads to decreased levels of ADC activity and thus avoids the accumulation of Put (Tiburcio et al., 1994).

POLYAMINES AND CROP IMPROVEMENT

It is interesting to note that PAs have been put to practical use in agriculture and tissue culture. For instance, Okii et al. (1980) have taken a patent for protection of crops against frost damage, loss of chlorophyll, photochemical oxidation, and wilting by using long-chain alkylenediamines such as octa- and decamethylenediamine. It has been shown that the unique and novel PAs, such as thermo-Spm and caldopentamine, confer thermoprotection during in vitro protein synthesis (Oshima,

1983). This suggests that crop plants could be protected against high temperature stress by using such compounds (Flores et al., 1989). Since PAs are antisenescence compounds, they may also be useful in retarding senescence and aging in plants. PAs and ethylene are antagonistic, and hence PAs may be used for delaying fruit ripening during storage, as exogenous PAs can lower ethylene formation. Wada et al. (1994) have demonstrated that the induction of flowering in seedlings of morning glory *(Pharbitis nil)* can be achieved by using PAs (particularly diamines—Put and Cad) and related compounds (e.g., unnatural 1,6-diaminohexane). This, too, may find practical application in agriculture, such as inducing flowering in nonflowering genotypes and in genotypes that do not flower because of new geographical locations. The exogenous application of Put has resulted in ovule longevity, effective pollination, and increased fruit set in Comice pear (Crisosto et al., 1988). PAs may be useful for the improvement of fruit quality during storage, probably through strengthening of the cell wall (Wang et al., 1993). Since PAs are involved in plant stress responses, they may be useful for protecting crops against abiotic stresses, especially salinity, low pH, low oxygen concentration, extreme temperatures, and drought (Flores, 1990; Flores et al., 1989; Rodriguez-Garay et al., 1989). Further, they may be useful as radical scavengers and protectants against ozone damage in plants (Bors et al., 1989). PAs can also protect plants against atrazine, a potent herbicide (Zheleva et al., 1994). PA conjugates are antimicrobial in nature, and hence they could be used as plant protectants against fungal and viral pathogens (Martin-Tanguy, 1985). The selective inhibition of fungal PA biosynthesis would be a novel and promising method of crop protection (Rajam, 1993; Rajam et al., 1985; Walters, 1995). It is noteworthy that paraquat resistance in *Hordeum glaucum* has been conferred by PAs (Preston et al., 1992).

PAs also have potential applications in plant tissue culture. The inclusion of PAs (Put and/or Spd) in the culture medium along with phytohormones results in the promotion of various developmental events such as somatic embryogenesis, floral buds, and shoot and root formation (Galston and Flores, 1991; Minocha and Minocha, 1995; Rajam and Subhash, 1987; Sharma and Rajam, 1995; Sharma et al., 1995). Recently, we have found it possible to increase the frequency of somatic embryogenesis five- to sixfold in eggplant tissue cultures amended with Put (Yadav and Rajam, unpublished results). This would be quite useful in micropropagation of plants and in the preparation of artificial seeds. The improvement of the viability of oat protoplasts (which show rapid accumulation of Put as a result of osmotic shock and do not divide in culture) by blocking Put accumulation using DFMA has been demonstrated (Tiburcio et al., 1986b). This may be useful in protoplast regeneration of certain cereal crops. We were able to improve and/or restore plant regeneration in long-term callus cultures of rice (which show massive accumulation of PAs, particularly Put, altered Put/Spd ratios, and near loss of plant regeneration) by manipulating endogenous PA levels either by using exogenous Spd or DFMA (Bajaj and Rajam, 1995, and unpublished results). These results would be helpful for genetic transformation (as efficient plant regeneration is a prerequisite for plant transformation) of rice, as it shows poor regeneration and a gradual loss of morphogenetic capacity with aging of the cultures (Bajaj and Rajam, 1995).

POLYAMINE BIOSYNTHESIS INHIBITION AND PLANT CHEMOTHERAPY

Plants are attacked by various pathogens and insect pests which upset the crop yields and quality. Therefore, crop protection has received major attention, and it has been achieved mainly through adjusting cultural practices, breeding resistant varieties, applying chemicals, or introducing genes. The chemical protectants against plant infections and pests have been most commonly used for crop protection. Unfortunately most of these agrochemicals are hazardous, cause environmental pollution, and damage wildlife (Rajam, 1993). These limitations have prompted researchers to look for novel alternative strategies for the protection of crop plants. One such strategies is the targeting of certain metabolic processes that occur in microbes, but not in higher plants or animals. Well-known examples include the specific inhibition of chitin biosynthesis (chitin is a major component of cell walls in fungi and also in insects but not in plants and animals) and lysine biosynthesis in fungi (in fungi lysine biosynthesis is peculiar and proceeds via an α-aminoadipic acid pathway rather than via diaminopimelic acid as in higher plants) by using a specific inhibitor of some step(s) in the biosynthesis. This approach has been proved to be a viable and novel approach for crop protection (Brent, 1983).

More recently, the specific inhibition of PA biosynthesis in fungi using PA biosynthesis inhibitors (particularly ODC inhibitors) has been demonstrated to be a useful and promising approach for control of fungal plant infections (reviewed by Rajam, 1993; Walters, 1995). In other words, PA biosynthetic enzymes, particularly ODC, have been targets for plant fungal disease control. This approach was based on the fact that fungi, with few exceptions, possess only an ODC pathway for PA formation (Khan and Minocha, 1989; Zarb and Walters, 1994a, b), while their host plants can synthesize PAs via both ODC and ADC pathways (the latter being more predominant than the former) (Rajam, 1993; Rajam and Galston, 1985). Further, PAs are absolutely required for fungal growth and differentiation; hence the inhibition of PA synthesis in fungi is lethal (Rajam and Galston, 1985). Rajam and Galston (1985) and Rajam et al. (1985) were first to report the control of several phytopathogenic fungi in vitro, and protection of bean plants against bean rust fungus, *Uromyces phaseoli,* respectively, using ODC inhibitors, notably DFMO, without affecting the host plant (Rajam et al., 1985, 1986). Further, DFMO was persistant, fast acting, and translocatable (Rajam et al., 1985, 1986). Since then, the fungicidal effects of DFMO as well as inhibitors of other PA biosynthetic enzymes like MGBG (an inhibitor of SAMDC) have been demonstrated by several researchers on different kinds of plant fungal pathogens in vitro and in vivo (particularly rusts and mildews, which were found to be very sensitive to inhibitors of PA synthesis) (Galston and Weinstein, 1988; Kepczynska, 1995; Rajam, 1993; Singhania et al., 1991; Walters, 1995) (Table 11-2). However, some fungi are relatively insensitive to PA inhibitors (Smith et al., 1992). Fungal spore germination (Garcia et al., 1991; Khurana et al., 1995; Rajam et al., 1989) and sporulation (Birecka et al., 1986; Machatschke et al., 1990; Singh et al., 1989) were also inhibited by DFMO. It is interesting to note that DFMO had no deleterious effects on the host plant

TABLE 11-2 Examples of Some Plant Fungal Pathogens Controllable by DFMO Under in Vitro and in Vivo Conditions

Inhibition of Plant Pathogenic Fungi in Vitro

Botrytis cinerea; Monilinea fructicola; Rhizoctonia solani; Helminthosporium maydis; H. carborum; H. orzae; Verticillium dahliae; Pyrenophora teres; P. avenae; Gaeumannomyces graminus; Fusarium culumorum; F. oxysporum f. sp. *lycopersici; Septoria nodorum; S. tritici; Ceratocystis ulmi; Neovossia indica; Postia placenta; Sclerotium rolfsii; Dreschslera carborum; Phytophthora infestans; Crinipellis perniciosa; Bipolaris maydis; Humicola lanuginosa; Mucor pusillus;* and *Tararomyces emersonii*

Control of Plant Pathogenic Fungi in Vivo

Uromyces phaseoli (French bean rust)[a]; *U. viciae-fabae* (broad bean rust); *Puccinia recondita* (wheat leaf rust); *P. graminis* f. sp. *tritici* (wheat stem rust); *P. graminis* f. sp. *avenae* (oat stem rust); *P. sorghi* (corn common rust); *Erysiphe graminis* (wheat powdery mildew); *E. graminis* f. sp. *hardei* (barley powdery mildew); *E. polygoni* (bean powdery mildew); *Podosphaera leucotricha* (apple powdeery mildew); *Verticillium dahliae* (tomato wilt); *Helminthosporium maydis* (corn leaf blight); and *Postia placenta* (wood decay)

[a] Disease name has been given in parentheses.

(Bharti and Rajam, 1995; Machatschke et al., 1990; Rajam et al., 1985, 1986, 1991; Weinstein et al., 1987; West and Walters, 1988), and this could be due to the paradoxical stimulation of ADC by DFMO and/or presence of two pathways (ODC and ADC) for PA biogenesis (Bharti and Rajam, 1995; Zarb and Walters, 1994a, b).

Mycorrhizal fungi, which are closely associated with many plants and affect their growth, have been shown to respond differently to PA inhibitors. For instance, the growth of *Laccaria proxima* (ectomycorrhizal fungus) was only marginally reduced by DFMO, while there was an increase in fungal growth with DFMA treatment; this is because of the presence of both ODC and ADC pathways for PA formation in this fungus (Zarb and Walters, 1994a). In contrast, the growth of *Paxillus involutus* was substantially reduced by DFMO with no change in cellular PA titers (Zarb and Walters, 1994b). Curiously, Put analogues (e.g., keto-Put and N-acetyl-Put), aliphatic Put analogues (e.g., E-1, 4-diaminobut-2-ene, E-(N,N,N', N'-tetraethyl)-1,4-diaminobut-2-ene), and alicylic Put analogues (e.g., 1,2-bis (aminomethyl)-4,5 dimethylcyclohexa-1,4-diene dihydrochloride) have also been used as potent fungicidal agents in plants (Walters, 1995). This approach has been extended to control some zoopathogenic fungi such as *Candida* spp. (Pfaller et al., 1990), *Microsporum, Trichophyton,* and *Aspergillus* (Boyle et al., 1988; Sinha and Rajam, 1992). Curiously, PA-oxidase-PA system has been used to inhibit and kill fungi like *Cryptococcus neoformans* (Levitz et al., 1990).

In general, bacteria possess both ODC (most predominant) and ADC pathways for PA biosynthesis, and therefore it has proven very difficult to manipulate PA levels in the bacterial cells (Pegg et al., 1989). In spite of this fact, the in vitro inhibition of some bacteria using PA synthesis inhibitors has been attempted (Rajam, 1993). MGBG and CHA have shown to be inhibitory for the growth of *Esche-*

richia coli and/or *Pseudomonas aeruginosa,* respectively (Pegg et al., 1989). The best inhibition of growth with a combination of PA inhibitors has been observed in *E. coli, P. aeruginosa,* and *Serratia marcescens* (Bitonti and McCann, 1987; Bitonti et al., 1982), but the growth inhibitory effects were by no means as pronounced as in a number of fungi. Nevertheless, it may be possible to eradicate a bacterial pathogen on a crop plant without affecting the growth and development of the host plant by taking a high concentration of ODC inhibitor (e.g., DFMO or MFMO) and a low concentration of ADC inhibitor (e.g., DFMA or MFMA) or by combining inhibitors of ODC, ADC, and SAMDC/Spd synthase (e.g., MGBG or CHA) enzymes in the appropriate concentrations (Rajam, 1993). In an interesting study, Ozawa and Tsuji (1993) have shown that soybean plants accumulate Put and Spd in their nodules to suppress the growth of bacteroids of *Bradyrhizobium japonicum* strain 138 NR; in fact the exogenous PAs at higher concentrations were antibacterial, and Spd at 0.5 mM (pH 7.0) inhibited more than 95% of bacteroids, which were not able to produce colonies on Spd-containing agar plates.

More recently, the influence of DFMO and MGBG on viral infections (TMV and cucumber mosaic virus) on tobacco has been examined in our laboratory, and we found that these inhibitors have significantly reduced the severity of these viral infections (Rao and Rajam, unpublished results).

The insect pests of plants may have only an ODC pathway for PA biogenesis (Joseph and Baby, 1988), and therefore it should be possible to control insect pests by specific inhibition of insect PA biosynthesis, without affecting the host plants (Rajam, 1991, 1993). Indeed, the growth and development of tobacco caterpillar (*Spodoptera litura* Fb.) has been suppressed by using DFMO, MGBG, or CHA; they also strongly retarded the growth and development of mosquito larvae with high mortality (Rajam, 1991).

CONCLUDING REMARKS AND FUTURE PROSPECTS

It is clear from the above discussion that PAs are intricately involved in important functions of plant metabolism and development. Most of the plants, especially cereals, accumulate huge quantities of Put as a response to abiotic stress, which is accompanied by an increase in ADC activity. This prompted Galston and Kaur-Sawhney (1990) to call ADC a "general stress enzyme" in cereals. The uncommon PAs (e.g., non-Spd, non-Spm, caldopentamine, and homocaldopentamine) seem to be produced by plants in response to abiotic stresses (Kuehn et al., 1990). Viruses and fungi can also elicit PA concentrations, particularly PA conjugates with hydroxycinnamic acids and PA enzymes as a consequence of plant-pathogen interactions (Martin-Tanguy, 1985). However, there is no general agreement concerning the physiological significance of Put accumulation. It could be the mechanism of plant protection against the imposed stress or it could also represent the means by which the stress injures the cell (Galston, 1989). Another hypothesis suggests that the primary injury-producing biochemical process is the liberation of ammonia, and that Put accumulation may be one of its major metabolic sinks (Slocum and

Weinstein, 1990). Although protective action of PAs in plants under stress is not clearly understood, it is suggested that they may stabilize the membranes, including thylakoid membrane activity (Besford et al., 1993; Yordanov and Goltsev, 1990), stabilize the structure of macromolecules, and prevent the breakdown of macromolecules (Altman et al., 1982) because of the unique properties of these small amines as polycations, and their strong binding to acidic sites of nucleic acids and cell membrane phospholipids (Tiburcio et al., 1993). It has also been suggested that excess synthesis of Put via the ADC pathway under stress conditions might be useful to maintain the ionic balance in the cell (Galston and Kaur-Sawhney, 1990; Tiburcio et al., 1993). Further, the mechanism of PA action in such varied functions could be due to their free radical scavenging properties and their effects on the RNase, protease, and other enzymes and on secondary messengers, and their interaction with ethylene formation (Tiburcio et al., 1993).

To understand the precise mechanism of action of PA functions in higher plants, the use of molecular approaches, such as cloning of PA biosynthetic genes, the production of transgenic plants with over- and underexpression of PA biosynthesis genes, and analysis of gene promoters fused with reporter genes would be appropriate (Rastogi et al., 1993). However, the roles of PAs in various cellular and molecular processes are better understood in animal systems (Heby and Persson, 1990). Unfortunately, almost nothing is known about the mechanism of action of PAs in plants, and only recently have some studies on molecular analysis of PA biosynthesis in plants been initiated. For example, the gene encoding ADC has been cloned in oat (Bell and Malmberg, 1990) and tomato (Rastogi et al., 1993). The ADC of oat is clipped from a precursor into two polypeptides and is synthesized as 16 kDa proenzyme, which undergoes proteolytic cleavage for activation (Malmberg and Cellino, 1994; Malmberg et al., 1992). It is noteworthy that transgenic tobacco plants overexpressing yeast ODC (in roots) were produced (Hammill et al., 1990). Interestingly, callus raised from tobacco transgenics expressing mouse ODC cDNA (driven by 35S promoter of cauliflower mosaic virus) have exhibited higher levels of Put (4- to 12-fold) and ODC activity than the controls exhibited (DeScenzo and Minocha, 1993). Subsequently, the expression of a human SAMDC cDNA (driven by 35S promoter) in transgenic tobacco has been achieved, and such transgenics have shown the increased SAMDC activity and Spd titer (2–3 times) compared with the control tissues; however Put levels were significantly reduced while Spm content remained unchanged (Noh and Minocha, 1994). Therefore, modulation of cellular PAs by genetic engineering (metabolic engineering) techniques should provide insights into the roles of PAs in plant metabolism and development. Recently, we have undertaken such studies to address some of the vexing questions pertaining to functions of PAs in higher plants, using rice and eggplant as model plants.

In the future, PA research field would provide potential applications in agriculture and tissue culture, especially protecting our crop plants against senescence, abiotic and biotic stresses and improving crop yield and quality, and morphogenesis in in vitro cultures.

ACKNOWLEDGMENTS

Our research on polyamines has received generous financial support from the Council of Scientific and Industrial Research, University Grants Commission, and Department of Science and Technology, New Delhi. I am grateful to my former graduate student Dr. Pankaj Sharma, Department of Genetics, University of Delhi, South Campus for his critical comments on the manuscript and help in the literature survey. I thank my graduate students Dr. Bharti, Mr. Shivendra Bajaj, and Mr. J. S. Yadav for their help, especially for literature survey, and Miss Ratna Kumria for critically reading the manuscript.

REFERENCES

Adiga, P. R. and G. L. Prasad (1985), *Plant Growth Regul., 3,* 205–226.

Altamura, M. M., P. Torrigiani, G. Falasca, P. Rossini, and N. Bagni (1993), *J. Plant Physiol., 142,* 543–551.

Altman, A. and U. Bachrach (1981), in *Advances in polyamine research* (C. M. Caldarera, V. Zappia, and U. Bachrach, Eds.), Vol. 3, Raven Press, New York, pp. 365–375.

Altman, A. and N. Levin (1993), *Physiol. Plant, 89,* 653–658.

Altman, A., R. Kaur-Sawhney, and A. W. Galston (1977), *Plant Physiol., 60,* 570–574.

Altman, A., R. Friedman, D. Amir, and N. Levin (1982), in *Plant growth substances* (P. F. Wareing, Ed.), Academic Press, London, pp. 483–494.

Altman, A., R. Friedman, and N. Levin (1983), in *Advances in polyamine research* (U. Bachrach, A. Kaye, and R. Chayen, Eds.), Vol. 4, Raven Press, New York, pp. 395–408.

Angelini, R., M. Bragaloni, R. Federico, A. Infantino, and A. Porta-Puglia (1993), *J. Plant Physiol., 142,* 704–709.

Apelbaum, A., I. Icekson, A. C. Burgoon, and M. Liberman (1982), *Plant Physiol., 88,* 996–998.

Abelbaum, A., A. Goldlust, and I. Icekson (1985), *Plant Physiol., 79,* 635–640.

Apelbaum, A., Z. N. Canellakis, P. B. Applewhite, R. Kaur-Sawhney, and A. W. Galston (1988), *Plant Physiol., 88,* 996–998.

Aurisano, N., A. Bertani, M. Mattana, and R. Reggiani (1993), *Physiol. Plant., 89,* 687–692.

Bagga, S., A. Dharma, G. C. Phillips, and G. D. Kuehn (1991), *Plant Cell Rep., 10,* 550–554.

Bagni, N. (1966), *Experentia, 22,* 732.

Bagni, N. and R. Pistocchi (1990), in *Polyamines and ethylene: Biochemistry, physiology and interactions* (H. E. Flores., R. N. Arteca, and J. C. Shannon, Eds.), Amer. Soc. Plant Physiol., Rockville, MD, pp. 62–72.

Bailey, J. P., A. J. Bower, and D. H. Lewis (1987), *Trans. Brit. Mycol. Soc., 89,* 83–87.

Bajaj, S. and M. V. Rajam (1995), *Plant Cell Rep., 14,* 717–720.

Balasundarum, D. and A. K. Tyagi (1991), *Mol. Cell Biochem., 100,* 129–140.

Basso, L. C. and T. A. Smith (1974), *Phytochemistry, 13,* 875–883.

Basu, R. and B. Ghosh (1991), *Physiol. Plant., 82,* 575–581.

Basu, R., N. Maitra, and B. Ghosh (1988), *Aust. J. Plant Physiol., 15,* 777–786.

Bell, E. and R. L. Malmberg (1990), *Mol. Gen. Genet., 224,* 431–436.

Belles, J. M., J. Carbonell, and V. Conejero (1991), *Plant Physiol., 96,* 1053–1059.

Belles, J. M., M. A. Perez-Amador, J. Carbonell, and V. Conejero (1993), *Plant Physiol., 102,* 933–937.

Bertrand, S., P. Nadeau, D. Dostaler, and A. Gosselin (1990), in *Polyamines and ethylene: Biochemistry, physiology and interactions* (H. E. Flores, R. N. Arteca, and J. C. Shannon, Eds.), Am. Soc. Plant Physiol., Rockville, MD, pp. 401–404.

Besford, R. T., C. M. Richardson, J. L. Campos, and A. F. Tiburcio, (1993), *Planta, 189,* 201–206.

Bey, P., C. Danzin, V. Vandorsselaer, P. Mamont, M. Jung, and C. Tardif (1977), *J. Med. Chem., 21,* 50–55.

Bey, P., C. Danzin, and M. Jung (1987), in *Inhibition of polyamine metabolism: Biological significance and basis for new therapies* (P. P. McCann, A. E. Pegg, and A. Sjoerdsma, Eds.), Academic Press, San Diego, CA, pp. 1–32.

Bharti and M. V. Rajam (1995), *Ann. Bot., 76,* 297–301.

Bird, C. R. and T. A. Smith (1983), *Phytochemistry, 22,* 2401–2403.

Birecka, H., T. S. Dinolfo, W. B. Martin, and M. W. Frolich (1984), *Phytochemistry, 23,* 991–994.

Birecka, H., M. O. Garraway, R. I. Baumann, and P. P. McCann (1986), *Plant Physiol., 80,* 798–800.

Birecka, H., K. P. Ireton, A. J. Bitonti, and P. P. McCann (1991), *Phytochemistry, 30,* 105–108.

Bitonti, A. J. and P. P. McCann (1987) in *Inhibition of polyamine metabolism* (P. P. McCann, A. E. Pegg, and A. Sjoerdsma, Eds.), Academic Press, San Diego, CA, pp. 259–275.

Bitonti, A. J., P. P. McCann, and A. Sjoersma (1982), *Biochem. J., 208,* 435–441.

Bitonti, A. J., P. J. Casara, P. P. McCann, and P. Bey (1987), *Biochem. J., 242,* 69–74.

Bors, W., C. Langebartels, C. Michel, and H. Sandermann, Jr. (1989), *Phytochemistry, 28,* 1589–1595.

Botha, M.-L. and C. S. Whitebread (1992), *Planta, 188,* 478–483.

Boyle, S. M., N. Sriranganathan, and D. Cordes (1988), *J. Med. Vet. Mycol., 26,* 227–236.

Brent, K. J. (1983), in *Biochemical plant pathology* (J. A. Callow, Ed.), Wiley, New York, pp. 435–452.

Bueno, M., D. Garrido, and A. Matilla (1993), *Physiol. Plant, 87,* 381–388.

Carbonell, J. and J. L. Navarro (1989), *Planta, 178,* 482–487.

Casas, J. L., M. Acosta, J. A. Delrio, and F. Sabater (1990), *Plant Growth Regul., 9,* 89–96.

Chen, C. T. and C. H. Kao (1993a), *Plant Growth Regul., 13,* 197–202.

Chen, C. T. and C. H. Kao (1993b), *Plant Growth Regul., 13,* 133–136.

Cheng, S. H., Y. Y. Shyr, and C. H. Kao (1984), *Bot. Bull. Acad. Sin., 25,* 191–196.

Coghlan, S. E. and D. R. Walters (1990), *New Phytol., 116,* 417–424.

Corey, K. A. and A. V. Barker (1989), *J. Am. Soc. Hort. Sci., 114*, 651–655.

Crisosto, C. H., P. B. Lombard, D. Sugar, and V. S. Polito (1988), *J. Am. Soc. Hort. Sci., 113*, 708–712.

Das, S., R. Basu, and B. Ghosh (1987), *Plant Physiol. Biochem., 14*, 108–116.

Del Duca, S., U. Tidu, R. Bassi, C. Esposito, and D. Serafini-Fracassini (1994), *Planta, 193*, 283–289.

DeRosa, M., S. DeRosa, A. Gambacorta, M. Carteni-Farina, and V. Zappia (1978), *Biochem. J., 176*, 1–7.

DeScenzo, R. A. and S. C. Minocha (1993), *Plant Mol. Biol., 22*, 113–127.

Dibble, A. R. G., P. J. Davies, and M. A. Mutschler (1988), *Plant Physiol., 86*, 338–340.

Dinella, C., D. Serafini-Fracassini, B. Grandi, and S. Del Duca, (1992), *Plant Physiol. Biochem., 30*, 531–539.

DiTomaso, J. M., T. L. Rost, and F. M. Ashton (1988), *Plant Cell Physiol., 29*, 1367–1372.

DiTomaso, J. M., J. E. Shaff, and L. V. Kochian (1989), *Plant Physiol., 90*, 988–995.

DiTomaso, J. M., J. J. Hart, and L. V. Kochian (1992), *Plant Physiol., 98*, 611–620.

Downs, C. G. and P. H. Lovell (1986), *Physiol. Plant, 66*, 679–684.

Evans, P. T. and R. L. Malmerg (1989), *Annu. Rev. Plant Physiol. Plant Mol. Biol., 40*, 235–269.

Even-Chen, Z., A. K. Mattoo, and R. Goren (1982), *Plant Physiol., 69*, 385–388.

Faure, O., M. Mengoli, A. Nougarede, and N. Bagni (1991), *J. Plant Physiol., 138*, 545–549.

Federico, R. and R. Angelini (1986), *Planta, 167*, 300–302.

Feng, J. and A. V. Barker (1993), *Hort. Sci., 28*, 109–110.

Flores, H. E. (1990), in *Stress responses in plants: Adaptation and acclimation mechanisms* (R. G. Alscher and J. R. Comming, Eds.), Wiley-Liss, New York, pp. 217–239.

Flores H. E. and P. Filner (1985), *Plant Growth Regul., 3*, 277–291.

Flores, H. E. and A. W. Galston (1982), *Science, 217*, 1259–1261.

Flores, H. E. and A. W. Galston (1984a), *Plant Physiol., 75*, 102–109.

Flores, H. E. and A. W. Galston (1984b), *Plant Physiol., 75*, 110–113.

Flores, H. E., N. D. Young, and A. W. Galston (1985), in *Cellular and molecular basis of plant stress* (J. L. Key and T. Kosuge, Eds.), Vol. 22, Liss, New York, pp. 93–114.

Flores, H. E., C. M. Protacio, and M. W. Signs (1989), in *Plant nitrogen metabolism. Recent advances in phytochemistry* (J. E. Poulton, J. T. Romex, and E. E. Cohn, Eds.), Vol. 23, Plenum Press, New York, pp. 329–393.

Foster, S. A. and D. R. Walters (1991), *Physiol. Plant., 82*, 185–190.

Foster, S. A. and D. R. Walters (1992), *J. Plant Physiol., 140*, 134–136.

Friedman, R., A. Altman, and N. Levin (1989), *Physiol. Plant, 76*, 295–302.

Fuhrer, J., R. Kaur-Sawhney, L. M. Shih, and A. W. Galston (1982), *Plant Physiol., 70*, 1597–1600.

Funayama, S., G. Zhang, and S. Nozoe (1995), *Phytochemistry, 38*, 1529–1531.

Galston, A. W. (1983), *BioScience, 33*, 382–387.

Galston, A. W. (1989), in *The physiology of polyamines* (U. Bachrach and Y. M. Heimer, Eds.), Vol. 2, CRC Press, Boca Raton, FL, pp. 99–105.

Galston, A. W. and H. E. Flores (1991), in *Biochemistry and physiology of polyamines in plants* (R. D. Slocum and H. E. Flores, Eds.), CRC Press, Boca Raton, FL, pp. 175–186.

Galston, A. W. and R. Kaur-Sawhney (1987) in *Plant senescence: Its biochemistry and physiology* (W. Thompson, E. A. Nothnagel, and R. C. Huffaker, Eds.), Am. Soc. Plant Physiol., Rockville, MD, pp. 167–181.

Galston, A. W. and R. Kaur-Sawhney (1990), *Plant Physiol., 94,* 406–410.

Galston, A. W. and A. F. Tiburcio (Eds.) (1991), *Lecture course on polyamines as modulators of plant development,* Vol. 257, Fundacion Juan March, Madrid.

Galston, A. W. and L. H. Weinstein (1988), in *Progress in polyamine research* (V. Zappia and P. E. Pegg, Eds.), Plenum, New York, pp. 589–599.

Garcia, J. I., G. Nicolas, and T. Valle (1991). *Plant Sci., 77,* 131–136.

Goldberg, R. and E. Perdrizet (1984), *Planta, 161,* 531–533.

Greenland, A. J. and D. H. Lewis (1984), *New Phytol., 96,* 283–291.

Guye, M. G., L. Vigh, and J. M. Wilson (1986), *J. Exp. Bot., 37,* 1036–1043.

Hamana, K. and S. Matsuzaki (1985), *J. Biochem., 97,* 1595–1601.

Hamana, K., K. Masahiro, H. Onishi, T. Akazawa, and S. Matsuzaki (1985), *J. Biochem., 94,* 1653–1658.

Hammill, J. D., R. J. Robins, A. J. Parr, D. M. Evans, J. M. Furze, and M. J. C. Rhodes (1990), *Plant Mol. Biol., 15,* 27–38.

Haslam, S. C. and T. W. Young (1992), *Phytochemistry, 31,* 4075–4079.

Hauschild, M. and S. Jacobsen (1990), in *Polyamines and ethylene: Biochemistry, physiology and interactions* (H. E. Flores, R. N. Arteca, and J. C. Shannon, Eds.), Am. Soc. Plant Physiol., Rockville, MD, pp. 405–407.

Heby, O. and L. Persson (1990), *Trends Biochem. Sci., 15,* 153–158.

Hiatt, A. C. and R. L. Malmberg (1988), *Plant Physiol., 86,* 441–446.

Hibasami, H., M. Tanaka, J. Nagai, and T. Ikeda (1980), *FEBS Lett., 116,* 99–101.

Huang, W. Y., Y. L. Wang, and L. J. Yuan (1990), *Acta Bot. Sin., 32,* 125–132.

Hughes, P. R., L. H. Weinstein, S. H. Wettlaufer, J. J. Chiment, G. J. Doss, T. W. Culliney, W. H. Gutenmann, C. A. Bache, and D. J. Lisk (1987), *J. Agric. Food. Chem., 35,* 50–54.

Hurng, W. P. and C. H. Kao (1993), *Plant Sci., 91,* 121–125.

Hurng, W. P., H. S. Lur, C. K. Liao, and C. H. Kao (1994), *J. Plant Physiol., 143,* 102–105.

Joseph, K. and T. G. Baby (1988), *Insect Biochem., 18,* 807–810.

Kakkar, R. K. and K. V. Rai (1993), *Phytochemistry, 33,* 1281–1288.

Kallio, A., P. P. McCann, and P. Bey (1981), *Biochemistry, 20,* 3163–3166.

Kallio, A., P. P. McCann, and P. Bey (1982), *Biochem. J., 204,* 771–775.

Kandpal, R. P. and N. A. Rao (1985), *Biochem. Int., 11,* 365–370.

Kaur-Sawhney, R. and P. B. Applewhite (1993), *Plant Growth Regul., 12,* 223–227.

Kaur-Sawhney, R. and A. W. Galston (1979), *Plant Cell Environ., 2,* 189–196.

Kaur-Sawhney, R., and A. W. Galston (1991), in *Biochemistry and physiology of polyamines in plants* R. D. Slocum and H. E. Flores, Eds.), CRC Press, Boca Raton, FL, pp. 201–211.

Kaur-Sawhney, R., H. E. Flores, and A. W. Galston (1980), *Plant Physiol., 65,* 368–371.

Kaur-Sawhney, R., L. Shih, T. Cegielska, and A. W. Galston (1982a), *FEBS Lett., 145,* 345–349.

Kaur-Sawhney, R., L.-M. Shih, H. E. Flores, and A. W. Galston (1982b), *Plant Physiol., 69,* 405–410.

Kaur-Sawhney, R., N. S. Shekhawat, and A. W. Galston (1985), *Plant Growth Regul., 3,* 329–337.

Ke, D. and R. J. Romani (1988), *Plant Physiol. Biochem., 26,* 109–116.

Kepezynska, E. (1995), *Plant Growth Regul., 16,* 263–266.

Khan, A. J. and S. C. Minocha (1989), *Life Sci., 44,* 1215–1222.

Khurana, N., R. K. Saxena, R. Gupta, and M. V. Rajam (1996), *Microbiology, 142,* 517–523.

Klein, H., A. Priebe, and H. J. Jager (1979), *Physiol. Plant, 45,* 497–499.

Kolb, M., C. Danzin, J. Barth, and N. Claverie (1982), *J. Med. Chem., 25,* 550–556.

Konigshofer, H. (1989), *J. Plant Physiol., 134,* 736–740.

Kotzabasis, K., C. Fotinou, K. A. Roubelakis-Angelakis, and D. Ghanotakis (1993), *Photosyn. Res., 38,* 83–88.

Kramer, G. F. and C. Y. Wang (1989), *Physiol. Plant, 76,* 479–484.

Kramer, G. F. and C. Y. Wang (1990), *J. Plant Physiol., 136,* 115–119.

Krishnamurthy, R. (1991), *Plant Cell Physiol., 32,* 699–703.

Krishnamurthy, R. and K. A. Bhagwat (1989), *Plant Physiol., 91,* 500–504.

Kuehn, G. D., S. Bagga, B. Rodriguez-Garay, and G. C. Phillips (1990), in *Polyamines and ethylene: Biochemistry, physiology and interactions* (H. E. Flores, R. N. Arteca, and J. C. Shannon, Eds.), Am. Soc. Plant Physiol., Rockville, MD, pp. 190–202.

Kulpa, J. M., A. G. Galsky, P. Lipetz, and R. Stephens (1985), *Plant Cell Rep., 4,* 81–83.

Kyriakidis, D. A., C. A. Panagiotidis, and J. G. Georgatsos (1983), *Methods Enzymol., 94,* 162–166.

Lahiri, K., S. Chattopadhyay, and B. Ghosh (1992), *Phytochemistry, 31,* 4087–4090.

Langebartels, C., K. Kerner, S. Leonardi, M. Schraudner, M. Trost, W. Heller, and H. Sandermann, Jr. (1991), *Plant Physiol., 95,* 882–889.

Le Dily, F., J.-P. Billard, and J. Boucaud (1991), *Plant Cell Environ., 14,* 327–332.

Lee, T. M. and C. Chu (1992), *Plant Physiol., 100,* 238–245.

Levitz, S. M., D. J. DiBenedetto, and R. D. Diamond (1990), *Antonie van Leeuwenhoek, 58,* 107–114.

Li, Z. C. and J. W. McClure (1989), *Phytochemistry, 28,* 2255–2259.

Lin, C. C. and C. H. Kao (1995), *Plant Growth Regul., 17,* 15–20.

Machsatschke, S., C. Kamrowski, B. M. Moerschbacher, and H.-J. Reisener (1990), *Physiol. Mol. Plant Pathol., 36,* 451–459.

Malmberg, R. L. and M. L. Cellino (1994), *J. Biol. Chem., 269,* 2703–2706.

Malmberg, R. L., K. E. Smith, E. Bell, and M. L. Cellino (1992), *Plant Physiol., 100,* 146–152.

Mariani, P., D. D'Orazi, and N. Bagni (1989), *J. Plant Physiol., 135,* 508–510.

Martin, C. and J. Martin-Tanguy (1981), *CR Acad. Sci., 292,* 249–251.

Martin-Tanguy, J. (1985), *Plant Growth Regul., 3,* 381–399.

Martin-Tanguy, J., C. Martin, and M. Gallet (1973), *CR Acad. Sci., 276,* 1433–1435.

Martin-Tanguy, J., C. Martin, M. Gallet, and R. Vernoy (1976), *CR Acad. Sci., 282,* 2231–2234.

Matsuzaki, S., K. Hamana, K. Isobe, M. Niitsu, and K. Samejima (1990), in *The biology and chemistry of polyamines* (S. H. Goldenberg and I. D. Algranati, Eds.), IRL Press, Oxford, UK, pp. 159–167.

McDonald, R. E. and M. M. Kushad (1986), *Plant Physiol., 82,* 324–326.

Mehta, A. M., R. A. Saffner, G. W. Schaffer, and A. K. Mattoo (1991), *Plant Physiol., 95,* 1294–1297.

Melnichuk, Y. P., A. K. Lishko, and F. L. Kalini (1983), *Fiziol. Biokhim. Kul't. Rast., 14,* 47–50.

Melnichuk, Y. P., A. S. Sobolev, and F. L. Kalinin (1984), *Fiziol. Biokhim. Kul't. Rast., 16,* 572–577.

Metcalf, B. W., P. Bey, C. Danzin, M. Jung, P. Casara, and J. P. Vevert (1978), *J. Am. Chem. Soc., 100,* 2251–2553.

Minocha, S. C. and R. Minocha (1995), in *Biotechnology in agriculture and forestry* (Y. P. S. Bajaj, Ed.), Vol. 30, Springer-Verlag, Berlin, pp. 53–70.

Muhitch, M. J., L. A. Edwards, and J. S. Fletcher (1983), *Plant Cell Rep., 2,* 82–84.

Nadeau, P. (1990), in *Polyamines and ethylene: Biochemistry, physiology and interactions* (H. E. Flores, R. N. Arteca, and J. C. Shannon, Eds.), Am. Soc. Plant Physiol., Rockville, MD, pp. 387–389.

Nadeau, P. and R. Paquin (1988), *Can. J. Plant Sci., 68,* 449–456.

Nadeau, P., S. Delaney, and L. Chouinard (1987), *Plant Physiol., 84,* 73–77.

Naik, B. I. and S. K. Srivastava (1978), *Phytochemistry, 17,* 1885–1887.

Negrel, J. and T. A. Smith (1984), *Phytochemistry, 23,* 739–741.

Negrel, J., J. C. Vallee, and C. Martin (1984), *Phytochemistry, 23,* 2747–2751.

Negrel, J., M. Paynot, and F. Javelle (1992), *Plant Physiol., 98,* 1264–1269.

Nielsen, K. A. (1990), *J. Exp. Bot., 41,* 849–854.

Noh, E. W. and S. C. Minocha (1994), *Transgenic Res., 3,* 26–35.

Okii, M. et al., (1980), U.S. Patent No. 4,231,789.

Oshima, T. (1983), in *Advances in polyamine research* (U. Bachrach, A. Kaye, and R. Chayen, Eds.), Vol. 4, Raven Press, New York, pp. 479–487.

Ozawa, T. and T. Tsuji (1993), *Plant Cell Physiol., 34,* 899–904.

Panagiotidis, C. A. and D. A. Kyriakidis (1985), *Plant Growth Regul., 3,* 247–255.

Peeters, K. M. U., J. M. C. Geuns, and A. J. Van Laere (1993), *J. Exp. Bot., 44,* 1709–1715.

Pegg, A. E. and H. G. Williams-Ashman (1987), in *Inhibition of polyamine metabolism: Biological significance and basis for new therapies* (P. P. McCann, A. E. Pegg, and A. Sjoerdsma, Eds.), Academic Press, San Diego, CA, pp. 33–48.

Pegg, A. E., P. P. McCann, and A. Sjoerdsma (1989), in *Enzymes as targets for drug design* (M. G. Palfreyman, P. P. McCann, W. Lovenberg, J. G. Temple, Jr., and A. Sjoerdsma, Eds.), Academic Press, San Diego, CA, pp. 157–183.

Pfaller, M. A., J. Riley, and T. Gerarden (1990), *Mycopathologia, 112,* 27–32.

Pfosser, M. (1993), *J. Plant Physiol., 142,* 605–609.

Pistocchi, R. and N. Bagni (1990), *J. Plant Physiol., 136,* 728–735.

Pistocchi, R., F. Keller, N. Bagni, and P. Matile (1988), *Plant Physiol., 87,* 514–518.

Pistocchi, R., F. Antognoni, N. Bagni, and D. Zannoni (1990), *Plant Physiol., 92,* 690–695.

Ponchet, M., J. Martin-Tanguy, A. Marias, and C. Martin (1982), *Phytochemistry, 21,* 2865–2869.

Popovic, R., D. Kyle, A. Cohen, and S. Zalik (1979), *Plant Physiol., 64,* 721–726.

Poso, H., P. P. McCann, R. Tanskanen, P. Beyt, and A. Sjoerdsma, (1984), *Biochem. Biophys. Res. Commun., 125,* 337–343.

Prakash, L. and G. Prathapasenan (1988), *Aust. J. Plant Physiol., 15,* 761–767.

Prakash, L., M. Dutt, and G. Prathapasenan (1988), *Aust. J. Plant. Physiol., 15,* 769–776.

Prasad, G. L. and P. R. Adiga (1986). *J. Biosci., 10,* 373–391.

Prasad, G. L. and P. R. Adiga (1987), *J. Biosci., 11,* 571–579.

Predieri, S., D. T. Krizek, C. Y. Wang, R. M. Mirecki, and R. H. Zimmerman (1993), *Physiol. Plant, 87,* 109–117.

Preston, C., A. Holtum, and S. Powels (1992), *Photosyn. Res., 34,* 193 (abstract).

Priebe, A., H. Klein, and H. I. Jager (1978), *J. Exp. Bot., 29,* 1045–1050.

Rajam, M. V. (1989), *Plant Sci., 59,* 53–56.

Rajam, M. V. (1991), *Indian J. Exp. Biol., 29,* 881–882.

Rajam, M. V. (1993), *Curr. Sci., 65,* 461–469.

Rajam, M. V. and A. W. Galston (1985), *Plant Cell Physiol., 26,* 683–692.

Rajam, M. V. and K. Subhash (1987), in *Proc. Symp. on Plant Cell and Tissue Culture of Economically Important Crop Plants* (G. M. Reddy, Ed.), Osmania University, Hyderabad, India, pp. 223–227.

Rajam, M. V., L. H. Weinstein, and A. W. Galston (1985), *Proc. Natl. Acad. Sci. USA, 82,* 6874–6878.

Rajam, M. V., L. H. Weinstein, and A. W. Galston (1986), *Plant Physiol., 82,* 485–487.

Rajam, M. V., L. H. Weinstein, and A. W. Galston (1989), *Plant Cell Physiol., 30,* 37–41.

Rajam, M. V., L. H. Weinstein, and A. W. Galston (1991), *Curr. Sci., 60,* 178–180.

Rastogi, R., J. Dulson, and S. J. Rothstein (1993), *Plant Physiol., 103,* 829–834.

Reddy, G. N., R. N. Arteca, Y. R. Dai, H. E. Flores, F. B. Negm, and E. J. Pell (1993), *Plant Cell Environ., 16,* 819–826.

Reggiani, R. (1994), *Plant Cell Physiol., 35,* 1245–1249.

Reggiani, R. and A. Bertani (1990), in *Polyamines and ethylene: Biochemistry, physiology and interactions* (H. E. Flores, R. N. Arteca, and J. C. Shannon, Eds.), Am. Soc. Plant Physiol., Rockville, MD, pp. 390–393.

Reggiani, R., A. Hochkoeppler, and A. Bertani (1989), *Plant Cell Physiol., 30,* 893–898.

Reggiani, R., S. Bozo, and A. Bertani (1994), *Plant Sci., 102,* 121–126.

Richards, F. J. and R. G. Coleman (1952), *Nature, 170,* 460.

Roberts, D. R., M. A. Walker, J. E. Thompson, and E. B. Dumbroff (1984), *Plant Cell Physiol., 25,* 315–322.

Roberts, D. R., E. B. Dumbroff, and J. E. Thompson (1986), *Planta, 167,* 395–401.

Rodriguez, M. T., M. P. Gonzalez, and J. M. Linares (1987), *J. Plant Physiol., 129,* 369–374.

Rodriguez-Garay, B., G. C. Phillips, and G. D. Kuehn (1989), *Plant Physiol., 89,* 525–529.

Roux, S. J. (1993), *Plant Growth Regul.*, *12*, 189–193.

Rowland-Bamford, A. J., A. M. Borland, P. J. Lea, and T. A. Mansfield (1989), *Environ. Pollut.*, *61*, 95–106.

Samborski, D. J. and R. Rohringer (1970), *Phytochemistry*, *9*, 1939–1945.

Santanenan, A. and L. K. Simola (1994), *Physiol. Plant*, *90*, 125–129.

Sarjala, T. and E.-M. Savonen (1994), *J. Plant Physiol.*, *144*, 720–725.

Schoofs, G., S. Teichmann, T. Hartmann, and M. Wink (1983), *Phytochemistry*, *22*, 65–69.

Schraudner, M., M. Trost, K. Kerner, W. Heller, S. Leonardi, C. Langebartels, and H. Sandermann, Jr. (1990), in *Polyamines and ethylene: Biochemistry, physiology and interactions* (H. E. Flores, R. N. Arteca, and J. C. Shannon, Eds.), Am. Soc. Plant Physiol., Rockville, MD, pp. 394–396.

Seiler, N. (1987), in *Inhibition of polyamine metabolism: Biological significance and basis for new therapies* (P. P. McCann, A. E. Pegg, and A. Sjoerdsma, Eds.), Academic Press, Orlando, FL, pp. 49–77.

Sen, K. and B. Ghosh (1984), *Phytochemistry*, *23*, 1583–1585.

Serafini-Fracassini, D., N. Bagni, P. G. Cionini, and A. Bennici (1980), *Planta*, *148*, 332–337.

Serafini-Fracassini, D., S. Del Duca, and D. D'Orazi (1988), *Plant Physiol.*, *87*, 757–761.

Sharma, P. and M. V. Rajam (1995), *J. Plant Physiol.*, *146*, 658–664.

Sharma, P., J. S. Yadav, and M. V. Rajam (1995), *Plant Physiol. Suppl.*, *108*, 48 (Abstract).

Shevyakova, N. I., B. P. Strogonov, and I. G. Kiryan (1985), *Plant Growth Regul.*, *3*, 365–369.

Shih, L., R. Kaur-Sawhney, J. Fuhrer, S. Samanta, and A. W. Galston, (1982), *Plant Physiol.*, *70*, 1592–1596.

Sinclair, C. (1969), *Plant Soil*, *30*, 423–438.

Sindhu, R. K. and S. S. Cohen (1984), *Plant Physiol.*, *76*, 219–223.

Singh, P. P., S. Singh, A. S. Basra, and P. S. Bedi (1989), *Phytoparasitica*, *17*, 323–326.

Singhania, S., T. Satyanarayana, and M. V. Rajam (1991), *Mycol. Res.*, *95*, 915–917.

Sinha, M. and M. V. Rajam (1992), *Indian J. Exp. Biol.*, *30*, 538–540.

Skinner, W. A. and J. G. Johannsen (1972), *J. Med. Chem.*, *15*, 427–428.

Slocum, R. D. and L. H. Weinstein (1990), in *Polyamines and ethylene: Biochemistry, physiology and interactions* (H. E. Flores, R. N. Arteca, and J. C. Shannon, Eds.), Am. Soc. Plant Physiol., Rockville, MD, pp. 157–165.

Slocum, R. D., R. Kaur-Sawhney, and A. W. Galston (1984), *Arch. Biochem. Biophys.*, *235*, 283–303.

Smith, T. A. (1973), *Phytochemistry*, *12*, 2093–2100.

Smith, T. A. (1983), *Rec. Adv. Phytochemistry*, *18*, 7–54.

Smith, T. A. (1985), *Annu. Rev. Plant Physiol.*, *36*, 117–193.

Smith, T. A. and C. Sinclair (1967), *Ann. Bot.*, *31*, 103–111.

Smith, T. A. and P. J. Davies (1985), *Plant Physiol.*, *79*, 400–405.

Smith, T. A., J. Negrel, and C. R. Bird (1983), *Adv. Polyamine Res.*, *4*, 347–370.

Smith, T. A., J. H. A. Barker, and W. J. Owen (1992), *Mycol. Res.*, *96*, 395–400.

Songstad, D. D., D. R. Duncan, and J. M. Widholm (1990), *J. Exp. Bot.*, *41*, 289–294.

Speranza, A. and N. Bagni (1978), Z. *Planzenphyiol.*, *88*, 163–168.

Srivastava, S., D. Vashi, and B. Naik (1983), *Phytochemistry*, *22*, 2151–2154.

Srivenugopal, K. S. and P. R. Adiga (1981), *J. Biol. Chem.*, *256*, 9532–9541.

Stoessl, A. (1965), *Phytochemistry*, *4*, 973–976.

Stoessl, A. and C. H. Unwin (1978), *Can. J. Bot.*, *48*, 465–470.

Suttle, J. C. (1981), *Phytochemistry*, *20*, 1477–1480.

Tiburcio, A. F. and A. W. Galston (1986), *Phytochemistry*, *25*, 107–110.

Tiburcio, A. F., M. A. Masdeu, F. M. Dumortier, and A. W. Galston (1986a), *Plant Physiol.*, *82*, 369–374.

Tiburcio, A. F., R. Kaur-Sawhney, and A. W. Galston (1986b), *Plant Physiol.*, *82*, 375–378.

Tiburcio, A. F., J. L. Campos, X. Figueras, and R. T. Besford, (1993), *Plant Growth Regul.*, *12*, 331–340.

Tiburcio, A. F., R. T. Besford, T. Capell, A. Borrell, P. S. Testillano, and M. C. Risueno (1994), *J. Exp. Bot.*, *45*, 1789–1800.

Torrigiani, P., D. Serafini-Fracassini, S. Biondi, and N. Bagni, (1986), *J. Plant Physiol.*, *124*, 23–29.

Triantaphylides, C., L. Nespoulous, C. Chervin, and A. Rosset (1990), in *Polyamines and ethylene: Biochemistry, physiology and interactions* (H. E. Flores., R. N. Arteca, and J. C. Shannon, Eds.), Am. Soc. Plant Physiol., Rockville, MD, pp. 383–386.

Turner, L. B. and G. R. Stewart (1986), *J. Exp. Bot.*, *37*, 170–177.

Turner, L. B. and G. R. Stewart (1988), *J. Exp. Bot.*, *39*, 311–316.

Upfold, S. J. and J. van Staden (1991), *Plant Cell Tiss. Org. Cul.*, *10*, 355–362.

Venis, M. A. (1969), *Phytochemistry*, *8*, 1193–1197.

Vernon, D. M., M. C. Tarczynksi, R. G. Jensen, and H. J. Bohnert (1993), *Plant J.*, *4*, 199–205.

Wada, N., M. Shinozaki, and H. Iwamura (1994), *Plant Cell Physiol.*, *35*, 469–472.

Walters, D. R. (1989), *Plants Today*, *2*, 22–26.

Walters, D. R. (1995), *Mycol. Res.*, *99*, 129–139.

Walters, D. R. and M. A. Shuttleton (1985), *New Phytol.*, *100*, 209–214.

Walters, D. R. and M. A. Wylie (1986), *Physiol. Plant*, *67*, 630–633.

Walters, D. R., P. W. F. Wilson, and M. A. Shuttleton (1985), *New Phytol.*, *101*, 695–705.

Wang, C. Y. (1987), *Physiol. Plant*, *69*, 253–257.

Wang, C. Y., W. S. Conway, J. A. Abbott, and G. F. Kramer (1993), *J. Am. Soc. Hort. Sci.*, *118*, 801–806.

Weinstein, L. H., R. Kaur-Sawhney, M. V. Rajam, S. H. Wettlaufer, and A. W. Galston (1986), *Plant Physiol.*, *82*, 641–645.

Weinstein, L. H., J. F. Osmeloski, S. H. Wettlaufer, and A. W. Galston (1987), *Plant Sci.*, *51*, 311–316.

West, H. M. and D. R. Walters (1988), *New Phytol.*, *110*, 193–200.

Williams-Ashman, H. G. and A. Schenone (1972), *Biochem. Biophys. Res. Commun.*, *46*, 288–295.

Yamanoha, D. and S. S. Cohen (1985), *Plant Physiol.*, *78*, 784–790.

Ye, X. S., S. A. Avdiushko, and J. Kuc (1994), *Plant Sci.*, *97*, 109–118.

Yordanov, I. and V. Goltsev (1990), *Plant Physiol. (Sofia), 4*, 42–51.

Young, N. D. and A. W. Galston (1983), *Plant Physiol., 71*, 767–771.

Young, N. D. and A. W. Galston (1984), *Plant Physiol., 76*, 331–335.

Zarb, J. and D. R. Walters (1994a), *Lett. Appl. Microbiol., 18*, 5–7.

Zarb, J. and D. R. Walters (1994b), *New Phytol., 126*, 99–104.

Zheleva, D. I., V. S. Alexieva, and E. N. Karanov (1993), *J. Plant Physiol., 141*, 281–285.

Zheleva, D. I., T. Tsonev, I. Sergiev, and E. Karanov (1994), *Plant Growth Regul., 13*, 203–211.

12

Air Pollutants

Bruce N. Smith and C. Mel Lytle

INTRODUCTION

Prehistoric air pollution came from volcanoes and wildfires. Man became even more intimately involved when controlled fire was used for cooking and warmth. For thousands of years man used dung, wood, and peat for fuel. The Chinese were the first to use coal, but eventual use and the resultant pollution became widespread, so much so that King Edward of England decreed death to anyone who burned coal while Parliament was in session. One unfortunate soul was hanged in 1307 for doing so (Kogan, 1970).

London smog persisted as a dangerous nuisance, as shown by these rhymes from a book of collected verse published in about 1663 (Nicolson, 1965):

> *He shewes that 'tis the seacoale smoake*
> *That allways London doth Inviron,*
> *Which doth our Lungs and Spiritts choake*
> *Our hangings spoyle, and rust our Iron.*
> *Lett none att Fumifuge be scoffing*
> *Who heare att Church our Sunday's Coughing.*

The industrial revolution had a great impact on air quality. The effects included dust and soot as well as smaller particles and oxides of sulfur, primarily from burning fossil fuel. "Cleopatra's Needle" refers to one of two Egyptian obelisks. Neither stone monument is in Egypt any more. One rests on the Thames embankment in London, the other in New York City's Central Park. Made in 1460 B.C., the obelisk in New York was given to the city in 1881 by Ismail Pasha, the former khedive of Egypt. In less than a century, New York smog has done what 33 centuries of

Egyptian climate did not do—obscure the hieroglyphics on the obelisk (Kogan, 1970).

Charles Dickens opened his novel *Bleak House* (1852) with this description of nineteenth-century London:

> Smoke lowering down from chimney-pots, making a soft black drizzle, with flakes of soot in it as big as full-grown snow flakes—gone into mourning . . . for the sun. . . . Fog everywhere. . . . Fog in the eyes and throats of ancient Greenwich pensioners wheezing by the firesides.

The dense fog of December 1952 killed more people in London than did the great cholera epidemic of the nineteenth century. Widespread use of internal combustion engines in the twentieth century has created additional problems not only for human health but for the entire biosphere. In many urban areas automobile exhaust accounts for at least 60% of the total air pollution, including much of the carbon dioxide, carbon monoxide, nitrogen oxides, and hydrocarbons (Boughey, 1975).

It is estimated that the burning of fossil fuels has released carbon dioxide into the atmosphere at a rate far above the capacity of plants to absorb it in photosynthesis. The content of this gas in air has risen from 290 to 330 parts per million (ppm) in the past century. Present measurements at the Mauna Loa observatory in Hawaii show that atmospheric carbon dioxide levels continue to increase at a rate of 0.72 ppm per year. If all of the known coal and oil deposits were to be burned, the CO_2 content of the atmosphere would increase an estimated seventeen times.

Of the solar energy striking the surface of the earth, 80% is visible light. Some of the energy absorbed is reemitted as infrared radiation. Carbon dioxide, methane, and water vapor in the atmosphere preferentially absorb infrared radiation and serve as a thermal blanket for the earth. Glass windows allow light to enter but do not lose much heat. Atmospheric carbon dioxide similarly acts to keep the earth warmer than it would otherwise be. The very high surface temperature of the planet Venus is thought to be due to an atmosphere 75 times as dense as that on earth, largely composed of CO_2—a runaway greenhouse effect. A number of people are now convinced that a greenhouse effect due to fossil fuel combustion is already apparent on earth. Clearing and burning of major forest areas in the world, particularly in the tropics, not only adds to atmospheric carbon dioxide but decreases the potential for carbon dioxide fixation in photosynthesis. Atmospheric levels of carbon dioxide have certainly changed during the history of the earth. Concern today is that human activity may negatively affect the self-regulating balances between atmosphere, climate, and the biosphere.

Some people hope that plants themselves may purify the air, not only of carbon dioxide but of a variety of pollutants, through the magic of bioremediation. There is little evidence to support that notion. Today there is worldwide concern for the effects of air pollution on plants. The literature from the last seven decades includes thousands of papers. Air pollutants may affect plant tissues either directly (leaf necrosis) or indirectly (acid rain). Oxides of nitrogen and sulfur combine with water to form strong acids that increase the hydrogen ion concentration in precipitation, which in turn changes soil and water chemistry. Desirable nutrients may be leached from the soil and toxic elements (e.g., aluminum) may be made available. Several pollutants may attack the plant at the same time. Plants exposed to air pollutants may be more susceptible to damage from pathogens and herbivores as well as drought, thermal variations, and so on. Multiple stresses often occur and may result in complex plant behavior. This chapter is a summary of information that may prove useful since air pollution will be with us for many decades in the future.

TYPES AND SOURCES OF AIR POLLUTANTS

Atmosphere

The lower atmosphere (up to a height of about 80 km) is well-mixed and differs in composition little from air at sea level. The approximate composition of unpolluted dry air is given in Table 12-1. Since the mass of the atmosphere is so great, the total mass of even a trace gas can be quite large. While man-made pollution can be distributed throughout the atmosphere, it is usually highly concentrated in the bottom 10 meters of the air column. With temperature inversions and still air, very high concentrations of pollutants may come in contact with plants.

Carbon Monoxide and Carbon Dioxide

While carbon monoxide is toxic to mammals and is produced in large amounts by inefficient combustion, it is not generally considered to be toxic to plants. Both CO and CO_2 can be taken up in photosynthesis and incorporated into plant carbon. Carbon monoxide rapidly disappears from the atmosphere due to photosynthesis. Increases in CO_2 concentration due to human activity may result in global warming. Considerable increases in ambient CO_2 are not harmful to plants (Hodges, 1977).

TABLE 12-1 Composition of Unpolluted Dry Air

Molecule	Formula	Fraction of the Atmosphere
Nitrogen	N_2	78.09%
Oxygen	O_2	20.94%
Argon	Ar	0.93%
Carbon dioxide	CO_2	0.032%
Neon	Ne	18 ppm
Helium	He	5.2 ppm
Methane	CH_4	1.3 ppm
Nitrous oxide	N_2O	0.25 ppm
Carbon monoxide	CO	0.1 ppm
Ozone	O_3	0.02 ppm
Sulfur dioxide	SO_2	0.001 ppm
Nitrogen dioxide	NO_2	0.001 ppm

Source: After Hodges (1977).

Particles

Dust and soot come from incomplete combustion and any process that releases fine particulate matter. In the nineteenth and early twentieth centuries, the most noticeable component of air pollution was smoke. The great reduction in smoke in American cities started in the 1930s and 1940s and has been due to the replacement of coal by fuel oil and natural gas and the control of particulate matter emissions on installations that burn coal (Hodges, 1977). Today over 80% of particulate matter in the air comes from coal-fired electric power generation plants, coking operations of steel mills, and other industrial processes. Particles fall to the earth near the pollution source with the larger particles settling out first. However, once small particles reach high altitudes they may be distributed by the prevailing winds over the globe and must be considered air pollutants. Small particles, including fungal spores, have been shown to have worldwide distribution and are found in ice in both Greenland and Antarctica (Weiss et al., 1971).

Heavy metals such as lead, cadmium, and manganese are usually distributed on particles. Since small particles have diverse chemical composition and a very large surface area, they are very biologically active.

Sulfur Dioxide

Sulfur is released into the atmosphere from volcanoes, anaerobic decay processes (H_2S), and aquatic and terrestrial plants (dimethyl sulfoxide) (Chrominski et al., 1989). A large amount of sulfur oxide is also released from human activity. Ores of copper, zinc, lead, nickel, and iron often form as sulfides in minerals, which may contain as much as 10% sulfur. During smelting SO_2 is produced. Today much of the SO_2 released in smelters is recovered by scrubbers and sold as sulfuric acid. Petroleum and coal also contain appreciable amounts of sulfur that is liberated upon combustion. Oxides of sulfur (and nitrogen) combine with water to form acid rain (Likens and Bormann, 1974), which may precipitate at great distances from the pollution source.

Fluorides

Industrial processes such as phosphate fertilizer production, aluminum reduction, steelmaking, ceramic firing, brick kilns, and so on, release hydrogen fluoride and other fluorides into the atmosphere both as gases and attached to particles. Fluorides are extremely phytotoxic, often showing effects on plants at concentrations less than one part per billion (ppb), and may have great impact close to a pollution source.

Photochemical Smog

In the 1930s and 1940s it became apparent that automobiles were a major source of air pollution. However, direct exposure to automobile exhaust was less toxic to plants than exposure to smog itself. Haagen-Smit (1952) then embarked on research that culminated in a description of the interaction of exhaust from internal combustion engines, oxygen, and sunlight to produce photochemical oxidants. Those of greatest importance were ozone, peroxyacetyl nitrate (PAN), and nitrogen oxides.

IMPACT OF AIR POLLUTANTS ON PLANTS

Particles

Scanning electron micrographs showed stomates of plants growing close to pollution sources clogged with particles. However, since particles carry many toxic substances bound to their surfaces, chemical pollutants are probably more important than the relatively few plants close enough to a pollution source to be impacted in a major way through physical obstruction by the particles themselves.

Manganese and lead from automobile exhaust are found in highest concentrations near roadways. Concentrations have been shown to diminish with distance from the edge of the pavement and to correlate with traffic volume (Lytle et al., 1995). Soil profiles taken 0.25 m from the edge of the pavement were sampled to a depth of 20 cm (Fig. 12-1). Manganese concentrations were an order of magnitude

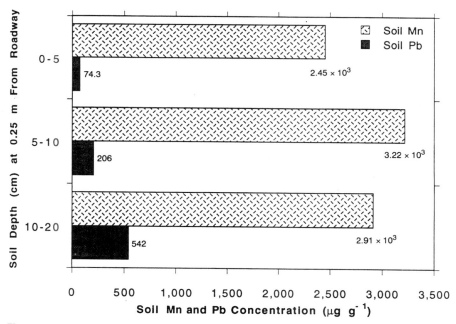

Figure 12-1. *Median roadside soil manganese and lead concentrations (μg g^{-1}) at different soil profile depths. Soil samples were taken at depths of 0–5, 5–10, and 10–20 cm at 0.25 m from moderately high traffic volume roadways. (After Lytle et al., 1995.)*

higher than lead concentrations at all depths. Lead concentrations increased with depth in the soil profile while manganese concentrations did not. This may be because of decreased use of leaded fuel in the United States during the past 20 years. Plant tissue concentrations of lead and manganese are proportional to soil concentrations and decrease with distance from the roadway. Metal concentrations in shoot tissues increase during the growing season and are highest in the fall (Lytle et al., 1995).

Heavy metals such at Cd, Mn, and Pb at high concentrations are toxic to plants (Carlson and Bazzaz, 1977). Usually, however, since they are rarely metabolized or excreted, a trophic channeling effect occurs in which, as plants are consumed by herbivores, the metals accumulate at higher concentrations at each trophic level, often showing toxicity in top carnivores (including man) when plant concentrations are still quite low and the plant tissues show no ill effects (Nordberg 1974). While metal-containing particles are distributed as air pollution (Pakarinen and Tolonen, 1976), much of the uptake into plants is from the soil (Van Hook et al., 1977)—where it can remain in the surface layers for several years (Lytle et al., 1995). Exposure (Table 12-2) to even low levels of heavy metals for a long period of time can result in gradual accumulation of the metal until the tissue concentration is high enough to negatively influence plant growth.

Cd, Mn, and Pb can accumulate to high levels in plant tissue when the ambient concentration is very low. The effect of cadmium on plant growth differs a great deal between species, but ambient concentrations of 0.2 to 9 μg mL^{-1} (with correspondingly higher tissue concentrations) were sufficient to greatly reduce plant growth (Page et al., 1972). Similarly lead in culture solution (10 mg L^{-1}) inhibited growth, photosynthesis, and respiration in a marine diatom (Woolery and Lewin, 1976). Once leaf tissue concentrations of manganese reached 1000 ppm, growth of several species of crop plants was diminished (Horiguchi, 1987). At very low ambient concentrations heavy metals accumulate in plant tissues. Manganese seems to be most available for plant uptake when roots are under low oxygen tension and the manganese is in the 2+ form. At higher oxidation states the manganese is much less available for plant uptake. As the tissue concentration of heavy metals increases, respiration and photosynthesis may be stimulated (Vallee and Ulmer, 1972). At still higher concentrations, metabolism is decreased, growth ceases, and eventually the tissue dies.

TABLE 12-2   Minimum Pollutant Doses for Supression of Photosynthesis

	Concentration (Dry Weight Leaf)			
Pollutant	μg m^{-3}	μg g^{-1} dry wt. leaf	Species	Reference
SO$_2$	262		White fir	Keller (1978)
O$_3$	294		Pines	Barnes (1972)
F		30	Pines	Keller (1977)
Cd		10	Sycamore	Carlson and Bazzaz (1977)
Pb		10	Sycamore	Carlson and Bazzaz (1977)

Sulfur Dioxide

Plants absorb sulfur dioxide (SO_2) mainly by gaseous diffusion through the stomata. Early techniques for monitoring SO_2 concentrations and for experimental fumigation of plants with SO_2 were developed by Thomas (Thomas et al., 1950). It has been suggested that SO_2 injury depends on the rate of uptake (Malhotra and Khan, 1984). Species of plants (Table 12-2) that are most sensitive to SO_2 absorbed greater amounts of gas than did those species that are resistant to it (Furukawa et al., 1980). Much less variation was found within a species. Resistant plants had more rapid closure of stomata and correspondingly high levels of abscisic acid (Kondo et al., 1980). At high concentrations, gaseous SO_2 inside the cell dissolves in water on moist cell surfaces, lowering the pH. Acid conditions may directly weaken walls and membranes. More likely the acid conditions promote free radical formation. The free radicals then peroxidize membrane fatty acids, destroying membrane integrity. For 70 years scientists have looked for structures and enzymes uniquely sensitive to sulfur dioxide poisoning. None have been found. The first symptom of sulfur dioxide damage is membrane destruction. Sulfate uptake from the soil, transport, and subsequent reduction require a great deal of energy at each stage, and thus usually do not result in plant damage (Steffens et al., 1986). In plants with C_4 photosynthesis, the assimilation of sulfur occurs in the bundle-sheath chloroplasts; the mesophyll chloroplasts are inactive (Gerwick et al., 1980). SO_3^{2-} can, in some cases, compete with HCO_{3-} on rubisco and thus inhibit photosynthesis (Zeigler, 1972).

Oxides of Nitrogen

Plants fumigated with low levels of NO_{2-} have been shown to form both nitrate and nitrite, which can then be reduced and incorporated into amino acids and proteins. At higher concentrations, toxicity is evident (Table 12-3). NO_2 is absorbed much more readily than NO since the latter is almost insoluble in water. NO_x may inhibit photosynthesis by competition with CO_2 for photoreductive power. Oxides of nitrogen given as air pollutants also stimulate activity of glutamine synthetase and glutamate synthetase systems. However, since activation of glutamine synthetase and induction of nitrate reductase depend on high concentrations of nitrate and low concentrations of ammonia, competition with photosynthesis would be a very

TABLE 12-3 Threshold Doses and Injury Responses to Photochemical Oxidants

Pollutant	Concentration (μg m^{-3})	Symptoms
NO_2	1,900	Necrosis
PAN	989	Bronzing, chlorosis
O_3	157	Necrosis, stunting

Source: After Smith (1981).

minor consideration. Dark respiration was more severely inhibited by high levels of NO_2 than was photosynthesis (Srivastava et al., 1975). As with sulfur dioxide, oxides of nitrogen probably damage plant tissue by acidification of walls and membranes, but most importantly by the formation of free radicals, which directly damage membranes.

Acid Rain

Oxides of sulfur and nitrogen combine with water vapor in the atmosphere to form acid droplets that can make rain and snow abnormally acidic. The effect of lower pH values in aquatic ecosystems has received a great deal of attention. Soils in northeastern North America and western Europe have become more acidic in recent decades. Acid rain has been blamed in Europe for the recent well-publicized forest decline. Acid soils make aluminum and other potentially toxic metals more soluble and available for plant uptake. Conversely, acid soils may allow leaching of essential and desirable mineral nutrients out of the soil (Likens and Bormann, 1974). Acid soils can also negatively impact soil microorganisms, including symbiotic fungi and bacteria. Soil is well buffered and complex, and the soil-microorganism-plant interfaces less well understood than we pretend. Acid soils may weaken plant defenses and make them more vulnerable to pathogens. Despite a great deal of effort and study, the main impact of acid rain is still not known (Treshow and Anderson, 1989). It may be (Lange et al., 1989) that the interaction of acid soil, gaseous air pollutants entering the stomates, and changes in plant defenses may combine to give the noted decline of forests in western Europe and elsewhere.

Fluorides

Gaseous fluoride (HF) is absorbed through the stomates while fluoride attached to particles falls on plant surfaces and is less well absorbed and consequently less harmful. HF or NaF is toxic at ppb levels while most air pollutants show damage only at the ppm level or greater (Table 12-2). Fluoride shows transient effects on several biochemical and physiological processes. However, no discernable effect could be detected before the appearance of necrosis. Fluoride ions may either act as free radicals or induce free radical formation, resulting in peroxidation of fatty acids in diacylglycerides in the membrane, leading to damage and death.

Peroxyacetyl Nitrate

Peroxyacetyl nitrate (PAN) is a major air pollutant produced from the interaction of automobile exhaust, oxygen, and sunlight (Haagen-Smit et al., 1952). A relatively small molecule, it can be distributed over great distances. Like other gaseous pollutants, PAN enters the leaf through the stomates. Herbaceous plants exposed to PAN typically show bronze, silver, or glaze on the undersurface of the leaf. It is the young, actively expanding tissue that is most sensitive to PAN. Chloroplasts are particularly sensitive to PAN, which may act to oxidize sulfhydryl groups on proteins (Table 12-3). PAN is a strong oxidant and may either act directly or induce free radical formation to oxidize fatty acids in the membranes, resulting in damage and death.

Ozone

Ozone (O_3) has probably received more detailed attention than any other gaseous air pollutant (Table 12-4). Since O_3 is a strong oxidant and thus easy to detect, it is often used as an index for air quality. It is always a gaseous pollutant and can be carried great distances from the point of origin. Ozone is taken up through the stomates and may react indirectly on membranes through the formation of free radicals and subsequent lipid peroxidation (Runeckles and Chevone, 1992). Ozone can either stimulate or inhibit respiration and photosynthesis, depending on concentration and exposure time (Barnes, 1972). Fumigation with O_3 inhibited photosynthesis, though this was reversed about 24 hours after termination of the experiment (Pell and Reddy, 1991). Ozone caused a decrease of chlorophyll in chloroplasts, probably by blocking synthesis. Maize leaves exposed to O_3-enriched air showed specific increase in some thylakoid membrane-associated proteins, while the D_1 protein in the thylakoid of photosystem II was greatly decreased (Pino et al., 1995). This may be a general phenomenon, allowing fluorescence of photosystem II chlorophyll *a* to be a sensitive indicator of environmental stress in plants (Ball et al., 1995; Srivastava et al., 1995). Plants experiencing moisture deprivation are generally less sensitive to ozone than plants receiving ample water, due to stomatal closure (Chevone et al., 1990). Nonetheless most O_3 entering the stomate oxidizes the plasmalemma and causes progressive collapse of cells in the substomatal cavity (Matyssek et al., 1995).

TABLE 12-4 Percentage Yield Loss Based on Seasonal Average 12-Hour Ozone Concentration

Crop	3-Month, 12-Hour Average Ozone Concentrations			Ambient Ozone
	40 ppb	50 ppb	60 ppb	
Lemons	12.7	20.8	22.9	28.3
Dry beans	10.5	16.9	22.7	27.2
Onions	14.2	20.8	22.9	23.2
Grapes	9.4	15.2	19.5	20.8
Cotton	6.6	11.1	15.3	19.6
Oranges	8.9	14.7	18.2	19.3
Rice	6.8	9.2	10.2	10.4
Alfalfa	4.3	7.4	7.5	7.6
Sweet corn	3.8	5.4	6.1	6.1
Tomatoes	0.6	1.6	2.6	4.5
Wheat	0.8	1.4	1.7	1.7

Source: After Treshow and Anderson (1989).

IMPACT OF AIR POLLUTION ON METABOLISM

Stomatal Opening

Particulate pollution may physically clog the stomatal opening, impeding normal gas exchange. Many particles, including those containing heavy metals, settle to the earth and the plant root system absorbs and distributes them. Gaseous air pollutants, however, enter the stomates to have their effect. If conditions are such that stomates are open, air pollutants can be taken up and have impact on plant metabolism. Thomas and Hill (1935) showed that absorption of sulfur dioxide was correlated with humidity. Low concentrations of SO_2 increased stomatal conductance in *Vicia faba*, possibly by damaging epidermal cells adjacent to the stomata (Black and Unsworth, 1980). Some species of plants appear to avoid injury by closing the stomates more rapidly than other species. Furukawa et al. (1980) found highly significant correlations between foliar injury and the amount of SO_2 absorbed. Resistance of Norway spruce *(Picea abies)* seedlings to SO_2 when stomata were open was attributed to cellular mechanisms for detoxifying SO_2 (Oku et al., 1980). Pollution-resistant species rapidly form abscisic acid (ABA) in response to the pollutant, probably due to a decrease in pH on the surface of guard cells. The ABA in turn causes the stomates to close (Kondo et al., 1980). Fumigation with O_3 increased ABA content in *Oryza sativa;* plants resistant to O_3 had higher ABA content than did sensitive ones (Jeong et al., 1980). NO_2 inhibited transpiration in illuminated leaves, apparently through causing partial stomatal closure (Srivastava et al., 1975). Poovaiah and Wiebe (1973) demonstrated stomatal closure in *Glycine max* in response to low levels of HF. Stomatal opening did not recover completely in clean air until the following day.

Biochemistry

Sulfur dioxide rapidly becomes phytotoxic sulfite upon contact with cells. Oxidation to the much less toxic sulfate greatly reduces damage to plant cells (Thomas et al., 1943). In some instances reduction of sulfite to sulfide in the chloroplast occurs. As either sulfate or sulfide, the absorbed sulfur dioxide has a substantial degree of mobility (Jensen and Kozlowski, 1975). In the presence of atmospheric sulfur dioxide, enough can be taken up through the stomates to completely provide the mineral sulfur requirement for growth and development. In fact, plants exposed to injurious concentrations of SO_2 emit considerable amounts of H_2S; a positive correlation between such emissions and SO_2 resistance was shown by Sekiya et al. (1980).

Photosynthesis

Decrease in photosynthetic rate may be simply due to stomatal closure. Ziegler (1972) observed with spinach that sulfite inhibited ribulose bisphosphate carboxylase by competing with CO_2 for the binding site. In the C_4 plant *Zea mays* (Ziegler, 1973) sulfite competed with bicarbonate for the active site on phosphoenolypyruvate carboxylase. Sulfur dioxide has also been shown to inhibit the oxidizing side of photosystem II (Shimazaki and Sugahara, 1980). Ozone fumigation reversibly inhibited photosynthesis (Hill and Littlefield, 1969). Ozone altered photosystem II fluorescence characteristics in beans long before any visible symptoms of injury appeared (Schreiber et al., 1978). Decreased photosynthesis has been demonstrated upon exposure of plants to gaseous NO and NO_2, even at concentrations that do not produce visible injury (Hill and Bennett, 1970). The inhibition may simply be due to competition between nitrite and carbon dioxide for reducing power in chloroplasts. Exposure of leaves to HF caused a reversible decline in photosynthesis (Thomas and Hendricks, 1956). Little is known about the effect of fluoride on enzymes or processes in photosynthesis.

Pigments

Fumigation with low concentrations of sulfur dioxide destroys chlorophyll in lichens (Rao and LeBlanc, 1965). Acids can also give the same result. Chlorophyll in pine needles was not affected by low levels of sulfur dioxide but was destroyed by high levels (Malhotra, 1977). A decrease in chlorophyll was noted in *Phaseolus vulgaris* following exposure to ozone (Beckerson and Hofstra, 1979). NO_2 fumigation caused a significant reduction in the chlorophyll content of various lichen species. It has been suggested that fluoride caused a reduction in chlorophyll synthesis (McNulty and Newman, 1961). It is probable that pigment destruction is due to free radical activity.

Respiration

Sulfur dioxide fumigation inhibited photorespiration in rice through the formation of glyoxylate bisulfite, which inhibits glycolate oxidase (Tanaka et al., 1972). Dark respiration either showed no response to SO_2 or showed a stimulation probably due to uncoupling (Furukawa et al., 1980). Ozone has been reported to stimulate (Todd, 1958) or inhibit (MacDowell, 1965) plant respiration. Dark respiration in primary leaves of *Phaseolus vulgaris* was more severely inhibited by increasing doses of NO_2 than was photosynthesis, and the inhibition was not reversed quickly upon removal of NO_2 from the atmosphere (Srivastava et al., 1975). Depending on the duration of exposure, HF caused either a stimulation or a depression in intact tissue respiration (Miller and Miller, 1974).

Amino Acids and Proteins

A decrease in the total protein content after fumigation with sulfur dioxide has been reported for a number of plants (Godzik and Linskens, 1974). Such a decrease could be attributed to breakdown of existing proteins and reduced de novo synthesis. In general, SO_2 fumigation results in an increase specifically in sulfur-containing amino acids (Ziegler, 1975). Ozone can cause either an increase (Tingey et al., 1973) or a decrease (Ting and Mukerji, 1971) in amino-acid content of plants.

Similarly, the protein content of some plants increased following ozone fumigation while the protein content of other plants decreased. Since plants are nearly always nitrogen deficient, oxides of nitrogen have been shown to stimulate amino acid synthesis (Matsushima, 1972). Nitrate reductase is inducible by NO_2 fumigation, while nitrite reductase is not inducible but utilizes reductive power from the photosystems (Zeevaart, 1976). Following nitrite reduction, ammonia is incorporated into amino acids via the glutamine synthetase and glutamate synthetase system. A marked increase in free amino acids following HF treatment of spruce needles was noted (Jäger and Grill, 1975). It is not known if this increase in amino acids was due to increased breakdown of cellular proteins or to inhibition of protein synthesis, or both.

Fatty Acid and Lipid Metabolism

Sulfur dioxide induced a decrease in chloroplast membrane galactolipids—possibly by increasing sulfolipids (Khan and Malhotra, 1977). Treatment of *Pinus banksiana* needles with SO_2 produced a marked reduction in the content of linolenic acid and an increase in palmitic acid. In tobacco leaves, ozone caused a decrease in all fatty acids, probably by inhibition of fatty acid synthesizing and desaturating enzymes (Tomlinson and Rich, 1969). Direct peroxidation of fatty acids by ozone also occurs. In *Chlorella pyrenoidosa,* nitrite markedly inhibited lipid biosynthesis (Yung and Mudd, 1966). This was also the case for *Pinus banksiana* seedlings. Fluoride caused a decrease in fatty acids in *Sphagnum fimbriatum* (Simola and Koskimies-Soininen, 1980).

Carbohydrates and Organic Acids

Fumigation with sulfur dioxide increased soluble sugars due to breakdown of polysaccharides (Khan and Malhotra, 1977). Organic acids, including shikimic acid, decreased with SO_2 treatment (Sarkar and Malhotra, 1979). In *Pinus ponderosa,* exposure to ozone resulted in an increase in the content of soluble sugars, starch, and phenols in the foliage, and a decrease in their contents in the roots (Tingey et al., 1976). Plants fumigated with HF show changes in concentration of sugars, polysaccharides, and organic acids (Adams and Emerson, 1961). The mechanism is still not understood.

PLANT RESPONSES TO POLLUTION

Wound Respiration

High concentrations, or long exposure to lower concentrations, of most air pollutants will produce irreversible visible damage and eventually necrosis to plant tissues, including a decline in photosynthesis and respiration rates. Lower concentrations and shorter exposure times may have reversible effects such as wound respiration (Vallee and Ulmer, 1972). Wound respiration is a reversible increase in respiration rate, less often in photosynthetic rate, in response to sublethal concentrations of a toxic substance. Higher concentrations of the toxin, of course, result in death.

Metabolic increase in response to pollutants may reflect a mobilization of plant defense mechanisms. The alternative pathway of respiration may act to maintain respiration during environmental stress and as an energy overflow (Weger and Dasgupta, 1993). Part of the metabolic increase may represent activation of mechanisms to repair cellular damage (Shirazi and Fuchigami, 1993). Examples may include synthesis of phenylalanine ammonia lyase, ethylene, or salicyclic acid, as well as activation of phenol oxidase systems and ascorbic acid oxidase.

Stress

Decline of biodiversity in forests of western Europe and North America has been attributed to acid rain and other effects of air pollution. Most forests downwind from population centers have relatively few species compared with ancient forests. In addition the remaining species are not replacing themselves. Careful management is required in order to maintain green belts and areas for recreation as well as production of wood and paper. This is expensive. Study of forests impacted by man (most of those in the world) has shown that trees are stressed and thus slower growing, less productive, and more susceptible to disease organisms. Species most susceptible to stress are those that have disappeared from the forests. Even resistant species showed signs of stress. A sensitive indicator of environmental stress in plants is photosystem II activity, which can be assayed by chlorophyll fluorescence (Ball et al., 1995). Careful monitoring of plant stress may allow protective measures to be taken before permanent damage occurs.

Free Radicals

For a long while, people looked for the first site of damage due to the presence of an air pollutant (Heath, 1980). Usually a sensitive enzyme or substrate was thought to be involved. Despite much effort not one key site of damage for any of the pollutants has been found. In addition, for each pollutant it is known that certain plant species are much more susceptible to damage at low concentrations, while other plants will show necrosis and other damage only at much higher concentrations. Thus pollution-resistant plants may simply have better defense mechanisms.

There is a growing consensus that air pollution, and other forms of environmental stress to which plants may be subjected, result in the formation of free radicals (superoxide, superhydroxide, hydrogen peroxide, and to a degree ozone and oxygen), which can oxidize the fatty acids in membrane lipids, which leads to membrane leakage and eventual death (Tanaka, 1994; Willekens et al., 1994). Mechanisms for free radical formation may differ in different compartments of the cell (Elstner, 1991). Plant lipoxygenases stimulate the synthesis of jasmonic acid, polyamines, and ABA in response to environmental stress (Geerts et al., 1994). Acclimation can help, in some cases, prepare for more intense environmental stress (Badiani et al., 1993). Plants resistant to air pollutants (and free radicals) have better defense mechanisms than do susceptible plants (Bowler et al., 1992). Plants subjected to stress produce elevated levels of ethylene and salicylic acid. Included in the defense mechanisms are enzymes such superoxide dismutase, catalase, peroxidase, ascorbic acid oxidase, polyphenol oxidase, and phenylalanine ammonialyase (Scandalios 1993; Foyer et al., 1994). In addition, free radical scavengers such as glutathione and other reduced compounds may play a role.

Antioxidants may not always be available to some of the sites where they are most needed in times of stress. For example, ascorbate is limited in the apoplastic space during attack by ozone (Luwe et al., 1993). Increased production of toxic oxygen derivatives is considered to be a universal or common feature of stress conditions (Foyer et al., 1994). Plants and other organisms have evolved a wide range of mechanisms to contend with this problem. The antioxidant defense system of the plant comprises a variety of antioxidant molecules and enzymes. Studies of transformed plants expressing increased activities of single enzymes of the antioxidative defense system indicate that it is possible to confer a degree of tolerance to stress by this means (Willekens et al., 1994). However, attempts to increase stress resistance by simply increasing the activity of one of the antioxidant enzymes have not always been successful, presumably because of the need for a balanced interaction of protective enzymes.

SUMMARY

Automobiles and industrial processes, including production of electricity using fossil fuels, have created the worldwide problem of air pollution. Small particles tend to be deposited relatively close to the pollution source, but some are carried by wind all over the world. The particles have bound to their surfaces a variety of metals (especially Cd, NaF, Mn, Pb) and other toxic substances. While stomatal uptake may occur, most particle-borne substances tend to be taken up by the root systems of plants from the soil.

Gaseous pollutants generally enter via stomates so uptake can be controlled by stomatal closure. Coal and oil combustion and the smelting of ore liberates SO_2, which is toxic to many plant tissues at the ppm level. HF and NaF are extremely phytotoxic (ppb level) and produced by several industrial processes.

Automobile exhaust includes CO, CO_2, Pb, and Mn. Exhaust fumes interact with sunlight and oxygen to produce the photochemical pollutants: NO_x, PAN, and O_3. With increased automobile use in recent decades, these pollutants are the chief offenders in many urban areas.

Oxides of sulfur and nitrogen combine with water vapor in the atmosphere to form acid rain, which lowers the pH of lakes, streams, and soil. Low pH may result in leaching from the soil of minerals required for plant growth and simultaneously increase availability and uptake of toxic heavy metals including aluminum.

After 70 years of research on the effects of air pollutants on plants, prelethal and reversible indexes of damage include the increase of metabolism known as wound respiration, which probably indicates synthesis of plant defense mechanisms and increase in variable fluorescence of chlorophyll *a* in photosystem II. The various air pollutants, whether taken up by stomates or by roots from the soil, do not target a key enzyme or metabolite. The first indication of damage is destruction of membranes starting with those first encountered by the pollutants.

Membranes are destroyed by free-radical peroxidation of fatty acids in the diacylglycerides of membranes. Damaged membranes rapidly lead to death of cells. Air pollutants induce the formation of superoxide, superhydroxide, and hydrogen peroxide. PAN and ozone are themselves strong oxidants and may act as free radicals. Plant defenses against free radicals include enzymes: superoxide dismutase, catalase, peroxidases, phenol oxidases, ascorbic acid oxidase, and so on. A number of compounds act as free radical scavengers: glutathione, ascorbic acid, cysteine, polyamines, and so on. Plant species poor in free radical defenses are very susceptible to air pollution damage. Plants with robust defenses against free radicals are pollution-resistant.

REFERENCES

Adams, D. F. and M. T. Emerson (1961), *Plant Physiol., 36,* 261–265.

Badiani, M., A. R. Paolacci, A. D'Annibale, and G. G. Sermanni (1993), *J. Plant Physiol., 142,* 18–24.

Ball, M. C., J. A. Butterworth, J. S. Roden, R. Christian, and J. J. G. Egerton (1995), *Aust. J. Plant Physiol., 22,* 311–319.

Barnes, R. L. (1972), *Environ. Pollut., 3,* 133–138.

Beckerson, D. W. and G. Hofstra (1979), *Can. J. Bot., 57,* 1940–1945.

Black, V. J. and M. H. Unsworth (1980), *J. Exp. Bot., 31,* 667–677.

Boughey, A. S. (1975), *Man and the environment,* 2nd ed., Macmillan, New York, pp. 371–407.

Bowler, C., M. Van Montagu, and D. Inze (1992), *Annu. Rev. Plant Physiol. Plant Mol. Biol., 43,* 83–116.

Carlson, R. W. and F. A. Bazzaz (1977), *Environ. Pollut., 12,* 243–253.

Chevone, B. I., J. R. Seiler, J. Melkonian, and R. G. Amundson (1990), in *Stress responses in plants: Adaptation and acclimation mechanisms* (R. G. Alscher and J. R. Cumming, eds.), Wiley-Liss, New York, pp. 311–328.

Chrominski, A., D. J. Weber, B. N. Smith, and D. F. Hegerhorst (1989), *Naturwissenschaften, 80,* 473–475.

Dickens, C. (1852), *Bleak House,* Chapman and Hall, London, 838 pp.

Elstner, E. F. (1991), in *Active oxygen/Oxidative stress and plant metabolism* (E. J. Pell and K. L. Steffen, Eds.), Am. Soc. Plant Physiol., Rockville, MD, pp. 13–25.

Foyer, C. H., P. Descourvieres, and K. J. Kunert (1994), *Plant, Cell Environ., 17,* 507–523.

Furukawa, A., O. Isoda, H. Iwaki, and T. Totsuka (1980), *Studies on the effects of air pollutants on plants and mechanisms of phytotoxicity: Res. Rep. Natl. Inst. Environ. Stud.*, Japan, *11*, 113–126.

Geerts, A., D. Feltkamp, and S. Rosahl (1994), *Plant Physiol.*, *105*, 269–277.

Gerwick, B. C., S. B. Ku, and C. C. Black (1980), *Science, 209*, 513–515.

Godzik, S. and H. F. Linskens (1974), *Environ. Pollut.*, *7*, 25–38.

Haagen-Smit, A. J. (1952), *Ind. Eng. Chem., 344*, 134.

Haagen-Smit, A. J., E. F. Darley, M. Zaitlin, H. Hull, and W. Noble (1952), *Plant Physiol., 27*, 18–34.

Heath, R. L. (1980), *Annu. Rev. Plant Physiol., 31*, 395–431.

Hill, A. C. and J. H. Bennett (1970), *Atmos. Environ., 4*, 341–348.

Hill, A. C. and N. Littlefield (1969), *Environ. Sci. Technol., 3*, 52–56.

Hodges, L. (1977), *Environmental pollution,* 2nd ed., Holt, Rinehart and Winston, New York, pp. 46–143.

Horiguchi, T. (1987), *Soil Sci. Plant Nutrit., 33*, 595–606.

Jäger, H. J. and D. Grill (1975), *Eur. J. For. Path., 5*, 279–286.

Jensen, K. F. and T. T. Kozlowski (1975), *J. Environ. Qual., 4*, 379–382.

Jeong, Y. H., H. Nakamura, and Y. Ota (1980), *Jap. J. Crop Sci., 49*, 456–460.

Keller, T. (1978), *Photosynthetica, 12*, 316–322.

Khan, A. A. and S. S. Malhotra (1977), *Phytochemistry, 16*, 539–543.

Kogan, B. A. (1970), *Health: Man in a changing environment,* Harcourt, Brace and World, New York, pp. 56–74.

Kondo, N., I. Maruta, and K. Sugihara (1980), *Studies on the effects of air pollutants on plants and mechanisms of phytotoxicity, Res. Rep. Natl. Inst. Environ. Stud.,* Japan, *11*, 127–136.

Lange, O. L., U. Heber, E.-D. Schulze, and H. Zeigler (1989), in *Forest decline and air pollution* (E.-D. Schulze, O. L. Lange, and R. Oren, Eds.), Springer-Verlag, Berlin, pp. 238–273.

Likens, G. E. and F. H. Bormann (1974), *Science, 184*, 1176–1179.

Luwe, M. W. F., U. Takahama, and U. Heber (1993), *Plant Physiol., 101*, 969–976.

Lytle, C. M., B. N. Smith, and C. Z. McKinnon (1995), *Sci. total environ., 162*, 105–109.

MacDowell, F. D. H. (1965), *Can. J. Bot., 43*, 419–427.

Malhotra, S. S. (1977), *New Phytol., 78*, 101–109.

Malhotra, S. S. and A. A. Khan (1984), in *Air pollution and plant life* (M. Treshow, Ed.), Wiley, New York, pp. 113–157.

Matsushima, J. (1972), *Bull. Fac. Agr., Mie University (Japan), 44*, 131–139.

Matyssek, R., P. Reich, R. Oren, and W. E. Winner (1995), in *Ecophysiology of coniferous forests* (W. K. Smith and T. M. Hinckley, Eds.), Academic Press, New York, pp. 255–308.

McNulty, I. B. and D. W. Newman (1961), *Plant Physiol., 36*, 385–388.

Miller, J. E. and G. W. Miller (1974), *Physiol. Plant, 32*, 115–121.

Nicolson, M. H. (1965), *Pepys' diary and the new science,* University of Virginia Press, Charlottesville, VA, p. 149.

Nordberg, G. F. (1974), *Ambio, 3*, 55–66.

Oku, T., K. Shimazaki, and K. Sugahara (1980), *Studies on the effects of air pollutants on plants and mechanisms of phytotoxicity: Res. Rep. Natl. Inst. Environ. Stud.*, Japan, *11*, 151–154.

Page, A. L., F. T. Bingham, and C. Nelson (1972), *J. Environ. Qual.*, *1*, 288–291.

Pakarinen, P. and K. Tolonen (1976), *Ambio*, *5*, 38–40.

Pell, E. J. and G. N. Reddy (1991), in *Active oxygen/Oxidative stress and plant metabolism* (E. J. Pell and K. L. Steffen, Eds.), Am. Soc. of Plant Physiol., Rockville, MD, pp. 67–75.

Pino, M. E., J. B. Mudd, and J. Bailey-Serres (1995), *Plant Physiol.*, *108*, 777–785.

Poovaiah, B. W. and H. H. Wiebe (1973), *Plant Physiol.*, *51*, 396–399.

Rao, D. N. and B. F. LeBlanc (1965), *Bryologist*, *69*, 69–75.

Runeckles, V. C. and B. I. Chevone (1992), in *Surface level ozone exposures and their effects on vegetation* (A. S. Lefohn, Ed.), Lewis, Chelsea, MI, pp. 189–270.

Sarkar, S. K. and S. S. Malhotra (1979), *Biochem. Physiol. Pflanzen*, *174*, 438–445.

Scandalios, J. G. (1993), *Plant Physiol.*, *101*, 7–12.

Schreiber, U., W. Vidaver, V. C. Runeckles, and P. Rosen (1978), *Plant Physiol.*, *61*, 80–84.

Sekiya, J., L. G. Wilson, and P. Filner (1980), *Plant Physiol.*, *Suppl. 65*, Abs. 407.

Shimazaki, K. and K. Sugahara (1980), *Studies on the effects of air pollutants on plants and mechanisms of phytotoxicity: Res. Rep. Natl. Inst. Environ. Stud.*, Japan, *11*, 79–89.

Shirazi, A. M. and L. H. Fuchigami (1993), *Oecologia*, *93*, 429–434.

Simola, L. K. and K. Koskimies-Soininen (1980), *Physiol. Plant*, *50*, 74–77.

Smith, W. H. (1981), *Air pollution and forests*, Springer-Verlag, New York.

Srivastava, A., H. Greppin, and R. J. Strasser (1995), *Plant Cell Physiol.*, *36*, 839–848.

Srivastava, H. S., P. A. Joliffe, and V. C. Runeckles (1975), *Can. J. Bot.*, *53*, 466–474.

Steffens, J. C., D. F. Hunt, and B. G. Williams (1986), *J. Biol. Chem.*, *261*, 13879–13882.

Tanaka, K. (1994), in *Causes of photooxidative stress and amelioration of defense systems in plants* (C. H. Foyer and P. M. Mullineaux, Eds.), CRC Press, Boca Raton, FL, pp. 365–378.

Tanaka, H., T. Takanishi, and M. Yatazawa (1972), *Water, Air, Soil Pollut.*, *1*, 205–211.

Thomas, M. D. and R. H. Hendricks (1956), in *Air pollution handbook* (P. L. Magill, F. R. Holden, and C. Ackley, Eds.), McGraw-Hill, New York, p. 44.

Thomas, M. D. and G. R. Hill (1935), *Plant Physiol.*, *10*, 291–307.

Thomas, M. D., R. H. Hendricks, T. R. Collier, and G. R. Hill (1943), *Plant Physiol.*, *18*, 343–371.

Thomas, M. D., R. H. Hendricks and G. R. Hill (1950), *Ind. Eng. Chem.*, *42*, 2231–2235.

Ting, I. P. and S. K. Mukerji (1971), *Am. J. Bot.*, *58*, 497–504.

Tingey, D. T., R. C. Fites, and C. Wickliff (1973), *Physiol. Plant.*, *29*, 33–38.

Tingey, D. T., R. G. Wilhour, and C. Standley (1976), *Forest Sci.*, *22*, 234–241.

Todd, G. W. (1958), *Plant Physiol.*, *33*, 416–420.

Tomlinson, H. and S. Rich (1969), *Phytopathology*, *59*, 1284–1286.

Treshow, M. and F. K. Anderson (1989), *Plant stress from air pollution*, Wiley, New York, 283 pp.

Vallee, B. L. and D. D. Ulmer (1972), *Annu. Rev. Plant Physiol.*, *41*, 91–128.

Van Hook, R. I., W. F. Harris, and G. S. Hendereson (1977), *Ambio*, *6*, 281–286.

Weger, H. G. and R. Dasgupta (1993), *J. Phycol.*, *29*, 300–308.

Weiss, H. V., M. Koide, and E. D. Goldberg (1971), *Science*, *172*, 261–263.

Willekens, H., W. Van Camp, M. Van Montagu, D. Inze, C. Langebartels, and H. Sandermann, Jr. (1994), *Plant Physiol.*, *106*, 1007–1014.

Woolery, M. L. and R. A. Lewin (1976), *Water, Air Soil Pollut.*, *6*, 25–31.

Yung, K. H. and J. B. Mudd (1966), *Plant Physiol.*, *41*, 506–509.

Zeevaart, A. J. (1976), *Environ. Pollut.*, *11*, 97–108.

Zeigler, I. (1972), *Planta*, *103*, 155–163.

Zeigler, I. (1973), *Phytochemistry*, *12*, 1027–1030.

Zeigler, I. (1975), *Residue Rev.*, *56*, 79–105.

13

Carbon Dioxide

Kazimierz Strzałka and Pieter Ketner

INTRODUCTION

Carbon constitutes only a small proportion of all elements present in the earth's crust. However, due to its unique (special) properties, it is one of the fundamental elements forming living matter (Lehninger, 1982). Carbon skeleton is a basic structural motif of all organic molecules including biomolecules, and in fact all life processes are the reactions occurring between derivatives of carbon compounds. After subtracting water, carbon is the major component of the dry residue of all living organisms. Apart from the biosphere, carbon is also present in atmosphere, mostly as CO_2, in hydrosphere, as carbonate ions, and in soil and rocks as inorganic carbon compounds, for example, limestone or organic carbon deposits in the form of coal, oil, and gas. There exists a constant interchange of carbon between these environments, forming a global cycle (Schlesinger, 1991).

Cycling of carbon is more or less strictly connected with the cycles of oxygen, hydrogen, nitrogen, and others. Parallel to carbon cycling there is a flow of carbon from fossil fuels into the atmosphere and other environments. This flow is unidirectional since present-day photosynthesis cannot restore the deposits of coal, oil, and gas, which were formed over the course of more than 600×10^6 years before present during Carbonaceous period of Paleozoic era. As a matter of fact, the biogeochemical cycle of carbon is not in an equilibrium. Proportions of carbon amounts between different constituents of the cycle are changing: there are short-term oscillations connected with time of the year, geographical latitude, and state of vegetation, as well as gradual changes over longer periods of time—decades and centuries— mainly due to human activity (Trabalka et al. 1985). Finally, there were also fluctuations in global carbon distribution in the past geological eras, to which important contribution was made by the evolution of photosynthesis and photoautotrophs.

From all the constituents of the biogeochemical cycle of carbon, the content and significance of CO_2 in the atmosphere and hydrosphere is the most important subject to study. Level of atmospheric CO_2 is one of the major factors influencing photosynthetic activity of plants and hence their biomass production (Bolin et al., 1979). This has important implications, not only for an increase in food production for growing human population, but also for applying biomass as an alternative energy source as fossil fuels are finite. On the other hand, CO_2 is one of the greenhouse gases, so the difference between the rate of its emission and fixation may have strong impact on global climate changes. Therefore, the effects of CO_2 on plant functioning, on both the molecular and the organismal levels, are discussed. Current models of atmospheric CO_2 changes and their implication for greenhouse effect and possible consequences for climate changes are also dealt with.

ORIGIN OF ATMOSPHERIC CO_2

It is generally accepted that the composition of the atmosphere, which was gradually formed on the primordial earth, used to be different from its present state. The primitive atmosphere was of a rather reducing character (Holland, 1984), with abundant nitrogen and lesser proportions of water vapors. CO_2 and other gases were introduced into the atmosphere by volcanic activity. It is, however, almost certain that the primordial atmosphere did not contain oxygen. Volatile elements, characteristic of the present day atmosphere, were probably brought to the earth's surface by meteorites of the "carbonaceous chondrite" type (Anders, 1989) at the early phases of formation of the globe. These volatile elements were subsequently released by the degassing processes of the earth's crust, due to high temperature in the crust and mantle. These volatile elements thus formed the primordial atmosphere.

Abiotic factors exclusively created the early earth atmosphere, but were supplemented with biotic ones once living organisms had formed and began to evolve. The reducing character of the atmosphere began to change to a more oxidizing one with the advent of the oxygenic type of photosynthesis (Fig. 13-1), which dates back to about 2.9 billion years before the present. Fossil cyanobacteria-like organisms have been discovered in Australian sediments dating from about 3.5 billion years ago (Awramik et al., 1983). Emergence of dioxygen in the atmosphere enabled the evolution of an oxygen-dependent respiratory type of metabolism, and as a consequence, living organisms became important constituents in the cycling of carbon between biosphere, atmosphere, and the earth's crust.

A combination of several direct and indirect analytical methods, such as geological carbon cycle models, analysis of air bubbles trapped in ice cores, determination of ^{13}C isotope, and modern spectroscopic methods, has enabled us to determine variations of atmospheric CO_2 level over the past 100 million years. In spite of the fact that the degree of uncertainty increases with the time distance, all the available data indicate that in the mid-Cretaceous period (10^8 years ago) the atmospheric level of CO_2 was significantly higher than its present concentration and reached a value of several thousands of ppm (Berner et al., 1983; Budyko and Ronov, 1979).

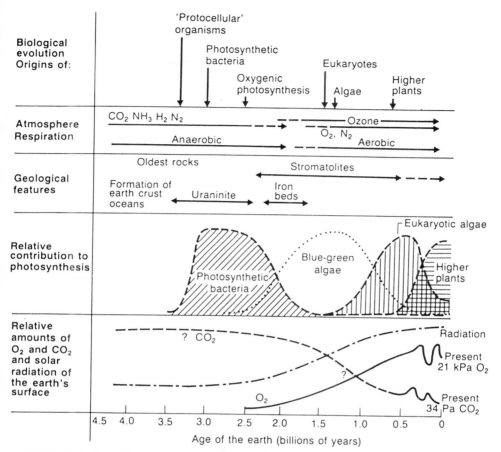

Figure 13-1. *Probable sequence of events in evolution of photosynthesis and their impact on some other global processes. (From Lawlor, 1990; with the permission of Addison Wesley, Longman.)*

Since that time, the level of CO$_2$ in the atmosphere has decreased to reach a value between 200 and 300 ppm, that is, close to contemporary values (Fig. 13-2).

According to Gammon et al. (1985), three different periods can be distinguished in the variations of the CO$_2$ content during the evolution of earth's atmosphere in the past 100 million years:

1. The first period, covering the time 10^8–10^4 years ago, that is, from the mid-Cretaceous period to the end of the last glacial period, characterized by a gradual drop in CO$_2$ concentration from several thousand ppm to a 200–300 ppm level. These variations parallel the glacial and interglacial cycles.

2. The second period, from the last postglacial period to the beginning of the nineteenth century, is characterized by rather constant CO$_2$ concentration in

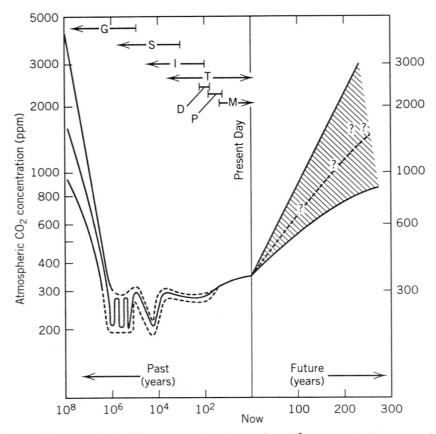

Figure 13-2. *Atmospheric CO_2 concentration changes from 10^8 years ago to the present day (log-log scale) and possible future CO_2 levels in the next few centuries (linear time scale). The presented data are based on analytical methods appropriate to the respective time scales and are indicated by the letters and arrows: G = geological carbon cycle models; S = ocean sediment cores; I = trapped air bubbles in ice cores; T = ^{13}C isotopic studies of tree rings; D = direct chemical measurements of the past century; P = spectroscopic plates from Smithsonian Solar Constant Program; M = Mauna Loa record and subsequent precise atmospheric CO_2 measurements by nondispersive infrared spectroscopy. The hatching and dashed lines indicate the general level of uncertainty about the estimated CO_2 level in each time interval. The ice-age cycling is only representative of the range of CO_2, not the specific number of glaciations. (Adapted from Gammon et al., 1985; with permission.)*

air, estimated to be in the range of 260–290 ppm. Human impact on the atmospheric CO_2 content was rather insignificant during that time.

3. The third period, which started with the beginning of the nineteenth century and lasts until the present. Here one can observe a gradual increase in the atmospheric CO_2 level due to human activity, mainly deforestation, fossil fuel combustion, and land use changes for modern agriculture.

Since the time of precise measurements of CO_2 level in the atmosphere, started by Keeling in 1957 in Hawaii, its content has increased from about 315 ppm to the value of 360 ppm (July 1993) (Boden et al., 1994). It shows further distinct rising tendency with no sign of change in the predictable future. According to various estimations, the present rate of CO_2 increase ranges between 1.5–2.5 ppm a year. Such a high concentration of CO_2 in the earth's atmosphere has not been observed since several million years ago.

Present-day atmospheric CO_2 concentration is the result of its evolution into the atmosphere from the one side, and its removal by different processes from the other side. The atmospheric pool of CO_2, which is estimated to account at present for about 768×10^{15} g of carbon, is sustained by such processes as organic matter decomposition and respiratory metabolism, human activity, and abiotic sources. Until modern times, the respiratory metabolism of heterotrophic organisms and the decomposition of organic matter were and still are the main path of CO_2 flux into the atmosphere, and they were nearly balanced by photosynthesis (Table 13-1). CO_2 input from abiotic processes (earth's crust degassing via volcanic emission) is at present times negligible for the flow of CO_2 into the atmosphere, and is estimated to be $0.01–0.05 \times 10^{15}$ g of carbon per year (Baes et al., 1976), which constitutes only about 0.004–0.02% of the total CO_2 flux into atmosphere. Since the beginning of the industrial era in the middle of the nineteenth century, human activity has begun to significantly contribute to the atmospheric CO_2 enrichment, the process being directly correlated with the increasing combustion of fossil fuels. According to recent estimations, annual release of carbon from fossil fuels into the atmosphere accounts for approximately 7×10^{15} g of carbon.

Flux of the carbon into the atmosphere is not balanced by its uptake due to photosynthesis and oceanic deposition, which results in a net increase of the atmospheric CO_2 concentration of about 1.5 ppm a year. Precise continuous monitoring of atmospheric CO_2 level shows characteristic short-term variations in the global CO_2 concentration (Fig. 13-3). These variations are dependent on the time of the year and are connected with seasonal changes in plant photosynthetic activity, especially in temperate climate zones of the Northern Hemisphere, as well as with different amounts of fuel used for heating purposes throughout the year.

TABLE 13-1   Carbon Flux into the Atmosphere

Source	Amount (10^{15} g C yr^{-1})	%
Respiratory processes and decomposition of organic matter:		
On land	120	51
In ocean	105	44.9
Human Activity:		
Combustion of fossil fuels, cement industry	7	3
Changes in vegetation	2	0.85
Abiotic sources, volcanic activity	0.01–0.05	0.004–0.02
Total	~234	~100

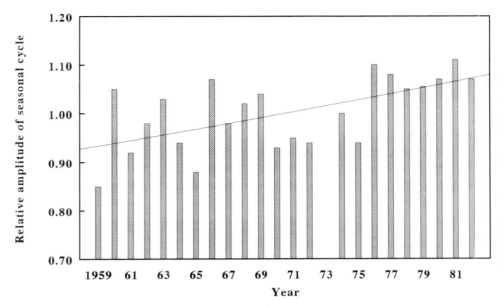

Figure 13-3. *Increasing CO₂ concentration and its oscillations at Mauna Loa, HI. (Redrawn from Bacastow et al., 1985.)*

Carbon dioxide is not the only form of carbon occurring in the atmosphere. Other carbon-containing gases like methane and carbon monoxide are also released into the atmosphere, although at much lower rates. Concentration of methane in atmosphere is estimated to be about 1.7 ppm, which is about 200 times less than the present amount of CO_2 (~ 360 ppm). Globally, the amount of CH_4 in the atmosphere is estimated to be in a range of about 5×10^{15} g and its content is increasing at the rate of about 1%, which is considerably faster than the rise in CO_2 level. Although the concentration of methane in the atmosphere is much lower than that of CO_2, its effect on global warming is significant. Due to its molecular structure, methane is about 20–25 times more effective than CO_2 in absorption of infrared radiation, which gives rise to the greenhouse effect (Lashof and Ahuja, 1990).

There are many sources for methane influx into the atmosphere. In the total flux, which is estimated to account for about 540×10^{12} g of CH_4 a year, natural sources like wetlands and paddy fields dominate (about 50%), while anthropogenic factors are estimated to account for about 30%. CH_4 is relatively unstable and readily reacts with hydroxyl radicals, giving formaldehyde, which is subsequently converted to CO and finally oxidized to CO_2. This reaction constitutes the major sink for atmospheric CH_4 (about 90%). Methanotrophic bacteria use only about 2% of the total flux of CH_4 into the atmosphere.

Analysis of gases trapped in ice cores revealed that the level of atmospheric methane was rather constant during last several thousand years and started to rise at the beginning of nineteenth century. From that time its amount in the atmosphere has doubled (Fig. 13-4).

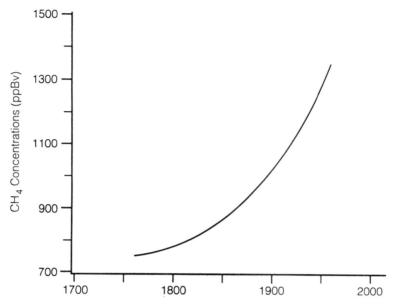

Figure 13-4. *Concentrations of atmospheric CH_4 from ice cores; ppBv = parts per billion volume. (From Walker, 1992; with the permission of Richard Walker.)*

Release rate of carbon monoxide into the atmosphere is comparable with that of methane and accounts for approximately 640×10^{12} g CO yr^{-1}. However, unlike methane, almost half of it results from the combustion of fossil fuels and organic matter (Warneck, 1988); thus its increasing release rate, estimated as 1.5–2% a year (Khalil and Rasmussen, 1988), is directly connected with human activity. CO does not contribute significantly to the global greenhouse effect and it is removed from the atmosphere mainly by reaction with OH radicals.

CARBON DIOXIDE FIXATION

Photosynthesis

Photosynthesis is the most important among the processes removing CO_2 from atmosphere as well as from hydrosphere. The amount of carbon fixed by photosynthesizing organisms on a global scale is estimated to be in a range of 100^{15} g annually, which is about 15 times more than the use of fossil fuels by humans. Thus photosynthesis plays a fundamental role in the global carbon cycle and in maintaining atmospheric level of CO_2, while at the same time coupling the cycling of carbon with cycling of oxygen. CO_2 fixation also occurs in chemosynthetic organisms; however, only the photoautotrophs, which can utilize light energy for producing organic compounds from CO_2 and water, have a real impact on global carbon budget. All forms of life depend directly or indirectly on the energy provided by photosynthesis; therefore the process is of the most fundamental importance for the life on our planet.

There is a mutual interdependence between plant metabolism and CO_2 level. Photosynthesizing organisms play an important role in setting the actual concentration level of CO_2 in the atmosphere and hydrosphere but, on the other hand, availability of CO_2 for plants is one of the most important factors influencing their photosynthetic activity and biomass production. Among the photosynthesizing organisms two different groups can be distinguished: (1) those carrying on the oxygenic type of photosynthesis, and (2) those that do not liberate oxygen and use substrates other than water as a source of electrons for reduction of CO_2. All photosynthesizing eukaryotic organisms perform an oxygenic type of photosynthesis, and among the prokaryota only the blue-green algae (cyanobacteria) produce oxygen. All other photosynthetic bacteria carry on an anoxygenic type of photosynthesis.

Oxygenic photosynthesis in eukaryotic organisms takes place in chloroplasts, which are the only type of plastids capable of performing this process. Plant cells may contain different numbers of chloroplasts, from a single chromatophore (chloroplast), as is frequently found in many algal species, to more than 100, depending on plant species as well as on growth conditions. Typically, a chloroplast is a discoidal body of a diameter of $4-10$ μm, and a thickness of 1 μm. A chloroplast is surrounded by two membranes forming a chloroplast envelope, which houses two distinct compartments: the stroma and a system of membranes called the lamellar system. The lamellar system consists of numerous vesicles called thylakoids, which exhibit a continuity of their internal space and are arranged into stacks of grana. Different grana stacks are connected by membrane bridges called stroma lamellae. Thylakoids contain all chloroplast chlorophyll and electron carriers, and this is where the light energy is converted into chemical energy. This process is described as a light phase of photosynthesis.

Stroma is an aqueous medium containing, among other compounds, soluble proteins and enzymes engaged in CO_2 fixation, which takes place during dark phase of photosynthesis at the expense of ATP and NADPH produced by light phase. Chloroplasts are semiautonomous organelles and they contain 200–300 copies of their own, stroma-localized circular DNA, and protein-synthesizing machinery of procaryotic type. Since the coding capacity of chloroplast DNA is limited to about 120 proteins and ribonucleic acids (Shinozaki et al., 1986), a majority of chloroplast proteins are coded by nuclear genes, translated on cytoplasmic 80 S ribosome-based protein-synthesizing machinery, and transported into these organelles in the form of a precursor.

Light Phase of Photosynthesis   During the light phase of photosynthesis energy of light quanta is captured by photosynthetic apparatus and converted into high energy phosphodiester bonds of ATP as well as the reducing power of NADPH. The essence of the light reactions is the use of energy of photons for transport of electrons from an electron source (water in the case of oxygenic photosynthesis, or other donors like hydrogen sulfide, simple organic compounds, and others, in the case of anoxygenic bacterial photosynthesis) to $NADP^+$. Electrons of NADPH have sufficient electronegative potential to be able to provide reducing power during reduction of CO_2 to the level of sugars.

In higher plants two photosystems (PSI and PSII) cooperate in the transfer of electrons from water to $NADP^+$. Each photosystem is composed of photosynthetic pigments (chlorophylls and carotenoids), proteins, and lipids. Of all pigments only one pair of chlorophyll a, forming a reaction center, is directly engaged in electron transport and charge separation. In PSI such a pair is designed as P-700 and in PSII as P-680. The numbers indicate their absorption maxima in the red region. All other pigments form various antenna complexes transferring the captured light energy to reaction center by resonance energy transfer. PSI and PSII differ in their composition and spectral properties. PSII, which mediates the electron transfer from water to PSI, contains about 250–300 chlorophyll molecules per reaction center. PSI comprises about 200 chlorophyll molecules per reaction center and contains less chlorophyll b than PSII, so it absorbs more long-wavelength radiation. Their schematic organization is depicted in Fig. 13-5. P-680 in its excited state becomes a powerful reductant, able to transfer its electron in a few picoseconds to pheophytin (Barber and Andersson, 1994). This electron moves to other carriers localized down on the redox potential scale, that is, more electropositive (Fig. 13-6). This move-

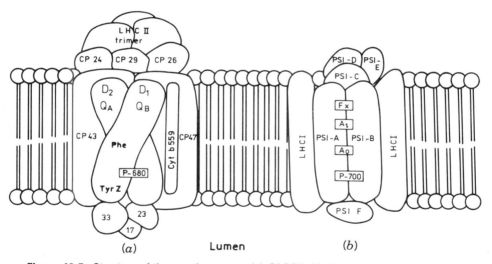

Figure 13-5. Structure of the reaction center. (a) Of PSII. (b) Of PSI. Abbreviations: D_1 and D_2 = polypeptides of the reaction center, binding chlorophyll P-680 and pheophytin; Phe = pheophytin; Q_A and Q_B = plastoquinone binding sites; Tyr Z = electron donor to P-680; Cyt b_{559} = cytochrome b_{559}; CP43 and CP47 = chlorophyll-protein complexes serving as immediate antenna for PSII reaction center; LHCII = chlorophyll a/b-protein complex serving as a main antenna for PSII; CP 24, CP 26, and CP 29 = chlorophyll-protein complexes transferring excitations from LHCII to CP 43 and CP 47; PSI-A and PSI-B = main polypeptides of reaction center of PSI, binding chlorophyll P-700; A_0 = molecule of chlorophyll; A_1 = vitamin K_1; F_x = iron-sulfur center; PSI-C = a polypeptide binding another iron-sulfur center (not indicated); PSI-D and PSI-E = ferredoxin-binding polypeptides; PSI-F = plastocyanin-binding polypeptide; LHCI = chlorophyll-protein complexes serving as antenna to PSI. Some polypeptides of PSI with still unknown function are not shown.

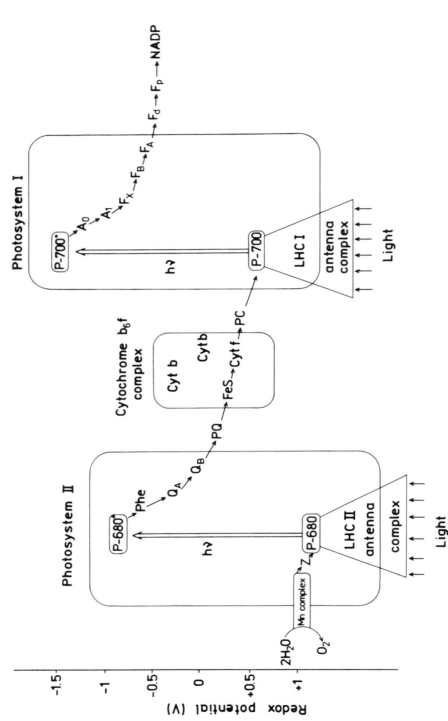

Figure 13-6. Schematic representation of photosynthetic electron transport chain. Abbreviations: P-680 and P-700 = chlorophyll a molecules in the reaction centers of PSII and PSI, respectively (an asterisk indicates an excited state); Q_A and Q_B = bound plastoquinone molecules; PQ = pool of free plastoquinone; FeS = iron-sulfur center; Cyt b and Cyt f = cytochromes b and f; PC = plastocyanin; A_0 = molecule of chlorophyll; A_1 = vitamin K_1; F_x, $F_{B'}$ F_A = iron-sulfur centers; F_d = ferredoxin; F_p = ferredoxin-NADP reductase.

ment is driven by free energy changes and may also occur in darkness. The next electron carrier, Q_A, is a plastoquinone molecule bound to a reaction center polypeptide D_2. Q_A is reduced within several hundred picoseconds to a semiquinone form (Babcock, 1993) and is the first stable electron acceptor on the microsecond time scale. The next plastoquinone molecule occupying the Q_B site on D_1 reaction center polypeptide forms a two electron gate, accepting from Q_A two electrons. After protonation by protons deriving from stromal side, the formed plastoquinol is released into membrane and its place is occupied by another plastoquinone molecule from the membrane pool. The plastoquinol, being the actual mobile electron and proton carrier, is oxidized by cytochrome b_6/f complex, acting as a plastoquinol-plastocyanin oxidoreductase. Only electrons are taken by the complex and protons are liberated on the lumenal side of the thylakoid membrane. Additional protons are transported from stromal side to lumenal side as a result of activity of a Q cycle operating at the level of cytochrome b_6/f complex (Cramer and Knaff, 1991).

Plastocyanin is another mobile electron carrier connecting supramolecular complexes of cytochrome b_6/f and that of PSI. It is a copper-containing peripheral protein localized on the lumenal side of thylakoid membrane, which feeds electron to the oxidized reaction center of PSI—the P-700$^+$. Another light absorption act by PSI is necessary to excite the P-700 and subsequent charge separation. The excited P-700 gains electronegative potential low enough to be able to reduce NADP$^+$. The NADP$^+$, however, is not the immediate electron acceptor from P-700, and there is a series of relatively poorly characterized compounds, several of them iron-sulfur proteins, which mediate the electron transfer between the primary electron acceptor of PSI (a chlorophyll molecule) and NADP$^+$.

The P-680$^+$ cation formed in PSII is reduced by electrons coming from water via the manganese cluster of the water-splitting complex and tyrosine Z (161) of D_1 polypeptide. Four electrons are transferred from the manganese cluster to P-680$^+$, before two molecules of water are split to molecular oxygen, four protons, which accumulate in lumenal space, and four electrons, which reduce the manganese cluster. The half-life of the chemical process of water splitting and oxygen liberation has been estimated by electron paramagnetic resonance (EPR) oximetry (Strzałka et al., 1986) and was found to occur in the microsecond time scale (400–500 μs) (Schulder et al., 1992; Strzałka et al., 1990). The concerted action of simultaneous decomposition of two water molecules is important, since the detachment of single electrons from water would lead to the formation of free radical species, dangerous for the photosynthetic apparatus.

Light-driven electron transport from water to NADP is coupled to a formation of ATP, which, together with NADPH, provides a necessary energy input for photosynthetic CO_2 fixation. Transfer of electrons generates a proton gradient across the thylakoid membrane as the result of: (1) water splitting, (2) proton transport at the level of plastoquinone, including the Q cycle, and (3) the NADPH formation on the stromal side of the membrane. The movement of accumulated protons through the transmembrane complex of ATP synthase provides a driving force for ATP synthesis from ADP and inorganic phosphate, according to the chemiosmotic mechanism proposed by Mitchell (1979).

Transport of a single electron from water to NADP requires two light quanta absorbed by PSI and PSII, which, as mentioned above, differ in their spectral properties. Thus for maximal electron transport to occur both photosystems must be active; hence they must be provided with the light of appropriate spectrum, matching their spectral properties. The proportion of both photosystems in thylakoid membrane and the amount of chlorophyll depend on many factors, among them the plant species and the light conditions. Plants growing in low intensity light conditions have relatively more PSII in relation to PSI than plants growing in a high light intensity. Low light plants also have relatively higher chlorophyll b to a ratio and have bigger antennae (i.e., more chlorophyll molecules per reaction center) than plants grown in full sun.

Dark Phase of Photosynthesis   During the dark phase of photosynthesis CO_2 is converted into organic compounds according to the general formula:

$$CO_2 + H_2O + h\nu \rightarrow CH_2O + O_2$$

where CH_2O represent organic matter formation. Organic matter is, in turn, a substrate for energy-generating respiratory processes, which result in CO_2 release according to the reaction:

$$CH_2O + O_2 \rightarrow CO_2 + H_2O + energy$$

CO_2 fixation occurs in the stroma of chloroplast in a series of reactions known as the Calvin–Benson cycle. The key enzyme of this cycle is ribulose-1,5-bisphosphate carboxylase (rubisco) composed of eight large (56 kDa) subunits coded for by chloroplast genes and eight small (14 kDa) subunits coded by family of genes located in nuclear DNA. Catalytic sites are localized on the large subunits only, but the small subunits exert a regulatory effect on the enzyme activity (Andrews and Lorimer, 1987). It is estimated that rubisco, accounting for about 50% of leaf soluble protein is the most abundant protein in the biosphere. Rubisco catalyzes binding of CO_2 to an acceptor molecule, ribulose 1,5-bisphosphate, forming a transient 6-carbon compound 3-keto-2-carboxy arabinitol 1,5-bisphosphate. This molecule is hydrolyzed immediately when still present at the active site of enzyme, yielding two molecules of 3-phosphoglycerate.

Reaction of carboxylation occurs with a relatively high drop of free energy ($\Delta G^{0'} = -12.4$ kcal mol^{-1}), which makes it irreversible even at low concentration of CO_2. After the phase of carboxylation a phase of reduction follows. Under action of such enzymes as phosphoglycerate kinase and glyceraldehyde 3-phosphate dehydrogenase and at the expense of ATP and NADPH, the 3-phosphoglycerate is phosphorylated to 1,3-diphosphoglycerate and then reduced to glyceraldehyde phosphate, that is, to sugar level (Fig. 13-7). The remaining series of the reactions constitute a regenerative phase in which action of such enzymes as adolases, transketolases, phosphatases, isomerases, epimerase, and kinase leads to regeneration of the acceptor molecule, ribulose 1,5-bisphosphate. After fixing six CO_2 molecules, one molecule of hexose or two molecules of phosphotriose may leave the cycle,

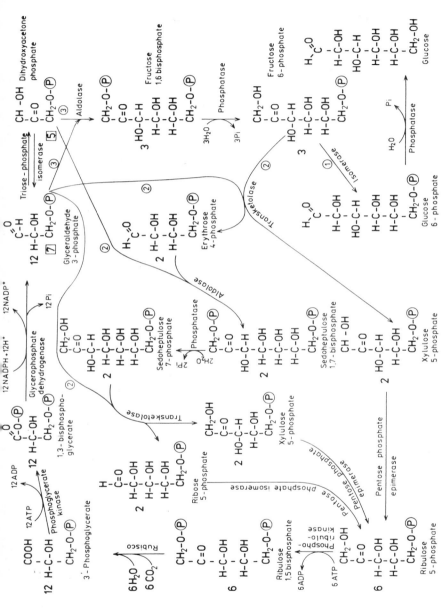

Figure 13-7. Reactions of the Calvin–Benson cycle. Circled numbers are the number of reacting molecules; numbers in squares indicate state of established equilibrium between number of glyceraldehyde 3-phosphate and dihydroxyacetone phosphate molecules. For six CO_2 molecules introduced into the cycle, one molecule of glucose can be withdrawn.

and this constitutes the net gain of reduced carbon compounds from the cycle. For each molecule of CO_2 assimilated, 3 ATP and 2 NADPH molecules are consumed in the cycle, so that fixation of 6 CO_2, which yields 1 molecule of glucose, needs an energy input of 18 ATP and 12 NADPH. Hydrolysis of 18 moles of ATP to ADP brings about change in free energy equal to 18×-7.3 kcal $= -131$ kcal. Similarly, oxidation of 12 molecules of NADPH to $NADP^+$ results in a free energy change of 12×-52.5 kcal $= -630$ kcal, so the total energy used for production of 1 mole of glucose from 6 moles of CO_2 and 6 moles of water accounts for -761 kcal. When comparing this value with the energy released from glucose during its complete oxidation to CO_2 and water (-668 kcal mol^{-1}), it is evident that about 90% of the energy released from ATP and NADPH during the Calvin–Benson cycle is retained in the sugar produced. This efficiency is much lower, however, when compared the amount of energy stored in glucose with the energy input of light quanta necessary to its synthesis.

The minimum number of photons required for fixation of one molecule of CO_2 is eight (four by each photosystem). Therefore assimilation of 6 moles of CO_2 (i.e., formation of 1 mole of glucose) requires energy input of 48 moles of photons. In the case of red light photons the energy content is 48×42 kcal $= 2\ 016$ kcal, from which only about 33% is retained as energy content of glucose molecules. When plants are illuminated with more energy carrying blue light, the energy loss is still higher and photosynthetic efficiency reaches about 20% only.

In comparison with a high quantum yield of photochemical reactions, exceeding the value of 0.9, the quantum yield of the whole process of photosynthesis is relatively low. In optimal conditions it accounts for the value of 0.1 only and in normal conditions for majority of plant species it accounts for about 0.05.

The overall rate of CO_2 assimilation in the Calvin–Benson cycle depends on the activity of the participating enzymes. Several of them are subject to precise multifactor regulation. Activity of rubisco is stimulated by alkalization of stroma and by magnesium influx, which occur during light reaction of photosynthesis. An important step in rubisco activation, both in the case of carboxylation and oxygenation reactions, is binding an activator CO_2 molecule (which is different from the substrate CO_2) to an amino group of lysine residue of the large subunits of the enzyme (Badger and Lorimer, 1976; Lorimer et al., 1976). The carbamylation of rubisco by the activator CO_2 is catalyzed by an enzyme rubisco activase in an ATP-dependent manner (Salvucci et al., 1985; Streusand and Portis, 1987; Werneke et al., 1988).

In many plants light exerts its regulatory effect on the rubisco activity also through an inhibitory effect of 2-carboxy arabinitol 1-phosphate. The inhibitor accumulates and binds to the enzyme in the dark, inhibiting carboxylation (Berry et al., 1987; Gutteridge et al., 1986). The binding is facilitated by a structural similarity of this compound to the first transient intermediate of CO_2 fixation, 3-keto-2-carboxy arabinitol 1,5-bisphosphate. Illumination of plants results in degradation of the inhibitor, which is accompanied by the changes in rubisco activity (Kobza and Seemann, 1988).

Several other enzymes are regulated by the redox state of their sulfhydryl groups, which is also under the control of light. An important role in this control mechanism is played by the ferredoxin/thioredoxin system (Cseke and Buchanan, 1986). Another important factor that has a great impact on the rate of CO_2 fixation is the level of intermediate metabolites of the cycle. In dark-adapted plants the level of these metabolites is rather low; therefore, upon illumination the rate of CO_2 fixation has a low value, increasing gradually within a period of a few minutes. During this time, which is referred to as the induction period, the level of the Calvin–Benson cycle metabolites gradually increases, which results in the acceleration of the whole cycle.

Fixed CO_2 is exported from chloroplasts to cytoplasm, mainly in a form of phosphotrioses, via the phosphate translocator carrier system, which shuttles one phosphotriose from chloroplast to cytoplasm in exchange for one molecule of inorganic phosphate, which is being transported in the reverse direction (Heldt and Flügge, 1987). The phosphate translocator is localized in the chloroplast envelope, but immunologically related proteins have also been detected in the envelope of nonphotosynthetic plastids (Ngernprasirtsiri et al., 1988).

Chemosynthesis

Chemosynthetic organisms are a class of autotrophs whose metabolism is not dependent on the energy of light, and they fix CO_2 using energy generated by oxidation of various inorganic or simple organic compounds. Occurrence of chemosynthesis is limited to microorganisms, and generally two groups of chemosynthesizing organisms can be distinguished: chemolithotrophs and chemoorganotrophs. Chemolithotrophs use inorganic substrates as a source of energy, and they are the sole organisms, apart from organisms carrying out photosynthesis, that can grow in complete absence of organic matter in the environment. Chemoorganotrophs generate energy for driving their metabolic processes by oxidation of simple organic compounds such as methane, methanol, or formate, and only some of them can fix CO_2, while the remaining require simple organic compounds as a source of carbon. Thus chemoorganotrophs do not fulfill the classical definition of autotrophy.

There is a great metabolic diversity among chemolithotrophs (Fig. 13-8), and according to the substrate that is used as an energy source, they are classified into the following main groups: sulfur bacteria, hydrogen-oxidizing bacteria, nitrifying (ammonium and nitrite oxidizing) bacteria, and iron-oxidizing bacteria.

Most chemolithotrophs can grow autotrophically, fixing CO_2 via the Calvin–Benson cycle, although some of them are facultative chemolithotrophs and may use simple organic compounds if they are present in the medium. Energetic efficiency of chemosynthesis is low, and usually only a few percent of energy liberated by substrate oxidation is used for CO_2 fixation. On a global scale, the biomass production via chemosynthesis is rather low, although in certain local habitats, for instance sulfur springs or acid-polluted environments, chemolithotrophs may make a pronounced contribution to organic matter production. On the other hand, metabolic activity of chemoorganotrophs leads to release of high amounts CO_2 to the atmosphere as the result of decomposition of organic matter.

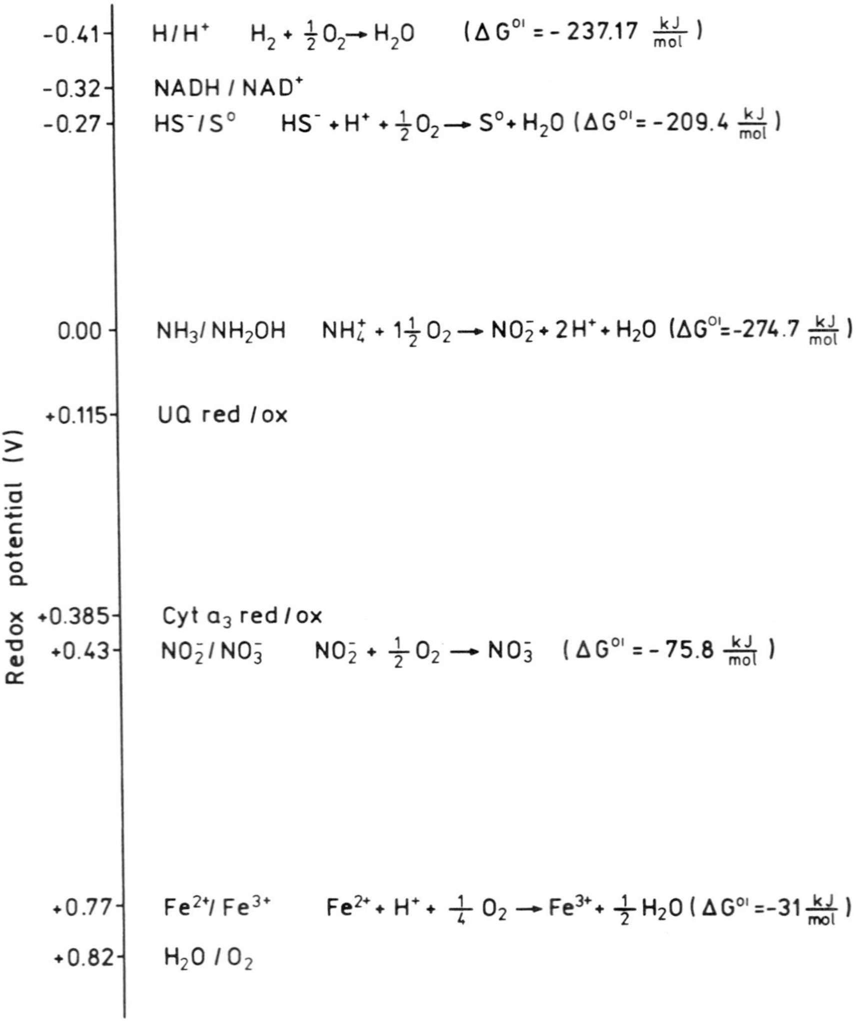

$$-0.41 \quad H/H^+ \quad H_2 + \tfrac{1}{2}O_2 \rightarrow H_2O \quad (\Delta G^{0'} = -237.17 \tfrac{kJ}{mol})$$

$$-0.32 \quad NADH / NAD^+$$
$$-0.27 \quad HS^-/S^0 \quad HS^- + H^+ + \tfrac{1}{2}O_2 \rightarrow S^0 + H_2O \ (\Delta G^{0'} = -209.4 \tfrac{kJ}{mol})$$

$$0.00 \quad NH_3/NH_2OH \quad NH_4^+ + 1\tfrac{1}{2}O_2 \rightarrow NO_2^- + 2H^+ + H_2O \ (\Delta G^{0'} = -274.7 \tfrac{kJ}{mol})$$

$$+0.115 \quad UQ \ red / ox$$

$$+0.385 \quad Cyt \ a_3 \ red / ox$$
$$+0.43 \quad NO_2^-/NO_3^- \quad NO_2^- + \tfrac{1}{2}O_2 \rightarrow NO_3^- \quad (\Delta G^{0'} = -75.8 \tfrac{kJ}{mol})$$

$$+0.77 \quad Fe^{2+}/Fe^{3+} \quad Fe^{2+} + H^+ + \tfrac{1}{4}O_2 \rightarrow Fe^{3+} + \tfrac{1}{2}H_2O \ (\Delta G^{0'} = -31 \tfrac{kJ}{mol})$$

$$+0.82 \quad H_2O / O_2$$

Redox potential (V)

Figure 13-8. Redox potentials of some substrates used by chemosynthetic bacteria and amounts of energy released during their oxidation.

CO$_2$ CONCENTRATION MECHANISMS

C$_4$ Plants

Plant species that assimilate CO$_2$ from the atmosphere directly via the Calvin–Benson cycle are referred to as C$_3$ plants because the first product of CO$_2$ binding is a three carbon compound, 3-phosphoglycerate. About 90–95% of the plants are known to use the C$_3$ photosynthetic pathway. Yet in some other groups of plants, the first product of CO$_2$ fixation is a four carbon compound—oxaloacetate—resulting from the binding of CO_2 to an acceptor molecule phosphoenol pyruvate, by the enzyme phosphoenol pyruvate carboxylase. These plants are called C_4 plants. Many tropical grasses like sugar cane and maize belong to the C_4 category. They have characteristic Kranz-type leaf anatomy, where two functionally different group of cells can be distinguished: mesophyll cells and bundle sheath cells (Fig. 13-9). Phosphoenol pyruvate carboxylase is present in cytosol of mesophyll cells and the product of its activity, oxaloacetate, is subsequently reduced to malate by NADPH-dependent malate dehydrogenase, which is abundant in the mesophyll cells. Malate thus formed is transported via plasmodesmata to bundle sheath cells, where it undergoes decarboxylation by malic enzyme, yielding CO_2 and pyruvate (Fig. 13-10). Liberated CO_2 is fixed again by the typical Calvin–Benson C_3 cycle occurring in the bundle sheath cell chloroplasts. Pyruvate moves back to mesophyll cells where it undergoes a specific reaction catalyzed by pyruvate phosphate dikinase, which phosphorylates at the expense of ATP pyruvate to phosphoenol pyruvate, and at the same time inorganic phosphate is phosphorylated to pyrophosphate, so that two high energy bonds of the ATP molecule are used. Thus the fixation of CO_2 in C_4 plants requires more energy per one assimilated CO_2 molecule than in the case of C_3 plants.

There exist several variations in the C_4 pathway, concerning mainly the form of CO_2 transport from mesophyll cells to bundle sheath cells. In some species oxaloacetate, instead of being reduced to malate, is transaminated to aspartate and the aspartate is subsequently transported to bundle sheath cells. The further fate of aspartate differs depending on the plant species. In some plant species aspartate is converted into oxaloacetate in mitochondria and then it is reduced and decarboxylated by a NAD-dependent malic enzyme. Yet some other species carry the aspartate transamination in cytoplasm and oxaloacetate formed undergoes decarboxylation due to the action of the enzyme phosphoenol pyruvate carboxykinase. Three-carbon compounds resulting from the decarboxylation reactions are converted into alanine and in such form transported back to mesophyll cells where phosphoenol pyruvate is regenerated. Depending on the type of decarboxylation reactions and enzymes engaged, the C_4 plants are classified into three categories. In two of these, NADP-ME and NAD-ME, decarboxylation occurs by the action of NADP- and NAD-dependent malic enzymes, respectively, while in the third, PEP-CK, it occurs by action of phosphoenol pyruvate carboxykinase. Details of these reactions can be found elsewhere (Hatch, 1987).

Division of labor between bundle sheath and mesophyll cells in C_4 plants is reflected by their enzyme composition. Phosphoenol pyruvate carboxylase and pyruvate phosphate dikinase, an unusual chloroplast enzyme, are abundant in the mesophyll cells, whereas decarboxylating enzymes are localized predominantly in the bundle sheath cells. Also rubisco (Huber et al., 1976) and most of the Calvin–Benson cycle enzymes exist only in the chloroplasts of the bundle sheath cells. Accordingly, malate and aspartate are formed in mesophyll cells, whereas 3-phosphoglycerate and starch are produced mainly in the bundle sheath cells. Similarly to the Calvin–Benson cycle, the activity of the key enzymes of C_4 pathway, such as phosphoenol pyruvate carboxylase, malate dehydrogenase, and pyruvate phosphate

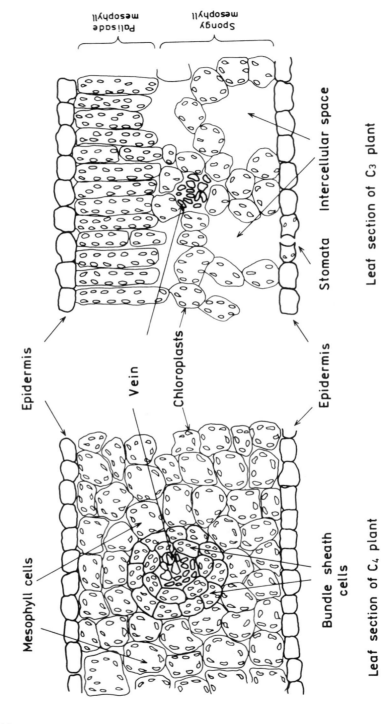

Epidermis

Vein

Chloroplasts

Epidermis

Palisade mesophyll

Spongy mesophyll

Intercellular space

Stomata

Leaf section of C₃ plant

Mesophyll cells

Bundle sheath cells

Leaf section of C₄ plant

Figure 13-9. Comparison of leaf anatomy of C_3 and C_4 plants.

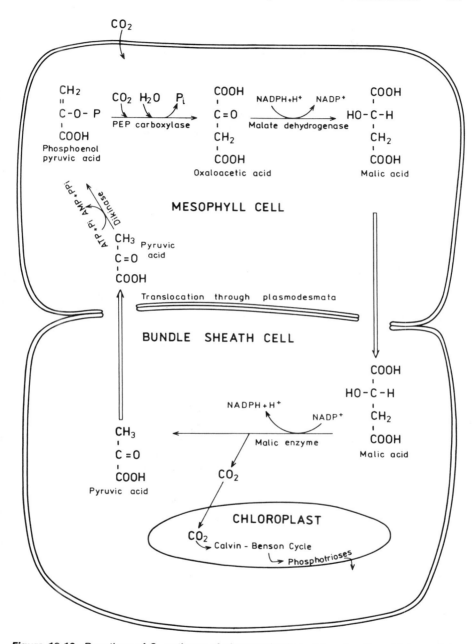

Figure 13-10. Reactions of C_4 pathway of photosynthesis and translocation of intermediate products.

dikinase, is indirectly regulated by light via reversible reduction of specific sulfhydryl groups or phosphorylation of threonine residues of respective enzymes.

The process of fixation of CO_2 into C_4 compounds, called the C_4 pathway or the Kortschak–Hatch–Slack pathway from the names of its discoverers (Hatch and Slack, 1970; Kortschak et al., 1965), represents an adaptation of plants to environmental conditions. C_4 plants usually grow in hot climates, and in order to limit the water loss due to transpiration, they close their stomata during the hottest parts of the day. Avoiding water loss, the plants limit at the same time CO_2 influx into intercellular spaces of leaf, as the stomatal conductance represents the main path for gas exchange processes. When stomata are closed during daytime, minimizing transpiration, inter- and intracellular CO_2 concentration falls considerably. The main C_3 carboxylating enzyme, rubisco, has rather low affinity for CO_2; therefore, it cannot fix it efficiently at its low concentrations. On the other hand, in such conditions a competitive process of oxygenation of 1,5-diphosphoribulose occurs. C_4 plants circumvent this problem by preliminary fixation of CO_2 to oxaloacetate by phosphoenol pyruvate carboxylase. This enzyme has much higher affinity for CO_2 (O'Leary, 1982) and can fix it efficiently even at its low concentration. Also of importance there is activity of carbonic anhydrase in mesophyll cells, which converts CO_2 to HCO_3^-, the latter being a preferable substrate for the carboxylase. Malate undergoing decarboxylation in the bundle sheath cells provides the local concentration of CO_2 high enough to ensure its efficient binding by rubisco. At the same time CO_2 loss due to photorespiration is greatly reduced. It has been estimated (Furbank and Hatch, 1987) that concentration of CO_2 in bundle sheath cells can be up to 100 times higher than in the leaf of C_3 plants, which is close to saturation value for rubisco in bundle sheath chloroplasts.

In ambient air, CO_2 is a mixture of stable carbon isotopes ^{12}C and ^{13}C and the $^{13}CO_2$ constitutes about 1.1% of the total CO_2 present in atmosphere. Rubisco and phosphoenol pyruvate carboxylase discriminate to a different extent the CO_2 molecules carrying the carbon isotopes ^{12}C and ^{13}C. Rubisco appears to be more selective; therefore it fixes relatively less $^{13}CO_2$ than does phosphoenol pyruvate carboxylase. As a consequence, the ratio of $^{13}C/^{12}C$ in dry organic matter of C_3 plants is lower than in C_4 plants. Relative content of ^{13}C in plant organic matter is a convenient measure of the type of photosynthetic processes and is usually described by a parameter $\delta^{13}C$, calculated from the following formula:

$$\delta^{13}C = \frac{^{13}C/^{12}C \text{ (sample)} - {}^{13}C/^{12}C \text{ (standard)}}{^{13}C/^{12}C \text{ (standard)}} \times 1\,000$$

Typically, the C_3 terrestrial plants exhibit values for the $\delta^{13}C$ parameter in the range of -23 to -36, whereas in C_4 plants the respective values are higher and lie in the range of -10 to -18 (Rundel et al., 1988).

It is generally agreed that a relatively low atmospheric CO_2 concentration (below its present level) during the Paleocene and Miocene was the major driving force in evolution of C_4 type of photosynthesis (Ehleringer et al., 1991) and the oldest fossil C_4 plants are known from the Miocene and Pliocene, that is, from 5–7 million years ago (Nambudiri et al. 1978; Thomasson et al., 1986). Development of a CO_2 concentrating mechanism ensures that the CO_2 level inside bundle sheath cells corresponds to the one present in atmosphere during the Cretaceous Era (Berner et al., 1983; Gammon et al., 1985).

The C_4 pathway of photosynthesis is now known to occur in a number of species, belonging to at least 18 different families of both monocotyledonous and dicotyledonous plants. They are present predominantly in families, namely Amaranthaceae, Chenopodiaceae, Cyperaceae, Euphorbiaceae, Poaceae, Portulacaceae (Downton, 1975; Raghavendra and Das, 1978). Out of about 10 000 grass species known worldwide, about 50% are C_4 species (Hattersley, 1987; Hattersley and Watson, 1992). Because of its distribution among many unrelated families, it is generally agreed that the C_4 photosynthesis evolved independently in several plant families (Smith et al., 1979).

C_4 plants require high light intensities and their photosynthetic activity does not get saturated even at full sunlight. Because of their high efficiency of carboxylation at high light intensity, the C_4 plants are more advantageous than the C_3 plants, especially in hot climate with limited water availability. At a temperature of about 30°C, the C_4 plants are about twice as efficient as C_3 plants in organic matter production. They are also more efficient (on average two times) in use of nitrogen as compared with C_3 plants (Brown, 1978; Sage et al. 1987), which makes them capable of growing on nitrogen-poor soils. A greater biomass production per unit of nitrogen in C_4 as compared with C_3 plants is connected with relatively smaller investment of nitrogen in the protein of rubisco.

Another characteristic feature of C_4 plants is that they are more tolerant to saline habitats. One more important factor determining geographical distribution of C_4 plants is temperature (Stowe and Teeri, 1978). Optimal temperature for the majority of C_4 plants lies around 35°C, as compared with about 25°C for C_3 plants. Therefore, C_4 plant species are predominantly present in arid and semiarid environments of tropical and subtropical regions. The number of species declines with increasing latitude (Fig. 13-11) as well as increasing altitude (Fig. 13-12). The flora of the temperate areas consists mostly of C_3 plant species (Black, 1971; Osmond et al., 1982). Only a relatively few C_4 species can grow in temperate and wet environments, of which some grow in saline habitats. The majority of C_4 plants would have a low productivity in such conditions, sometimes even lower than that of C_3 species. One of the reasons for this can be sensitivity of pyruvate phosphate dikinase to low temperature. Below 15–10°C the enzyme becomes inactive, due to its dissociation to a monomeric form (Hull et al., 1990). Another reason for low productivity of C_4 plants at low temperatures can be inefficient photosynthate translocation (Potvin et al., 1985).

A number of species are known to exhibit features intermediate between C_4 and C_3 plants. They are usually characterized by a value of CO_2 compensation point ranging from 5 to 35 ppm, which lies between the ranges typical of C_4 plants (0–10 ppm) and C_3 plants (30–50 ppm). Another characteristic feature of the C_4–C_3 intermediate species is their lower photorespiration rate and lower degree of inhibition of photosynthesis by atmospheric O_2, as compared to typical C_3 species. Fur-

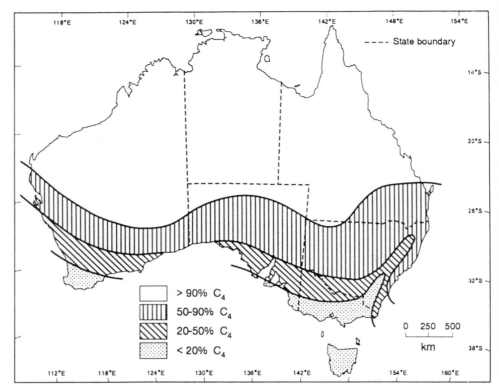

Figure 13-11. *Map of Australia showing 90, 50, and 20% isolines of C_4 plants. Percent C_4 refers to the percentage of native species in a regional grass flora that is C_4. (After Hattersley, 1983; courtesy, Springer-Verlag).*

ther, the Kranz-type leaf anatomy and the degree of compartmentalization of C_4 enzymes are intermediate between those of C_4 and C_3 plants.

CAM Plants

Crassulacean acid metabolism (CAM) plants are characteristic of hot, arid areas like deserts and semideserts. Many of them are succulents with characteristic thick cuticle, large vacuoles, and thin layer of cytoplasm.

Of thousands of species exhibiting CAM syndrome, the majority belong to 26 families of monocotyledonous and dicotyledonous Angiospermae, like Agavaceae, Bromeliaceae, Cactaceae, Orchidaceae and several others, and some of them belong to Pteridophyta (Szarek and Ting, 1977; Griffiths, 1989). Apart from deserts and semideserts, several species of CAM plants grow also on salt marsh areas and epiphytic sites, and about half of epiphytic species from the families Bromeliaceae and Orchidaceae are the CAM plants.

Because of scarce water availability CAM plants developed a specific type of

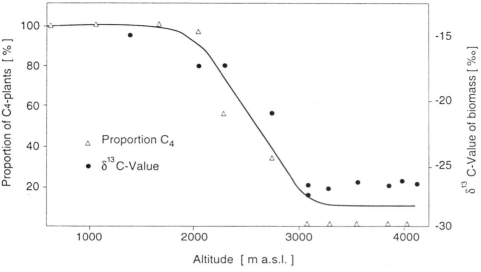

Figure 13-12. *Decrease in C₄ plants in the vegetation of Kenya (as percent of biomass) and increasingly negative isotope ratios ¹³C/¹²C (δ¹³C value) with altitude. (From Tieszen et al., 1979; courtesy, Eugen Ulmer).*

metabolism that permits them to fix CO_2 with a minimal loss of water due to transpiration (Table 13-2). The metabolic strategy that has evolved among CAM plants is directly related to the environmental conditions they experience. During the daytime, when the temperature is high, they keep their stomata closed, to minimize water loss due to transpiration. However, at the same time penetration of atmospheric CO_2 into the assimilatory organs is greatly reduced. At night, when temperature drops significantly and relative humidity increases, which is characteristic for desert and semidesert areas, the CAM plants open their stomata, making possible free influx of CO_2, and in such conditions water loss due to transpiration is low.

Available CO_2 cannot be used for fixation in the Calvin–Benson cycle, because during night time the products of the light phase of photosynthesis (ATP, NADPH) are not made. Therefore, CAM plants preliminarily fix the CO_2 available during the

TABLE 13-2 Cost of Assimilation of One CO_2 Molecule in C_3, C_4, and CAM Plants

Photosynthetic Pathway	Number of Molecules Used		Number of Lost H_2O Molecules
	ATP	NADPH	
C_3	3	2	~500
C_4	4–5	2	~250
CAM	5.5–6.5	2	~50

Source: Anderson and Beardall (1991); Taiz and Zeiger (1991).

night and store it in a form of malate until light and light phase products are available.

At night, similarly to C_4 species, the CAM plants fix CO_2 in the form of HCO_3^- by the enzyme phosphoenol pyruvate carboxylase to an acceptor, which is phosphoenol pyruvate, yielding a C_4 compound, oxaloacetate (Fig. 13-13). The source of phosphoenol pyruvate is starch degradation and the subsequent glycolytic pathway. Glycolysis also provides reducing power (NADH) for conversion of oxaloacetate to malate. Malate is transported to the vacuole, where it accumulates in high amounts (over 0.2–0.3 M), bringing about its acidification, and the pH value may drop down to 4–3, giving the plant an acidic taste. In some plants citrate and isocitrate also accumulate to a considerable extent.

During the day, the stomata are closed and accumulated malate is the source for CO_2 needed for photosynthesis. As depicted in Fig 13-13, malate leaves the vacuole and undergoes the decarboxylation in the cytosol by malic enzyme, which results in liberation of CO_2 and formation of pyruvate. Due to closed stomata CO_2 cannot diffuse out of the leaves and is fixed by rubisco in a typical Calvin–Benson cycle, whereas pyruvate is phosphorylated by pyruvate phosphate dikinase to form phosphoenol pyruvate. Phosphoenol pyruvate enters the gluconeogenesis pathway and finally starch is regenerated.

Thus CO_2 fixation in CAM plants resembles in several aspects CO_2 concentrating mechanism of C_4 plants; first product of the fixation is four-carbon oxaloacetate, subsequently reduced to malate, which serves as a CO_2 source for rubisco. There also exist, however, important differences. In C_4 plants the synthesis of malate and its decarboxylation occur at the same time but in different cells, thus being separated spatially. On the other hand, in CAM plants, both processes take place in the same cell but at different periods of the day, thus being separated temporally.

As a consequence of having both carboxylation and decarboxylation reactions occur in the same cell, a precise control mechanism for the activities of carboxylating and decarboxylating enzymes must have been developed, to avoid simultaneous action of both enzymes, which would result in a futile cycle of fixing of CO_2 only to have it released instantly. The regulatory mechanism involved has not been thoroughly elucidated; however some aspects have already been described (Brulfert et al., 1986). The activity of the key enzyme for C_4 carboxylation, phosphoenol pyruvate carboxylase, is a subject of regulation via phosphorylation and dephosphorylation of an essential serine residue. During the day the enzyme exists in a nonphosphorylated form, which is susceptible to inhibition by malate; thus its carboxylating activity will be blocked. At night the enzyme is phosphorylated and due to this phosphorylation it becomes insensitive to malate inhibition. Carboxylation can then proceed efficiently in spite of accumulation of large amounts of malate.

CAM metabolism is expressed to a various extent in plants belonging to different groups, Similar to the C_3–C_4 intermediate species, there exists also a variety of C_3–CAM intermediate plants. In several families, like Cactaceae, Bromeliaceae and others, there are many species that constitutively express CAM photosynthesis, regardless of environmental conditions, whereas in some other species, especially those belonging to Crassulaceae family and others, plants develop the CAM syn-

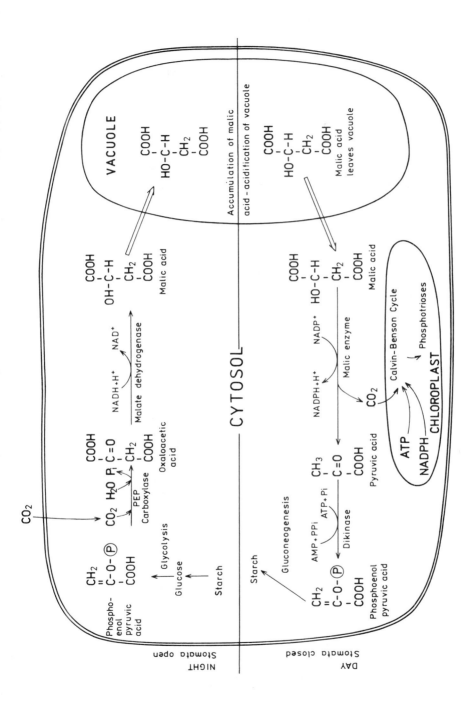

drome as a response to stress situations (drought or salinity), but in normal conditions they carry typical C_3 photosynthesis.

PHOTORESPIRATION

Photorespiration is frequently defined as a process of a light-stimulated CO_2 release by plants. Photorespiration is inevitably connected with the intrinsic property of rubisco which, depending on the local environment, may act as a carboxylase, binding CO_2 to ribulose bisphosphate, which yields two molecules of phosphoglycerate, or may carry out oxygenation of ribulose bisphosphate molecule with molecular oxygen, which results in its splitting to phosphoglycerate and phosphoglycolate (Fig. 13-14). Oxygen and CO_2 compete for the same active sites in the enzyme molecule, and the actual proportion of the both gases at the reaction sites decides which type of reaction prevails.

Rubisco exhibits a higher degree of affinity toward CO_2 than toward O_2 molecules, and in an equimolar mixture of both gases, the carboxylation rate would be about 80 times higher than the rate of oxygenation. At the proportion at which both gases exist in the atmosphere, the oxygenation rate would be considerably higher, and it may account for as much as about one-fourth to one-third of carboxylation rate in C_3 plant species. The dual enzymatic activity is an intrinsic universal property of all types of rubisco, even those present in anaerobic photosynthetic microorganisms. It is assumed that the enzyme structure and function have evolved in past geological eras when the atmosphere was enriched with high CO_2 levels and only traces or no oxygen. Thus an evolutionary pressure to eliminate the oxygenation activity did not exist.

One of the products of the oxygenation of ribulose 1,5-bisphosphate, the phosphoglycolate, is the substrate for photorespiration (Bowes et al., 1971). During photorespiration, in which such cellular organelles as chloroplasts, peroxisomes, and mitochondria participate (Tolbert, 1981), two phosphoglycolate molecules are converted into one molecule of 3-phosphoglycerate in a series of reactions (Fig. 13-15). Oxidation of glycolate to glyoxalate occurs in peroxisomes; glyoxalate is then transaminated to glycine and two molecules of glycine undergo an oxidative decarboxylation in mitochondria, giving rise to serine, NADH, CO_2, and ammonia. This decarboxylation process is central to photorespiratory metabolism (Husic et al. 1987; Ogren, 1984; Tolbert, 1980). During photorespiration intensive recycling of nitrogen takes place, and the intensity of this process may be several times higher than net assimilation of inorganic nitrogen (Wallsgrove et al., 1983). Photorespiratory CO_2 release under moderate to high irradiance and ambient CO_2 may exceed that of "dark" respiration and is accompanied by energy consumption. Due to photorespiration, the energy requirement (ATP + NADPH) for fixation of one mole of CO_2 is increased from 521 kJ to 867 kJ, and the thermodynamic efficiency of this process is reduced from 90% to about 50% (Taiz and Zeiger, 1991). Thus in contrast to carboxylation reaction, the oxygenase activity of rubisco and the process of photorespiration are considered to be wasteful processes resulting in decrease in

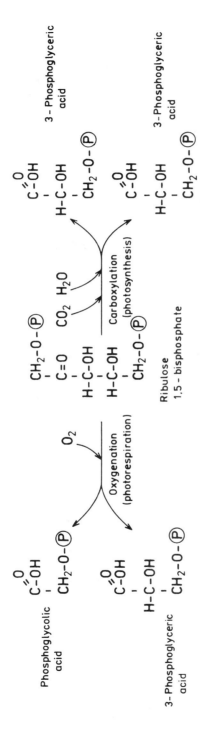

Figure 13-14. Oxygenation and carboxylation activities of rubisco.

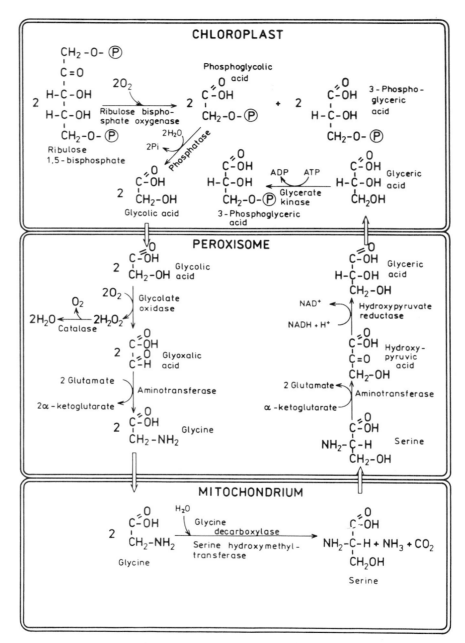

Figure 13-15. Reactions of photorespiration and cooperation of chloroplasts, peroxisomes, and mitochondria in this process.

accumulation of assimilated CO_2, and hence they decrease the photosynthetic efficiency and biomass production.

Photorespiration is strongly reduced or even totally absent in C_4 and CAM species. These plants have evolved a CO_2 concentration mechanism that ensures a high level of CO_2 at the site of rubisco activity. Thus oxygenation of ribulose 1,5-bisphosphate does not occur to significant degree. All eventually released CO_2 is refixed in mesophyll cells and returned to bundle sheath chloroplasts; therefore C_4 plants do not release CO_2 in light. A characteristic anatomical feature of C_4 leaves, the Kranz-type structure, also prevents atmospheric oxygen diffusion into chloroplasts of bundle sheath cells where rubisco is located and photorespiration could eventually occur. The C_4 plants (but not the CAM plants) are thus benefitted in spite of a higher energy cost for CO_2 fixation; they grow faster and exhibit substantially higher biomass production rate than C_3 species, which are devoid of such CO_2 concentration mechanism and are prone to photorespiratory processes.

Photorespiration occurring in C_3 plants presents a serious agricultural problem. In field conditions with C_3 crops, during a hot sunny day with limited wind, the ambient concentration of CO_2 may decrease severalfold, causing the CO_2/O_2 proportion to be changed in favor of O_2. This, in turn, would intensify photorespiratory processes and significantly decrease biomass production. Under conditions of ambient CO_2 concentration, strong irradiation, and the temperature about 25°C, carbon loss due to photorespiration in C_3 plants may exceed even 50% of the assimilated CO_2 (Sharkey, 1988).

The physiological significance of photorespiration is not understood thoroughly and is still a debatable issue. There is apparently no evident benefit for plants from this process, but rather waste of assimilated CO_2 and energy. Photorespiration can, however, be regarded as a process leading to the dissipation of the excess of the energy, which would be harmful to the photosynthetic apparatus in the situation of low CO_2 level in the ambient atmosphere. Moreover it is connected with metabolism of such compounds as lipids and amino acids.

DARK RESPIRATION

The direction of the flow of carbon between intercellular leaf spaces and the ambient atmosphere depends on the relative intensity of two concurrent processes, photosynthesis and respiration. Apart from light-stimulated CO_2 release during photorespiration, which contributes especially in C_3 plants, to the flux of carbon into the atmosphere, there exist in plants as well as in heterotrophic organisms a variety of energy generating processes coupled to CO_2 liberation. These processes, which on diagrams depicting biogeochemical cycle of carbon are usually covered by a general term "respiration," are in fact various reactions with various, very different natures, such as mitochondrial oxygen-dependent respiration (in the case of plants frequently called a dark respiration, to differentiate it from photorespiration), anaerobic respiration (where inorganic compounds like nitrate or sulfate serve as electron acceptors instead of oxygen), and different kinds of fermentation. These processes play a fundamental role in recycling of carbon between the biosphere and the atmosphere.

As substrates these processes use primary and secondary products of photosynthesis, most frequently sugars and fats but in many cases also proteins and other compounds. The sources of liberated CO_2 on the molecular level are various decarboxylation reactions occurring, for instance, in the course of conversion of pyruvate to acetyl CoA, during tricarboxylic acid cycle and pentose phosphate pathway and accompanying ethanolic fermentation.

In plants, apart from typical, cyanide-sensitive respiration, in certain circumstances another type of respiration, insensitive to cyanide and carbon monoxide, occurs. During this process, a large part of the energy released during electron transport is dissipated as heat. Activation of this cyanide-resistant respiration results in enhanced CO_2 production (Lambers, 1982).

Respiratory processes in plants liberate substantial amounts of CO_2, which may account for more than half of its net (less photorespiration) assimilation (Amthor, 1995). There exists also an interrelationship between photosynthesis and respiration, and it has been found that in a mature leaf, a night respiration activity is positively related to the net photosynthesis of the preceeding day (Amthor, 1995; Azcon-Bieto and Osmond, 1983). However, on the global scale, the most important CO_2 flux into the atmosphere is not from plant respiration but results from microbial decomposition of humus and other forms of dead organic materials.

CO_2 EXCHANGE IN PLANTS

Availability of CO_2 to plants is of a fundamental importance for the effectiveness of photosynthetic processes and biomass production. Concentration of CO_2 at the site of carboxylation by rubisco is of special significance. This concentration may vary over a broad range, depending on external factors as well as on the type of photosynthetic processes involved. In plant species that have evolved CO_2 concentration mechanisms (C_4, CAM), the actual level of CO_2 may exceed by severalfold the ambient CO_2 concentration. In plants devoid of CO_2 concentration mechanisms (such as C_3 plants), the actual concentration of CO_2 inside photosynthesizing tissues is lower than the atmospheric one. This is due to the fact that in the process of CO_2 influx into assimilatory organs several barriers have to be overcome. In principle, in the CO_2 uptake two phases can be distinguished; a gas phase diffusion followed by a liquid medium diffusion. The gas phase diffusion consists of several steps, leading to the penetration of atmospheric CO_2 into intercellular spaces. The first barrier CO_2 encounters on its path into the leaf is the thin layer of relatively immobile air facing the leaf surface. This so-called boundary layer resistance is bigger when the leaves are large, especially when they are covered with hairs. The magnitude of this barrier is also dependent on air movement and wind can decrease it significantly.

The cuticle covering the leaf surface is hardly permeable for CO_2 uptake; hence CO_2's main path into the leaf leads through the pores of the stomata, which can be either on the underside of the leaf (hypostomatous leaves), on both surfaces (amphistomatous leaves), or on the upper surface only, as is the case in emerged leaves of aquatic plants. The total area of the pores, in a majority of plants, may account for 0.5–3% of the total leaf surface, depending on plant species. However, it can be much lower in succulent species and scleromorphic leaves (Larcher, 1995). Stomatal diffusion resistance is a major factor controlling the gas exchange processes between the interior of the leaf and the ambient atmosphere, and the stomatal response to various factors has a decisive effect on this process. CO$_2$ that has passed through the stomata enters the substomatal cavity and then equilibrates with the gas phase, filling the intercellular spaces.

The resistance to the diffusion of CO$_2$ within the leaf consists of several other barriers, including intercellular air space resistance and a resistance connected with the diffusion in liquid phase, which encompasses the CO$_2$ penetration through cell wall, plasma membrane, and cytosol into the chloroplast, where it is finally fixed by rubisco in the Calvin–Benson cycle. All these barriers are, however, much lower than the barrier of stomatal resistance, and they account maximally for about 10% of it, when stomata are open. Therefore the degree of stomatal opening has a decisive effect on the influx of CO$_2$ into the leaf. The actual concentration of CO$_2$ in the intercellular spaces depends on the rate of its consumption by photosynthesis and the rate of its uptake via stomata. Under the conditions of a high photosynthesis accompanied by an intensive consumption of CO$_2$ by rubisco, the stroma level of this gas drops significantly, creating a steeper concentration gradient of CO$_2$ between chloroplast and the ambient atmosphere. The steeper the gradient, the larger the influx of CO$_2$ into the leaf.

Assuming that barriers for CO$_2$ diffusion account for less than 10% of the stomatal resistance, a concentration of CO$_2$ inside leaf can be calculated from the following formula:

$$C_i = C_a - \frac{1.6 \times A \times P}{g}$$

where C_i and C_a are CO$_2$ partial pressure in intercellular leaf spaces and ambient atmosphere, respectively; A is intensity of photosynthesis; P is atmospheric pressure; and g is conductance of water vapors through stomata (reciprocal of stomatal resistance); the factor 1.6 is used to compensate for the difference in diffusion coefficients between CO$_2$ and H$_2$O.

Stomata affect not only an influx of atmospheric CO$_2$ into the assimilatory organs during photosynthesis, but they also control the general gas exchange processes in plants, such as respiratory CO$_2$ output during dark respiration, oxygen exchange, and transpiration. Since the CO$_2$ and water vapor share the same path through the stomata, plants must cope with two counteracting goals; to increase the influx of CO$_2$, concentration of which is a limiting factor for photosynthesis in a majority of plant species, and at the same time to decrease water loss due to transpiration.

Degree of stomatal opening depends on several external and internal factors, such as light intensity, availability of water, CO$_2$ concentration, and some plant hormones (Raschke, 1979; Schulze and Hall, 1982). In general light, low intercellular CO_2 concentration and availability of water promote stomata opening, while darkness, a high CO_2 concentration inside the leaf and shortage of water result in closure of stomatal pores. There are, however, exceptions to these rules as well; for example, CAM plants open their stomata at night and close them during daytime, but during a period of severe drought, stomata also remain closed at night, which protects the plant from water loss, and at the same time CO_2 released by respiratory processes is retained within the plant and is reutilized by carboxylating enzymes. Moreover, the action of light on the stomatal opening can be both direct and indirect, through a light-induced sink of CO_2 due to increased photosynthesis (Morison, 1987; Sharkey and Ogawa, 1987). The response of the stomata to changing CO_2 concentration in the air is also light dependent. Stomata of *Vicia faba* responded significantly to a CO_2 level equal to half of its normal value (177 ppm) only in light, but not in darkness (Kappen et al., 1987). There is also an interplay between the influence of CO_2 and humidity on stomatal opening while the response to CO_2 deficiency may overrule the effect of low air humidity (Kappen and Hager, 1991).

The mechanism by which guard cells detect changes in the above-mentioned factors and respond by changing the stomatal aperture is not thoroughly understood yet. Turgor-dependent changes in the shape of the guard cells are the driving forces for variations in stomatal aperture. Upon illumination, an ATP-dependent proton pump extrudes protons from guard cells to the surrounding subsidiary cells of epidermis, which is accompanied by reverse flux of K^+ ions. The concentration of K^+ in guard cells may reach a value of 0.5 M, and this creates a negative osmotic potential causing absorption of water and swelling. Due to a special arrangement of cellulose microfibrilles in the cell wall, the swelling results in a change in the shape of guard cells that is directly related to the degree of stomata opening. In darkness or during water deficit, K^+ level in the guard cells diminishes with concomitant loss of water and turgor pressure drops down. This leads to the closure of stomata. Calmodulin-mediated Ca^{2+} influx, and abscissic acid, which is produced by plants in response to a water stress, have a role in the mechanism of stomatal closure (Schurr, 1992; Zhang and Davies, 1989).

ENVIRONMENTAL FACTORS INFLUENCING CO_2 FIXATION

Various external environmental conditions affect the rate of CO_2 fixation by plants. Among the most important external factors that influence CO_2 assimilation are light, CO_2 level, water and nutrient availability, level of pollutants, and temperature.

Light

Light is a very variable factor; its intensity depends not only upon geographical latitude and the season of the year, but it may also change depending on the time of day, cloud cover, and the spot where the plant grows (see also the Chapter 1). There exist a whole range of environments differing in the intensity of incident irradiance, from habitats exposed almost constantly to full sunlight, to environments like the bottom of certain types of forest, where there is not enough light even for the shade-loving plants to sustain photosynthesis.

Dependence of CO_2 fixation on light intensity is of a complex nature. In darkness, when photosynthesis does not occur, CO_2 release by plants is observed as a result of dark respiration. After illumination of plants with light of increasing intensity, a concurrent process of CO_2 fixation by photosynthesis occurs and its intensity rises in parallel to the intensity of irradiance (Fig. 13-16).

At a certain light intensity the amount of released CO_2 is balanced by CO_2 uptake due to assimilation, so that the net CO_2 flux between atmosphere and leaf is equal to zero. The intensity of irradiance at which this phenomenon occurs is defined as the light compensation point of photosynthesis. At the ambient CO_2 (360 ppm) and temperature of 20°C, the light compensation point in many plant species occurs at the light intensities in the range of 8–15 μmol m^{-2} s^{-1}. For sun-adapted plants these values usually lie between 10 and 20 μmol m^{-2} s^{-1}, whereas for shade plants these values are considerably lower, accounting for 1–5 μmol m^{-2} s^{-1} only. Low light compensation point in shade plants is related to a low rate of respiration characteristic for those plants.

At low and moderate light there is a direct relationship between its intensity and the rate of photosynthesis, and the photon flux density is the only factor limiting CO_2 assimilation. However, for higher light intensities, saturation of the process finally occurs and factors other than light eventually limit photosynthetic performance.

Very high light intensities may be deleterious to the photosynthetic apparatus, leading to its destruction. This phenomenon, termed photoinhibition, occurs when plants absorb more light than they are able to use for photosynthesis. Overreduction of the acceptor side of PSII triggers the photoinhibitory processes, whose detailed

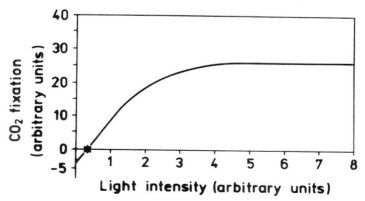

Figure 13-16. Dependence between light intensity and CO_2 assimilation. Asterisk indicates the light compensation point of photosynthesis.

molecular mechanism still has not been elucidated and which is manifested, among other signs, by a rapid increase in turnover rate of D_1 protein of PSII reaction center and damage of the oxygen evolving complex.

Various adaptation mechanisms at different organizational levels have evolved to adjust an optimal performance of the photosynthetic apparatus to actual light conditions. At a molecular level sun and shade plants differ in structure of the photosynthetic apparatus, shade plants having a higher chlorophyll *b* to *a* ratio, a higher PSII to PSI proportion, and higher antenna size than sun plants. Adaptations on a cellular level concern the rearangement of chloroplasts within the cell depending on light intensity. At low light, chloroplasts gather at the cell walls, oriented perpendicularly to the direction of the incident light, and expose their flat surfaces in such a way as to ensure maximum light absorption. At high light intensity, chloroplasts move to the cell walls that are parallel to the direction of incident light, thereby minimizing light absorption (Zurzycki, 1955). Shade plants usually have shorter palisade mesophyll cells and thinner leaves than sun plants.

Another adaptation to light conditions is heliotropism and some plant species perform so-called solar tracking, keeping their leaves oriented perpendicularly to the sun's rays from dawn to dusk.

CO_2 Level

In the absence of CO_2 no photosynthesis can occur and CO_2 is released as an effect of catabolic processes. A rise in ambient CO_2 results in an increase in the intensity of photosynthesis, and at a certain CO_2 concentration its fixation and liberation are balanced, so there is no net flux between leaf and the atmosphere. The concentration of CO_2 at which this balance occurs is referred to as the CO_2 compensation point of photosynthesis (by analogy to the light compensation point of photosynthesis). At concentrations exceeding the CO_2 compensation point, fixation of CO_2 prevails over its liberation, so that the net influx of CO_2 into leaves can be observed. C_3 and C_4 plants differ significantly in their values of CO_2 compensation point (Fig. 13-17). For C_3 plants it is in the range of partial pressure 5–10 Pa (at 25°C) and may be doubled, when the temperature rises by 10°C. C_4 plants have considerably lower CO_2 compensation point, in the range of 0.5–2 Pa, and it approaches 0 when intraleaf CO_2 concentration is concerned. This is because the C_4 plants may effectively fix the CO_2 released during catabolic processes.

In natural conditions atmospheric level of CO_2 is the factor limiting photosynthesis in a majority of plant species. There is a direct relationship between CO_2 level and efficiency of its assimilation over some range of concentrations; therefore CO_2 is frequently used as a fertilizer in greenhouse plant production. There is, however, a limitation in this relationship and at certain concentrations saturation of CO_2 fixation occurs. C_4 plants get saturated at CO_2 concentrations considerably lower (20–30 Pa) than do C_3 plants (about 100 Pa). Rise in CO_2 concentration above saturation level does not lead to the increase of photosynthesis but, on the contrary, may be harmful to plants. Other aspects concerning the importance of ambient CO_2 level

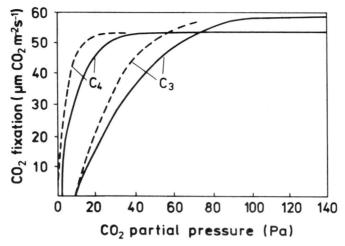

Figure 13-17. *Relationship between concentration of CO₂ and intensity of photosynthesis in C₃ and C₄ plants. Solid lines refer to the ambient CO₂ and dashed lines to actual intracellular level of CO₂. (From Berry and Downton, 1982, with the permission of Academic Press.)*

are discussed in the sections devoted to influence of increased level of CO_2 on photosynthesis as well as in the section on CO_2 exchange in plants.

Water Availability

Water is not an immediate regulatory factor of photosynthetic processes. Of all the water that passes from the soil through the plant to the leaves, only a very small percentage (about 1%) is used in the photosynthesis. However, water availability is directly related to optimal photosynthetic performance of plants. Efficiency of CO_2 assimilation is connected with a general water status of plants, and many species developed various mechanisms and adaptations to limit water loss due to transpiration, the process which, on the other hand, is important for the plants as it leads to the decrease in leaf temperature in hot climate areas.

Plants subjected to water deficit lose their turgor, and basic metabolic processes, including photosynthesis, are impaired. During prolonged drought CO_2 fixation initially slows down and finally it is totally blocked. Tolerance of drought varies among plant species. The majority of plants lose their turgor and decay after losing 30–50% of water, whereas some species (resurrection plants) can survive almost total dehydration.

Nutrient Availability

Nutrients are of vital importance for the processes of photosynthesis and biomass production. For plant growth a whole range of minerals are required, of which nitrogen, phosphorus, and potassium are the most important. Essential elements for chlorophyll biosynthesis are nitrogen and magnesium. When these minerals become limiting, leaves turn yellow and the rate of photosynthesis will be lowered and eventually stop.

Level of Pollutants

Photosynthetic processes are very vulnerable to a wide range of air pollutants, such as sulfur dioxide, fluoride, nitrogen oxides, and peroxyacyl nitrate. Sulfur dioxide results in a breakdown of chlorophyll (chlorosis) and subsequently cell destruction (necrosis). The sensitivity varies per species. The loss of chlorophyll causes a decrease in the rate of photosynthesis. Chlorosis and, at the higher concentration, necrosis is caused also by fluoride, which enters leaves through stomata and accumulates in chloroplasts. A harmful effect on photosynthesis is also exerted by ozone, which is produced by interaction of light and N_xO present in the atmosphere.

Temperature

Photosynthetic processes may occur over a broad range of temperatures, from sub-zero to about 50°C; however in some plants the temperature limit may reach even 65–70°C. Various plant species may adapt to different temperatures; this adaptation involves, among others, changes in the organization of the photosynthetic apparatus. The influence of temperature on the photosynthetic process is of a complex nature and depends also on other factors such as ambient CO_2 concentration, light intensity, and so on.

Two categories of processes occurring during photosynthesis can be distinguished; temperature-independent physical phenomena (as, e.g., energy transfer between photosynthetic pigments) and temperature-dependent biochemical reactions (e.g., the Calvin–Benson cycle reactions). According to van't Hoff's rule, in a certain range of temperatures, increase of temperature by 10°C results in a doubling of the rate of biochemical reaction involved. As far as CO_2 assimilation is concerned, the rise in temperature accelerates the rubisco action, but on the other hand an affinity of CO_2 for the enzyme diminishes, hence photorespiration is greatly enhanced. As a consequence of the increased temperature C_3 plants have an increased CO_2 compensation point and lower quantum yield of photosynthesis.

In temperatures above 40°C, CO_2 assimilation slows down and eventually ceases, as in majority of plant species inactivation of oxygen-evolving complex occurs in the temperature range of 45–50°C, followed by denaturation of chlorophyll protein complexes at about 60–70°C.

At low temperature extremes, a low intrachloroplastic phosphate level, resulting from diminished synthesis of sucrose and starch, frequently becomes one of the main reasons of decreased CO_2 assimilation. Low temperature also decreases fluidity of thylakoid membranes (Gruszecki and Strzałka, 1991), and as a consequence, diffusion of such electron carriers as plastoquinone and plastocyanin is hampered, which lowers the yield of the light phase of photosynthesis.

ASSIMILATION OF CO$_2$ IN AQUATIC PLANTS

Roughly four categories of vascular water plants may be distinguished:

1. Those with emergent foliage
2. The rooted floating leaved species
3. The free-floating macrophytes, which live within and upon the water and obtain most of their CO$_2$ from the atmosphere
4. The submerged aquatic plants, which are exposed to different forms of dissolved inorganic carbon and which have morphological and physiological adaptations for carbon uptake

Aquatic plants occur in freshwater and under saline conditions.

The majority of plants with emerged foliage perform typical C$_3$ photosynthesis, and their gas exchange processes and CO$_2$ compensation points are similar to those of their C$_3$ terrestrial counterparts. Another carbon source for rooted aquatic plants can be sedimentary CO$_2$ directly absorbed by the roots. A few free-floating species and heterophyllous aquatic species also utilize CO$_2$ dissolved in water.

The leaves of submerged vascular plants have extreme reduction or elimination of mesophyll and only a few stomata, since gas diffusion takes place through the epidermal cell walls, which have a very thin cuticle and contain numerous chloroplasts. For many phanerophytes and macroalgae the sparingly soluble nature and slow diffusion rate of CO$_2$ in water (the respective diffusion coefficient of CO$_2$ in water being about 10,000 times lower than in air) will limit its uptake and photosynthesis. The rate at which it is limiting depends upon differences in water chemistry, physical mixing, and rates of CO$_2$ utilization. Concentration of CO$_2$ in water at equilibrium with air accounts for approximately 10 μM; however in many habitats concentration of inorganic carbon is either higher (some marine habitats, alkaline bicarbonate lakes) or lower (waters with high CO$_2$ consumption by phytoplankton). Accordingly there may occur a net efflux or influx of CO$_2$ from these habitats.

Submerged vascular plants need high CO$_2$ concentrations to saturate photosynthesis. In most of them kinetic properties of rubisco are different from those in terrestrial plants, and due to lower affinity the concentration of CO$_2$ at the site of rubisco activity should be high enough to ensure efficient carboxylation. Many species are able to utilize bicarbonate ions as a carbon source when free CO$_2$ is very low and HCO$_3^-$ is abundant (Wetzel and Grace, 1983). Thus in case of limited CO$_2$, an additional carbon source may be used, the availability of which is dependent on the chemical composition of the water. A slower CO$_2$ diffusion in water and the presence of numerous interconnected air chambers in stems and roots may help the refixation of respired and photorespired CO$_2$ and are adaptations to limited CO$_2$ conditions (Sondergaard and Wetzel, 1980).

A majority of aquatic plants have developed complex mechanisms of inorganic carbon acquisition. In a green alga *Udotea* an intracellular CO$_2$ concentration mechanism of C$_4$ type was postulated (Reiskind et al., 1988). There exist also a variety of transport systems allowing accumulation of dissolved inorganic carbon in the cells of cyanobacteria and unicellular algae. Due to such active uptake, the concentration of inorganic carbon within the cell may be over 1000-fold that of the external concentration (Ogawa and Kaplan, 1987). However, the internal CO_2 level may exceed the external concentration up to 17 000 times, depending on the intracellular pH, and on whether CO_2 and HCO_3^- are in equilibrium (Canvin, 1990). At such a high concentration of CO_2 in the cell, oxygenation processes by rubisco do not occur; thus photorespiration is suppressed. On the other hand, carboxylation activity is greatly enhanced. The presence of proteins acting as CO_2/HCO_3^- pumps is dependent upon the external concentration of CO_2; they are not present in an environment with a high level of CO_2 but are induced at low CO_2 concentration (Taiz and Zeiger, 1991)

There are different mechanisms governing the uptake of CO_2 and HCO_3^- in cyanobacteria and unicellular algae. Cyanobacteria do not have an externally localized carbonic anhydrase, a zinc-containing enzyme, that maintains equilibrium between CO_2 and HCO_3^-; however its activity has been demonstrated in the cytoplasm of these organisms (Badger et al., 1985).

It seems that cyanobacteria employ independent mechanisms for uptake of CO_2 and HCO_3^-, although there are also reports indicating conversion of CO_2 to HCO_3^- at the transport site (Canvin, 1990; Reinhold et al., 1987; Volokita et al., 1984). Ability of the cells to take up exogenous inorganic carbon also depends on whether the cells were grown previously in a medium with low or high CO_2 concentration. Low CO_2 cells demonstrate the presence of active transport systems for both CO_2 and HCO_3^-, whereas cyanobacteria grown at a high CO_2 concentrations possess an active transport for CO_2 but their capacity for HCO_3^- uptake is very limited if not absent (Badger and Gallagher, 1987; Miller and Canvin, 1987).

Unicellular algae grown on low CO_2 medium exhibit an active transport of both CO_2 and HCO_3^-. However, in contrast to cyanobacteria, their transport system, in addition to an intracellular carbonic anhydrase, also includes an external carbonic anhydrase. The systems for inorganic carbon transport in unicellular algae may have different location, as for example, plasma membrane or chloroplast envelope (Moroney et al., 1987; Williams and Turpin, 1987), whereas in cyanobacterial cell these systems are located in the plasma membrane. Values for the CO_2 compensation point in submerged angiosperms are very variable in comparison to terrestrial species, and may differ in different seasons.

THE ROLE OF VEGETATION IN CARBON BIOGEOCHEMICAL CYCLE

The carbon reservoirs and fluxes in the biosphere play an important role in the global biogeochemical cycle of carbon. Not only are assimilation and respiration major fluxes between atmosphere and biosphere, but so are the fluxes due to fossil fuel burning, deforestation, and intensive land use.

Since the Industrial Revolution, atmospheric CO_2 has risen from about 290 ppm (around 1860) to about 360 ppm ($=$ca 768 Gt C) in 1993 (Boden et al., 1994), with the fastest increase in the past 30 years. Almost all life on earth depends on the uptake of CO_2 from the atmosphere by green plants, whereby CO_2 is converted into carbohydrates (mainly glucose), the foundation stone for any further biochemical synthesis of new complex compounds. This uptake is called gross primary photosynthesis or gross primary production (GPP). Some of these carbohydrates are used up by the plants for processes of maintenance, active absorption of nutrients, and the production of biomass, whereby CO_2 is released back to the atmosphere. This is called autotrophic respiration (R_A). What is left from the GPP after respiration is the net primary production (NPP) $=$ GPP $-$ R_A.

This is the increase in biomass if no plant material will disappear.

Plant materials form the food base for animals, which can also be a resource for other animals (food chains). Of the total plant material ingested by animals, some parts are not used, that is, rejected, others are used for respiration ($R_H =$ heterotrophic respiration), and a part is converted into new biomass, as growth and reproduction. Dead plant and animal material is finally decomposed through complex decomposition processes by microbial activities, also forming a part of the heterotrophic respiration. Heterotrophic respiration causes the main flux of carbon from the biosphere to the atmosphere. The differences between NPP and total R_H is called the net ecosystem production (NEP) (Fig. 13-18):

$$NEP = NPP - R_H \quad \text{or} \quad NEP = GPP - (R_A + R_H).$$

In climax ecosystems the net primary production and the total heterotrophic respiration are in equilibrium and total biomass remains relatively constant over a longer period of time. There is no absolute increase although fluctuations with the different pools may take place. In young growing ecosystems the NEP is greater than 0, production is higher than total respiration, and carbon is accumulated in one of the pools of living biomass, dead biomass, or soil organic matter. Ecosystems where total respiration is higher than production are degrading systems; total biomass decreases over time.

Next to the natural flux of carbon from the terrestrial ecosystems to the atmosphere, there is an extra flux as an indirect result of human activities such as deforestation, burning of savannas, and the exploitation of peat. Large quantities of CO_2 are released through these activities. Whether the biosphere as a whole is a net source or a net sink for carbon is an intriguing question.

In view of the increasing atmospheric CO_2, the global carbon cycle has attracted much attention over the past decades (Bolin et al., 1979, 1986; Clark, 1982; Olson et al. 1985; Woodwell and Pecan, 1973). Numerous simulation models have been developed to give insight into the mechanism of the carbon cycle, to test sensitivity, and to predict future development (Berner and Lasaga, 1989; Björkström, 1979; Goudriaan and Ketner, 1984; Kohlmaier et al. 1981; Oeschger et al., 1975).

Various authors have made attempts to quantify the carbon reservoirs and fluxes of the biosphere by estimating the surface areas of different types of ecosystems

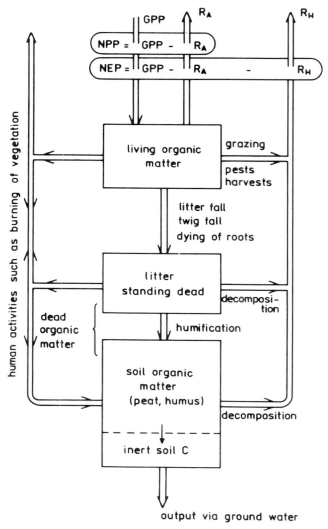

Figure 13-18. Diagram of the most important carbon pools and fluxes in terrestrial ecosystems. On the right the natural fluxes from biota to atmosphere are indicated, on the left the fluxes as a result of human activities. Abbreviations: GPP = gross primary production; R_H = respiration of the plants; R_A = respiration of man, animal, and microorganisms; NPP = net primary production (GPP-R_A); NEP = net ecosystem production [GPP-($R_A + R_H$)]. (From Gezondheidsraad, 1983, with permission.)

and biomass and net primary productivity data from field studies both for land biota and for aquatic environments (Ajtay et al., 1979; Bazilevich et al., 1971; De Vooys, 1979; Moiseev, 1969; Rodin and Bazilevich, 1967; Schlesinger, 1977; Whittaker and Likens, 1973, 1975).

Pools (Reservoirs)

The atmospheric carbon pool is the only one that can be measured accurately; it now contains approximately 768 Gt C (value for 1993, Boden et al., 1994). Almost all the carbon of the biosphere is stored on land, as wood, in leaves, and also in dead organic matter in soils, such as humus, and as inert elemental carbon. Each pool has its own residence time. Living biomass is estimated to be about 600 Gt C, of which only a very small part concerns animal and microorganisms (Ajtay et al., 1979; Gezondheidsraad, 1983, 1987); on earth this is decreasing. Litter is estimated to be 50 Gt C, standing dead wood 30 Gt C, humus 700–1300 Gt C, and inert carbon in the soil amounting to 700 Gt C.

A large pool of carbon, which originates from plant growth, is present in the form of fossil fuels such as coal, oil, and natural gas. The content of this pool is estimated at 6 000–10 000 Gt C. Total living biomass of aquatic systems is much smaller, being about 3 Gt C, mainly present in the superficial waters of oceans (2.2 Gt C).

Dead organic matter in the oceans can be divided into particulate organic carbon (POC) and dissolved organic carbon (DOC) and amounts to 30 (20 for oceans and 10 for shelfs and estuaries) and 700 Gt, respectively, divided over a vertical gradient from the surface downwards. Besides DOC and POC, the oceans contain a huge amount of dissolved inorganic carbon (DIC), in the form of CO_2, HCO_3^-, and CO_3^{2-}, which is estimated at about 39 000 Gt C. Particulate inorganic carbon is very small, namely 1 Gt. Thus the total amount of oceanic carbon is about 40,000 Gt.

Fluxes

Net primary production of all land biota of the world is estimated at close to 60 Gt C (Ajtay et al., 1979). Assuming that NPP equals autotrophic respiration, the gross primary production flux is 120 Gt Cyr^{-1}. However, it should be noted that plant respiration varies per species, season, and age of the plant.

It is far more difficult to assess the carbon fluxes between the different compartments such as litter fall (flux from living to dead material) or total heterotrophic respiration (R_H, respiration of animals, humans, and through decomposition of organic matter by microbial activities), the major flux from biosphere to atmosphere. Both are estimated to be in the order of 40–50 Gt C^{-1}.

Oceans Net primary production of the oceans amounts to 45 Gt C per year, which constitutes about 98% of total aquatic production of the earth. The majority of the produced material is decomposed again rapidly in the upper layer of the oceans, during which process CO_2 is released into the ocean. In addition to this net inorganic carbon flux from atmosphere to the ocean, there is a nonneglectable constant flux of particulate and dissolved organic carbon from the upper layer of the ocean downwards, as a result of vertical mixing and sedimentation, unobstructed by the thermal stratification, escaping recycling in the upper productive layer. On its way downwards this organic material is decomposed, thus increasing the carbon content of ocean water. Only a minor part of it is deposited at the ocean bottom in the form of calcium bicarbonate skeletons and organic matter (Kempe, 1979). The amount of carbon being transferred in this way is difficult to assess and depends on the primary production per site. Estimates of the flux vary from 1 to 15% of total net production.

The main absorber of carbon emissions to the atmosphere is the ocean (about 2 Gt C yr^{-1}. This uptake is based on the fact that in an undisturbed situation there exists a physicochemical equilibrium between the atmosphere and the ocean. With increasing CO_2 pressure of the atmosphere, the ocean absorbs more carbon till a new equilibrium is established. However, uptake by the ocean is limited by: (1) the chemical buffer characteristics, and (2) the slow rate of mixing between the different ocean layers.

Through its chemical composition seawater functions as a buffer solution. Most of the ocean carbon is present in the form of bicarbonate (89%) and locked up beyond chemical equilibrium, the rest as carbonate ions (CO_3^{2-}, 10%) and CO_2 (1%). There exists a chemical equilibrium in the chain CO_2 <↔> HCO_3^- <↔> (CO_3^{2-}). With increasing atmospheric CO_2, more CO_2 dissolves in the ocean. The equilibrium shifts to HCO_3^-; pH will decrease, which in turn will limit CO_2 uptake. The integral buffer factor ($C_{dic}/C_{co_3}^{-2}$) is about 10, and increases further with increasing CO_2 uptake. The net effect is that the relative change in carbon content of the ocean is only one-tenth of the relative change in atmospheric CO_2 partial pressure. The effective pool size is about 4000 Gt, concentrated in the top layers.

The carbon that is absorbed by the ocean is mainly stored in the upper water layers where mixing takes place. This mixed layer is thinner at lower latitudes than at higher latitudes. Exchange between the mixed layer and the deep water layers takes a very long time because of thermal stratification. Only at higher latitudes, near the Antarctic and in the North Atlantic Ocean, where thermal stratification is interrupted, does transport of carbon to deep layers take place by downward water movements (advection), as cooled water sinks down. The downwelling water transports some of the extra atmospheric CO_2. In the tropics there are upward flows in upwelling areas, which when reaching the upper layer, release CO_2, and supply nutrients, leading to an increase of primary production. This means extra uptake of CO_2 and downward flow of particulate organic carbon. Only over a very long period of time (hundreds of years) will the deeper layers be able to absorb more atmospheric CO_2. Potentially, the ocean is able to absorb 85% of CO_2 emitted to the atmosphere but because of the high residence time of the deeper layers, the present uptake is 40% of the ejected CO_2 (Gezondheidsraad, 1983, 1987; Goudriaan, 1987, 1993). It has been calculated that of the total CO_2 emitted in the period 1860–1980 some 60 Gt C has been absorbed by the oceans, of which 12 Gt C was moved downwards by water movements, 40 Gt diffused to a depth of some 450 m, and 8 Gt is found in the mixed layer (MacIntyre, 1980).

Fossil Fuel Burning By fossil fuel combustion considerable amounts of carbon are injected into the atmosphere, enhancing the natural carbon cycle. Global CO_2 emissions from fossil fuel burning, cement production, and gas flaring amounted to

6.2 Gt C in 1991 (Boden et al., 1994). Not all of the injected carbon remains in the atmosphere, but is redistributed over the other carbon reservoirs (e.g., the oceans as described above). The so-called airborne CO_2 fraction (=ratio of the annual atmospheric CO_2 increase to the annual industrial output) is about 58% (Freyer, 1979).

Land Use Changes   Next to the release of CO_2 from fossil fuel burning, there is additional flux caused by human activities like deforestation, burning of grassland vegetation, and other land use changes such as reclamation of peatlands. An estimate of this flux from deforestation is $1-2$ Gt C yr^{-1}, including loss of soil carbon through increased oxidation. Higher estimates are also reported (Wiersum and Ketner, 1989).

Total carbon flux due to burning is about $4-7$ Gt C yr^{-1}, of which a main part is attributed to tropical grassland burning (Hall and Scurlock, 1991). It should be noted that with deforestation only a part of the total phytomass is being burned. The stems are taken away and used elsewhere, for example, as durable wood products, being a reservoir with relatively long turnover time or, as paper, with a short turnover. The remaining phytomass is often insufficiently burned and converted into charcoal, ashes, or hardly burned stems and tree stumps. The latter are gradually decomposed by microbial activities.

Burning efficiency in savannas is much higher, but in fact does not contribute to enhanced CO_2 in the atmosphere, since the phytomass would have decomposed naturally anyway and since regrowth takes place. Burning will lead to a gradual increase of inert soil carbon. As a result of this incomplete burning of phytomass there is an increase of inert carbon in the soil, estimated at $0.5-1.7$ Gt C yr^{-1}. With increasing population pressure there is an increase in area being burned as well as an increased frequency of burning, decreasing total phytomass.

Oxidation of Soil Organic Matter   Enhanced decomposition of soil organic matter takes place in the case of shifting cultivation, in the case of selective or total logging, and in the case of peat burning and reclamation of peat areas. The latter occurs at an increasing rate, both in northern European countries, and in southeast Asia. The released carbon is estimated to be on the order of $0.6-4.6$ Gt C (Hampicke, 1979). As the oceans can only absorb about 40% of the CO_2 emissions resulting from fossil fuel burning, there must be additional sinks to compensate for the second main flux caused by land use changes. If there were no such sinks, atmospheric CO_2 content would now be much higher.

The Biosphere as a Sink for Carbon   There is sufficient experimental evidence that with increased CO_2 levels photosynthesis and net primary production are stimulated, not only under optimal conditions but also in natural vegetation (Lemon, 1983; Strain and Cure 1985, 1994). Each increment in atmospheric CO_2 will give rise to an increment of biomass production. The relative response (biota growth factor, β) varies between 0.7 and 0 (Gezondheidsraad, 1983, 1987; Goudriaan, 1993). For model studies often the value of 0.5 is assumed, meaning that with each percent of relative increment of CO_2 content, growth increases by 0.5 percent.

The total net flux as a result of stimulation of growth can only be assessed by using a dynamic model (Goudriaan 1993; Goudriaan and Ketner, 1984) and lays in the order of 1.1 Gt C yr^{-1}. About 1/14 of the actual rate of increase of agricultural productivity worldwide can be ascribed to the fertilizing effect of rising atmospheric CO_2. However, it is difficult to detect because of constantly changing farming practices and variations in yield due to interannual climatic variations.

Despite decreasing total biomass as a result of deforestation and burning and other human activities, there still exist ecosystems where a net carbon accumulation takes place. To mention some, there are reforested areas, regrowth in formerly exploited temperate forests, secondary forest development (particularly in the tropics), peat formation in nonexploited areas, and finally, managed temperate grasslands, originally derived from moist temperate forests, where remnants of leaves, stolons, roots, and so on, gradually raise the soil organic matter. In all these ecosystems net ecosystem production (NEP) exceeds zero, because a climax state has not yet been reached. Reforestation and regrowth in temperate areas has been estimated at 1.0–1.2 Gt C yr^{-1} (Armentano and Ralston, 1980). Also, growth of secondary forests on partly or totally exploited plots in the tropics sequesters carbon. Total carbon accumulation through processes of regrowth in land biota is estimated to be 1.3–2 Gt C yr^{-1}. It is, however, uncertain whether this regrowth is greater than the decomposition, particularly in the first years after exploitation. Total peat formation may be in the order of 0.01–0.15 Gt C yr^{-1} (Bramryd, 1979;) or even higher (Moore and Bellamy, 1973).

Net Result of the Various Fluxes   The net results of the source and sink functions of the biosphere can only be studied through making use of dynamic simulation models, whereby each of the natural fluxes is separated from the different human-induced fluxes and whereby also for each compartment of an ecosystem different turnover rates are used. In a first attempt Goudriaan and Ketner (1984) found that the balance of natural fluxes amounted to a total net ecosystem production of 7.25 Gt C yr^{-1}, while the net flux as a result of human disturbances (from biosphere to atmosphere) is on the order of 6.91 Gt C. The net increase of the total terrestrial biosphere is thus 0.34 Gt C yr^{-1}, which consists of an increase in soil carbon (humus and charcoal) on the one hand and a net decrease of living phytomass on the other hand.

The same model indicated that about 20% of the total primary production of the oceans moves down toward the ocean bottom, that is, 8 Gt C. This amount is needed in order to maintain the gradient of carbon concentration with depth (Goudriaan, 1987; Goudriaan and Ketner, 1984). This value is much higher than those found in literature (De Vooys, 1979). Later Goudriaan (1990) estimated this flux at 3–4 Gt C. If this downward "bump" were to stop (no plankton growth), there would be an increase of atmospheric CO_2 by 50% within about 200 years. A doubling of plankton growth would lead to a reduction of atmospheric CO_2, to below 200 ppm (Goudriaan, 1987). It can be concluded that the biosphere is one of the major components, the major "driving force" (Goudriaan, 1987), in the global carbon cycle (Fig. 13-19). Without the primary production processes in land biota and oceans,

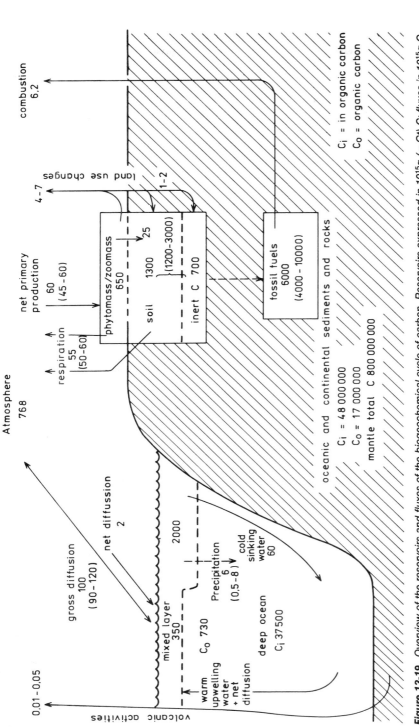

Figure 13-19. Overview of the reservoirs and fluxes of the biogeochemical cycle of carbon. Reservoirs expressed in 10^{15}g (=Gt) C; fluxes in 10^{15}g C yr^{-1}; the values in brackets give range of uncertainty. The atmospheric carbon pool was calculated assuming its actual concentration of about 360 ppm. (Figure adapted from Gezondheidsraad, 1983; data from Berner and Lasaga, 1989; Bolin et al., 1979; Goudriaan, 1993; Goudriaan and Ketner, 1984; with the permission of Gezondheidsraad.)

437

the carbon content of the atmosphere would be much larger and would cycle much more slowly.

INFLUENCE OF INCREASED CO_2 ON PHOTOSYNTHESIS

Primary physiological effects of increased CO_2 levels include changes in molecular structure of the photosynthetic apparatus, increased photosynthesis in C_3 plants, reduced photorespiration, changes in dark respiration, and reduced stomatal conductance (Gifford, 1988; Gifford et al., 1985). Secondary effects include decreased transpiration, increased water use efficiency, alteration in composition, concentration, and translocation of photosynthate; increased tolerance of air pollution; and increase in the carbon/nitrogen ratio of plant tissue (Strain, 1987).

Numerous studies with field and laboratory experiments, as well as application of simulation models, have been carried out on the effects of elevated atmospheric CO_2 on photosynthesis, plant growth, and ecosystems processes, also in combination with climate change.

At the molecular level several phenomena associated with increased CO_2 concentration can be observed. Tobacco plants grown with a CO_2 content in the atmosphere of 700 ppm show numerous changes in the composition of the photosynthetic apparatus and its activity as compared with plants grown at ambient (350 ppm) CO_2. Among others they exhibit quantitative changes in composition of PSI (decrease in core peptides with concomitant increase of the light-harvesting complex of PSI) and changes in the proportion between chlorophylls, carotenoids, and proteins, as well as other structural modifications. PSI-mediated electron transport activity increased by 18% (He et al. 1995; Makewicz et al., 1995).

Exposure of tomato plants to a high CO_2 level brought about changes of abundance of several transcripts of nuclear genes coding for chloroplast proteins. Some of them were decreased as *cab-7, cab-3,* and *Rca,* coding for the chlorophyll *a*/*b* binding proteins of the light-harvesting complexes I and II and rubisco activase, respectively, whereas others were either not affected (*psa A-B, psb A,* encoding the proteins of the core complex of PSI and D_1 protein of PSII) or increased (transcript for the B subunit of ADP-glucose phosphorylase), thus mimicking an effect of glucose or sucrose supplied to the leaf tissue (Van Oosten et al., 1994). In addition, a deactivation of rubisco is several plant species was reported as an effect of both short-term and long-term responses to high concentrations of CO_2 (authors Makino, 1994; Sage et al., 1989). There are also data on decrease in activity of other CO_2-fixing enzymes with concomitant increase in enzyme activities linked to the respiratory process as a result of a long-term exposure to an elevated CO_2 level (Van Oosten et al., 1992). Short-term perturbations, as well as long-term growth under elevated CO_2 concentrations, were found to result in an enhanced ethylene response from leaf tissue (Woodrow and Grodzinski, 1993).

At the whole-plant level, it has been reported that enhanced CO_2 may increase net production, cause changes in leaf area and leaf structure, modify growth forms, and alter developmental phenology, reproduction, and senescence (Bazzaz et al.,

1985; Garbutt et al. 1990; Kramer, 1981; Oechel and Strain, 1985; Strain, 1987; Strain and Armentano, 1980). An interesting observation is a response of stomatal density to CO_2 level. As found in herbarium specimens of leaves of eight temperate arboreal species collected over the last 200 years, the stomata density decreased by 40%, which coincides well with the rise in atmospheric CO_2 concentration during this period. This finding was confirmed by an experiment in which CO_2 concentration was increased from 280 ppm to 340 ppm, resulting in the decline of stomatal density by 67% (Woodward, 1987).

Table 13-3 gives an enumeration of the various changes in molecular, physiological, and morphological as well as phenological characteristics of plants as a result of elevated CO_2.

TABLE 13-3 Impact of Enhanced Atmospheric CO_2 Concentration on Molecular, Physiological, and Morphological, as Well as Phenological, Characteristics of Plants

Molecular

Changes in the abundance of transcripts of nuclear genes coding for chloroplast proteins
Changes in proportion between chlorophylls, carotenoids, and proteins
Quantitative changes in PSI peptide composition
Deactivation of rubisco
Decrease in specific activity of photorespiratory enzymes
Enhanced ethylene production

Physiological

Increase in: photosynthesis, dark CO_2 fixation, CO_2 fixation in CAM plants, dark respiration, and growth rate (dry weight)
Decrease in: photorespiration, transpiration, stomatal conductance
Change in: photosynthate composition; photosynthate allocation; carbon partitioning; photosynthetic pathway ($C_3 \rightarrow$ CAM)

Morphological

Root/shoot ratio increase
Lateral branches increase in number
Leaf area increase
Flower size decrease
Fruit size increase
Increase root length
Longer leafstalks
Cytological changes

Phenological

Leaf senescence delayed
Flowers produced earlier
Crop maturity accelerated
Number of seeds per plant increase
Fruit number increase
Induction of germination

Source: Makewicz et al., 1995; Makino, 1994; Strain and Armentano, 1980; Strain and Cure, 1994; Van Oosten et al., 1994; Woodrow and Grodzinski, 1993.

It is now generally accepted that increased concentrations of atmospheric CO_2 will increase photosynthesis as well as net primary production in most C_3 plants, for which current CO_2 levels are suboptimum. These effects, however, will be less, if they occur at all, in C_4 plants. This is in agreement with the practice of CO_2 enrichment in greenhouses, which has already been widely practiced for a long period to raise crop production. The utilization of other growth factors may also be enhanced, such as water and nutrients.

Increase in production is usually greater under experimental conditions than under field conditions and not so strong if nutrient supply is insufficient (Goudriaan and De Ruiter, 1983). However, great differences in growth response between species of the same photosynthetic pathways have been found, between herbaceous and woody species as well as between annuals and perennials (Patterson and Flint, 1990). Significant CO_2- species interactions have been found for leaf area, leaf biomass, biomass of reproductive parts, and seed biomass, indicating species-specific responses for these characters (Garbutt et al. 1990; Strain and Cure, 1994). However, no generalizations can be made as there is a large interspecific variation in the individual response of plant species to increased CO_2 concentration, as was shown in a review by Poorter (1993).

Much experimental work on the impact of increased atmospheric CO_2 has been carried out with plants of single species under controlled environments, as have some studies of whole-crop responses over the entire life cycle. Under optimal conditions (optimum water and nutrient supply) primary production of certain crops may increase by up to 50%, experimentally shown by studies in open top chambers, enhancing water use efficiency (WUE), but total seasonal water use did not change significantly because of a higher (transpiring) biomass. This increased WUE can be explained by the fact that a higher ambient CO_2 pressure will lead to an increased rate of CO_2 diffusion through the stomata, even when the stomata are partly closed if water loss exceeds the uptake from the roots. The ability to conserve water may benefit both C_3 and C_4 species, if water and nutrients are not in optimal supply, as is the case in natural communities.

There is a strong interaction with nutrient and water shortage. The positive effect of elevated CO_2, observed under optimal conditions, is maintained under water shortage because of increased WUE. In pot experiments nutrient limitation counteracts with CO_2 enrichment, particularly in the case of phosphorus or potassium shortage (Goudriaan and De Ruiter, 1983), but in the field there may be a positive interaction as a result of enhanced root growth.

Nitrogen is an important element for the carboxylation enzyme rubisco. With an increased rate of photosynthesis under high CO_2 levels, resulting in starch accumulation, an intensification of nitrogen metabolism occurs, decreasing rubisco in the leaf. If a nitrogen shortage occurs, there can still be a positive effect, as plant tissue with a lower nitrogen concentration is produced. It appears that species adapted to nutrient-limited conditions (slow growing species) respond less to elevated CO_2 than faster growing species adapted to nutrient availability (Hunt et al., 1991; Poorter, 1993).

Arp and Berendse (1993) came to the conclusion that the response to elevated

CO_2 levels of natural, nutrient-limited ecosystems depends on the long-term availability of nutrients to support an increase of biomass. Increased root growth resulting from higher CO_2 level may result in a higher uptake of nutrients as well as in increased availability through a stimulation of microbial activity and recycling of microbial nitrogen.

A change in the carbon/nitrogen ratio in the plant tissues may lead to an increase of the carbon/nitrogen ratio of the litter, if no reallocation takes place before the plant parts are shed. A high carbon/nitrogen ratio of litter usually leads to a reduction in the decomposition rate, and thus to a reduction of the availability of nutrients in the soil. Any change in chemical composition (e.g., lower inorganic content, higher secondary chemical content) of plant tissue will have an impact on herbivory. Consumption may increase as the compensation for lower nitrogen content. This in turn could speed up nutrient turnover and increase supply. Experiments with plants grown under elevated CO_2 have indicated that probability for insect infestation is higher. However no generalization can be made (Oechel and Strain, 1985).

It has been experimentally proved that increasing CO_2 will enhance nitrogen-fixing activity of nitrogen-fixing tree seedlings, but the magnitude of the response depends much on the plant and soil nutrient status (Thomas et al., 1991).

Experiments revealed that elevated CO_2 and pollutant combinations were often contradictory. It seems that the stimulating effect of increased CO_2 is reduced by the pollutants O_3, NO_x, and NH_3 (Van der Eerden et al., 1993). On the other hand, elevated CO_2 decreases O_3 damage. While NH_3 and CO_2 both stimulate growth or photosynthesis, in combination this stimulation may disappear. Because of the complexity of interactions between air pollutants and CO_2 and because of the individual response of plant species to pollutants as well as to elevated CO_2, it is not yet possible to assess the effects of both on natural vegetation.

For floating water plants and those with floating leaves it has been observed that they may have higher net photosynthesis, greater leaf area index, and increase in biomass under elevated CO_2 (Allen et al., 1988; Alpert et al., 1992). However, it was also found that acclimation takes place after longer period of exposure to high CO_2 (Spencer and Bowes, 1986), but the net result was still a gain in dry weight. Because of the high CO_2 requirements of submerged aquatic plants it is also very likely that they will respond to higher CO_2 levels by increased photosynthesis, as it is easier to assimilate free CO_2 than bicarbonate. The rate of response will depend on the water chemistry.

There are only a few, mainly short-term, experimental data about the effects of elevated CO_2 on tropical plants (Hogan et al., 1991; Strain and Cure, 1994), but no studies have been carried out in the field. Reekie and Bazzaz (1991) found no effect on photosynthesis or overall growth of individually grown plants species, but higher CO_2 did affect canopy architecture. Stomatal conductance decreased slightly, but had no effect on whole water use or leaf water potential. However, CO_2 had a significant effect on the competitive performance of the species grown together, through its effect upon canopy structure, thereby changing plant's competitive ability for light. Korner (1992) found in tropical plants no significant difference in stand biomass, leaf area index, or stomatal behavior between ambient and elevated CO_2, but a massive starch accumulation in the tops of the canopy occurred, as well as an increase in fine root production and a doubling of CO_2 evolution from the soil.

The individual responses of plant species to elevated CO_2 will have influence on plant competition and other ecological processes in ecosystems, managed as well as natural. Experiments with combinations of two or more species (combinations of C_3 species, and C_3 with C_4 species) grown under different levels of CO_2 have revealed a scale of response possibilities (Patterson and Flint, 1990).

Rapid downward acclimation of photosynthesis to high CO_2 has been observed. Increase as well as decrease in respiration may occur in individual plants and whole stands (Hilbert et al., 1991). Changes in the morphology of plants as a result of high CO_2 may eventually act as a disadvantage; for instance, a larger leaf area may result in more competition for light. A study using (mean) general circulation model output in a transient scenario showed that forest productivity will initially increase, but will then reduce if the CO_2 concentration is doubled, due to an increasing negative effect of temperature on forest productivity.

Information is rather meager about the response of intact ecosystems where natural interactions between the various components occur in the presence of increased CO_2. Studies have been carried out on the tussock tundra in Alaska (Hilbert et al., 1987; Oechel and Strain, 1985) and one has been done on salt marsh communities in the Chesapeake Bay (Arp 1991; Curtis et al. 1989a, b). The salt marsh contains C_3 and C_4 species. Increased CO_2 had no effects on the C_4 species *(Scirpus olneyi)*. The C_3 species *(Spartina patens)*, however, grown in pure stands or mixed, showed a substantial increase in shoot density and aboveground dry matter, and a delay in senescence of leaves and shoots. Differences in growth response of mixtures of plant species have been mentioned by various authors (Overdieck, 1986; Pickett and Bazzaz, 1978; Strain and Bazzaz, 1983).

The study on the salt marsh community showed that after four years of exposure to elevated CO_2 there was no downward regulation of photosynthesis, contrary to what is found in experiments with plants grown in small pots (Arp, 1991). In the mixed C_3/C_4 community a very large and statistically significant increase in biomass of the C_3 species was observed under elevated CO_2 in all four years, while in the fourth year the biomass of the C_4 species declined, probably because of competition for light and other resources.

In the tundra there was a low growth response to elevated CO_2, which might be attributed to the fact that this ecosystem type is generally nutrient-limited and has a low productivity. An increase in tillering has been observed in the dominant species, *Eriophorum vaginatum* (Hilbert et al., 1987; Oechel et al., 1992; Tissue and Oechel, 1987). At elevated temperature and high CO_2 the tundra had slightly higher mean respiration rates. Soil CO_2 efflux was not affected by CO_2 enrichment (Oberhauer et al., 1986).

Differential species growth and changes in other aspects of the plants' response (such as WUE, phenology, and competitive power) will eventually result in shifts, extensions, or contractions in the habitat range of the species or may lead to extinction. These changes will ultimately affect other components of the ecosystems including pollinators, dispersers, herbivores, and so on. In combination with the impacts of CO_2-induced climate changes the final implications are difficult to forecast. Total structure and functions of ecosystems will be modified. Even new ecosystem types may evolve. There will certainly be limits to the tolerance of plants to the increase of CO_2. Acclimation can take place, but also irreversible damage may occur at too high levels of CO_2 concentration.

GREENHOUSE EFFECT AND ITS IMPACT ON CLIMATE

One of the most important factors in both short- and long-term climate change is the composition of the atmosphere and how it affects the energy balance of the planet. The solar radiation reaching the atmosphere is primarily visible light. About 30% of this is reflected back into space by such features as clouds, snow, or ice-covered earth, and atmospheric particles. The rest of the radiation is absorbed by liquids, solids, and gases that constitute the planet. The energy absorbed must eventually leave the earth in the form of infrared radiation. The emission of this radiation is hampered by certain trace gases present in the atmosphere, CO_2 being one of the most important.

Besides CO_2 there are various other trace gases that function as so-called greenhouse gases in the atmosphere, for example, methane (CH_4), N_2O, NO_x, and aerosols such as chlorofluorocarbons (CFC-11 and CFC-12) and others such as halons, tropospheric O_3, and stratospheric water vapor. These gases, which characteristically contain three or more atoms per molecule, transmit incoming solar radiation, but absorb the outgoing terrestrial long-wave radiation, thus trapping heat. With a rise in concentrations of these gases, more radiation will be absorbed, resulting in a warming of the earth. It has been estimated that without the presence of the greenhouse gases in the atmosphere, the average surface temperature of earth would be below freezing.

Due to differing physical properties and atmospheric lifetimes, the various greenhouse gases have different greenhouse warming potentials. The relative contribution of CO_2 in the greenhouse effect is estimated to be about 63% (Burke and Lashof, 1990; Hansen et al., 1988) of methane 18%. The potential of the latter is much greater than that of CO_2 (Fig. 13-20).

The main cause of the increase in atmospheric CO_2 is the burning of fossil fuel, which, together with gas flaring and cement production, releases about 6.2 Gt C yr^{-1}. In addition to CO_2, CH_4 is also emitted. Agricultural activities (land clearing for agriculture, burning of agricultural wastes, grasslands, and forests) also contribute to an increase in the level of greenhouse gases. Burning of biomass releases predominantly amounts of CO_2, CH_4, N_2O, and NO_x. Methane emissions from anaerobic respiration in rice (*Oryza sativa* L.) paddies and domestic animal rumens account for 30–50% of the global total. With population growth, agriculture will grow, and subsequently these emissions will increase. Based on the present trends of emissions a doubling of CO_2 is likely to occur in the first half of the next century.

It is expected that, as a result of increased atmospheric CO_2 and other trace gases, the global climate will eventually change. Model calculations indicate that

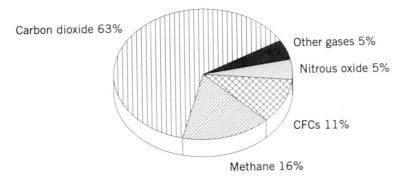

Carbon dioxide 63%

Other gases 5%

Nitrous oxide 5%

CFCs 11%

Methane 16%

Figure 13-20. *The major atmospheric gases contributing to "greenhouse effect." Solar energy enters the atmosphere unimpeded by CO_2. Light is absorbed by the earth. It is reradiated at longer heat wavelengths, some of which are captured by CO_2; the others escape into space. At the normal atmospheric CO_2 level the quantity of heat escaping into space determines the earth's temperature and climate. Higher concentrations of CO_2 trap more reradiated heat. Atmospheric and surface temperature increases, affecting weather and climate. A rise in global temperatures of about one degree celsius by the year 2025 is anticipated. The major greenhouse gases contributing to this warming are CO_2, CH_4, chloroflurocarbons (CFCs), nitrous oxide, and other gases. (Redrawn from "The Greenhouse Gases" Fact Sheet, Atmospheric Environment Service, Environment Canada, 1991.)*

with a doubling of preindustrial CO_2 the global mean temperature will rise by 2–4.5°C. This increase will be larger in the northern regions and less in the tropics, and will be larger in winter than in summer. The average global temperature has already increased by about 0.73°C since 1850 (Houghton et al., 1990). If climate models are correct, future warming will proceed more rapidly than in any warming period since the last ice age, which ended 10 000–12 000 years ago, and during which the average global temperatures at the surface were only about 5°C less than they are today. It seems that for the first time, unlike in the geological ages when natural cycles determined the climate, the influence of human activities will determine the climatic future, with some unexpected implications for life on earth. An average global surface temperature greater than 2°C over the current temperature would be unprecedented in human history.

A global intensification of the hydrological cycle (more precipitation and more evaporation) by 5–15% is also predicted. Increased evaporation might, in turn, lead to regional drying (Manabe et al., 1981; Taylor and McCracken, 1990). On the other hand, with increasing concentration of water vapor there will be an important positive feedback on temperature rise, as water vapor itself has a greenhouse effect.

This predicted intensification of the hydrological cycle under future weather conditions will lead to shifts in precipitation patterns. The geographical distribution of the predicted changes is, however, highly uncertain. As pointed out before, under higher CO_2 the water use efficiency will be higher; hence more dry matter will be produced for the same amount of water (Gifford, 1979; Goudriaan and Bijlsma, 1987). In parts of the present (semi)arid regions, the constraint on moisture availability might be to a certain extent decreased, leading to higher primary production, resulting in a denser vegetation cover and/or greater extent of vegetation cover. This in turn could reduce water runoff.

Very little is known about the possible change in direct and indirect (diffuse) radiation as a result of CO_2-induced climate change. In addition to temperature and soil moisture, the amount of radiation is also a determining factor for the duration of the growing season and the growth of plants. During the summer period a temperature rise, not accompanied by an increase in radiation energy, will more commonly diminish growth rate.

The most likely climate scenario for Europe is a rise in annual mean temperature that exceeds the rise in global mean, with uncertainty about rise in all seasons. There are indications that, due to a larger annual variability, with occasional occurrence of colder winter, the average winter temperature in Central and Western Europe might initially even decrease. An important effect of an increase in temperature will be, for the cooler climate regions, an increase in the length of the growing season (which equals the number of days per year with a mean temperature above 5°C). An increase of the growing season by 15 to 20 days per 1°C rise of annual mean temperature may occur in the temperate zone. However there may be limitations due to other growth factors, such as nutrients. A clearly negative effect could be a regional reduction of precipitation. Theoretical estimates indicate that the global net primary production may increase by 2–5% if the world average temperature rises 1°C.

As far as the precipitation pattern is concerned, Europe will follow the global tendency to more precipitation and more evaporation. However it is difficult to predict what the precipitation excess or deficit in the different seasons will be. In autumn and winter the already existing precipitation will probably increase, but the period of deficit (in Western Europe on average only a few months per year) could further extend from summer to spring.

On a global scale a shift in climate zones may result, whereby an increase of temperature by 3°C corresponds with a change in bioclimatic zone of about 600 km. Altitudinally this would mean a shift of about 150 m. These changes in climate zones will also have far-reaching effects on crop production (Goudriaan et al. 1990; Ominde and Juma, 1991; Parry, 1990; Van de Geijn et al., 1993). Concerning the impact of the greenhouse effect, a distinction should be made between:

- Direct effects as a result of increased CO_2 on, for example, physiological processes in plants and growth of plants.
- Indirect effects through climate change (change in temperature, precipitation, evaporation, and so on).
- From climate change derived effects such as sea level rise, changes in sedimentation and erosion patterns, and so on.
- These effects can be synergetic or counteracting. The indirect effects, global warming, and changes in precipitation and the frequency of weather extremes may have greater impact than the direct effects of CO_2 on physiological processes.

Changes in nature resulting from elevated CO_2 and climate changes will take place at three levels:

1. Direct response of individual species to changing circumstances
2. Response of two or more organisms that interact naturally
3. Response of whole vegetation/ecosystems

Of course these levels are not strictly separate.

It may be concluded that in general elevated atmospheric CO_2 *supra vide* may be beneficial for dry matter production and that it may enhance plant tolerance to stress, including low water availability, high or low temperature, and photoinhibition. Changes in competitive power, morphological, and phenological features may result in changes in species abundance and distribution patterns. For example, some species may be able to extend their range into physically less favorable sites. Biological interactions may become relatively more important in determining the distribution and abundance of species.

On the other hand, there will be species for which the temperature will be more advantageous than CO_2 enrichment. There are numerous examples of close relations between distribution patterns of plant species and temperature (Ketner, 1990a). Increased temperature may lead to changes in distribution patterns. Species with a southern distribution may spread northwards. C_4 species, particularly, may benefit from higher temperatures. They are predominantly thermophilous species and may extend into new areas. As several of these C_4 species are aggressive weeds, problems may arise in agriculture (Ketner, 1990b).

Two phenomena that are particularly sensitive to a temperature change are seed dormancy breaking and germination. Earlier emergence and a longer growing season may enhance seed production. In combination with stimulated seed production by enhanced CO_2, particular annual weedy species may become a plague.

Increasing atmospheric CO_2 concentrations will affect agricultural production via changes in photosynthesis and respiration rates as well as via a change in climate. In general the direct effects of CO_2 for crop production may be beneficial, but together with the indirect climatic effects it is difficult to predict the final outcome. For some areas of the world the effects together may be positive, elsewhere negative.

Many simulation studies have been carried out on the impact of CO_2-induced climate change and crop production (Goudriaan et al., 1990; Nonhebel, 1993; Parry, 1990; Van de Geijn et al., 1993). The implications of the temperature response (growing season response) for crop production depend on the growth form of the crop. That can be either determinate (e.g., cereals that have a discrete life cycle, ending when the crop is mature) or indeterminate (as in perennial grass or in potato, which continue growing and forming yield as long as the temperature is above the minimum threshold). For the first category of crops in temperate regions, higher temperature will be a disadvantage, while an increase in potential yield may be expected in the latter category. For C_3 cereals the higher CO_2 effect will thus be counteracted by a higher temperature.

Nonhebel (1993) found that, for potential production (optimal water) of wheat, a 3°C temperature rise caused a yield decline due to a shortening of the growing season. Doubling of the CO_2 concentration gave rise to an increase in yield of 40%, due to higher assimilation rates. Effects of higher temperature and higher CO_2 concentration were superimposed and the combination of both led to a yield increase of 1–2 ton ha^{-1}. Both rise in temperature and increase in CO_2 reduced water requirements of the crop.

For silage maize (C_4) in northwestern Europe, where it usually grows below optimal temperature, a longer growing season and higher temperature will be an advantage. In southern Europe and most regions of North America, as well as in tropical and subtropical regions, a higher temperature will lead to a loss in crop yields. This might be partly diminished by higher CO_2, but probably not if there is a nutrient shortage, as is often the case in the tropics. Tropical plants may be more narrowly adapted to prevailing temperature regimes than are temperate plants, so expected changes in temperature may be more important in the tropics. Reduced transpiration due to stomatal conductance could modify the effects of water stress as a cue for vegetative or reproductive phenology of plants of seasonal tropical areas (Hogan et al., 1991).

Despite numerous predictions about climate change from a large number of general circulation models and the almost uncountable number of experimental and field studies on impact of elevated CO_2, there are only indications of what might happen. There are still many physical phenomena that are not understood or that remain to be discovered. There are also many biological feedbacks, of which we have hardly any knowledge and which so far cannot be modeled. Also, findings from experiments on single plants or plant combinations under enhanced CO_2, often short-term and covering one phase of the plant's life, cannot be extrapolated to field conditions.

The counteracting effect of sulfate aerosols, which exert another large anthropogenic radiative forcing of the atmosphere, should be taken into account when constructing greenhouse gas models of climate changes. According to recent estimations (Mitchell et al., 1995) a future global mean warming of 0.3 K per decade is predicted if only greenhouse gases are taken into consideration, but it drops to 0.2 K when sulfate aerosol forcing is also included. Therefore great precautions should be taken in generalizing the ultimate effects of increased atmospheric CO_2. For example, sea level rise due to thermal expansions of the oceans and melting glaciers is one of the most probable aspects of a global temperature increase and most estimations predict that the sea level may rise a few centimeters to 1.5 meters. However, even in this case, a decrease in sea level cannot be ruled out.

Although the relative probability of events caused by increased atmospheric concentrations of CO_2 and other greenhouse gases is not known, it is generally agreed that nature's adaptability to global warming depends on the rate and magnitude of the climate change. If the change is slow enough, nature can adapt via such mechanisms like migration, evolutionary selection, and adaptation processes, whereas human society can adapt via infrastructure changes and new technologies (Abrahamson, 1989). The concern lies with the threat of a rapid climate change and those irreversible impacts to which society and nature cannot adapt.

CONCLUSIONS

As pointed out in the preceding sections, the importance of ambient CO_2 far exceeds its relative abundance in the atmosphere. Variations in the level of CO_2, being one of the trace gases, are much more significant for the biosphere than the changes of the same order in the level of the major gaseous constituents of the atmosphere. CO_2 concentration in the atmosphere is the result of many biotic and abiotic factors and processes and its actual level has, in turn, because of a feedback coupling, a great impact on the biosphere and environment. This concerns its effect on various structures, processes, and adaptations occurring in the biosphere, not only on a molecular level, but also on subcellular, cellular, whole organism, and ecosystem levels. Atmospheric CO_2 concentration is one of the global factors heavily influenced by human activity, with possible far-reaching consequences for the earth's climate and climate-related changes. The biogeochemical cycle of carbon, which is the fundamental element of living matter and all organic molecules, is tightly coupled with the cycles of other elements, biomass production, and energy flow through biosphere, all the constituents of this system being in a dynamic equilibrium. One of humanity's greatest challenges for the future is to manage the global carbon reservoirs and fluxes in such a way as to maintain this delicate balance, established during hundreds of millions of years of earth's evolution.

ACKNOWLEDGMENTS

The authors are thankful to Addison Wesley, Longman (Fig. 13-1); US DOE (CDRD) (Fig. 13-2); Richard Walker, Oxygraphic Ltd. (Fig. 13-4); Springer (Fig. 13-11); Eugen Ulmer (Fig. 13-12); Academic Press (Fig. 13-17); Gezondheidsraad (Figs. 13-18 and 13-19); and Atmospheric Environment Service, Canada (Fig. 13-20) for permission to use their copyrighted figures. Constructive criticism and help from Dr M. N. V. Prasad is gratefully acknowledged.

REFERENCES

Abrahamson, D. (Ed.) (1989), *The challenge of global warming*. Island Press, USA.

Ajtay, G. L., P. Ketner, and P. Duvigneaud (1979), in *The global carbon cycle. Scope report 13* (B. Bolin, E. T. Degens, S. Kempe, and P. Ketner, Eds.) Wiley, London, pp. 129–187.

Allen, S. G., S. B. Idso, B. A. Kimball, and M. G. Anderson (1988), *Agric. Ecosyst. Environ., 20*, 137–141.

Alpert, P., F. R. Warembourg, and J. Roy (1992), *Am. J. Bot., 78*, 1459–1466.

Amthor, J. S. (1995), in *Ecophysiology of photosynthesis* (E.-D. Schulze and M. M. Caldwell, Eds.), Springer-Verlag, Berlin, pp. 71–101.

Anders, E. (1989), *Nature, 342*, 255–257.

Anderson, J. W. and J. Beardall (1991), *Molecular activities of plant cells*. Blackwell Scientific, Oxford, UK.

Andrews, T. J. and G. H. Lorimer (1987), in *Rubisco: structure, mechanisms and prospects for improvement* (M. D. Hatch and N. K. Boardman, Eds.), Academic Press, San Diego, CA, pp. 131–218.

Armentano, T. V. and C. W. R. Ralston (1980), *Can. J. For.*, *10*, 53–60.

Arp, W. J. (1991), *Vegetation of a North American salt marsh and elevated atmospheric carbon dioxide*, Ph.D. Thesis, Free University, Amsterdam, 181 pp.

Arp, W. and F. Berendse (1993), in *Climate change; crops and terrestrial ecosystems* (S. C. van de Geijn, J. Goudriaan, and F. Berendse, Eds.), CABO-DLO, Agrobiologische Thema's No. 9, Wageningen.

Awramik, S. M., J. W. Schopf, and M. R. Walter (1983), *Precambrian Res.*, *20*, 357–374.

Azcon-Bieto, J. and C. B. Osmond (1983), *Plant Physiol.*, *71*, 574–581.

Babcock, G. T. (1993), *Proc. Natl. Acad. Sci. USA*, *90*, 10893–10895.

Bacastow, R. B., C. D. Keeling, and T. P. Whorf (1985), *J. Geophys. Res.*, *90*, 10529–10540.

Badger, M. R. and A. Gallagher (1987), *Aust. J. Plant Physiol.*, *14*, 189–201.

Badger, M. R. and G. H. Lorimer (1976), *Arch. Biochem. Biophys.*, *175*, 723–729.

Badger, M. R., M. Bassett, and H. N. Comins (1985), *Plant Physiol.*, *77*, 465–471.

Baes, C. F., H. E. Goeller, J. S. Olson, and R. M. Rotty (1976), *The global carbon dioxide problem*. ORNL-5194, 1–72, Oak Ridge National Laboratory, Oak Ridge, TN.

Barber, J. and B. Andersson (1994), *Nature*, *370*, 31–34.

Bazilevich, N. I., L. Y. Rodin, and N. N. Rozov (1971), *Sov. Geogr. Rev. Transl.*, *2*, 293–317.

Bazzaz, F. A., K. Garbutt, and W. E. Williams (1985), in *Direct effects of increasing carbon dioxide on vegetation* (B. R. Strain and J. D. Cure, Eds.), US Dept. of Energy, DOE/ER-0238, pp. 155–170.

Berner, R. A. and A. C. Lasaga (1989), *Sci. Am.*, March, 54–61.

Berner, R. A., A. C. Lasaga, and R. M. Garrels (1983), *J. Am. Sci.*, *283*, 641–683.

Berry, J. A. and J. A. Downton (1982) in *Photosynthesis, development, carbon metabolism and plant productivity* (Govindjee, Ed.), Vol. 2, Academic Press, New York, pp. 263–343.

Berry, J. A., G. H. Lorimer, J. Pierce, J. R. Seeman, J. Meeks, and S. Freas (1987), *Proc. Natl. Acad. Sci. USA*, *84*, 734–738.

Björkström, A. (1979), in *The global carbon cycle. Scope report 13* (B. Bolin, E. T. Degens, S. Kempe, and P. Ketner, Eds.), Wiley, London, pp. 411–457.

Black, C. C. (1971), *Adv. Ecol. Res.*, *7*, 87–113.

Boden, T. A., D. P. Kaiser, R. J. Sepanski, and F. W. Stoss (1994), *Trends '93. A compendium on data on global change*. Carbon Dioxide Information Analysis Center, Publication No. ORNL/CDIAC-65, ESD Publication No. 4195.

Bolin, B. (Ed.) (1981), *Carbon cycle modelling. Scope report 16*, Wiley, New York.

Bolin, B., E. T., Degens, S. Kempe, and P. Ketner (Eds.) (1979), *The global carbon cycle. Scope report 13*, Wiley, London, p. 491.

Bolin, B., B. R. Doos, J. J. Jäger, and R. A. Warrick (Eds.) (1986), in *The greenhouse effect, climate change and ecosystems*, John Wiley & Sons, New York.

Bowes, G., W. L. Ogres, and R. Hageman (1971), *Biochem. Biophys. Res. Comm.*, *45*, 716–722.

Bramyd, T. (1979), in *The global carbon cycle. Scope report 13* (B. Bolin, E. T. Degens, S. Kempe, and P. Ketner, Eds.) Wiley, London, pp. 181–218.

Brown, R. H. (1978), *Crop Sci., 18,* 93–98.

Brulfert, J., J. Vidal, P. Le Marechal, P. Gadal, and O. Queiroz (1986), *Biochem. Biophys. Res. Commun., 136,* 151–159.

Budyko, M. J. and A. B. Ronov (1979), *Geochem. Intern., 16,* 1–9.

Burke, L. M. and D. A. Lashof (1990), in *Impact of carbon dioxide, trace gases, and climate change on global agriculture* (B. A. Kimball, N. J. Rosenberg, and L. H. Allen, Eds.), ASA Special Publication no. 53, 27–43.

Canvin, D. T. (1990), in *Plant physiology, biochemistry and molecular biology* (D. T. Dennis and D. H. Turpin, Eds.), Longman Scientific & Technical, Burnt Mill, Harlow, England, pp. 253–273.

Clark, W. C. (1982), in *Carbon dioxide review,* Oxford University Press.

Cramer, W. A. and D. B. Knaff (1991) *Energy transduction in biological membranes,* Springer Verlag, New York.

Cseke, C. and B. B. Buchanan (1986), *Biochim. Biophys Acta, 853,* 43–64.

Curtis, P. S., B. G. Drake, and D. F. Whigham (1989a), *Oecologia, 78,* 97–301.

Curtis, P. S., B. G. Drake, P. W. Leadley, W. J. Arp, and D. F. Whigham (1989b), *Oecologia, 78,* 20–26.

De Vooys, C. G. N. (1979), in *The global carbon cycle. Scope report 13* (B. Bolin, E. T. Degens, S. Kempe, and P. Ketner, Eds.), Wiley, London, pp. 259–292.

Downton, W. J. S. (1975), *Photosynthetica, 9,* 96–105.

Ehleringer, J. R., R. F. Sage, L. B. Flanagan, and R. W. Pearcy (1991), *Trends Ecol. Evol., 6,* 95–99.

Freyer, H. D. (1979), in *The global carbon cycle. Scope report 13* (B. Bolin, E. T. Degens, S. Kempe and P. Ketner, Eds.), Wiley, London, pp. 79–99.

Furbank, R. T. and M. D. Hatch (1987), *Plant Physiol., 85,* 958–964.

Gallagher, A. (1987), *Aust. J. Plant Physiol., 14,* 189–201.

Gammon, R. H., E. T. Sundquist, and P. J. Fraser (1985), in *Atmospheric carbon dioxide and the global carbon cycle,* US Dept. of Energy, DOE/ER - 0239, December, pp. 25–62.

Garbutt, K., W. E. Williams, and F. A. Bazzaz (1990), *Ecology, 71,* 1185–1194.

Gezondheidsraad (1983), *Part I—Report on CO_2 probem* (English translation), Recommendation to the Ministers of Public Health and of Physical Environment issued by the Health Council of the Netherlands, Health Council, The Hague, Netherlands.

Gezondheidsraad (1987), *CO_2 problem, scientific opinions and impacts on society* (English translation), Second advice and final report, Health Council. The Hague, Netherlands.

Gifford, R. M. (1979), *Aust. J. Plant Physiol., 6,* 367–378.

Gifford, R. M. (1988), in *Greenhouse: Planning for climate change* (G. I. Pearman, Ed.), Brill, New York, pp. 506–519.

Gifford, R. M., H. Lambers, and J. I. L. Morison (1985), *Physiol. Plant., 63,* 351–356.

Goudriaan, J. (1987), *Neth. J. Agric. Sci., 35,* 177–187.

Goudriaan, J. (1990), in *The greenhouse effect and primary productivity in European agroecosystems* (J. Goudriaan, H. van Laar, and H. H. van Keulen, Eds.), Pudoc, Wageningen, Netherlands, pp. 23–25.

Goudriaan, J. (1993), in *Climate change: Crops and terrestrial ecosystems* (S. C. van de Geijn, J. Goudriaan, and F. Berendse, Eds.), *Agrobiologische Thema's, 9,* 125–138.

Goudriaan, J. and R. J. Bijlsma (1987), *Neth. J. Agric. Sci.*, *35*, 189–191.

Goudriaan, J. and H. E. De Ruiter (1983), *Neth. J.Agric. Sci.*, *31*, 157–169.

Goudriaan, J. and P. Ketner (1984), *Climate Change, 6*, 167–192.

Goudriaan, J. and M. H. Unsworth (1990), in *Impact of carbon dioxide, trace gases, and climate change on global agriculture* (B. A. Kimball, N. J. Rosenzweig, and L. H. Allen, Eds.), ASA Special Publication No. 353, Madison, WI, pp. 111–130.

Goudriaan, J., H. van Keulen, and H. H. Van Laar (Eds.) (1990), *The greenhouse effect and primary productivity in European agro-ecosystems*, Pudoc, Wageningen, Netherlands, 90 pp.

Griffiths, H. (1989), in *Vascular plants as epiphytes. Evolution and ecophysiology* (U. Luettge, Ed.), Springer, Berlin, pp. 42–46.

Gruszecki, W. I. and K. Strzałka (1991), *Biochim. Biophys. Acta, 1060*, 310–314.

Gutteridge, S., M. A. J. Parry, S. Burton, A. J. Keys, A. Mudd, J. Feeney, J. C. Servaites, and J. Pierce (1986), *Nature, 324*, 274–276.

Hall, D. O. and J. M. O. Scurlock (1991), *Annl. Bot., 67*, 49–55.

Hampicke, U. (1979), in *The global carbon cycle. Scope report 13*, (Bolin, B., E. T. Degens, S. Kempe, and P. Ketner, Eds.) John Wiley & Sons, London, pp. 219–236.

Hansen, J., I. Fung, A. Lacis, S. Lebedeff, R. Rind, G. Russel, and P. Stone (1988), *J. Geophys. Res., 93*, 9341–9364.

Hatch, M. D. (1987), *Biochim. Biophys. Acta, 895*, 81–106.

Hatch, M. D. and C. R. Slack (1970), *Annu. Rev. Plant Physiol., 21*, 141–162.

Hattersley, P. W. (1983), *Oecologia, 57*, 113–128.

Hattersley, P. W. (1987), in *Grass systematics and evolution* (T. R. Soderstrom, K. W. Hilu, C. S. Campbell, and M. E. Barkworth, Eds.), Smithsonian Institution Press, Washington DC, pp. 49–64.

Hattersley, P. W. and L. Watson (1992), in *Grass evolution and domestication* (G. P. Chapman, Ed.), Cambridge University Press, Cambridge, UK, pp. 38–116.

He, P., K. P. Bader, A. Radunz, and G. H. Schmid (1995), *Z. Naturforsch., 50C*, 781–778.

Heldt, H. W. and U. I. Fluegge (1987), *The biochemistry of plants*, Vol. 12, Academic Press, New York, pp. 49–85.

Hilbert, D. W., T. I. Prudhomme, and W. C. Oechel (1987), *Oecologia, 71*, 466–742.

Hilbert, D. W., A. Larigauderie, and J. F. Reynolds (1991), *Annl. Bot., 68*, 365–376.

Hogan, K. P., A. P. Smith, and L. H. Ziska (1991), *Plant Cell Environ., 14*, 763–778.

Holland, H. D. (1984), *The chemical evolution of the atmosphere and oceans*, Princeton University Press, Princeton, NJ.

Houghton, J. T., G. J. Jenkins, and J. J. Ephraums (1990), *Climate change scientific assessment*, IPCC Cambridge University Press, Cambridge, UK.

Huber, S. C., T. C. Hall, and G. E. Edwards (1976), *Plant Physiol., 57*, 730–733.

Hull, M. R., S. P. Long, and C. P. Raines (1990), in *Current research in photosynthesis* (M. Baltscheffsky, Ed.), Vol. 4, Kluwer Academic, Dordrecht, Netherlands, pp. 675–678.

Hunt, R., D. W. Hand, M. A. Hannah, and A. M. Neal (1991), *Funct. Ecol., 5*, 410–421.

Husic, D. W., H. D. Husic, and N. E. Tolbert (1987), *CRC Crit. Rev. Plant Sci., 5*, 45–400.

Kappen, L. and S. Hager (1991), *J. Exp. Bot., 42*, 979–986.

Kappen, L., G. Andresen, and R. Loesch (1987), *J. Exp. Bot., 38*, 126–141.

Kempe, S. (1979), in *The global carbon cycle. Scope report 13* (B. Bolin, E. T. Degens, S. Kempe, and P. Ketner, Eds.), Wiley, London, pp. 317–342.

Ketner, P. (1990a), in *Effect of climate change on terrestrial ecosystems. Report from a seminar in Trondheim* (J. I. Holten, Ed.), 10-1-1990, NINA Notat., 4, 47–60.

Ketner, P. (1990b), in *The greenhouse effect and primary productivity in European agro-ecosystems* (J. Goudriaan, H. van Keulen, and H. H. van Laar, Eds.), Pudoc, Wageningen, Netherlands, pp. 18–19.

Khalil, M. A. K. and R. A. Rasmussen (1988), *Nature, 332*, 242–244.

Kobza, I. and J. R. Seemann (1988), *Proc. Natl. Acad. Sci. USA, 85*, 3815–3819.

Kohlmaier, G. H., G. Kratz, H. Brohl, and E. O. Sire (1981), in *Energy and ecological modelling* (W. J. Mitsch, R. W. Bosserman, and J. W. Klopatek, Eds.), Elsevier, Amsterdam.

Korner, C. (1992), *Science, 257*, 1672–1675.

Kortschak, H. P., C. E. Hartt, and G. O. Burr (1965), *Plant Physiol., 40*, 209–213.

Kramer, P. (1981), *BioScience, 31*, 29–33.

Lambers, H. (1982), *Physiol. Plant., 55*, 478–485.

Larcher, W. (1995), *Physiological plant ecology.* Springer, Berlin.

Lashof, D. A. and D. R. Ahuja (1990), *Nature, 344*, 529–531.

Lawlor, D. W. (1990), *Photosynthesis: Metabolism, control and physiology,* 2nd ed., Longman, London.

Lehninger, A. L. (1982), in *Principles of biochemistry,* Worth Publishers, New York, pp. 45–94.

Lemon, E. R. (Ed.) (1983), *CO₂ and plants. AAAS Selected Symp. 84,* Westview Press, Boulder, CO, 280 pp.

Lorimer, G. H., M. R. Badger, and T. J. Andrews (1976), *Biochemistry, 15*, 529–536.

MacIntyre, F. (1980), *Oceanol. Acta, 3*, 505–506.

Makewicz, A., A. Radunz, and G. H. Schmid (1995), *Z. Naturforsch., 50C*, 511–520.

Makino, A. (1994), *J. Plant Res., 107*, 79–84.

Manabe, S., R. T. Wetherald, and R. J. Stouffer (1981), *Climate Change, 3*, 347–386.

Miller, A. G. and D. T. Canvin (1987), *Plant Physiol., 84*, 118–124.

Mitchell, P. (1979), *Science, 206*, 1148–1159.

Mitchell, J.F.B., T. C. Johns, J. M. Gregory, and S. F. B. Tett (1995), *Nature, 376*, 501–504.

Moiseev, P. A. (1969), *The living resources of the world ocean,* Moscow (English trans., Israel Programs of Scientific Translations Ltd., Jerusalem, 1971).

Moore, P. D. and D. J. Bellamy (1973). *Peatlands,* Elek Science, London, 221 pp.

Morison, J. I. L. (1987), in *Stomatal function* (E. Zeiger, G. D. Farquhar and I. R. Cowan, Eds.), Stanford University Press, Stanford, CA, pp. 229–252.

Moroney, J. W., M. Kitayama, R. K. Togosaki, and N. E. Tolbert (1987), *Plant Physiol., 83*, 460–463.

Nambudiri, E. M. V., W. D. Tidwell, B. N. Smith, and N. P. Hebbert (1978), *Nature, 276*, 816–817.

Ngernprasirtsiri, J., P. Harinasut, D. Macheral, K. Strzałka, R. Takebe, T. Akazawa, and K. Kojima (1988), *Plant Physiol., 87*, 371–378.

Nonhebel, S. (1993), *The importance of weather data in crop growth simulation models and assessment of climate change effects,* Doctoral dissertation, Wageningen Agricultural University, Netherlands, 144 pp.

Oberhauer, S. F., W. C. Oechel, and G. H. Riechers (1986), *Plant Soil, 96,* 145–148.

Oechel, W. C. and B. R. Strain (1985), in *Direct effects of increasing carbon dioxide on vegetation* (B. R. Strain and J. D. Cure, Eds.), US Dept. of Energy, DOE/ER-0238, pp. 117–154.

Oechel, W. C., G. H. Riechers, W. T. Lawrence, T. I. Prudhomme, N. Grulke, and S. J. Hastings (1992), *Funct. Ecol., 6,* 86–100.

Oeschger, H., U. Siegenthaler, U. Schotterrer, and A. Gugelmann (1975), *Tellus, 27,* 169–192.

Ogawa, T. and A. Kaplan (1987), *Plant Physiol., 83,* 888–891.

Ogren, W. L. (1984), *Annu. Rev. Plant Physiol., 34,* 415–442.

O'Leary, M. H. (1982), *Annu. Rev. Plant Physiol., 33,* 297–315.

Olson, J. S., R. M. Garrels, R. A. Berner, T. V. Armentano, M. I. Dyer, and D. H. Yaalon, (1985), in *Atmospheric carbon and the global carbon cycle* (J. R. Trabalka, Ed.), DOE/ER/-0239, Oak Ridge National Laboratory, Oak Ridge, TN, pp. 175–213.

Ominde, S. H. and C. Juma (1991), *A change in the weather. African perspectives on climate change,* African Centre for Technology Studies, Nairobi, Kenya, 210 pp.

Osmond, C. B., K. Winter, and H. Ziegler (1982), in *Encyclopaedia of plant physiology (NS),* Vol. 12B, Physiological plant ecology II (O. L. Lange, P. S. Nobel, C. B. Osmond, and H. Ziegler, Eds.), Springer Verlag, Berlin, pp. 479–547.

Overdieck, D. (1986), *Biometeorology, 30,* 323–332.

Parry, M. (1990), *Climate change and world agriculture,* Earthscan, London, 157 pp.

Patterson, D. T. and E. P. Flint (1990), in *Impact of carbon dioxide, trace gases, and climate change on global agriculture* (G. H. Heichel, C. W. Stuber, and D. E. Kissel, Eds.), ASA Special Publication no. 53, pp. 83–104.

Pickett, S. T. A. and F. A. Bazzaz (1978), *Ecology, 59,* 1248–1255.

Poorter, H. (1993), *Vegetatio, 104/105,* 77–97.

Potvin, C., B. R. Strain, and J. D. Goeschl (1985), *Oecologia, 67,* 305–309.

Raghavendra, A. S. and V. S. R. Das (1978), *Photosynthetica, 12,* 200–208.

Raschke, K. (1979), in *Encyclopedia of plant physiology (NS),* Vol. 7, Physiology of movement (W. Haupt and M. E. Feinleib, Eds.), Springer, Berlin, pp. 383–441.

Reekie, E. G. and F. A. Bazzaz (1991), *Can. J. Bot., 69,* 2475–2481.

Reinhold, L., M. Zviman, and A. Kaplan (1987), in *Progress in photosynthesis research* (J. Biggins, Ed.), Vol. 4, Martinus Nijhoff, Dordrecht, Netherlands, pp. 289–296.

Reiskind, J. B., P. T. Seamon, and G. Bowes (1988), *Plant Physiol., 87,* 686–692.

Rodin, L. E. and N. I. Bazilevich (1967), *Production and mineral cycling in terrestrial vegetation,* Oliver and Boyd, London, 288 pp.

Rundel, P. W., J. R. Ehleringer, and K. A. Nagy (1988), *Stable isotopes in ecological research,* Springer, Berlin.

Sage, R. F., R. W. Pearcy, and J. R. Seemann (1987), *Plant Physiol., 85,* 355–359.

Sage, R. F., T. D. Sharkey, and J. R. Seemann (1989), *Plant Physiol., 89,* 590–596.

Salvucci, M. E., A. R. Portis, and W. L. Ogren (1985), *Photosynth. Res., 7,* 193–201.

Schlesinger, W. H. (1977), *Annu. Rev. Ecol. Syst.*, *8*, 51–81.

Schlesinger, W. H. (1991), *Biogeochemistry. An analysis of global changes*, Academic Press, Harcourt Brace Jovanovich, San Diego, CA, 439 pp.

Schulder, R., K. Burda, K. Strzałka, K. P. Bader, and G. H. Schmid (1992), *Z. Naturforsch.*, *47C*, 465–473.

Schulze, E. D. and A. E. Hall (1982), in *Encyclopedia of plant physiology (NS)*, Vol. 12B, Physiological plant ecology II. Water relations and carbon assimilation (O. L. Lange, P. S. Nobel, C. B. Osmond, and H. Ziegler, Eds.), Springer Verlag, Berlin.

Schurr, U. (1992), *Plant Cell Environ.*, *15*, 561–567.

Sharkey, T. D. (1988), *Physiol. Plant*, *73*, 247–152.

Sharkey, T. D. and T. Ogawa (1987), in *Stomatal function* (E. Zeiger, G. D. Farquhar and I. R. Cowan, Eds.), Stanford University Press, Stanford, CA, pp. 195–208.

Shinozaki, K., M. Ohme, M. Tanaka, T. Wakasugi, N. Hayashida, T. Matsubayashi, N. Zaita, J. Hunwongse, J. Obokata, H. Yamaguchi-Shinozaki, C. Ohto, K. Torazawa, B. Y. Meng, M. Sugita, H. Deno, T. Kamogashira, K. Yamada, J. Kusuda, F. Takaiwa, A. Kato, N. Tohdo, H. Shimada, and M. Sugiura (1986), *EMBO J.*, *5*, 2043–2049.

Smith, B. N., G. E. Martin, and T. W. Boutton (1979), in *Stable isotopes: Proceedings of third international conference* (E. R. Klein and P. D. Klein, Eds.), Academic Press, New York, pp. 231–237.

Sondergaard, M. and R. G. Wetzel (1980), *Can. J. Bot.*, *58*, 591–598.

Spencer, W. and G. Bowes (1986), *Plant Physiol.*, *82*, 528–533.

Stowe, L. G. and J. A. Teeri (1978), *The American Naturalist 112*, 609–623.

Strain, B. R. (1987), *Tree*, *2*, 18–21.

Strain, B. R. and T. V. Armentano (1980), *Environmental and societal consequences of CO_2 induced climate change: Response of "unmanaged" ecosystems*, Position paper Am. Assoc. Advanc. Science, 47 pp.

Strain, B. R. and F. A. Bazzaz (1983), in *CO_2 and plants. The response of plants to rising levels of atmospheric carbon dioxide* (E. R. Lemon, Ed.), AAAS Selected Symposium 84, Westview Press, Boulder, CO, pp. 177–222.

Strain, B. R. and J. D. Cure (Eds.) (1985), *Direct effects of increasing carbon dioxide on vegetation*, US Dept. of Energy, DOE/ER-0238.

Strain, B. R. and J. D. Cure (1994), *Direct effects of atmospheric CO_2 enrichment on plants and ecosystems*, An updated bibliographic database, ORNL/CDIAC, Oak Ridge National Laboratory, Oak Ridge, TN, 301 pp.

Streusand, V. I. and A. R. Portis, Jr. (1987), *Plant Physiol.*, *85*, 152–154.

Strzałka, K., T. Sarna, and J. S. Hyde (1986), *Photobiochem. Photobiophys.*, *12*, 67–71.

Strzałka, K., T. Walczak, T. Sarna, and H. M. Swartz (1990), *Arch. Biochem. Biophys.*, *281*, 312–318.

Szarek, S. R. and I. P. Ting (1977), *Photosynthetica*, *11*, 330–342.

Taiz, L. and E. Zeiger (1991), *Plant Physiology*. Benjamin/Cummings, Redwood City, CA.

Taylor, K. E. and M. C. McCracken (1990), in *Impact of carbon dioxide, trace gases, and climate change on global agriculture* (B. A. Kimball, N. J. Rosenberg, and L. H. Allen, Eds.), ASA Special Publication No. 53, pp. 1–18.

Thomas, R. B., D. D. Richter, H. Ye, P. R. Heine, and B. R. Strain (1991), *Oecologia, 88,* 415–421.

Thomasson, J. R., M. E. Nelson, and R. J. Zakrzewski (1986), *Science, 233,* 876–878.

Tieszen, L. L., M. M. Senyimba, S. K. Jmbamba, and J. H. Troughton (1979), *Oecologia, 37,* 337–350.

Tissue, D. T. and W. C. Oechel (1987), *Ecology, 68,* 401–410.

Tolbert, N. E. (1980), in *Metabolism and respiration, Biochemistry of plants,* (D. D. Davies, Ed.), Vol. 2, Academic Press, New York, pp. 487–523.

Tolbert, N. E. (1981), *Annu. Rev. Biochem., 50,* 133–157.

Trabalka, J. R. (Ed.) (1985), *Atmospheric carbon dioxide and the global carbon cycle,* US Dept. of Energy, DOE/ER-0239, 316 pp.

Trabalka, J. R., J. A. Edmonds, M. Reilly, R. H. Gardner, and L. D. Voorhees (1985), *Atmospheric carbon dioxide and the global carbon cycle* (J. R. Trabalka, Ed.) US Dept. of Energy, DOE/ER-0239, pp. 247–287.

Van der Eerden, L., T. Dueck, and M. Perez-Soba (1993), in *Climate changes, crops and terrestrial ecosystems* (S. C. Van der Geijn, J. Goudriaan, and F. Berendse, Eds.), *Agrobiologische Thema's, 9,* CABO-DLO, Wageningen, Netherlands, pp. 59–70.

Van de Geijn, S. C., J. Goudriaan, and F. Berendse (Eds.), (1993), *Climate change, crops and terrestrial ecosystems, Agrobiologische Thema's, 9,* CABO-DLO, Wageningen, Netherlands.

Van Oosten, J.-J., D. Afif, and P. Dizengremel (1992), *Plant Physiol. Biochem., 30,* 541–547.

Van Oosten, J.-J., D. Wilkins, and R. T. Besford (1994), *Plant, Cell Environ., 17,* 913–923.

Volokita, M., D. Zenvirth, A. Kaplan, and L. Reinhold (1984), *Plant Physiol., 76,* 599–602.

Walker, D. (1992), *Energy, plants and man,* 2nd ed., Oxygraphic, Brighton, UK.

Wallsgrove, R. M., A. J. Keys, P. J. Lea, and B. J. Miflin (1983), *Plant Cell Environ., 6,* 301–309.

Warneck, P. (1988), *Chemistry of the natural atmosphere,* Academic Press, London.

Werneke, J. M., J. M. Chatfield, and W. L. Ogren (1988), *Plant Physiol., 87,* 917–920.

Wetzel, R. G. and J. B. Grace (1983), in *CO$_2$ and plants. The response of plants to rising levels of atmospheric carbon dioxide* (E. R. Lemon, Ed.), AAAS Selected Symposium 84, Westview Press, Boulder, CO, pp. 223–280.

Whittaker, R. H. and G. E. Likens (1973), in *Carbon and the biosphere* (G. M. Woodwell and E. V. Pecan, Eds.), AEC Symposium Series 30, NTIS, US Dept. of Commerce, Springfield, VA, pp. 281–302.

Whittaker, R. H. and G. E. Likens (1975), in *Primary productivity of the biosphere. Ecol. Stud. 14,* (H. Lieth and R. H. Whittaker, Eds.), Springer-Verlag, Berlin, Heidelberg, New York, pp. 305–328.

Wiersum, F. and P. Ketner (1989) in *Climate and energy. The feasibility of controlling CO$_2$ emissions* (P. A. Okken, R. J. Swart, and S. Zwerver, Eds.), Kluwer Academic, the Netherlands, pp. 107–124.

Williams, T. G. and D. H. Turpin (1987), *Plant Physiol., 83,* 92–96.

Woodrow, L. and B. Grodzinski (1993), *J. Exp. Bot., 44,* 471–478.

Woodward, F. I. (1987), *Nature, 327,* 617–618.

Woodwell, G. M. and E. V. Pecan (Eds.) (1973), *Carbon and the biosphere,* AEC Symposium Series 30, NTIS, US Dept. of Commerce, Springfield, VA.

Zhang, J. and W. J. Davies (1989), *Plant Cell Environ., 12,* 73–81.

Zurzycki, J. (1955), *Acta Soc. Bot. Pol., 24,* 27–63.

14

Radionuclides

Noriyuki Momoshima

INTRODUCTION

Radionuclides cause radiation exposure to plants; this arises from naturally occurring radionuclides ubiquitously found in soil, and from artificial radionuclides released to the environment from nuclear weapons testing, nuclear reactor accidents, nuclear facilities, and so on. The deposition of the artificial radionuclides from nuclear weapons testing occurred mainly in 1950s and 1960s on a global scale; we can still detect some long-lived artificial radionuclides in soil, plants, and other various environmental materials from these tests. Accidents at nuclear reactors resulted in radioactive contamination in the environment on a local or more extended scale. The Chernobyl accident, in 1986, caused a significant radioactive contamination in the Northern Hemisphere. In European countries the Chernobyl accident exceeded the nuclear weapons testing in radioactive fallout. Nuclear facilities continue to release radionuclides to the environment under normal operation; these are nuclear power stations and nuclear fuel reprocessing plants. However, their contribution to the environmental radioactive contamination is small compared to the former two large sources, and the radioactive contamination usually occurs in a quite limited local area. The radionuclides, whether naturally occurring and artificial, expose plants both externally and internally.

The basic chemical nature of radionuclides does not differ from stable isotopes of the same atomic number but the difference in mass causes chemical isotope effects, both equilibrium and kinetic. Hydrogen isotopes are expected to represent greater isotope effects than any other elements in the periodic table because of they have the largest difference in the relative mass weight in three isotopes, ^1H, ^2H, and ^3H. As the atomic number of the element increases, the difference in relative mass weight of the isotopes becomes smaller, yielding less isotope effect. The isotope effects do not affect biological availability of the isotopes; the radioactive and stable isotopes are both incorporated into the same metabolic processes and would behave almost similarly in plants. The isotope effect could appear as different absorption and metabolic rates of the isotopes in plants, and consequently may result in different specific activity of isotopes between plants and substrate; for example, the activity ratios of ^3H to ^1H may differ between plant and soil water. However, it would be difficult to detect and evaluate this isotope effect in a natural soil-to-plants system because of variable microenvironmental conditions and fluctuation of radionuclide concentrations in soil water.

In this chapter, several topics relating radioisotopes to plants are described. The concentration levels of naturally occurring and artificial radionuclides in plants are summarized, and various contamination pathways of radionuclides to plants are described. The pulse input event of radionuclides from the Chernobyl accident afforded a unique opportunity to examine behaviors of radionuclides in various soil-to-plant systems. The radial distribution of radionuclides in wood are discussed in association with that of stable isotopes. The radionuclides from the nuclear weapons testing, accumulated in the bole of a tree by root uptake, gave information about physiological behaviors of radionuclides in xylem. In the last section, different responses of plants exposed to irradiation are described in relation to the irradiation experiment done on natural plant population and community.

RADIONUCLIDES IN PLANTS

Naturally Occurring Radionuclides

Naturally occurring radioactivity can be found in all substances, living and nonliving. About 340 nuclides have been found in nature; about 70 are radioactive. The naturally occurring radionuclides can be divided into those that occur singly and those that construct a decay chain series.

Representative singly occurring natural radionuclides are listed in Table 14-1. All singly occurring natural radionuclides have stable isotopes of the same atomic number. Three kinds of decay chain series are found in nature:

1. The uranium decay series, which originates with ^{238}U
2. The thorium decay series, which originates with ^{232}Th
3. The actinium decay series, which originates with ^{235}U

In Table 14-1, representative radionuclides belonging to the uranium and thorium decay series are listed. The contribution of the actinium decay series to radioactivity in plants is considered to be very small compared to that of the uranium decay series because of small isotopic abundance of ^{235}U (0.72%). The singly occurring natural radionuclides include those both of cosmic and terrestrial origin, but the decay chain series are all of terrestrial origin. The radionuclides with an atomic number greater than 83 have no stable isotope; the isotopes of U, Th, Pa, Ra, Rn, At, and Po are,

TABLE 14-1   Naturally Occurring Radionuclides

Nuclide	Half-Life	Radiation	Category
^{3}H	12.28 yr	β	Single[a]
^{7}Be	53.44 d	β, γ	Single
^{14}C	5730 yr	β	Single
^{40}K	1.28×10^9 yr	β, γ	Single
^{87}Rb	4.73×10^{10} yr	β	Single
^{208}Tl	3.1 min	β, γ	$(4n)$
^{210}Bi	5.013 d	β	$(4n + 2)$
^{210}Pb	22.3 yr	β, γ	$(4n + 2)$
^{210}Po	138.4 d	α	$(4n + 2)$
^{212}Bi	60.6 min	α, β, γ	$(4n)$
^{212}Pb	10.64 h	β, γ	$(4n)$
^{212}Po	2.98×10^{-7} s	α	$(4n)$
^{214}Bi	19.9 min	β, γ	$(4n + 2)$
^{214}Pb	26.8 min	β, γ	$(4n + 2)$
^{214}Po	1.64×10^{-4} s	α	$(4n + 2)$
^{216}Po	0.145 s	α	$(4n)$
^{218}Po	3.05 min	α	$(4n + 2)$
^{224}Ra	3.62 d	α	$(4n)$
^{226}Ra	1600 yr	α, γ	$(4n + 2)$
^{228}Ra	5.75 yr	β	$(4n)$
^{228}Ac	6.13 h	β, γ	$(4n)$
^{228}Th	1.91 yr	α	$(4n)$
^{230}Th	7.7×10^4 yr	α	$(4n + 2)$
^{232}Th	1.41×10^{10} yr	α	$(4n)$[b]
^{234}Th	24.1 d	β, γ	$(4n + 2)$
^{234}U	2.45×10^5 yr	α	$(4n + 2)$
^{238}U	4.47×10^9 yr	α	$(4n + 2)$[c]

[a] Singly occurring radionuclide.
[b] Thorium decay series.
[c] Uranium decay series.

therefore, all radioactive. All kinds of naturally occurring radionuclides might occur in plants, but their levels are quite different.

Artificial Radionuclides

The first significant release of artificial radionuclides to the environment began with the discharge of effluent in 1944 from the Hanford Atomic Plants in the United States to the northeastern Pacific Ocean via the Columbia River. Soon a series of nuclear weapons testing programs started in the United States and the Soviet Union. The nuclear weapons testings done in the open air have brought a remarkable amount of fission products to the atmosphere. The fission products released to the atmosphere circled the earth, carried by the westerly wind in the Northern Hemisphere, and were deposited on the ground surface as well as in the ocean. The global deposition of the radioactive fallout occurred mainly in late 1950s and early 1960s, in accordance with nuclear weapons testing programs taking place on a

TABLE 14-2 Artificial Radionuclides

Nuclide[a]	Half-Life	Radiation
^{3}H	12.28 yr	β
^{14}C	5730 yr	β
^{60}Co	5.271 yr	β, γ
^{63}Ni	100.1 yr	β
^{90}Sr	28.6 yr	β
^{95}Zr	64.02 d	β, γ
^{99}Tc	2.13×10^5 yr	β
^{103}Ru	39.35 d	β, γ
^{106}Ru	368.2 d	β
^{129}I	1.57×10^7 yr	β
^{131}I	8.04 d	β, γ
^{134}Cs	2.062 yr	β, γ
^{137}Cs	30.17 yr	β, γ
^{237}Np	2.14×10^6 yr	α
^{238}Pu	87.75 yr	α
^{239}Pu	24 131 yr	α
^{240}Pu	6 569 yr	α
^{241}Pu	14.4 yr	β
^{241}Am	432.2 yr	α, γ

[a] ^{3}H and ^{14}C are also found as naturally occurring radionuclides.

dense schedule during 1957–1958 and 1961–1962 (UNSCEAR, 1982). Representative and important artificial radionuclides, both fission and activation products, are listed in Table 14-2.

Peaceful use of nuclear energy is another source of artificial radionuclides in the environment. Under normal operations, nuclear power plants and the fuel reprocessing plants release trace amounts of radionuclides to the air through stacks and to the ocean/river through outlet of effluents. The radionuclides released under normal operations would be observed in a limited area around the nuclear facility; however, an accidental release of radioactivity sometimes causes a serious and large-scale environmental contamination, as we have experienced in the accidents of the Three Mile Island Nuclear Power Plant in the United States in 1979, and more recently at the Chernobyl Nuclear Power Plant in the Soviet Union in 1986. A variety of fission products were released to the environment from the both accidents.

Radionuclides of moderate half-life would be important for a long-term assessment of radioactive contamination in plants, but those of short half-life yield an acute irradiation on plants. The importance of such nuclear facilities as a potential source of environmental radioactivity would increase with development of big nuclear power plants and fuel reprocessing plants in future.

Concentrations of Naturally Occurring Radionuclides in Plants

Concentrations of naturally occurring radionuclides are low in plants growing in the general environment compared to those growing at a place where the radioactivity level is enhanced, indicating a positive correlation in radionuclide concentrations between plant and soil (Table 14-3). The environments of high concentrations of naturally occurring radionuclides appear around uranium mines (Ibrahim and Whicker, 1988a, 1988b; Summerton, 1992), in a region being exposed to contaminated discharges from uranium mining (Pettersson et al., 1993), and in an area where soil contains high concentrations of ^{238}U and ^{232}Th and their daughter nuclides (Linsalata et al., 1989). The plants growing around the uranium mines show a significant variation in radionuclide concentrations in accordance with a spatial variation in soil originating from mining activities.

The concentrations of the singly occurring natural radionuclides can be estimated by the stable isotope concentrations in plants and a known isotopic abundance of the radionuclide. For example, for ^{40}K and ^{87}Rb, the isotopic abundances are 1.17 and 27.83%, respectively. Bowen (1979) has summarized ^{40}K concentrations of 20–260 Bq kg^{-1} dry for lichens, 30–750 Bq kg^{-1} dry for trees, and 30–2100 Bq kg^{-1} dry for vegetables, and for ^{87}Rb, concentrations of 2–22 Bq kg^{-1} dry for lichens, 3–9 Bq kg^{-1} dry for trees, and 0.5–18 Bq kg^{-1} dry for vegetables, based on stable isotope concentration data.

Radionuclide concentrations in plants are characterized by a plant–soil concentration ratio, because the activity of the radionuclide in a plant is positively related to that in the soil. The concentration ratio is called a transfer factor (TF), and is defined as:

$$TF = \frac{\text{activity in plant (Bq kg}^{-1}\text{ dry or wet weight)}}{\text{activity in underlying soil (Bq kg}^{-1}\text{ dry or wet weight)}} \quad (14\text{-}1a)$$

A translocation factor (TLF), which is similar to TF, is also used. The definition of the TLF is:

$$TLF = \frac{\text{activity in plant (Bq kg}^{-1}\text{ dry or wet weight)}}{\text{activity in underlying soil (Bq m}^{-2}\text{)}} \quad (14\text{-}1b)$$

The TF and TLF describe the amount of radionuclide expected to enter plants from soil under equilibrium conditions, and the TF and TLF approach requires us to assume that the plant and soil concentrations are linearly related (Sheppard and Sheppard, 1985). Compilation of the TF and TLF values of various species has a practical aspect for prediction of radionuclide concentrations in plants; the values provide a measurement of soil activity that enables us to calculate radionuclide concentrations in many kinds of plant species growing in a particular area.

TF values for 37 elements have been compiled and authorized by International Atomic Energy Agency (IAEA) (1994). However, a wide range of TF and TLF values have been observed experimentally even for the same element in the same species. Factors affecting the TF and TLF are the concentration of the element and/ or radionuclide in soil and the proportion of that concentration actually available to plants, the physical and chemical characteristics of soil, and the age and metabolic activity of plants.

TABLE 14-3 Concentrations of Naturally Occurring Radionuclides in Plants

Species	Nuclide	Concentration (Bq kg^{-1} dry wt)	Location	Reference
Pine needles	^{3}H	0.8–3.0[a]	General environment, Japan	Takashima et al. (1987)
		1.4–3.6[b]		
		3.0–80.5[a]	Tokaimura, Japan	Momoshima et al. (1991)
		6.4–47[b]		
Rice grain	^{3}H	1.1–3.8[a]	General environment, Japan	Hisamatsu et al. (1991)
		0.7–2.2[b]		
	^{14}C	251–261[c]		Shibata and Kawano (1994)
Cherry, leaf	^{40}K	388–847	Verona, Italy	Scotti and Silva (1992)
Peach, leaf		702–2 130		
Apple, leaf		384–635		
Pear, leaf		258–498		
Hickory and oak	^{210}Pb	0.4–13	Illinois, US	Holtzman (1970)
Shrub samphire	^{234}Th	3.3	General environment, Australia	Summerton (1992)
	^{226}Ra	0.95		
	^{227}Ac	<0.4		
	^{228}Ra	1.6		
	^{40}K	436		
	^{238}U	1.76		
	^{230}Th	2.76		
	^{232}Th	1.57		

	Nuclide	Value	Location	Reference
	^{234}Th	70	Uranium treatment plant, Australia	Summerton (1992)
	^{226}Ra	124		
	^{227}Ac	28		
	^{228}Ra	62		
	^{40}K	250		
	^{238}U	68.7		
	^{230}Th	1672		
	^{232}Th	58.4		
Grasses and forbs	^{238}U	4.4	General environment, Wyoming, US	Ibrahim and Whicker (1988a)
	^{228}Th	11		
	^{230}Th	9.3		
	^{232}Th	1.9		
	^{226}Ra	4–11		Ibrahim and Whicker (1988b)
	^{238}U	41–407	Uranium and mill complex, Wyoming, US	Ibrahim and Whicker (1988a)
	^{228}Th	9.6–19		
	^{230}Th	363–2849		
	^{232}Th	2.2–3.3		
	^{226}Ra	12–348		
Vegetables	^{232}Th	6.5–14.6	Pocos de Caldas Plateau, Brazil	Ibrahim and Whicker (1988b)
	^{226}Ra	2.1–10.6		Linsalata et al. (1989)
	^{228}Ra	8.4–24.7		

[a] Free water fraction (Bq L^{-1} water).
[b] Organic fraction (Bq L^{-1} combustion water).
[c] Bq kg^{-1} carbon.

TABLE 14-4   Transfer Factor (TF) Values for Naturally Occurring Radionuclides[a]

Nuclide	Port Pirie, Australia[b]	Wyoming, USA[c]	Pocos de Caldas Plateau, Brazil[d]	Alligator Rivers Region, Australia[e]
^{210}Pb				0.009–0.031
^{210}Po				0.0095–0.029
^{226}Ra	0.066 (0.031)	0.14–0.43 (0.07)	0.0466	0.012–0.027
^{228}Ra	0.075 (0.045)		0.0236	
^{228}Th	0.080 (0.034)	0.16–0.36 (0.23)		0.014–0.034
^{230}Th	0.061 (0.061)	0.05–2.9 (0.2)		0.008–0.017
^{232}Th	0.057 (0.053)	0.04–0.07 (0.04)	0.00011	0.0045–0.010
^{234}Th	0.039 (0.070)			
^{234}U				0.007–0.015
^{238}U	0.039 (0.041)	0.05–0.8 (0.1)		0.0065–0.015

[a] The values were obtained for an environment with high radioactivity, while those in parentheses are for the natural environment.
[b] Shrub samphire (*Halosarcia halocnemoides*); Summerton (1992).
[c] Grasses and forbs; Ibrahim and Whicker (1988a, b).
[d] Vegetables; Linsalata et al. (1989).
[e] Waterlily (*Nymphaea violacea*); Pettersson et al. (1993).

The TF values reported for naturally occurring radionuclides belonging to the uranium and thorium decay series are summarized in Table 14-4. Linsalata et al. (1989) examined TF values of several kinds of vegetables growing at the Pocos de Caldas Plateau, Brazil, where concentrations of radionuclides and rare earth elements are 5–20 times higher than those in the soil of the United States. For the same type of vegetables, the TF values observed appear to be in reasonable agreement with those obtained in New York, in spite of the different soil characteristics. The geometric mean TF values have decreased as ^{228}Ra $>$ ^{226}Ra $>$ La $>$ Nd $=$ Ce $>$ ^{232}Th, or simply as M (II) $>$ M (III) $>$ M (IV).

The TF values for uranium and thorium isotopes on plants growing at the evaporation pond in the uranium and monazite treatment facility at Port Pirie, Australia, were not so different from those growing in the general environment; only the concentration of radium isotopes seemed to be higher for the plants growing at the evaporation pond (Summerton, 1992).

On the other hand, Ibrahim and Whicker (1988a, b) have observed higher TF values for U, Th, and Ra isotopes on plants growing at the uranium mine and mill complex in Wyoming, compared to those from the general environment. The TF value was highest at the edge of the tailing impoundment and decreased with a distance from the tailing pond. The authors stated that a low acidity (about pH 1.8) and wet condition at the pond site, as well as deposition of pond water spray and dry particles on plants, are possible reasons for the high TF values in this case. Within the Pocos de Caldas Plateau, Brazil, Linsalata et al. (1989) observed high soil pH values, ranging from pH 4.6 to 7.2.

Concentrations of Artificial Radionuclides in Plants

The accident at the Chernobyl Nuclear Reactor occurred on April 26, 1986; it caused extensive contamination in plant populations and communities. Radioactive materials initially released spread with winds in a northerly direction and were subsequently released toward the west and southwest and other directions. Many kinds of fission and activation products, having different half-lives, were identified by filter sampling. Deposition of radioactive material on the ground was very inhomogeneous, mainly occurring in association with rain. Among the radionuclides deposited, much attention has been paid to ^{137}Cs and ^{134}Cs because of their long half-lives and a large amount of deposition. The concentrations of ^{137}Cs and ^{134}Cs in plants have been measured extensively in relation to plant physiology and their cycling in forest ecosystems, and the results are given in many publications. The activities of ^{137}Cs and ^{134}Cs, both gamma-emitting radionuclides, are detected relatively easily by gamma spectrometry, which makes it possible for us to measure many samples without complex chemical treatments.

Concentrations of ^{137}Cs driven from the Chernobyl fallout varied widely in plants, depending upon species, sampling location, and time of sampling, as shown in Table 14-5. The sampling location would be the primary factor controlling ^{137}Cs levels in plants because the deposition occurred in quite irregular patterns; differences of more than 2 orders in ^{137}Cs concentrations have been observed in plants, for example, in fern collected in Finland (Jasinska et al., 1990). Plants sampled at the same location but over several years after the Chernobyl accident showed an overall decline in ^{137}Cs concentration, as shown by Jackson (1989). Accumulation of ^{137}Cs and ^{134}Cs in plants can be considered to occur in two ways: one is interception and retention of fallout by plant surface, and the other is root uptake. A plant's ability to intercept and retain fallout is dependent upon the structure and surface condition of the leaf. Mechanical removal of radionuclides from the plant surface would begin soon after the interception, and is realized to be important during the relatively early period after atmospheric deposition. The root uptake would begin soon after atmospheric deposition on plants with shallow root system, such as grasses. Most of ^{137}Cs detected in annual plants after 1987, the year after the Chernobyl accident, must have been supplied by root uptake, though a small quantity may be absorbed from the leaves following resuspension of soil particles. The perennial plants showed higher activities than the annuals (Livens et al., 1991).

^{137}Cs in Plants from the Chernobyl Accident and the Nuclear Weapons Tests

Occurrence of ^{137}Cs in plants has resulted from the fallout due to the Chernobyl accident and from nuclear weapons testing. The relative contributions of ^{137}Cs from above two sources to the plants could be evaluated using ^{134}Cs because concentrations of that nuclide due to the nuclear weapons testing no longer remained in the

TABLE 14-5 Artificial Radionuclides in Plants

Species	Year	Nuclide	Concentration (Bq kg^{-1} dry wt)	Location	Reference
Spruce (Picea excelsa), 2 y-needles	1987	^{137}Cs	20–410	Southern Poland	Jasinska et al. (1990)
		^{134}Cs	6–160		
	1988	^{137}Cs	34–875	Finland	
		^{134}Cs	0–200		
Fern (Athyrium sp.)	1987	^{137}Cs	99–15890		
		^{134}Cs	47–5430		
Apricot, fruit	1987–1990	^{137}Cs	0.5–6	Northern Greece	Antonopoulos-Domis (1991)
Apricot, leaf			0.6–10.5		
Olive, fruit			1.4–13		
Olive, leaf			1.9–15		
Tea leaves	1986	^{137}Cs	4500–20 000	Turkey	Ünlü et al. (1995)
	1987		250–600		
Various kinds of plant species	1989	^{137}Cs	140–9140	Northeastern Italy	Livens et al. (1991)
		^{134}Cs	30–1720		
		^{137}Cs	100–52 500	Norway	
		^{134}Cs	18–10 500		
		^{137}Cs	23–5860	Scotland	
		^{134}Cs	38–1170		
Flowers (black locust, sweet chestnut, and white clover)	1987	^{137}Cs	9.3–64.3	Piedmontese, Italy	Bonazzola et al. (1991)
		^{134}Cs	3.1–23.1		
Grass (Trichophorum caespitosum)	1987	^{137}Cs	240–1800	Babaria, Germany	Bunzl and Kracke (1989)
Grass (Molinia coerulea)			320–4000		
Heather (Calluna vulgaris)			5000 and 10 000		
Ling (Calluna vulgaris)	1986	^{137}Cs	2310–8880	West Cumbria, England	Jackson (1989)
		^{134}Cs	1319–5510		
	1987	^{137}Cs	<275–2489		
		^{134}Cs	300–4180		

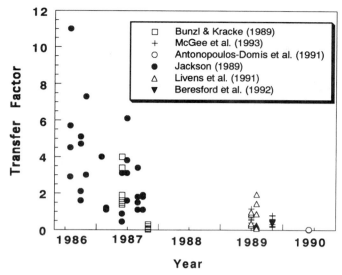

Figure 14-1. Change of ^{137}Cs transfer factor (TF) with time.

ground when the Chernobyl accident occurred in 1986. By using an activity ratio of ^{137}Cs/^{134}Cs, Beresford et al. (1992) estimated TLF values of 0.03 and 0.05 m^2 kg^{-1} for the radiocesium from the Chernobyl fallout, while TLF values were 0.007 and 0.013 m^2 kg^{-1} for the aged ^{137}Cs from the nuclear tests. The larger TLF values for the Chernobyl fallout suggest a preferable chemical and physical form of the young radiocesium in respect to root uptake. This is more clearly seen in Fig. 14-1, in which TF values for various plant species are plotted as a function of sampling time. The TF values apparently decreased with time and became similar to the aged ^{137}Cs a few years after the Chernobyl accident. Since no marked decline in ^{137}Cs concentration in soil was observed, the decrease in the TF values resulted from the decrease in ^{137}Cs availability in plants. Adsorption and fixation of radiocesium by soil minerals would be responsible for the decrease in availability.

PATHWAYS OF RADIONUCLIDES TO PLANTS

When radionuclides are released to the atmosphere from a nuclear facility, an initial contamination of plants occurs by interception and retention of atmospheric deposition, regardless of whether the release is being carried out under controlled or uncontrolled conditions. The efficiency of the interception and retention by plants varies with shapes, structure, and surface conditions of foliage. Needles of conifers have been realized to have more effective structure for trapping particles than the leaves of deciduous trees, which are usually broad, flat, and smooth (Ronneau et al., 1991). The foliage is considered to be a kind of filter that intercepts and retains atmospheric particles and the deposition of radionuclides to foliage would occur effectively in a well-developed forest. The radionuclides deposited are eventually removed from the surface of the foliage by meteorological factors, such as rain and wind, and the removal rate depends upon the surface microstructure and on the condition of the foliage, as well as on the physical and chemical characteristics of the radioactive particles. The removal rate, which is a measure of the retention capability of the foliage, is defined as weathering half-time. The weathering half-times reported have ranged from a few days to several hundreds of days.

The radionuclides trapped on the surface of foliage are called external contamination. If radionuclides are incorporated into the tissues of plants, that is called internal contamination. A portion of the radionuclides deposited on the surface of plants is absorbed into the tissue through the cuticle layer of the foliage, and it is often difficult to distinguish between the radionuclides that are tightly adsorbed on the surface of the foliage and those that are absorbed within the tissues of the leaf. Clear distinction of radionuclides has not been given in many reports between external and internal contaminations. The radionuclides on or within foliage ultimately accumulate on the ground, due to the weathering process and fallen leaves, and they are incorporated into a nutrient cycle where root uptake is the main process. Resuspension and rainsplash of soil particles can increase radionuclide concentrations in plants, but contribution of these processes would be small compared to root uptake. The resuspension process would become important for radionuclides existing in soil in a form less available to plants.

Contamination of Plants by Atmospheric Deposition and Weathering Half-Time

Interception and retention of atmospheric deposition causes external contamination of plants. Time course of radionuclide inventory on vegetation from atmospheric deposition is expressed using a equation of the form (Pinder, 1991):

$$\frac{dA(t)}{dt} = [1 - \exp{(-\mu(B(t))})]D - \lambda e A(t) \tag{14-2}$$

where $A(t)$ is the radionuclide concentration on vegetation at time t (Bq kg^{-1} or Bq m^{-2}), D is the deposition rate (Bq m^{-2}t^{-1}), $B(t)$ is the vegetation biomass (kg m^{-2}) at time t, μ is the absorption coefficient (m^2 kg^{-1}), and λe is the effective loss rate constant (t^{-1}). The effective loss rate constant is the sum of the loss rate constants due to weathering of radionuclides from plant surface and radioactive decay, and is expressed:

$$\frac{1}{Te} = \frac{1}{Tw} + \frac{1}{Tr} \tag{14-3}$$

where Te, Tw, and Tr are the effective half-time, the weathering half-time, and the radioactive half-life, respectively. It should be noted that the implication of the weathering half-time varies with time and duration of the measurement that was performed to evaluate the weathering half-time. If the weathering half-time is evaluated based on the measurement for a short-term period after atmospheric deposition, meteorological factors would be a main contributor. However, if the weathering half-time is evaluated based on a long-term measurement, the weathering half-time would include other environmental factors relating to root uptake, such as fixation of radionuclides on soil matrix and amendment of biological availability. Some authors, then, use the term "biological half-time" instead of "weathering half-time" (Uchida et al., 1991). From equation (14-2), the concentration of radionuclides on vegetation increases with time and approaches a steady-state value as the plant grows up under the condition of a constant deposition flux of radionuclides. This situation would occur around nuclear installations, those chronically releasing radionuclides to the atmosphere. For the case of accidental releases, the radionuclide concentration is expected to decrease exponentially with time after atmospheric deposition because interception and retention would occur one time.

In equation (14-2), a constant rate of the weathering half-time is assumed. However, a change of weathering half-time has been observed in relation to the Chernobyl accident. Koprdá (1991) measured concentrations of the mixture of radionuclides on various kinds of foliage during an 80 day period after the Chernobyl accident. The activities in the leaves, collected in Bratislava, Czechoslovakia, decreased with time. The estimated effective half-time ranged from 4 days when measured 10 days after the accident to 150 days at 60 days. The weathering half-time became longer as time passed. The author also observed different levels of contamination on foliage, depending on the surface structures of the leaves, hairy surface leaves having higher contamination than smooth ones. The change of the weathering half-time could be attributed to heterogeneous size composition of particles intercepted and retained on foliage. Larger particles tend to be removed faster than smaller ones.

Weathering half-time would be changeable with changing weather conditions during or soon after atmospheric deposition. Removal of ^{134}Cs was examined using wheat and tomato that were grown in an experimental field and contaminated by a spray of solution containing ^{134}Cs (Scotti et al., 1993). The sample groups sprinkled with water at 6 hours after contamination showed a great reduction in ^{134}Cs concentrations compared with those of the nonirrigated ones; the wheat grain and tomato fruit sprinkled showed 80 and 55% lower concentrations, respectively, than the nonirrigated ones. The shorter the time between contamination and sprinkling, the more radioactivity was removed (Scotti, 1994).

Weathering half-time is also changeable with growing states of plants. Big sagebrush *(Artemisia tridentata)* and squirreltail bottlebrush *(Elymus elymoides)* contaminated with ^{141}Ce and ^{134}Cs aerosols at different seasons showed a longer weathering half-time in summer than in spring (Fraley et al., 1993). The weathering half-time in summer ranged from 22 to 42 days and that in spring from 9 to 15 days. The authors pointed out a significant growth dilution at the spring experiments.

Absorption by Foliage

Radionuclides are certainly absorbed from leaf and translocated to other part of the plant. The translocation of radionuclides is known to be correlated with metabolism of plants. Absorption of ^{137}Cs by fruit leaves (cherry, peach, apple, and pear) and subsequent translocation and incorporation to the edible parts were investigated in association with the Chernobyl accident (Scotti and Silva, 1992). The fruits harvested in the autumn of 1986 showed less variable ^{137}Cs activities than the leaves, suggesting that absorption and translocation are related more to genetic factors of the individual species than to the amount of radioactivity on the foliage. Rapid absorption and translocation of ^{137}Cs from the Chernobyl fallout was observed in spruce tree. A considerable amount of ^{137}Cs was detected in newly formed needles whose buds had not yet been open at the time of the accident, suggesting rapid absorption of ^{137}Cs by older needles (Ronneau et al., 1991). Tent et al. (1993) carried out sequential extraction experiments using vegetables contaminated with different radionuclides to get information about absorption by leaf. The surfaces of lettuce plants were contaminated with ^{134}Cs, ^{85}Sr, and ^{110m}Ag solutions and washed successively with water and then chloroform, to differentiate fractions of water soluble, adhering to cuticle, and particulate. The authors observed quicker absorption of ^{134}Cs than of the other radionuclides, and all radionuclides examined showed an increase in uptake by leaf with time.

No measurable absorption and translocation of ^{238}Pu has been observed on orange trees growing at the H-Area nuclear fuel chemical separations facility of the Savannah River Plant, USA, at which ^{238}Pu-bearing particles, primarily 1μm agglomerates of plutonium metal or plutonium oxide, are released to the atmosphere (Pinder et al., 1987). The negligible transfer of plutonium from leaf to fruit would be a result of the insoluble nature of these particles, because the weathering half-time of ^{238}Pu was not so different from that of other radionuclides. The weathering half-times of ^{238}Pu observed ranged from 33 to 99 days. As far as contamination of fruit with atmospheric deposition of plutonium, the authors postulate the importance of the direct interception of ^{238}Pu by fruits rather than leaf absorption followed by transfer to fruits. Iodine-131 was known to be absorbed quickly from both leaves and stem. Singhal et al. (1994) examined uptake of ^{131}I by fenugreek *(Trigonella foenum-graecum)* and okra *(Hibiscus esculentus)*, both very common vegetables in India. The plants were contaminated with airborne ^{131}I and variations of ^{131}I activity in leaves and stem were measured. The absorbed fraction increased with time and more than 80% of ^{131}I was absorbed at 100 and 200 hours after the contamination.

Resuspension

Contamination of plants by resuspension of contaminated soil was evaluated in wheat, soybeans, and corn grown in experimental fields where the soil was contaminated with ^{238}Pu (Pinder and McLeod, 1989). The amount of soil on plant surfaces, estimated from the concentrations of ^{238}Pu, did not differ significantly among crops. The amounts of soil evaluated were 4.8, 2.1, and 1.4 mg g^{-1} for wheat, soybeans, and corn, respectively, and the mean inventories were 1.3, 1.4, and 1.2 g m^{-2}, respectively. Based on the above results, model calculations were carried out, using a hypothetical scenario in which a nuclear facility operated for 30 years with a constant deposition rate of 100 mBq m^{-2} d^{-1} of ^{239}Pu (Pinder et al., 1990), and the contributions of different pathways to crop contamination were evaluated. The ^{239}Pu concentration predicted for corn grains due to the three pathways were estimated to be 10 mBq kg^{-1} from atmospheric deposition, 0.05 mBq kg^{-1} from root uptake, and 0.005 mBq kg^{-1} from resuspension, respectively. When a contamination source is an atmospheric releases of radionuclides, atmospheric deposition on plant surfaces is a primary contributor of contamination to crops, while the contribution of resuspension is small, only a few percent, but that of root uptake would be variable, low in plutonium but certainly high in ^{137}Cs.

Root Uptake

Root uptake of radionuclides is a major process that causes internal contamination of plants. It is true that any kind of radionuclide can be taken up by plants, but the amounts vary according to the radionuclide. Many factors are known to influence root uptake: physical and chemical characteristics of the radionuclides; type and physicochemical properties of soil; interaction of radionuclides with soil; plant species; the relative position of rooting zone to the contaminated soil layer; and climate. The root uptake of ^{137}Cs has been examined widely under natural conditions with respect to the Chernobyl accident.

Ünlü et al. (1995) examined ^{137}Cs concentrations in newly formed fresh tea leaves, growing in the eastern Black Sea region in Turkey, at every April–May, June–July, and August–September from 1986 to 1992. A large portion of the ^{137}Cs (94.8% of the activity) had decreased with an effective half-time of 125 days in 1986, suggesting a rapid removal from the leaves by rain and wind. A longer effective half-time, 1114 days, was observed for the following years, indicating root uptake of ^{137}Cs and translocation to the leaves. The weathering half-time calculated from equation (14-2) is 3.4 years, suggesting that ^{137}Cs fixed in the soil matrix becomes a potential source for root uptake. Decreases in ^{137}Cs concentration in plants and milk were also observed in southern Chile over the years from 1982 and 1990 (Schuller et al., 1993). All of the ^{137}Cs in the soil was considered to be due to fallout from nuclear testing, because no significant increase in ^{137}Cs concentration was observed after the Chernobyl accident. The concentrations of ^{137}Cs in prairie plants and milk have decreased, showing very similar effective half-times, about 5 years, while the concentration of ^{137}Cs in soil has decreased at a much slower rate, which is consistent with the physical decay, 30 years. The consistent decline in ^{137}Cs concentration in soil with the physical half-life suggests a progressive irreversible immobilization of ^{137}Cs within soil.

Root uptake of ^{137}Cs is depressed by increasing the concentration of stable cesium (Shaw and Bell, 1991), suggesting no isotopic discrimination among cesium isotopes during root uptake. A similar effect has been observed for cations that have a chemical nature generally equivalent to Cs$^+$, such as K$^+$ and NH^{4+}, and the absorption rate of ^{137}Cs is known to obey Michaelis–Menten kinetics (Shaw and Bell, 1991). Plants such as *Agrostis capillaris* and *Calluna vulgaris,* growing widely in upland areas of Britain where high Chernobyl fallout occurred, were cultivated with varying concentrations of K$^+$ in the rooting medium and showed a negative relationship between the uptake rate of ^{137}Cs and the concentrations of K$^+$ in the plant tissue (Jones et al., 1991). The effect of fertilizers, K$^+$, N, and P, on ^{137}Cs uptake has been studied, and a decline of ^{137}Cs uptake following application of K$^+$ fertilizer has been shown (Belli et al., 1995). Sandalls and Bennett (1992) examined TLF values for ^{137}Cs on grasses growing at soils with different character-istics in west Cumbria and observed positive correlations of the TLF values for ^{137}Cs with the organic matter content and the amounts of exchangeable potassium and ^{134}Cs in the soil, although a negative correlation of the TLF values for ^{137}Cs with the ^{134}Cs and potassium concentrations in soil water were observed at field capacity. These authors' observations are consistent with those of other researchers mentioned above. High concentrations of exchangeable K and ^{134}Cs in soil suggest a high concentration of ^{137}Cs in exchangeable form in the same soil, and also a high concentration in soil water, leading to a positive correlation, while an increase in potassium and ^{134}Cs concentrations in soil water would suppress the absorption of ^{137}Cs by root uptake because of competition with ^{134}Cs or potassium. Robison and Stone (1992) examined the effect of K$^+$ on the uptake of ^{137}Cs in coconut growing at Bikini Atoll on coral soils contaminated by fallout from thermonuclear explo-sions; they concluded that periodic additions of K$^+$ at rates of about 1000 kg ha^{-1} would provide a feasible and highly effective means of reducting ^{137}Cs contamina-tion in coconut food products.

Pu isotopes and ^{241}Am showed very poor root uptake and translocation compared to ^{137}Cs (Coughtrey et al., 1984). Distribution of plutonium and americium in mango trees growing near a solid waste disposal area was generally in this order: leaves > cotyledon > shell of fruit > fruit flesh, suggesting that the least amounts of plutonium and americium are found in the edible portion of the mango fruit (Matkar et al., 1994).

Seasonal Variation of Radionuclide Concentration in Plants

Bunzl and Kracke (1989) determined ^{137}Cs, ^{134}Cs, and K concentrations in intervals of about 14 days from June to November 1987 on three plant species grown in a peat bog contaminated by the Chernobyl fallout. The purpose of these experiments was to examine the seasonal variability of K and ^{137}Cs between the evergreen plant (*Calluna vulgaris*) and the grass species (*Trichophorum caespitosum* and *Molinia coerulea*), which have perennial roots but sprout every year while the old leaves wither. The concentrations of ^{137}Cs and K in the grass species decreased consider-ably in a similar manner during the growing season, while that of the evergreen plant was rather constant during 1987. The constant concentration ratios ^{137}Cs/K, however, were observed in all plant species, in spite of significantly different sea-sonal concentrations of ^{137}Cs and K, suggesting similarities with respect to the behavior of ^{137}Cs and K in these plants. The authors explain that the significant decrease in ^{137}Cs and K concentrations in the grass species are due to the transloca-tion to the roots in autumn, while in the evergreen plant most of the nutrients can be stored in the leaves and stems during winter.

Seasonal variation of ^{137}Cs concentrations in grass plants have reported by Rudge et al. (1993). They observed an exponential decline in ^{137}Cs from June 1986 through to the summer of 1987 after the Chernobyl accident, attributing it partly to mechanical loss and partly to growth dilution, because the deposition occurred at the start of of the growing season. However, for the samples collected in late 1987 they observed an increase in ^{137}Cs concentrations, and the peak was followed by a decline during 1988. The tempory increase in ^{137}Cs in plants would be associated with translocation of elements within plants taking place prior to leaf fall or die-back. The senescent samples in November 1987 had markedly lower levels of ^{137}Cs than live ones, supporting the movement of ^{137}Cs from old to new tissue.

Declines in ^{137}Cs concentrations in fruits and their leaves were measured for fruit trees grown in northern Greece by Antonopoulos-Domis et al. (1991). The concentration of ^{137}Cs decreased exponentially from 1987 to 1990 with an effective half-time of 0.7 year, which is rather short compared to that observed in tea leaves (Ünlü et al., 1995) and prairie plants (Schuller et al., 1993). According to the authors, most of the ^{137}Cs accumulated in the trees was intercepted by leaves and absorbed through the leaves, not taken up by the roots, because the trees planted before the Chernobyl accident had a ^{137}Cs concentration in leaves about one-third that of the trees planted after the Chernobyl accident. The trees preplanted had deep rooting systems that were less likely to absorb ^{137}Cs, which accumulated mostly at the surface layer of the soil. The root systems of the postplanted trees, however, were distributed near the surface layer. The same authors described long-term ^{137}Cs contamination in trees by a model in which they considered a translocation of ^{137}Cs to fruits from a tree body (reservoir) and a conversion of ^{137}Cs from available form for fruit contamination to unavailable one (Antonopoulos-Domis et al., 1990). They estimated a very small flow of ^{137}Cs every year from the inventory to new tree prod-ucts, and therefore, a small rejection of ^{137}Cs from trees by annual removal of fruits.

RADIONUCLIDES IN XYLEM

Radial Distribution of Radionuclides in Annual Rings

Occurrence of radionuclides in wood can be classified into two types, based on the chemical forms: one is radionuclides incorporated into the structure of xylem by photosynthesis, such as ^{3}H, ^{14}C, and ^{35}S, and the other is those existing as cation or salt. The radionuclides in cationic form are fixed on negatively charged binding sites on the cell walls in consequence of the interaction between sap and binding site. Radial distribution of radionuclides in annual rings in aged tree trunks gives physiological information concerning migration and fixation of elements in xylem. Concentrations of radionuclides in xylem in various tree species are summarized in Table 14-6. The radioactive fallout due to nuclear weapons testing enables us to

TABLE 14-6 Radionuclide Concentrations in Xylem

Species	Nuclide	Concentration (Bq kg⁻¹ dry wt)	Location	Reference
Red spruce (*Picea rubens*)	^{90}Sr	0.3–18.0	Tennessee, USA	Momoshima and Bondietti (1994)
Eastern hemlock (*Tsuga canadensis*)		0–10.4	Tennessee, USA	
Eastern white pine (*Pinus strobus*)		1.4–3.3	Tennessee, USA	
Hickory (*Carya*)		0–2.0	Tennessee, USA	
Elm (*Ulmus*)		0.35–2.6	Tennessee, USA	
American beech (*Fagus grandifolia*)		0.45–8.1	Tennessee, USA	
Sugar maple (*Acer saccharum*)		3.9–7.3	Tennessee, USA	
Yellow poplar (*Liriodendron tulipifera*)		1.5–3.1	Tennessee, USA	
Red spruce (*Picea rubens*)		0–12	Tennessee, USA	Bondietti et al. (1990)
Japanese cedar (*Cryptomeria japonica*)		0–1.5	Tokyo, Japan	Chigira et al. (1988)
Red spruce (*Picea rubens*)	^{137}Cs	4.4–7.2	Tennessee, USA	Momoshima and Bondietti (1994)
Eastern hemlock (*Tsuga canadensis*)		1.4–4.8	Tennessee, USA	
Eastern white pine (*Pinus strobus*)		0.07–0.37	Tennessee, USA	
Hickory (*Carya*)		0.15–0.65	Tennessee, USA	
Elm (*Ulmus*)		0.09–0.28	Tennessee, USA	
American beech (*Fagus grandifolia*)		0.85–2.8	Tennessee, USA	
Sugar maple (*Acer saccharum*)		0–0.33	Tennessee, USA	
Yellow popular (*Liriodendron tulipifera*)		0.04–0.07	Tennessee, USA	
Japanese cedar (*Cryptomeria japonica*)		0.09–0.7	Tokyo, Japan	Chigira et al. (1988)
Cypress (*Chamaecyparis obtusa*)		0.45–0.73	Chiba, Japan	Kohno et al. (1988)
Japanese cedar (*Cryptomeria japonica*)		0.09–1.5	Ehime, Japan	
Japanese cedar (*Cryptomeria japonica*)		0.08–1.0	Nagasaki, Japan	Kudo et al. (1993)
Japanese cedar (*Cryptomeria japonica*)	$^{239+240}$Pu	0.4–27[a]	Nagasaki, Japan	Kudo et al. (1993)
Hickory and oak	^{210}Pb	0.4–13	Illinois, USA	Holtzman (1970)

[a] mBq kg⁻¹ dry weight.

study lateral movement of radionuclides in the xylem by comparing their radial distribution patterns and a known historical accumulation pattern at the ground.

Carbon-14 and ^{3}H incorporated into the cellulose structure in annual rings by photosynthesis are chemically stable and have quite slow metabolic turnover rates. Their radial distributions are known to follow the change of $^{14}CO_2$ concentrations in air (Cain and Suess, 1976; Grootes et al., 1989) and of ^3H_2O concentrations in rain (Kozák et al., 1993; Yamada et al., 1989). The atmospheric concentration of $^{14}CO_2$ in the Northern Hemisphere increased by about 200% in early 1960s, due to atmospheric nuclear testing, and the concentration of ^{3}H in rain also increased, sometimes a few orders higher level being observed. Radial distribution of ^{3}H in xylem of Japanese cedar *(Cryptomeria japonica)* grown in Fukuoka, Japan, is shown in Fig. 14-2, together with ^{3}H concentrations in rain recorded in Tokyo and in wine produced at Kofe. The historical change of ^{3}H concentration in rain—the initial increase in 1954 and the maximum in 1963 followed by decline—were well recorded in the annual rings of Japanese cedar without time lag. The ^{3}H concentrations in the annual rings were determined for isolated cellulose fractions from cell walls. Cellulose has 10 hydrogen atoms in a molecule in which 3 atoms in hydroxyl group are labile. Historical information on ^{3}H no longer remains in hydroxyl groups.

Radial distribution of ^{90}Sr in xylem of Japanese cedar *(Cryptomeria japonica)* grown in Fukuoka, Japan is shown in Fig. 14-3. The ^{90}Sr distribution almost follows the cumulative deposition of radioactive fallout at the ground. Similar results

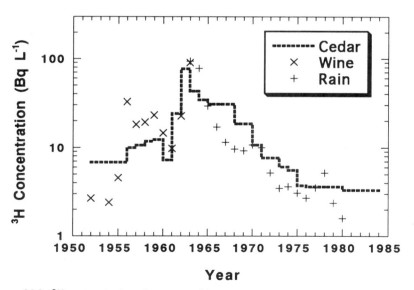

Figure 14-2. ^{3}H concentrations in annual rings of Japanese cedar (Cryptomeria japonica) felled at Fukuoka, Japan, in 1985. Cellulose fraction was isolated chemically and tritium activity was measured. For comparison concentrations in wine produced at Kofu, Japan (Takahashi et al., 1969) and in rain collected at Tokyo and Tsukuba, Japan (Katsuragi et al., 1983) are shown.

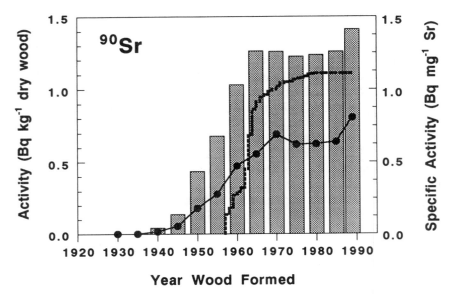

Figure 14-3. *Radial distribution of* [90]*Sr in Japanese cedar* (Cryptomeria japonica) *grown at Fukuoka, Japan, and felled in 1989. The concentrations of* [90]*Sr (Bq kg*[-1] *dry wood) are presented by bars and the specific activity (Bq mg*[-1] *Sr), by circles. The cumulative deposition of fallout in the Northern Hemisphere is shown by the broken line (UNSCEAR, 1982). Most nuclear explosions in the atmosphere occurred before 1963, and the atmospheric tests after 1962 were small in comparison with the earlier explosions; they ceased completely after 1980 (UN-SCEAR, 1988).*

have reported by Chigira et al. (1988) for Japanese cedar trees grown at different locations in Japan. Momoshima and Bondietti (1994) analyzed radial distributions of ^{90}Sr and ^{137}Cs in eight tree species grown in the United States; the radial distributions of ^{90}Sr were similar to Japanese cedar in all gymnosperms analyzed, red spruce *(Picea rubens)*, eastern hemlock *(Tsuga canadensis)*, and white pine *(Pinus strobus)*. Three of five angiosperms analyzed, hickory *(Carya)*, elm *(Ulmus)*, and American beech *(Fagus grandifolia)*, also showed radial patterns similar to Japanese cedar, but the highest ^{90}Sr concentration was observed in the 1970s, about a decade behind the historical maximum. The deep root systems of the angiosperms would correspond to the vertical migration time interval of ^{90}Sr, from the soil surface to the rooting zone. No correlation with the historical record of cumulative deposition was observed in two other angiosperms, yellow poplar *(Liriodendron tulipifera)* and sugar maple *(Acer saccharum)*, but the reason is not clear.

Radial distribution of ^{137}Cs in Japanese cedar xylem is shown in Fig. 14-4, which is the same sample analyzed for [90]Sr (Fig. 14-3). The distribution patterns are quite different from those of [90]Sr. In contrast to ^{90}Sr, ^{137}Cs in xylem does not present any correlation with deposition pattern of fallout. The concentration of ^{137}Cs in the heartwood was significantly higher than that in the sapwood; however the similar distribution pattern of stable cesium to ^{137}Cs results in relatively constant

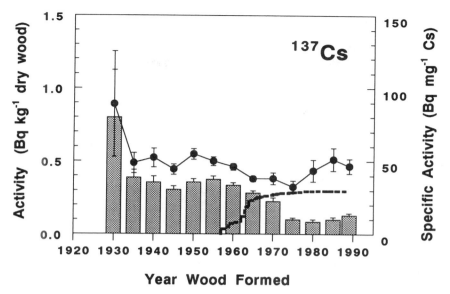

Figure 14-4. Radial distribution of ^{137}Cs in Japanese cedar (Cryptomeria japonica) grown at Fukuoka, Japan, and felled in 1989. The concentrations of ^{137}Cs (Bq kg^{-1}dry wood) are presented by bars and the specific activity (Bq mg^{-1} Cs), by circles. The cumulative deposition of fallout in the Northern Hemisphere is shown by the broken line (UNSCEAR, 1982). Most nuclear explosions in the atmosphere occurred before 1963, and the atmospheric tests after 1962 were small in comparison with the earlier explosions; they ceased completely after 1980 (UNSCEAR, 1988).

specific activities from the pith to the mostouter ring, suggesting a mixing of cesium isotopes and active lateral movement in the xylem. All of the radial distributions reported on ^{137}Cs do not follow the cumulative deposition of fallout at the ground (Brownridge, 1984; Chigira et al., 1988; Katayama et al., 1986; Kohno et al., 1988; Momoshima and Bondietti, 1994). Behavior of ^{40}K is supposed to resemble ^{137}Cs because of the similar radial distribution patterns in xylem to ^{137}Cs (Brownridge, 1984).

Radial distribution of $^{239+240}$Pu in a Japanese cedar tree grown in Nagasaki, Japan, was analyzed (Kudo et al., 1993). Nagasaki is the first city attacked by the plutonium atomic bomb in 1945. The 78-year-old tree growing 2.8 km east of the hypocenter, where the local fallout of the 1945 blast was highest, was harvested and analyzed at 3-year intervals. The radial distribution pattern of plutonium showed that the plutonium was immobile in the xylem and the tree recorded the history of plutonium in the surrounding environment of Nagasaki.

There are quite a few reports on the radial distribution of naturally occurring radionuclides except ^{40}K. Radial distribution of ^{210}Pb (half-life 22.3 years), measured in white oak (Holtzman, 1970), has decreased with an apparent half-time of about 22 years from the cambium toward the pith, suggesting negligible lateral movement of ^{210}Pb after assimilation.

Chemistry of the Binding Site for Radionuclides

The differences in the distribution patterns for ^{90}Sr and ^{137}Cs shown in Figs. 14-3 and 14-4 could be attributed to the different chemical and metabolic behaviors of strontium and cesium in trees. The radial distribution of ^{90}Sr suggests negligible lateral movement of strontium in xylem, especially gymnosperms, while there is rapid movement and mixing of cesium isotopes. Strontium and other cations are fixed on negatively charged binding sites in the cell wall by electrostatic interaction; the chemical species functioning at the binding site at the range of sap pH are probably carboxyl groups. Different pK values are observed on different types of carboxyl groups, but the carboxyl groups in cell wall probably have pK values between 3 and 4. The distribution of cations between sap and binding site is explained theoretically by the Donnan model (Momoshima and Bondietti, 1990). According to the Donnan model, the distribution changes with the chemical composition of cations and their concentrations in sap, sap pH, and concentration of the binding site.

The interaction of cations with binding site occur during upward movement of cations from base to top of the stem; cations absorbed by root ascend in the stem to the crown, interacting with binding sites in the cell walls. Chromatographic fractionations of cations based on charge density and cation concentration in stem were examined using 15 radioactive tracers in the laboratory (Wolterbeek et al., 1984), and an integrated result of this relationship was observed in red spruce grown in Tennessee (Momoshima and Bondietti, 1990). The concentrations of alkaline earth cations at different vertical positions in the stem were analyzed, and the relative concentrations of cations to Ca^{2+} were compared at different vertical positions. It was observed that the translocation rate of Mg^{2+} from the base to the top of the stem was faster than that of Ca^{2+}, while Sr^{2+} showed a translocation rate comparable to that of Ca^{2+}, and the translocation of Ba^{2+} was slower than Ca^{2+}. These observations coincided with a relationship expected from ion exchange theory.

The negatively charged binding site could react with positively charged ions; however the affinity of cations to the binding site varies from element to element, as observed on the fractionation of alkaline earth elements in the stem of red spruce. The most important factor producing the difference in the interaction is the charge density of the cation; divalents have larger affinity to the binding site than do monovalents. Hence a change in the cation composition of the sap, such as a change of the concentration ratio of monovalent to divalent, might appear in a more pronounced manner at the binding site. The change in the isotopic composition of the strontium in soil solution, the ratio of ^{90}Sr to Sr over time, due to fallout, was introduced into sap without isotopic fractionation, and also into the binding site because the chemical characteristics of ^{90}Sr and stable Sr are almost identical.

The distribution of cations between sap and cell wall will change with the binding capacity of the cell wall. The binding capacity would be proportional to the number of binding sites in the cell wall, which is correlated with the contents of uronic acid in the xylem (Crooke et al., 1964), and therefore, varies among species. The binding capacity of red spruce decreases almost linearly from the pith toward the cambium (Momoshima and Bondietti, 1990), suggesting physiological control in production of binding sites. The binding sites spread throughout the cell wall (Van de Geijn and Petit, 1979), and the portion of the cell wall where the immobile anions are distributed is known as the Donnan free space. Estimation of the volume of Donnan free space is difficult, but Bovis and Briggs (1968) estimated the volume of Donnan free space at 4% in yew wood or 800 meq L^{-1} xylem, using a very complex set of measurements and assumptions. Momoshima and Bondietti (1990) estimated the volume of Donnan free space to range from 400 in heartwood to 200 meq L^{-1} in sapwood in red spruce.

Trees conduct solutes in many rings, so that binding sites in the wood are exposed to changing sap concentrations. Conduction of sap water is carried out by tracheids or vessels in the annual rings in the sapwood. As an older tracheid or vessel loses its ability to conduct water, the equilibrium condition established between sap and binding site becomes isolated from the water conducting system, and will be preserved if lateral movement and mixing do not occur in future, thus changing the specific activity of ^{90}Sr observed in the wood (Fig. 14-4). The constant specific activity of ^{90}Sr from the 1960s to the 1980s, shown in Fig. 14-4, suggests a steady-state nutrient availability in this Japanese cedar for recent decades. As long as the sap chemistry is maintained at this steady-state condition, the distribution of cations between sap and binding sites will be constant. Bondietti et al. (1990) observed a rapid decline, after the 1950s, in ^{90}Sr specific activity and concentration in annual rings of red spruce grown in Tennessee. The decrease in ^{90}Sr specific activity in the wood was accompanied by a compensatory increase in strontium concentration in annual rings. The increase in strontium concentration, thus the decrease in ^{90}Sr specific activity in the wood, suggests a change of nutrient availability in recent decades.

METABOLISM OF TECHNETIUM IN PLANTS

The behavior of radionuclides in plants cannot be considered and explained without considering the behavior of stable isotopes as shown in the radial distribution of ^{137}Cs and ^{90}Sr in xylem (see the section on radionuclides in xylem, above). All chemical and physical reactions involved in the metabolism of plants occur almost equally on radioactive and nonradioactive isotopes of the same atomic number. This is the reason why radionuclides have been widely used as tracers in laboratory and field experiments. Metabolic processes concerning to radionuclides can thus be perceived from stable isotopes if the metabolism of the stable element is known. Metabolism of radionuclides that have no stable isotopes must be examined using the radionuclides themselves. However, our knowledge is quite limited; the metabolisms of all the elements and radionuclides have not yet been examined and details in the process are not yet clear.

Technetium-99 is believed to resemble rhenium in chemical nature. In field studies, ^{99}Tc was found migrating through soil as TcO_4^- anion in aerobic conditions and being absorbed in the soil matrix as TcO_2 or Tc_2S_7 in anaerobic conditions

(Masson et al., 1989). Pertechnetate anion is, therefore, the chemical species that exists in groundwater and is taken up by root (Murphy and Johnson, 1993). The uptake of ^{99}Tc by trees intercepting contaminated groundwater from a radioactive waste storage site in Oak Ridge National Laboratory, TN, was examined (Garten et al., 1986). Only pertechnetate is transported from root to shoot via the xylem, however. Technetium was not easily leached from the trees by rainfall, and was not readily extractable from forest floor leaf litter by water, indicating that metabolism following root uptake rendered it less biogeochemically mobile than expected. The low extractability of technetium incorporated in plant material has potential to cause technetium contamination in forest ecosystem to persist, due to a high soil-to-plant transfer of technetium and a continuous slow release of technetium through degradation of litter (Dehut et al., 1989).

Toxic effects of technetium have been reported, ranging from reduced yield of plants to macroscopic and microscopic morphological observations (Cataldo et al., 1989; Neel and Onasch, 1989). Plant toxicity resulting from micromolar levels of technetium are generally attributed to chemical rather than to radiation effects because of the low dose rate of ^{99}Tc associated with a low energy beta-ray emission. Characterization of the chemical form of technetium in maple leaves and wood growing at the radioactive waste storage site in Oak Ridge suggests that a part of the technetium in tree and leaf tissues homogenized and solubilized into phosphate buffer, and was associated with organic molecules greater than 10,000 amu (Garten and Lomax, 1989). Most of the technetium in wood was chemically removed by successive extractions with ethanol, water, and weak mineral acid; most of the technetium in leaves was speculated to be related to structural polysaccharides (hemicelluloses). The authors also demonstrated that incubations of leaf and wood homogenates with protease approximately doubled the amounts of technetium released from contaminated tissues. The above results strongly suggest incorporation of technetium into metabolic process in plant tissues.

Fixation and incorporation experiments of oxyanions TcO_4^- and SO_4^{2-} reveal a similar fate and distribution of technetium and sulfur in soybean plants (Cataldo et al., 1989). In vitro assay of chloroplast-based sulfur reduction and incorporation systems showed technetium to be reduced valence from VII to lower values and incorporated into amino nitrogen-containing products. The uptake of TcO_4^- is well explained by a mathematical model that assumes a continuous metabolization of TcO_4^- to TcX in the leaves (Van Loon et al., 1989). Accumulation experiments of ^{99}Tc on leaves of beans with or without cotyledons indicated that cotyledon excision substantially enhanced the toxic effects of ^{99}Tc on growth (Vázquez et al., 1990). In bean plants, cotyledons play an important metabolic role during the initial 12 days after sowing. Higher concentrations of ^{99}Tc in plants without cotyledons than in plants with cotyledons indicate that the active incorporation of technetium without cotyledons must occur in association with the shortage of organic substances, which should be provided from cotyledons. Above observations suggest that technetium toxicity is mainly associated with anabolic processes in developing tissues, a competitive assimilation of sulfur and technetium. The actively metabolized technetium would not be functional as an analog of sulfur, probably the reason for the high toxicity of technetium.

EXTERNAL AND INTERNAL IRRADIATION ON PLANTS

External Irradiation

The effect of ionizing radiation on plants has been studied in experimental fields, where natural plant populations and communities are irradiated with a radiation source under controlled conditions. Such an irradiation facility is usually provided with a strong external gamma-ray point source such as ^{137}Cs and ^{60}Co in the center of the irradiation field so that the radiation dose rate contour lines distribute from the strongest area, just near the source, to the level of almost background. One problem with such irradiation experiments on plants in the irradiation field is a rapid decrease in exposure intensity with distance from the source according to the inverse square law, due to absorption and scattering. The decrease in intensity is often so rapid that there may not be enough examples of the same plant species within the area where the same radiation dose has been achieved. The intensity may vary by a factor of a few orders from the source to a distance of 100 m. Exact dose rate of each tree is difficult to estimate because the trunks of other trees and other natural objects can produce significant shielding, especially near the source, although the dose rates at specific points in the field could be measured by dosimetoric techniques such as films, thermoluminescent dosimeters (TLD), radiation counters, and so on. A 10-cm diameter tree trunk may reduce the exposure by one-half. To minimize the above problems, radiation effects on plants have been treated statistically in relation to the average radiation dose. This approach would be more successful with plants than with animals due to the plants' immobility.

Response of Forest Ecosystem to Irradiation   Radiation exposure experiments for the terrestrial plant communities are well documented by IAEA (1992). These experiments could be categorized as either acute or chronic exposures, depending on the dose. However, no clear distinction has been drawn between acute and chronic exposures. In an acute exposure it is understood that the dose is delivered in a short time compared with the time over which any obvious biological response develops, while a chronic exposure is one in which the dose could continue over a large fraction of the natural life of the plant.

Death of the plant is the severest response to irradiation and could be used as an index of radiosensitivity of plant species. Among species examined pine trees (genus *Pinus*) are the most sensitive to irradiation. All specimens of *P. elliottii* died within a few month after an acute exposure of $> = 3$ Gy, and *P. palustris* less than five years old died at dose of $> = 8$ Gy and older ones > 28 Gy (IAEA, 1992). The condition of acute exposure would be satisfied only by a few plants near the irradiation source, and other plants at a distance from the source would experience chronic exposure. For chronic exposure the response of plants to irradiation is evaluated by mortality of species, a change in species composition, biomass above ground per unit ground area, and biomass deposited per unit ground area.

Response of Canadian boreal forest trees to chronic gamma irradiation has been examined during preirradiation, irradiation and postirradiation, spanning a period of 20 years (Amiro, 1994). The forest irradiated consisted of six major plant communities, and three tree communities have been measured for canopy cover percentage of each species in quadrates that were selected as permanent sampling stations before irradiation was started. The selected quadrates, 15×25 m in size, were distributed from near the ^{137}Cs source (370 TBq) to a distance of about 550 m, and the radiological dose rate ranged from 65 to 0.005 mGy hr^{-1}. The irradiation was started in 1973 and terminated in 1986, 13 years of irradiation. All trees were killed at a dose rate of >20 mGy hr^{-1}. Canopy cover of coniferous tree species, such as balsam fir, black spruce, and jack pine, decreased at dose rate of 0.5–10 mGy hr^{-1} and mortality of conifers increased with an increase of dose rate. All coniferous trees killed within 2–3 years at a dose rate of 5–10 mGy hr^{-1}. These gymnosperms were concluded to be more radiosensitive than angiosperms (willow, alder) because the canopy cover of the angiosperms was constant or increased at a dose rate of 0.5–10 mGy hr^{-1}, while that of gymnosperms decreased. No effect was observed on canopy level at dose rate of <0.1 mGy hr^{-1}. The damage to the forest canopy that was produced even at dose rates of 5 mGy hr^{-1} is recovering six years after termination of the irradiation.

Shrubs are less sensitive to radiation than trees, but more radiosensitive than herbs. Among shrubs, a great variation in radiosensitivity has been reported. Dugle and Mayoh (1984) examined 56 shrub taxa that were growing naturally in the gamma irradiation field in the Canadian boreal forest, and have reported that radioresistant species included *Rubus idaeus, Diervilla lonicera, Salix bebbiana, Prunus pensylvanica*, and *Ribes hirtellum,* while radiosensitive species included *Chimaphila umbellata, Lonicera villosa,* and *Vibrunum trilobum.* The LD_{50}, the dose rates at which 50% of the plant species were killed, was different by a factor of more than 30 between least and most radiosensitive species.

The different responses of plants to irradiation, radioresistant or radiosensitive, are considered to be correlated with the size of chromosomes. The plants with larger chromosomes have a higher chance that the radiation damages the DNA in chromosomes compared with the plants with small chromosomes. *Acer spicatum* was predicted to be more sensitive than *Alnus rugosa*, which is more sensitive than *Symphoricarpos albus,* based on the DNA content (El-Lakany and Dugle, 1972) and the same order of radiation sensitivity was confirmed (Dugle and Mayoh, 1984). The high radiation sensitivity of pine trees is also attributable to the larger chromosome size (Sparrow and Miksche, 1961). Such radiation sensitivities are predictable to within a factor of perhaps two from cellular characteristics, particularly the interphase chromosome volume (Sparrow et al., 1967).

Morphological and Physiological Change   Long-term irradiation of plants causes some noticeable effects on leaf morphology, phenology, and structure. The frequency of morphological changes increases with an increase in radiation dose. Dugle and Hawkins (1985) reported some teratological features, such as apparent fused leaves, alternate or missing leaflets, chlorophyll mottling and sectoring, abnormal venation, and irregular leaves, on ash trees irradiated at dose rates ranging from 1.83 to 62.48 mGy hr^{-1}, and the most severe effects were often associated with subsequent plant death. They speculated that an appearance of significant morphological effects is related to an overwinter irradiation on tissue cells of developing buds. Radiation affects meristems by slowing down or stopping mitosis and by causing untimely maturation and vacuolation of cells, pycnosis of nuclei, loss of cell content, disintegration and collapse of cells, and distortion of the apex (Lapins and Hough, 1970). The cause of the effects may or may not be transferred mitotically to other cells in the perennial plants, but they are not transferred genetically via the chromosomes of pollen and egg cells (Dugle and Hawkins, 1985).

Dose Rate Effect on Irradiation   The response of plants to irradiation changes with exposure dose rate even if the plants are irradiated with the same total radiation dose. This is clearly observed on the relationship between the total radiation dose and radiation dose rate for LD_{50}. Compiled LD_{50} data on pine trees indicates that the total accumulated radiation dose for LD_{50} increased as the radiation dose rate decreased (IAEA, 1992). To cause the same irradiation effect on plants, the lower radiation dose rate requires the larger total accumulated radiation dose. This relationship suggests a recovery of plants from the radiation damage during irradiation. The time for recovery, repair, or even modification of the radiation effect would be more available as the exposure dose rate decreases. The same relationship has been reported on black and green ash, deciduous tree with a slender trunk (Dugle and Hawkins, 1985); the radiation dose rate for LD_{50} decreased from > 60 mGy hr^{-1} at the first year of irradiation to 15.6 mGy hr^{-1} at 11th year of irradiation. The radiation dose rate for LD_{50} on balsam fir, one of the most radiosensitive tree species in the mixed boreal forest, also slowly decreased over the years, from 50 mGy hr^{-1} at the first year of irradiation to 1.5 mGy hr^{-1} at eleventh year of irradiation (Dugle, 1986).

Radiation Hormesis   The response of plants to chronic irradiation with very low radiation dose rate has been examined, covering radiation dose rates from slightly higher than the background to lower than any detrimental effect being observed. Sheppard et al. (1982) irradiated Scotch pine *(Pinus sylvestris)* with radiation dose rates from 0.002 to 7 mGy hr^{-1} to assess its use in revegetaing radioactive waste disposal areas. They measured fascicle needle growth, total plant biomass, and stem growth of one-year-old seedlings during 150-day exposure, and observed significant stimulation of both fascicle needle length and total biomass in seedlings exposed to 0.0025, 0.009, and 0.078 mGy hr^{-1} compared to those exposed to background level (0.002 mGy hr^{-1}). The mean shoot weights at the above levels were 183, 194, and 177% of the controls, suggesting an optimal stimulation at about 0.009 mGy hr^{-1}. No significant stimulation, however, was noted in the stem growth. Amiro (1986) observed no stimulation of stem growth of jack pine *(Pinus banksiana)* irradiated with radiation dose rate of 3.7 mGy hr^{-1}, which is, however, 50 times higher than the level of Sheppard's experiment mentioned above (Sheppard et al., 1982). The effect, stimulation of plant growth, is called "radiation hormesis," and many review articles have noted radiation hormesis at certain low levels of exposure (Luckey, 1982; Skok et al., 1965). No clear explanation has been given for the mechanism of such responses; the fact, however, that certain low-level exposures result in increased yield, is consistent with known mechanisms of cellular damage and known responses of plants to a compromise in apical dominance (Miller and Miller, 1987).

Internal Irradiation

In reality any contamination of a forest ecosystem with radioactive materials causes an internal irradiation on plants from alpha- and beta-emitting radionuclides that are accumulated in the tissue by root uptake as well as external irradiation from gamma-emitting radionuclides deposited on the ground and leaf surfaces. For the purpose of assessment of biological effect of ionizing radiation on plants, it should be necessary to evaluate radiation dose from both internal and external irradiation. However, calculations are only employed on evaluation of the internal radiation dose because of a technical difficulty in direct measurement of the radiation dose from alpha- and beta-emitting radionuclides in plants. In calculation a suitable biological effectiveness is taken for alpha and beta particles emitted in plants to compensate for the different degrees of damage to plant cells originating from different energy transfer coefficient per unit length. For alpha particles the quality factor may be taken to have a value of 20, and for beta particles and gamma rays a value of 1.

The response of plants to the internal irradiation was examined (Konoplyova et al, 1993) as seeds of orchard grass (*Dactylis glomerata* L.) were grown for 2 months in control and contaminated soils; those soils were collected from outside the 30-km zone of Chernobyl and within the 10-km zone around the Chernobyl Nuclear Power Plant, respectively. After the exposure, leaf segments were used for culture in vitro for 3 weeks. The radionuclides in the contaminated soil were ^{137}Cs, ^{134}Cs, ^{144}Ce, and ^{106}Ru at levels more than 10^5 Bq kg^{-1}, while that in the control soil was only ^{137}Cs of 30.0 Bq kg^{-1}. The total doses of external radiation during the 2-month exposure were 0.105 Gy for the contaminated and 0.194 mGy for the control soil; however no evaluation of the internal exposure was carried out. The somatic embryo formation and the average number of plantlets per explant were reduced 1.8 and 2 times, respectively, in the plants grown in the contaminated soil compared to the control one. The 50% reduction in the number of plantlets occurred at the external radiation dose of 0.105 Gy in the present experiment but the authors observed the same effect at a radiation dose of 6.0 Gy when seeds were irradiated with only an external ^{60}Co gamma-ray source (Tomaszewski et al., 1988). The severer response of the plants grown in the radioactive soil is attributable to the internal exposure originating from the radionuclides accumulated within the plant tissue.

What Is the Allowable Level of Radioactivity in Plants?

To protect natural plant populations from radiation exposure, reasonable radiation protection standards for plants are necessary. However, at present there is no authorized radiation protection standard for plants, or for animals except that for humans. The maximum radiation dose limit for the members of public is 1 mSv yr^{-1}.

All we can do is to evaluate an equilibrium radiation dose or equilibrium radionuclide concentration in plants growing in a contaminated environment where the radiation dose for the critical human population group is 1 mSv yr^{-1}. This evaluation would give a certain basis for calculation of the limitation on radioactivity in plants and would make it possible to compare that value with the radioactivity in plants listed in Tables 14-3 and 14-5.

Mathematical model calculations have been employed for estimation of the equilibrium radiation dose for plants (IAEA, 1992). The mathematical model employed was originally designed to evaluate the radiation dose to people who live in particular circumstances involving radioactivity. Among the circumstances, two of the model scenarios considered radioactive contamination of terrestrial plants:

1. Controlled releases of radionuclides to the atmosphere
2. Uncontrolled constant releases of radionuclides from a shallow land nuclear waste repository

To perform the model calculations, several assumptions are necessary as follows. The radionuclides released chronically from the nuclear installations cause a steady-state concentration of radionuclides in the air, and the shallow waste repository causes a steady-state concentrations of radionuclides in soil. These steady-state concentrations of radionuclides yield a radiation dose of 1 mSv yr^{-1} to the critical people living in the area. The plants are growing under the same conditions of radionuclide concentrations in air or soil, resulting in an equilibrium radiation dose to the plants.

The atmospheric releases cause external and internal contaminations of plants as already described in the section on pathways of radionuclides to plants: atmospheric deposition, root uptake, and resuspension. The model includes weathering effect on radionuclides deposited on the plant surface. The radiation dose on plants is given by internal and external exposures originating from alpha- beta-, and gamma-rays emitted from the radionuclides in plant tissues and from those deposited on the surface of plants, as well as from gamma-rays from radionuclides on the ground. The waste repository considers root uptake of radionuclides in the shallow landfill where the radioactive waste is homogeneously mixed with soil and distributed from the ground surface to a depth of several meters.

The model calculations were carried out for the important fission products ^{3}H, ^{90}Sr, ^{95}Z, ^{129}I, ^{131}I, and ^{137}Cs, and for long-lived alpha emitters ^{226}Ra, ^{235}U, ^{238}U, ^{239}Pu, and ^{241}Am (IAEA, 1992). The estimated equilibrium radionuclide concentrations in plants are distributed from 72 to 1800000 Bq kg^{-1} dry weight for the chronic atmospheric release scenario, the exact amount depending on the radionuclide, the lowest value being obtained on ^{239}Pu and a highest on ^{3}H, and the estimated equilibrium radiation dose rates range from 0.011 mGy d^{-1} for ^{239}Pu to 0.9 mGy d^{-1} for ^{95}Zr. On the other hand, the estimated radiation dose rates for the shallow landfill scenario are distributed from 0.009 mGy d^{-1} for ^{129}I to 0.7 mGy d^{-1} for ^{226}Ra. The differences and expanse of the equilibrium radionuclide concentrations and the radiation dose rates originate from a diversity of decay types of radionuclides (alpha or beta, with or without gamma-ray), their decay energy, and the physical, chemical, and biological properties of radionuclides.

For the case of the chronic atmospheric release, the equilibrium concentration of ^{137}Cs is estimated to be 1900 Bq kg^{-1} dry weight in plants, which yields an upper estimate of the radiation dose rate to plant tissue, 0.13 mGy d^{-1} (IAEA, 1992). If we simply use the above relationship between radiation dose rate and ^{137}Cs concentrations in plants, we can estimate a radiation dose rate on the plants from the Chernobyl accident as shown in Table 14-5. The highest ^{137}Cs activity observed on plants collected in 1989 in Norway (Livens et al., 1991) gives the radiation dose rate, 0.15 mBq hr^{-1}.

For the case of the shallow landfill scenario, the model calculation on ^{226}Ra gives an upper estimate of the radiation dose rate to the plant, 0.7 mGy d^{-1}; however, an equilibrium concentration of ^{226}Ra in plants is not presented. Thus we could not directly convert the ^{226}Ra concentrations to radiation dose rates on the plants listed in Table 14-3. The equilibrium concentration of ^{226}Ra in plants might be estimated using a TF value for ^{226}Ra such as that listed in Table 14-4, since the model calculation gave the equilibrium ^{226}Ra concentration in the soil as 300 Bq kg^{-1} dry. However, it would be better estimation to use the relationship between the ^{226}Ra concentration in soil and the radiation dose rate on plants because we do not know the TF value used for the model calculation. The highest ^{226}Ra concentrations on the plant in Table 14-3 are those in plants growing around uranium mines, where the highest concentrations of ^{226}Ra in soil occurred (Ibrahim and Whicker, 1988b). The ^{226}Ra concentration in the soil was 2200 Bq kg^{-1} dry, which is about seven times higher than the case in the model calculation. Thus the radiation dose rate for the most highly ^{226}Ra contaminated plant is estimated to be 0.21 mGy hr^{-1}.

The radiation dose rate of 0.42 mGy hr^{-1} has been understood as the critical level that yields a harmful effect on plants (IAEA, 1992). The radiation dose rates on the plants collected from the severest conditions in ^{137}Cs and ^{226}Ra contamination are both lower than the critical level, probably suggesting no apparent harmful effect on these plants.

REFERENCES

Amiro, B. D. (1986), *Environ. Exp. Bot.*, *26*, 253–257.

Amiro, B. D. (1994), *J. Environ. Radioactivity*, *24*, 181–197.

Antonopoulos-Domis, M., A. Clouvas, and A. Gagianas (1990), *Health Phys.*, *58*, 737–741.

Antonopoulos-Domis, M., A. Clouvas, and A. Gagianas (1991), *Health Phys.*, *61*, 837–842.

Belli, M., U. Sansone, R. Ardiani, E. Feoli, M. Scimone, S. Menegon, and G. Parente, (1995), *J. Environ. Radioactivity*, *27*, 75–89.

Beresford, N. A., B. J. Howard, and C. L. Barnett (1992), *J. Environ. Radioactivity*, *16*, 181–195.

Bonazzola, G. C., R. Ropolo, A. Patetta, and A. Manino (1991), *Health Phys.*, *60*, 575–577.

Bondietti, E. A., N. Momoshima, W. C. Shortle, and K. T. Smith (1990), *Can. J. For. Res.*, *20*, 1850–1858.

Bovis, C. P. and G. E. Briggs (1986), *Proc. Roy Soc. Lond. Biol. Sci.*, *169*, 379–397.

Bowen, H.J.M. (1979), *Environmental chemistry of the elements*, Academic Press, London, pp. 83–107.

Brownridge, J. D. (1984), *J. Plant Nutrit.*, *7*, 887–896.

Bunzl, K. and W. Kracke (1989), *Health Phys.*, *57*, 593–600.

Cain, W. F. and H. E. Suess (1976), *J. Geophy. Res.*, *81*, 3688–3694.

Cataldo, D. A., T. R. Garland, R. E. Wildung, and R. J. Fellows (1989), *Health Phys.*, *57*, 281–287.

Chigira, M., Y. Saito, and K. Kimura (1988), *J. Radiat. Res.*, *29*, 152–160.

Coughtrey, P. J., D. Jackson, C. H. Jones, P. Kane, and M. C. Thorne (1984), *Radionuclide distribution and transport in terrestrial and aquatic ecosystem, A critical review of data*, Vol. 4, A. A. Balkema, Rotterdam, Netherlands, pp. 19–74.

Crooke, W. M., A. H. Knight, and J. Keay (1964), *Forest Sci.*, *10*, 415–427.

Dehut, J. P., K. Fonsny, C. Myttenaere, D. Deprins, and C. Vandecasteele (1989), *Health Phys.*, *57*, 263–267.

Dugle, J. R. (1986), *Can. J. Bot.*, *64*, 1484–1492.

Dugle, J. R. and J. L. Hawkins (1985), *Can. J. Bot.*, *63*, 1458–1468.

Dugle, J. R. and K. R. Mayoh (1984), *Environ. Exp. Bot.*, *24*, 267–276.

El-Lakany, M. H. and J. R. Dugle (1972), *Evolution*, *26*, 427–434.

Fraley, L., Jr., G. Chavez, and O. D. Markham (1993), *J. Environ. Radioactivity*, *21*, 203–212.

Garten, C. T., Jr. and R. D. Lomax (1989), *Health Phys.*, *57*, 299–307.

Garten, C. T., Jr., C. S. Tucker, and B. T. Walton (1986), *J. Environ. Radioactivity*, *3*, 163–188.

Grootes, M., G. W. Farwell, F. H. Schmidt, D. D. Leach, and M. Stuiver (1989), *Tellus*, *41B*, 134–148.

Hisamatsu, S., T. Katsumata, and Y. Takizawa (1991), *J. Radiat. Res.*, *32*, 389–394.

Holtzman, R. B. (1970), *Environ. Sci. Tech.*, *4*, 314–317.

IAEA (1992), *Tech. Rep. Ser. No. 332*, IAEA, Vienna.

IAEA (1994), *Tech. Rep. Ser. No. 364*, IAEA, Vienna.

Ibrahim, S. A. and F. W. Whicker (1988a), *Health Phys.*, *54*, 413–419.

Ibrahim, S. A. and F. W. Whicker (1988b), *Health Phys.*, *55*, 903–910.

Jackson, D. (1989), *Health Phys.*, *57*, 485–489.

Jasinska, M., K. Kozak, J. W. Mietelski, J. Barszcz, and J. Greszta (1990), *J. Radioanal. Nucl. Chem., Lett.*, *146*, 1–13.

Jones, H. E., A. F. Harrison, J. M. Poskitt, J. D. Roberts, and G. Clint (1991), *J. Environ. Radioactivity*, *14*, 279–294.

Katayama, Y., N. Okada, Y. Ishimura, T. Nobuchi, and A. Aoki (1986), *Radioisotopes*, *35*, 636–638.

Katsuragi, Y., K. Kawamura, and H. Inoue (1983), *Pap. Met. Geophys.*, *34*, 21–30.

Kohno, M., Y. Koizumi, K. Okumura, and I. Mito (1988), *J. Environ. Radioactivity*, *8*, 15–19.

Konoplyova, A. A., L. V. Zseltonozskaya, B. V. Conger, and D. M. Grodzinsky (1993), *Environ. Exp. Bot.*, *33*, 501–504.

Koprdá, V. (1991), *J. Radioanal. Nucl. Chem.*, *Lett.*, *153*, 15–27.

Kozák, K., D. Rank, Biró, T., V. V. Rajner, F. Golder, and F. Staudner (1993), *J. Environ. Radioactivity*, *19*, 67–77.

Kudo, A., T. Suzuki, D. C. Santry, Y. Mahara, S. Miyahara, and J. P. Garrec (1993), *J. Environ. Radioactivity*, *21*, 55–63.

Lapins, K. and L. Hough (1970), *Radiat. Bot.*, *10*, 59–68.

Linsalata, R., R. S. Morse, H. Ford, and M. Eisenbud (1989), *Health Phys.*, *56*, 33–46.

Livens, F. R., A. D. Horrill, and D. L. Singleton (1991), *Health Phys.*, *60*, 539–545.

Luckey, T. D. (1982), *Health Phys.*, *43*, 771–789.

Masson, M., F. Patti, C. Colle, P. Roucoux, A. Grauby, and A. Saas (1989), *Health Phys.*, *57*, 269–279.

Matkar, Y. M., U. Narayanan, and I. S. Bhat (1994), *J. Radioanal. Nucl. Chem.*, *Articles*, *182*, 71–73.

McGee, E. J., H. J. Synnott, and P. A. Colgan (1993), *J. Environ. Radioactivity*, 18, 53–70.

Miller, M. W. and W. M. Miller (1987), *Health Phys.*, *52*, 607–616.

Momoshima, N. and E. A. Bondietti (1990), *Can. J. For. Res.*, *20*, 1840–1849.

Momoshima, N. and E. A. Bondietti (1994), *J. Environ. Radioactivity*, *22*, 93–109.

Momoshima, N., T. Okai, T. Kaji, and Y. Takashima (1991), *Radiochim. Acta*, *54*, 129–132.

Murphy, C. E. and T. L. Johnson (1993), *J. Environ. Qual.*, *22*, 793–799.

Neel, J. W. and M. A. Onasch (1989), *Health Phys.*, *57*, 289–298.

Pettersson, H.B.L., G. Hancock, A. Johnston, and A. S. Murray (1993), *J. Environ. Radioactivity*, *19*, 85–108.

Pinder, J. E., III (1991), *J. Environ. Radioactivity*, *14*, 37–53.

Pinder, J. E., III and K. W. McLeod (1989), *Health Phys.*, *57*, 935–942.

Pinder, J. E., III D. C. Adriano, T. G. Ciravolo, and A. C. Doswell (1987), *Health Phys.*, *52*, 707–715.

Pinder, J. E., III, K. W. McLeod, and D. C. Adriano (1990), *Health Phys.*, *59*, 853–867.

Robison, W. L. and E. L. Stone (1992), *Health Phys.*, *62*, 496–511.

Ronneau, C., L. Sombre, C. Myttenaere, P. Andre, M. Vanhouche, and J. Cara (1991), *J. Environ. Radioactivity*, *14*, 259–268.

Rudge, S. A., M. S. Johnson, R. T. Leah, and S. R. Jones (1993), *J. Environ. Radioactivity*, *19*, 173–198.

Sandalls, J. and L. Bennett (1992), *J. Environ. Radioactivity*, *16*, 147–165.

Schuller, P., C. Løvengreen, and J. Handl (1993), *Health Phys.*, *64*, 157–161.

Scotti, I. A. (1994), *Environ. Exp. Bot.*, *34*, 213–216.

Scotti, I. A. and S. Silva (1992), *J. Environ. Radioactivity*, *16*, 97–108.

Scotti, I. A., S. Silva, and F. Carini (1993), *J. Environ. Radioactivity*, *20*, 63–68.

Shaw, G. and J.N.B. Bell (1991), *J. Environ. Radioactivity, 13,* 283–296.

Sheppard, M. I. and S. C. Sheppard (1985), *Health Phys., 48,* 494–500.

Sheppard, M. I., D. H. Thibault, and J. E. Guthrie (1982), *Environ. Exp. Bot., 22,* 193–198.

Shibata, S. and E. Kawano (1994), *Appl. Radiat. Isot., 45,* 815–816.

Singhal, R. K., U. Narayanan, and I. S. Bhat (1994), *Health Phys., 67,* 529–534.

Skok, J., W. Cvhorney, and E. J. Rakosnik, Jr. (1965), *Radiat. Botany, 5,* 281–292.

Sparrow, A. H. and J. P. Miksche (1961), *Science, 134,* 282–283.

Sparrow, A. H., A. G. Underbrink, and R. C. Sparrow (1967), *Radiat. Res., 32,* 915–945.

Summerton, A. P. (1992), *J. Radioanal. Nucl. Chem., Articles, 161,* 421–428.

Takahashi, T., M. Nishida, S. Ohno, and T. Hamada (1969), *Radioisotopes, 18,* 560–563.

Takashima, Y., N. Momoshima, M. Inoue, and Y. Nakamura (1987), *Appl. Radial. Isot., 38,* 255–261.

Tent, J., M. Vidal, M. Llauradó, G. Rauret, J. Real, and P. Mischler (1993), *J. Radioanal. Nucl. Chem., Articles, 173,* 377–385.

Tomaszewski, Z., Jr., B. W. Conger, H. Brunner, and F. J. Novak (1988), *Environ. Exp. Bot., 28,* 335–341.

Tran Van, L. and T. Le Duy (1991), *J. Radioanal. Nucl. Chem., Lett., 155,* 451–458.

Uchida, S., Y. Muramatsu, M. Sumiya, and Y. Ohmomo (1991), *Health Phys., 60,* 675–679.

Ünlü, M. Y., S. Topcuoglu, R. Kücükcezzar, A. Varinlioglu, N. Güngör, A. M. Bulut, and E. Güngör (1995), *Health Phys., 68,* 94–99.

UNSCEAR (1982), *Inoizing radiation: Sources and biological effects,* United Nations, New York.

UNSCEAR (1988), *Source, effects and risks of ionizing radiation,* United Nations, New York, pp. 29–31.

Van de Geijn, S. C. and C. M. Petit (1979), *Plant Phys., 64,* 954–958.

Van Loon, L. R., G. M. Desmet, and A. Cremers (1989), *Health Phys.,* 57, 309–314.

Vázquez, M. D., A. Bennassar, C. Cabot, C. Poschenrieder, and J. Barceló (1990), *Environ. Exp. Bot., 30,* 271–281.

Wolterbeek, H. T., J. van Luipen, and M. de Bruin (1984), *Physiol. Plant, 61,* 599–606.

Yamada, Y., M. Itoh, N. Kiriyama, K. Komura, and K. Ueno (1989), *J. Radioanal. Nucl. Chem., 132,* 59–65.

15

Fire

P. S. Ramakrishnan, K. G. Saxena, A. K. Das, and K. S. Rao

INTRODUCTION

Fire is one of the early and continuing interests in ecology. The ecological significance of fire as a natural factor or as a management tool, determining the ecosystem
structure and processes, has been dealt in a number of studies in different parts of
the world (Kozlowski and Ahlgren, 1974; Mooney et al., 1981). However, ecological impacts of fire in the tropics are not as extensively studied as those in the temperate regions. Natural lightning-induced fires in the humid and subhumid tropics are
not as frequent as fire caused by humans, intentionally or unintentionally. Fire is a
management tool employed by the subsistence tribal farmers of the northeastern hill
region of India (Ramakrishnan, 1992), as well as elsewhere in the hot and humid
tropics (Gliessman et al., 1981; Nye and Greenland, 1960; Spencer, 1966; Watters,
1971). Radically different from the fire regime in the shifting agriculture are ground
fires set annually by local communities in order to improve upon the fodder availability from pine-dominated government reserve forests in the central and the northwestern Himalayas, and the burning of crop residues in agroecosystems based on
the intensive commercial production of sugarcane in the Indo-gangetic alluvial
plains. This chapter presents a review of the ecological role of fire in a shifting
agricultural system.

FIRE REGIME UNDER SHIFTING AGRICULTURE

Fire regime is determined by the frequency and intensity of burning in a given
ecosystem. In a shifting agricultural system, the frequency of fire depends upon the
length of the fallow period between two successive croppings on the same site. The

frequency of fire would increase with shortening of the cultivation cycle. Intensity of fire is a function of the fuel load, which in turn depends on biomass accumulation and flammability attributes of the secondary successional species during fallow regeneration. Longer shifting cultivation cycles thus impose a more intense but less frequent fire regime, compared to the shorter cycles where the disturbance of fire is less intense but more frequent. The length of the cultivation cycle depends upon the population pressure. In the northeastern Himalayas, the cultivation cycle, which used to be 20–30 years or more in the past, has now come down to an average of 4–6 years. The period of cultivation following burning till the site is fallowed for ecosystem recovery varies between 1 and 3 years, depending upon the site quality. In an extreme situation of land degradation and population pressure, the site may be burnt and cultivated annually until all the possibilities of cultivation are exhausted. Such a situation has also come up in the fringes of urban areas where farmers have access to modern agricultural inputs like chemical fertilizers and a switch over from subsistence-based shifting agriculture to cash crop-based agroecosystems is advancing (Ramakrishnan, 1992).

FIRE AND SOIL PHYSICOCHEMICAL PROPERTIES

Impacts of fire on soil physicochemical properties have important implications for ecophysiological adaptations of fire followers. The acidic tropical soils in northeastern India that are cultivated under a shifting agriculture regime exhibit a decline in soil organic carbon and nitrogen levels following the burn, along with an increase in available phosphorus as a result of the increase in soil pH (Fig. 15-1). The increase in soil pH is partly due to release of ash. These changes are more drastic under longer cycles with higher intensity of burning as compared to the shorter cycles. The intensity of physicochemical change also depends on the ecophysiological features of the preburn species. Thus burning of slash dominated by bamboos in a 10-year-old fallow field is characterized by release of greater quantities of potassium (Table 15-1), as this species is an accumulator of potassium. Fire in a 30-year fallow field releases larger quantities of calcium because bamboos are largely replaced over time by dicot trees, which accumulate calcium in higher concentrations than the bamboos (Ramakrishnan and Toky, 1981; Rao and Ramakrishnan, 1989).

The changes in soil physicochemical properties obviously influence the structure and function of soil microbial communities. High rates of nitrification after slashing the vegetation (Fig. 15-2) are recognized as due to more favorable microenvironmental conditions (Smith et al., 1968), but even low intensity burns, raising soil temperature to 35–58°C, cause severe damage to the thin-walled nitrifiers. The period of inhibition of nitrification following the burn depends upon the fire intensity and on subsequent environmental conditions that affect the recovery of the nitrifier population (Raison, 1979). On the other hand, heterotrophic microbial activity is stimulated by the heating effects (Simmon-Sylvestre, 1976). Although soil

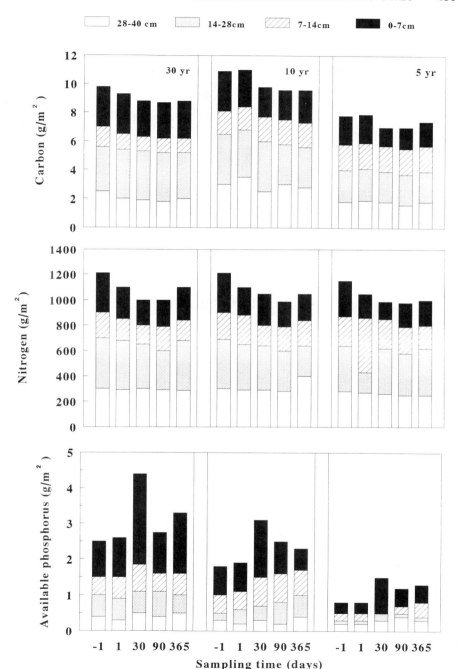

Figure 15-1. *Changes in total quantity of carbon, nitrogen, and available phosphorus within a soil column of 40 cm depth day before burn (−1) and after burn during cropping under 30-, 10-, and 5-year jhum (slash and burn) cycles at Burnihat in Meghalaya, India. (From Ramakrishnan and Toky, 1981.)*

TABLE 15-1   Total Quantity of Nutrients Liberated
Through Ash and the Amount Lost Through Blow-off
(kg ha^{-1}) in Agroecosystems Under Various Jhum[a]
Cycles at Burnihat, Meghalaya State, Northeastern India

	Jhum Cycle (yr)		
	30	10	5
Ash			
Released	17.4	13.8	6.9
Blown off	8.2	8.2	1.9
Phosphorus			
Released	313.0	262.2	150.7
Blown off	147.1	155.6	42.7
Potassium			
Released	1739.0	2070.0	685.0
Blown off	817.0	1228.5	194.0
Calcium			
Released	956.5	193.2	116.5
Blown off	449.4	114.7	33.0
Magnesium			
Released	208.7	151.8	113.7
Blown off	98.0	90.1	32.2

[a]Slash and burn agriculture in northeastern India is referred to as
"jhum."

Source: From Toky and Ramakrishnan (1981).

physicochemical properties become more favorable for nitrifiers following the burn
(Woodmansee and Wallach, 1981), the low residual viable nitrifier population
could, perhaps, be responsible for a steep decline in nitrification soon after the burn.
A decrease in nitrification with increasing length of fire cycle is related to structural
variation in microbial communities before the burn and differential heating regimes
(Saxena and Ramakrishnan, 1986). Competitive interactions between the hetero-
trophs and autotrophs seem to have a stronger effect on nitrogen mineralization
than on that of allelopathic inhibition of nitrification in the hot and humid tropical
ecosystems (Lamb, 1980; Ramakrishnan, 1992; Vitousek, 1982). The minimum
period required for recovery of nutrient losses due to fire varies depending upon
the fire regime, soil type, and climatic conditions, together with ecophysiological
attributes of early successional species (Eden et al., 1991; Mishra and Ramakrishnan, 1984; Pivello and Coutinoh, 1992; Ramakrishnan and Toky, 1981; Uhl and
Murphy, 1981). The success of a species in a mixed-plant community would depend on its ability to capitalize upon the spatiotemporal heterogeneity in light and
soil resources (Aerts et al., 1992; Ramakrishnan, 1992; Tilman, 1985), and the way
it spends the accumulated nutrients and biomass for different life purposes, together
with its regenerative strategy.

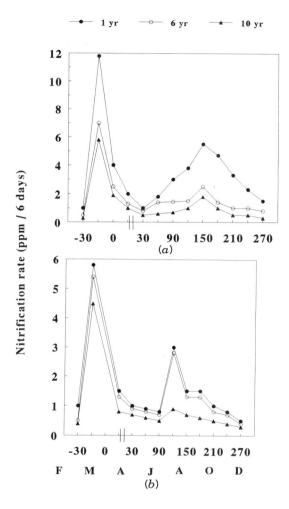

Figure 15-2. *Changes in nitrification rate during cropping at Burnihat, Meghalaya, India. (a) Under different jhum cycles. (b) Under jhum fallows. Time period in days after the burn, − 30 being 30 days before the burn and 0 being the day of the burn. (From Saxena and Ramakrishnan, 1986.)*

POSTFIRE SUCCESSION

Ecosystem resilience to disturbance of fire varies, depending on the bioclimate and vegetation type (Tables 15-2 and 15-3). Use of fire (i.e., prescribed burning) is an ages-old management tool intended, in the tropics, to manipulate the natural course of succession so as to serve the interests of human beings in the form of fast growth of desired species, and, in mediterranean and temperate regions, to avoid the possi-

TABLE 15-2   Principal Shrub Vegetation Types, Estimated Natural Fire Frequency, and a Rough Estimate of the Minimum and Maximum Fire-Free Interval to Which They Are Resilient

Vegetation Type	Modal Fire Frequency (yr)	Resilience	
		Minimum Fire-Free Interval (yr)	Maximum Fire-Free Interval (yr)
Mediterranean-climate, evergreen shrub	20–50	10	100–200 (?)
Mediterranean-climate, deciduous shrub	30–100	10	100 (150?)
Humid evergreen shrub	20–30	2–5	200 (?)
Arid (desert shrub) temperate forest	50–100	10–20	(?)
Successional shrubs:			
Western forest shrubs	20–100	5	300–400 (?)
Eastern forest shrubs	100–150	5	——
Shrub steppe	100–300	30	——
Tropical rain forest understory shrubs	——	——	——

Source: From Keeley (1981).

TABLE 15-3   Principal Tree Vegetation Types, Estimated Natural Fire Frequency, and a Rough Estimate of the Minimum and Maximum Fire-free Interval to Which They Are Resilient

Vegetation Type	Modal Fire Frequency (yr)	Minimum Fire-Free Interval (yr)	Maximum Fire-Free Interval (yr)
Temperate Coniferous Forests			
Sequoia mixed coniferous forests	10–100	1–3	600
Lodgepole forest, Sierra Nevada	100–300	?	——
Lodgepole forest, Rocky Mountains	40–80	15	200–300
Southeastern pine savanna	20–50	0	200
Pine barrens	10–20	1–3	200
Chaparral conifers			
Cupressus	50–100	20	200?
Pinus	20–40	10	100
Pinyon-juniper woodland	100–300	100	——
Boreal forest	20–300	?	——
Temperate deciduous forest	100–500	25	——
Tropical rain forest	——	——	——

Source: From Keeley (1981).

bilities of catastrophic fires (Komarek, 1974; Pyne, 1982; Waldrop et al., 1992). The successional changes following slash and burn depend upon the fire adaptations of the preburn species and the new arrivals from the surrounding vegetation. Early successional herbaceous species establishing following slash and burn compete with the planted crops for the available environmental resources and are weeds from an agroecosystem perspective. In-situ regeneration potential affected by the fire is further influenced by the weeding operations undertaken during the cropping phase (Swamy and Ramakrishnan, 1987a). Postfire succession in the ecosystems faced with periodic disturbance of fire alone (Christensen and Kimber, 1975; Gill, 1975; Purdie and Slatyer, 1976; Zammit and Zedler, 1988) tends to follow an "initial floristic composition" model (Egler, 1954). In shifting agriculture, where disturbance of fire is accompanied by the weeding disturbances, the initial floristic composition model was found to operate under more frequent perturbations under short cultivation cycles of 4–6 years, whereas the classical "relay floristics" model (Odum, 1969) explained the vegetational change under less frequent perturbations under longer cycles of 10–20 years (Saxena and Ramakrishnan, 1984a). Continuous slash-burn at 4–6-year intervals leads to "arrested succession," where the community is exclusively composed of herbaceous weedy species. In contrast, under the longer slash-burn cycles of 10–20 years, these shade-intolerant weedy species are replaced by the shade-tolerant herbaceous species. Longer fire cycles also allow regeneration of many shrubs, bamboos, and tree species. Species diversity was very low in the first 5 years of fallow development, increased considerably between 5 and 10 years, and showed further but slower increase in the next 10 years in a low elevation typical shifting agriculture area in the northeast of India (Fig. 15-3). The period of stabilization of species diversity may vary, depending upon the land use history and ecological conditions (Aweto, 1981).

REGENERATIVE STRATEGIES

Success of a species in fire-prone ecosystems depends upon the ecophysiological adaptations of heat tolerance and the thermal regime created by the fire. Air temperature following fire may be as high as 840°C, depending on the flammability and environmental conditions (Miranda et al., 1993). In fire-prone tropical forest ecosystems, insulation provided by thick bark is an adaptation rendering fire resistance to many tree species (Miranda et al., 1993; Uhl and Kauffman, 1990). Physicochemical changes, such as increases in nitrate and ammonium nitrogen concentrations following fire, have been found to stimulate germination of many fire-adapted species (Keeley, 1991; Keeley and Keeley 1988; Thanos and Rundel, 1995). Germination of some species is shown to be stimulated by short-term heat shocks, but very high temperatures could be lethal, resulting in long-delayed recovery in terms of colonization by the vascular plants (Clement and Touffet, 1990). There is also evidence of statistically insignificant differences in soil seed banks of burnt and unburnt sites (Valbuena and Trabaud, 1995). The success of a species following fire regimes would depend on the extent of insulation provided by the soil to the

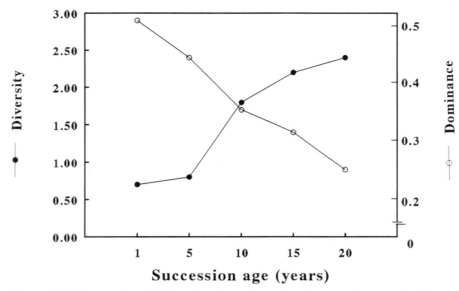

Figure 15-3. *Species diversity (closed circles) and species dominance (open circles) in successional communities up to 20 years at Burnihat, Meghalaya, India. (From Toky and Ramakrishnan, 1983.)*

belowground regenerating organs and the seed rain from the surrounding areas (Legg et al., 1992; Mallik et al., 1984).

Reproduction and regeneration of a species following fire (Table 15-4) may occur from soil-insulated underground vegetative organs (sprouting regenerative strategy) or through viable sexual propagules present in the soil seed bank/arrived from the neighboring areas (nonsprouting regenerative strategy). Obligate seeding and resprouting have been considered as alternative strategies for coping with an evolutionary history of recurrent disturbance by fire (Keeley, 1977; Keeley and Zedler,

TABLE 15-4   Reproductive and Regenerative Options in the Reproductive Cycle of Plants

Reproduction		Regeneration	
Seed[a]	Vegetative	Belowground Parts	Aboveground Parts
Disperse	"Runners" above- and belowground	Resprouts from stem or roots	Resprouts from epicormic buds
Remain in situ in the soil	Layering		
Remain in situ on the plant	Gradual spread by repeated resprouting from basal parts		

[a] Sexual and asexual.

Source: From Keeley (1981).

1978; Zammit and Westoby, 1988). In fire-prone chaparral ecosystems, obligate seeders are relatively short-lived, and their disappearance from long unburned stands led Zammit and Zedler (1988) to speculate that shrub longevity has been molded by the fire frequency during the course of evolution. Nonsprouting species are considered to have evolved from their sprouting ancestors owing to the loss of sprouting ability under frequent fire regimes (Stebbins, 1974).

These evolutionary trends in demographic and reproductive divergence and their relevance in explaining the shrub species dominance in ecosystems subjected to fire disturbance alone may not be equally applicable to the situations under shifting agriculture, where disturbance of fire gets modulated by other disturbances of slashing, weeding, and cropping and their ecological consequences. Successful colonization by both nonsprouting and sprouting species even when fire cycle is as short as 3–4 years under shifting agriculture in the northeastern India (Table 15-5) supports this contention. Many species, however, have a combination of nonsprouting and sprouting regenerative strategies. *Eupatorium* spp., apart from heavy seed production (Kushwaha et al., 1981; Ramakrishnan and Mishra, 1981), also sprout through a root stock. *Imperata cylindrica*, although largely reproducing through under-

TABLE 15-5 Photosynthetic Pathway and Regenerative Strategy of the Species Constituting Early Successional Herbaceous Communities

Species	Photosynthetic Pathway	Regenerative Strategy[a]
Ageratum conyzoides L.	C_3	NS
Borreria articularis (L.f.) Willd.	C_3	NS
Brachiaria distachya (L.) Stapf.	C_4	NS
Cassia tora L.	C_3	NS
Costos sp.	C_3	SP
Crossocephalum crepidioides (Benth.) S.	C_3	NS
Digitaria adscendens (H.B.K.) Henr.	C_4	NS
Erigeron linifolius Willd.	C_3	NS
Eupatorium odoratum L.	C_3	NS
Euphoribia hirta L.	C_4	NS
Grewia elastica Royle	C_3	SP
Imperata cylindrica Beauv. Var. Major	C_4	SP
Manisuris granularis L.f.	C_4	NS
Mimosa pudia L.	C_3	SP
Mollugo stricta L.	C_3	NS
Panicum khasianum Munro.	C_4	NS
P. maximum Jacq.	C_4	NS
Paspalidium punctatum (Burm.) A. Camus.	C_4	NS
Rottboelia goalparensis Bor.	C_4	NS
Saccharum arundinaceum Hook f.	C_4	SP
Setaria palmifolia (Koen.) Stapf.	C_4	NS
Thysanolaena maxima (Roxb.) O. Ketze	C_4	SP

[a] NS = nonsprouting; SP = sprouting.

Source: From Saxena and Ramakrishnan (1988).

ground extensive rhizomes, also produces seeds under frequent perturbations (Kushwaha et al., 1983b). *Mikania micrantha* multiplies rapidly through vegetative means by rooting at nodes (through ramets), and at the same time has heavy seed production that is fire-regulated (Swamy and Ramakrishnan, 1987a, b). There are also studies suggesting that an evolutionary divergence of nonsprouting and sprouting strategy may not hold true in all ecosystem types. Olson and Platt (1995) found that some shrubs of all species resprouted from underground organs; none regenerated from soil seed banks in longleaf pine savannas. With fire disturbance being followed by perturbations during cropping under shifting agriculture, both sprouting and nonsprouting strategies could be useful in efficient utilization of a heterogeneous microclimate and soil fertility status. Often, fire and stresses interact to modify the occurrence and intensity of competitive interactions in the ecosystem (Inchausti, 1995).

The course of succession changes when the site is fallowed after 1–3 years of cropping, partly because of depletion in soil nutrient pools and regeneration potential as a result of crop harvests, nutrient losses through leaching and runoff, and weeding. Lower reproductive potential and dispersal ability of sprouting species leads to a drastic reduction in their abundance when a burnt site is fallowed after cropping. Nonsprouting species exhibiting high reproductive potential and efficient dispersal traits become more successful. Apart from regenerative mechanism, adaptive traits related to photosynthetic pathway, growth behavior, resource allocation pattern, and nutritional relationships would be vital in determining the success of a given species in perturbed environments.

RESOURCE ALLOCATION PATTERNS AND THEIR ADAPTIVE SIGNIFICANCE

Since organisms have limited quantities of resources to expend for various life purposes such as growth, maintenance, and reproduction, the manner in which accumulated resources are allocated for different purposes would reflect the fitness of organism in a given environment (Cody, 1966). For an organism, the ecological implications of such partitioning would be critical when the given resource is limited and plant components being considered for allocation represent different alternatives, so that an increase in proportion allocated to one component would be linked with a decrease in others (Harper, 1977). The identification of biomass or energy as crucial resource has always been a critical assumption in evaluation of adaptive strategies in plants. Reproduction also requires mineral nutrients, and reproductive structures can make no contribution in this regard. Allocation of nutrients could be more crucial than biomass or energy under the situations of nutrient stresses such as those related to nitrogen and phosphorus in the tropics. The importance of nutrient allocations in gaining a better understanding of adaptive strategies of plants has been emphasised by many workers (Abrahamson and Caswell, 1982; Aerts et al., 1992; Fitter and Setters, 1988; Ohlson and Malmer, 1990; Saxena and Ramakrishnan, 1984b).

Evaluation of resource-allocation patterns of nonsprouting and sprouting herbaceous species establishing after fire and late successional species in northeastern India showed early successional nonsprouting species generally as having r-strategy with heavy seed production and/or vegetative reproduction, and the late successionals generally as following k-strategy, giving more emphasis to the allocation to nonreproductive organs (Fig. 15-4). The dominant sprouting species, such as *Imperata cylindrica, Thysanolaena maxima, Saccharum arundinaceum,* and *Grewia elastica,* direct the resources for underground vegetative organs of regeneration by economizing on the sexual reproductive growth. This preferential allocation of resources to underground organs was more marked for nitrogen and phosphorus compared to potassium. This could be related to more limited availability of nitrogen and phosphorus compared to that of potassium (Ramakrishnan, 1992).

Distinct compromises in resource partitioning may occur within a particular type of life history; for example, an annual weed species like *Ageratum conyzoides* achieves a high reproductive potential by partitioning its limited reproductive resources into a higher number of propagules instead of increasing the total cost of production (Saxena and Ramakrishnan, 1983). Reproductive strategies thus need to be evaluated in terms of number of seeds produced in addition to the bulk allocation of biomass and nutrients (Kushwaha et al., 1981; Ramakrishnan and Mishra, 1981). If one were to look at three kinds of strategies, namely, r- and k-strategies of MacArthur and Wilson (1967) and the additional competetive strategy occupying an intermediate position between r- and k-strategies (Grime, 1979), sprouting species, by virtue of their ability to grow fast soon after the burn and their lower sexual reproductive allocations, seem to be more competetive than the nonsprouting species. Individuals establishing through sprouts are more aggressive than those establishing through seeds. In mixtures, the plants arising from rhizomes in case of *I. cylindrica* had a distinct advantage over those arising from seeds, as seen from the higher relative yield for the former (Table 15-6). However, this initial advantage gradually tended to level off over a period of time (Kushwaha et al., 1983a). In sprouting species that primarily regenerate by vegetative means, the small number of new recruits that may arise from seeds are extremely important in maintaining the genetic diversity of the population (Saxena and Ramakrishnan, 1983).

Comparison of growth and allocation of four dominant postfire perennial species

TABLE 15-6 Relative Yield Total (RYT), Relative Yield Quotient (RYQ), and Relative Replacement Rate (RRR) at Three Harvests of *Imperata cylindrica* Raised from Rhizome and Seeds (20 Individuals in a Replacement Series of Pot Culture)

| | Relative Yield | | | | RRR |
Harvests	Rhizome (r_1)	Seed (r_2)	RYT $(r_1 + r_2)$	RYQ (r_1/r_2)	Rhizome with Respect to Seed
H_1	0.69	0.31	1.00	2.22	
H_2	0.53	0.47	1.00	1.12	0.51
H_3	0.51	0.49	1.00	1.04	0.92

Source: From Kushwaha et al. (1983a).

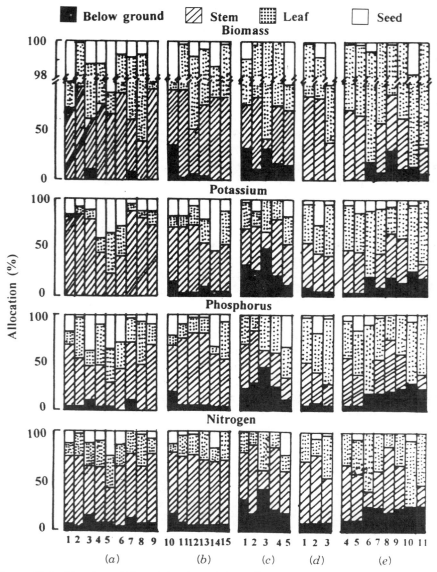

Figure 15-4. *Allocation of biomass/nutrients (%) to different components of early versus late successional species. (a) Early successional nonsprouting annuals. (b) Early successional nonsprouting perennials. (c) Early successional sprouting perennials. (d) Late successional nonsprouting annuals. (e) Late successional sprouting perennials. Each bar represents a different species out of a total of 31 important species studied. (From Saxena and Ramakrishnan, 1982.)*

in the early stages of secondary succession showed that survival of nonsprouting *E. odoratum* is ensured through rapid growth rates and allocation of many resources to the shoot system. On the other hand, the slow growth rates in sprouting species, namely, *G. elastica, I. cylindrica,* and *T. maxima,* may result from initial allocation of biomass and nutrients to belowground parts of regeneration. The disadvantage of slow growth rates in these sprouting species was to some extent compensated for by the transfer of underground resource reserves to support a vigorous shoot growth in early stages of regrowth. Thus the persistence of nonsprouting species that establish with low initial resources in the seeds would depend upon their nutrient uptake efficiency and use together with fast growth in the early stages of succession. The persistence of sprouting species would depend upon their ability to rebuild adequate belowground growth of regenerating organs before the next fire event. The abundance of *I. cylindrica* was explained as due to its rapid growth and multiplication through underground rhizomes (Saxena and Ramakrishnan, 1983).

ALLOCATION STRATEGY OVER A SUCCESSIONAL GRADIENT

A comparative study of populations of *E. odoratum* in seral environments following fire revealed sharp plasticity changes in response to the changing ecological conditions as a result of vegetation regeneration. The growth was more marked in recently fallowed sites (Fig. 15-5). The increase in biomass with fallow age was because of the age of this perennial plant in older fallows. Allocation of a high

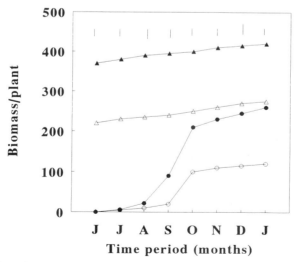

Figure 15-5. *Growth curves of* Eupatorium odoratum *in different jhum fallows at Burnihat, Meghalaya, India. Closed circle = recent fallow but uncropped; open circle = recent fallow but cropped; closed triangle = 2-year-old fallow; open triangle = 6-year-old fallow. Vertical bars represent LSD (p = 0.05). (From Saxena and Ramakrishnan, 1984a.)*

proportion of the available resources to the supporting organs but a lower proportion to the photosynthetic and reproductive structures in older fallows than in the recent ones (Fig. 15-6) is related to the tendency to grow taller and avoid shading in the more mature habitats where light could be limiting (Abrahamson, 1979; Newell and

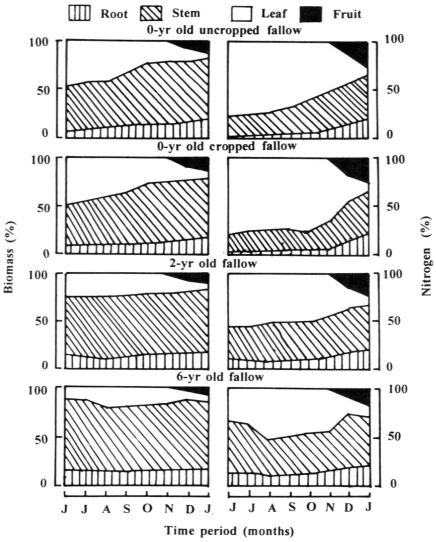

Figure 15-6. *Allocation of biomass and nutrients to different component organs (expressed as percentage of total biomass or nutrients) in* Eupatoriaum odoratum *in different jhum fallows. (From Saxena and Ramakrishnan, 1984a.)*

TABLE 15-7 Nutrient Uptake by *Eupatorium odoratum* in Different Jhum Fallows at Burnihat, Meghalaya, India

Fallow Field	Nitrogen	Nutrient Uptake (mg g^{-1} root biomass)	
		Phosphorus	Potassium
F$_1$	71.02	9.21	72.40
F$_2$	57.60	6.30	60.80
2-year-old	1.82	1.30	5.74
6-year-old	1.40	0.72	4.70
LSD ($p = 0.01$)[a]	9.48	2.26	9.20

[a] LSD = least significant difference.

Source: From Saxena and Ramakrishnan (1984a).

Tramer, 1978; Saxena and Ramakrishnan, 1982; Swamy and Ramakrishnan, 1987a).

Quantification of reproductive cost in the perennials is complicated because of production of the previous year(s) is added to that of the next year (Hickman, 1975; Ogden 1974). In most of the field studies on reproductive strategy analysis, reproductive allocation has been measured as a proportion of existing biomass. The importance of underground organs in the resource dynamics is also often ignored (Chapin, 1980; Mooney and Billings, 1960; Saxena and Ramakrishnan, 1983). The cost to a perennial plant of reproduction may be more accurately calculated as the proportion of the current increment in available resources devoted to reproductive units (Harper and Ogden, 1970). Such a measure of reproductive allocation (cf. Harper and Ogden, 1970; net reproductive effort expressed as the ratio of reproductive biomass to net primary production) was substantially lower in older fallows than in the recent ones (Table 15-7). Reproductive allocation of nitrogen and phosphorus (nitrogen in 2-year-old fallow field and both nitrogen and phosphorus in 6-year-old fallow field) exceeded the uptake during the current growing season, indicating an extreme ruderal strategy of this species relying on seeds for multiplication. Increasing nutrient stresses during fallow development, together with emphasis on sexual reproductive growth, results in gradual replacement of *E. odoratum* after 5–6 years of fallow development (Toky and Ramakrishnan, 1983). Fire, by improving the soil nutrient levels and insolation, favors persistence of *E. odoratum.*

The effect of fire on plant strategies could also be illustrated in the case of *M. micrantha,* which relys on both sexual and vegetative means of propagation (Swamy and Ramakrishnan, 1988). Its fire dependence is shown by faster growth in the burnt successional fallows of all ages compared to unburnt fallows (Table 15-8). This effect was greater in 8-year-old fallows, suggesting that the species rapidly declines in vigor in the absence of recurrent fire disturbance. The ruderal nature and fire dependence of *M. micrantha* is also reflected in its nutrient allocation strategy; it is able to flower better and at lower nutrient cost in burnt than in unburnt conditions (Table 15-9).

TABLE 15-8 Mean (±SE) Values of Relative Growth Rate (RGR), Net Assimilation
Rate (NAR), and Leaf Area Ratio (LAR) Growth Functions of *Mikania micrantha* in
Burnt and Unburnt Early Successional Fallows at Lailad, Meghalaya, India

Growth Function	Fallow Age (yr)	Burnt	Unburnt
RGR (mg mg^{-1} d^{-1})	2	0.0235 ± 0.0005	0.0148 ± 0.0003
	4	0.0247 ± 0.0004	0.0153 ± 0.0002
	8	0.0257 ± 0.0004	0.0070 ± 0.0004
NAR (mg cm^{-2})$^{-1}$ d^{-1})		20.171 ± 0.003	0.154 ± 0.004
	4	0.184 ± 0.003	0.163 ± 0.003
	8	0.197 ± 0.006	0.088 ± 0.001
LAR (cm^2 mg^{-1})	2	0.195 ± 0.006	0.139 ± 0.005
	4	0.254 ± 0.008	0.214 ± 0.005
	8	0.293 ± 0.009	0.010 ± 0.005

Source: From Swamy and Ramakrishnan (1988).

FIRE ADAPTATION AND WEED POPULATION STRUCTURE

In an ecosystem subjected to frequent perturbations, the effects of these perturbations would first be reflected at the population level. The manner in which fire alters
population density is illustrated here through the case of *M. micrantha,* studied
before and after fire in 4-, 8-, and 12-year-old fallow fields. The higher density in
burnt plots (Table 15-10), when compared to unburnt plots of the same fallow,
could be related to increased insolation, enriched soil, and possible release from the
allelopathic effects of its own individuals (Wong, 1964) and from other species
such as *E. odoratum* (Yadav and Tripathi, 1981). Competition may not be a factor
in increase in density of this species in burnt plots, as the densities of most associated species are either unaffected or increased after the burn. Heavy mortality and
consequent reduced density giving way to another wave of heavy recruitment could
explain double peaking in net population size in burnt sites during May and August
(Fig. 15-7). A relatively slower recruitment in unburnt plots may result in a slow
increment in net population size and one peak attained during July–August. The
total absence of seedling recruitment in unburnt plots of 2- and 4-year-old fallows
may be due, in part, to rapid vegetative regeneration that would reduce available
space and light penetration at ground level. Seedling recruitment in unburnt plot of
an 8-year field could be accounted for by vertical growth of the already existing
fewer *M. micrantha* individuals, which would not interfere with the light penetration to the ground level. Rosette formation took place only in burnt plots. Such a
fire-related seedling recruitment strategy is in spite of profuse seed production by
this weed (Swamy and Ramakrishnan, 1988).

PHOTOSYNTHETIC PATHWAYS AND NUTRITIONAL RELATIONSHIPS

The ecological advantages and disadvantages of sprouting and nonsprouting regenerative strategies interact with photosynthetic pathways and associated nutritional

TABLE 15-9 Reproductive Effort of *Mikania micrantha* in Burnt and Unburnt Successional Fallows at Lailad, Meghalaya, India

Reproductive Effort[a]	Fallow Age (yr)	Burnt	Unburnt
Biomass	2	24	23
	4	25	22
	8	27	22
LSD[b] at 5% level		5.2	6.4
Nitrogen	2	57	53
	4	60	61
	8	64	88
LSD[b] at 5% level		8.4	9.2
Phosphorus	2	51	46
	4	53	49
	8	56	80
LSD[b] at 5% level		7.4	7.3
Potassium	2	31	33
	4	31	39
	8	33	83
LSD[b] at 5% level		9.2	8.4

[a] Reproductive effort calculated as biomass or nutrient allocation to the seed as the percentage of the net primary production or nutrient uptake during the growing season.
[b] LSD = least significant difference.
Source: From Swamy and Ramakrishnan (1988).

attributes. Most of the native early successional grasses that come up largely through sprouts following fire are C_4, though several less abundant C_3 species do occur. These C_4 species have lower reproductive potential and so limited capacity of invading new sites, but higher efficiency of nutrient uptake and use. The dominant exotic weeds are all nonsprouting C_3 species with high reproductive potential, and so a high capacity for invading open sites created by fire, but low nutrient uptake and use efficiency (Fig. 15-8). Some C_3 exotic weeds like *M. micrantha* and *E. odoratum* grow fast in resource-rich postfire environments by maintaining a high

TABLE 15-10 Reproductive Potential of *Imperata cylindrica* at Clipped and Burnt Plots Under 3-Year Jhum Fallow at Burnihat, Meghalaya, India

	Clipped	Burnt
Tillers (m^{-2})	390	413
Spikes (m^{-2})	130	160
Seeds spike^{-1}	291	306
Seeds (m^{-2})	37,830	48,960
Seeds tiller^{-1}	97	119
Reproductive capacity (m^{-2})	34,047	44,064

Source: From Kushwaha et al. (1983b).

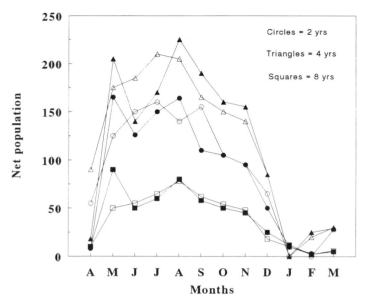

Figure 15-7. *Net population size of* Mikania micrantha *in 2-year-, 4-year-, and 8-year-old jhum fallows at Lailad, Meghalaya, India. Closed symbols are burnt sites; open symbols are unburnt sites. (From Swamy and Ramakrishnan, 1987a.)*

leaf area ratio (Saxena and Ramakrishnan, 1983; Swamy and Ramakrishnan, 1987a). C_3 species responded more to soil nutrient and moisture stresses compared to the C_4 species in experimental conditions, suggesting the latter to be more stress tolerant. Variability in spatiotemporal gradients in soil nutrient levels as determined by the fire regime and water stress, and in light regimes as determined by the climatic conditions, reproductive strategies, growth behavior, and nutritional requirements together would influence the relative abundance of different species or functional groups during the course of succession following fire. The C_3 and C_4 species would be expected to coexist as long as severe nutrient stresses are lacking. The C_3 species, with large utilization of nutrients for a given dry matter production, are suited to occupy nutrient-rich microsites, while C_4 species, with higher nutrient use efficiency, can successfully colonize nutrient-poor microsites. Interacting characteristics of plant strategies that are adaptive under varied levels of resource availability and changes in these strategies that occur as a consequence of fire and succession are summarized in Fig. 15-9 (Chapin and Van Cleve, 1981).

Divergence in phenological patterns of growth of C_3 and C_4 species during the year (Fig. 15-10) in a seasonal climate could also favor their coexistence. The early part of the growing season (April–June in northeastern India) is characterized by intense solar radiation, warmer temperatures, and low rainfall; it is the period of the year when water stress is most likely to occur. Nutrient stress, particularly that related to nitrogen and phosphorus, is also most likely to be present during these months, due to volatilization of these nutrients during fire and slow nitrification

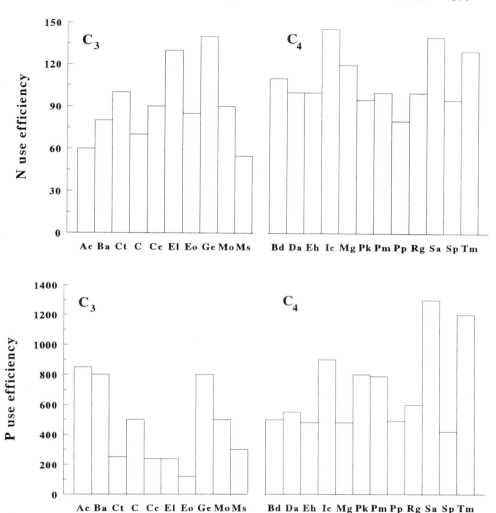

Figure 15-8. *Nutrient-use efficiency (expressed as mg dry matter production per mg nutrient absorbed) of different species coming after jhum at Burnihat, Meghalaya, India. Different columns from left to right are: C₃ species;* Ageratum conyzoides, Borreria articularis, Cassia tora, Costus *spp.,* Crossocephalum crepidioides, Erigeron linifolius, Eupatorium odoratum, Grewia elastica, Mimosa pudica, Mollugo stricta; *C₄ species:* Brachiaria distachya, Digitaria adscendens, Euphorbia hirta, Imperata cylindrica, Manisuris granularis, Panicum khasianum, P. maximum, Paspalidium punctatum, Rottboelia goalparensis, Saccharum arundinaceum, Setaria palmifolia, Thysanolaena maxima. *(From Saxena and Ramakrishnan, 1984c.)*

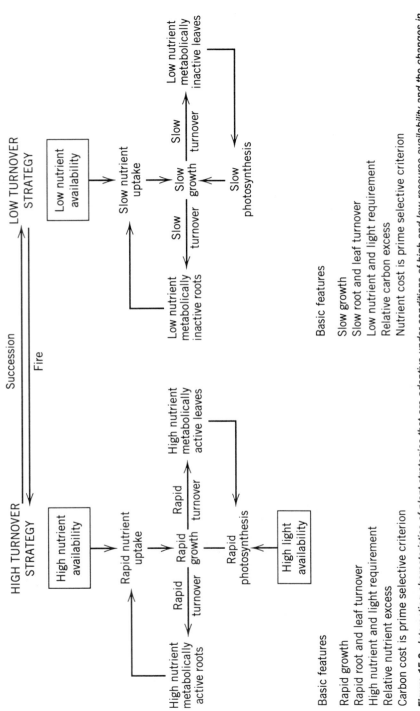

Figure 15-9. *Interacting characteristics of plant strategies that are adaptive under conditions of high and low resource availability and the changes in these strategies that occur as consequences of fire and succession. (From Chapin and Van Cleve, 1981.)*

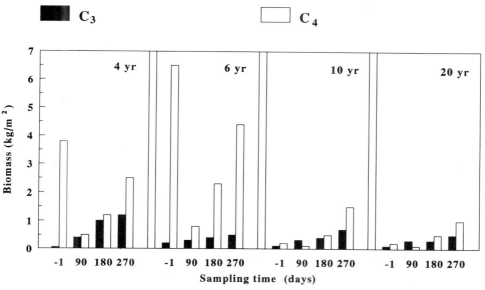

Figure 15-10. *Changes in biomass of C_3 and C_4 species on plots under 4-, 6-, 10-, and 20-year jhum cycles, a day before the burn after slashing (-1) and at different times after the burn. (From Saxena and Ramakrishnan, 1984c.)*

following fire. These ecological conditions would favor the performance of C_4 pathway from the point of physiological adaptations over that likely from C_3 pathway (Ehleringer and Bjorkman, 1977). Irradiance is less intense, and consequently cooler temperatures characterize the later part of the growing season (October–December). There is some improvement in soil nutrient availability also, as a result of recovery in nutrient release processes adversely affected by the fire during March–April (Ramakrishnan and Toky, 1981; Saxena and Ramakrishnan, 1986). These environmental conditions during the later part of the growing season would favor the performance of C_3 species over the C_4 ones. Thus divergence in responses to light and soil fertility regimes associated with C_3 and C_4 photosynthetic pathways, in conjunction with the spatiotemporal variability in these resources created by seasonality of the climate and fire, may help foster the coexistence of C_3 and C_4 species. Such a niche divergence cannot be generalized too far in view of the studies suggesting similarity in growth responses of cooccurring C_3 and C_4 species (Caldwell et al., 1977).

Many exotic C_3 weeds, despite of their high reproductive potential and long-range dispersal capacity, fail to invade the extremely nutrient-poor ecosystems at Cherrapunji, Meghalaya, India, (Ram and Ramakrishnan, 1988, 1992; Ramakrishnan and Ram, 1988; Ramakrishnan and Saxena, 1984). In these ecosystems, soil nutrient pool has been depleted to a much greater extent because of more frequent fire in a very high rainfall regime and consequently high rates of soil erosion and nutrient losses. Species constituting the seral grasslands in these extreme ecosys-

TABLE 15-11   Nutrient Use of Different Species at Cherrapunji, Meghalaya, India

Species	Nutrient Use Efficiency (mg dry matter produced per mg nutrient absorbed)		
	Nitrogen	Phosphorus	Potassium
Arundinella bengalensis	149.9 ± 12.5	1020.4 ± 95.4	136.4 ± 12.5
	(101.7 ± 9.7)	(1010.1 ± 91.7)	(158.7 ± 14.5)
Carex cruciata	108.1 ± 9.1	980.4 ± 75.9	91.7 ± 8.5
	(124.8 ± 11.1)	(740.4 ± 61.4)	(89.2 ± 7.1)
Chrysopogon gryllus	109.1 ± 9.4	595.2 ± 41.5	158.7 ± 13.7
	(77.9 ± 5.4)	(657.9 ± 51.3)	(1220.0 ± 10.5)
Ischaemum goeblii	117.6 ± 9.7	877.2 ± 73.4	169.5 ± 14.3
Eragrostiella leoptera	150.9 ± 12.7	943.4 ± 81.4	188.7 ± 17.1
LSD (p = 0.01)[a]	3.99	31.1	4.86

[a] LSD = least significant difference.

Source: From Ram and Ramakrishnan (1988).

tems are adapted for survival under nutrient stress through their allocation strategy being geared for clonal propagation and through a greater emphasis on C_4 species, which are more nutrient-use efficient (Table 15-11). Significance of nutrient use efficiency in explaining the development and distribution of vegetation has been illustrated in other ecosystem types also (Gray, 1983; Vitousek, 1982).

CONCLUSIONS

The impacts of fire in an ecosystem depend upon the frequency and intensity of burning. The adaptations toward fire in a species could be looked at on a range of levels, namely germination responses, regenerative/reproductive strategy, resource allocation patterns, growth strategies, and nutritional relationships. The response of a species to the perturbations of fire is determined by a number of adaptive features. Adaptive strategy analysis becomes a useful tool for predicting the ecosystem responses to fire and thereby improvement in fire management. The three hierarchial levels of adaptive strategy analysis, namely physiological, population, and community level, need to be looked at in an integrated perspective.

REFERENCES

Abrahamson, W. G. (1979), *Am. J. Bot., 66,* 71–79.

Abrahamson, W. G. and H. Caswell (1982), *Ecology, 63,* 982–991.

Aerts, R., H. De. Caluwe, and H. Konings (1992). *J. Ecol., 80,* 653–664.

Aweto, A. O. (1981). *J. Ecol., 69,* 601–607.

Caldwell, M. M., R. S. White, R. T. Moore, and L. B. Camp (1977), *Oecologia, 29,* 265–300.

Chapin, F. S. (1980), *Annu. Rev. Ecol. Syst., 11,* 233–260.

Chapin, F. S. and K. Van Cleve (1981), in *Fire regimes and ecosystem properties* (H. A. Mooney, T. M. Bonnicksen, N. L. Christensen, J. E. Lotan, and W. A. Reiners, Eds.), USDA For. Serv. Gen. Tech. Rep., WO 26, pp. 301–321.

Christensen, N. L. and P. C. Kimber (1975), *Proc. Ecol. Soc. Austral.*, *9*, 85–106.

Clement, B. and J. Touffet (1990). *J. Veg. Sci.*, *1*, 195–202.

Cody, M. L. (1966), *Evolution, 20*, 174–184.

Eden, M. J., P. A. Furley, W. Milliken, and J. A. Ratter (1991), *For. Ecol. Manage.*, *38*, 283–290.

Egler, F. E. (1954), *Vegetatio, 4*, 412–417.

Ehleringer, J. R. and O. Bjorkman (1977), *Plant Physiol.*, *59*, 86–90.

Fitter, A. H. and N. L. Setters (1988), *J. Ecol.*, *76*, 617–636.

Gill, A. M. (1975), *Austr. For.*, *38*, 1–25.

Gliessman, S. R., E. R. Garcia, and A. M. Amadore (1981), *Agroecosystems, 7*, 173–185.

Gray, J. T. (1983). *J. Ecol.*, *71*, 21–41.

Grime, J. P. (1979), *Plant strategies and vegetation processes*, Wiley, New York.

Harper, J. L. (1977), *Population biology of plants*. Academic Press, London.

Harper, J. L. and J. Ogden (1970), *J. Ecol.*, *58*, 681–698.

Hickman, J. C. (1975), *J. Ecol.*, *63*, 691–698.

Inchausti, P. (1995), *J. Ecol.*, *83*, 231–243.

Keeley, J. E. (1977), *Ecology, 58*, 820–829.

Keeley, J. E. (1981), in *Fire regimes and ecosystem properties* (H. A. Mooney, T. M. Bonnicksen, N. L. Christensen, J. E. Lotan, and W. A. Reiners, Eds.), USDA For. Serv. Gen. Tech. Rep., WO 26, pp. 231–277.

Keeley, J. E. (1991), *Bot. Rev.*, *57*, 81–116.

Keeley, J. E. and S. E. Keeley (1988), in *North American terrestrial vegetation* (M. G. Barbour and W. D. Billings, Eds.), Cambridge University Press, Cambridge, UK, pp. 65–84.

Keeley, J. E. and P. H. Zedler (1978). *Am. Midl. Nat.*, *99*, 142–161.

Komarek, E. V. (1974), in *Fire and ecosystems* (T. T. Kozlowski and C. E. Ahlgren, Eds.), Academic Press, New York, pp. 251–277.

Kozlowski, T. T and C. E. Ahlgren (Eds.) (1974), *Fire and ecosystems*, Academic Press, New York, 542 pp.

Kushwaha, S. P. S., P. S. Ramakrishnan, and R. S. Tripathi (1981), *J. Appl. Ecol.*, *18*, 529–535.

Kushwaha, S. P. S., P. S. Ramakrishnan, and R. S. Tripathi (1983a), *Trop. Plant Sci. Res.*, *1*, 53–57.

Kushwaha, S. P. S., P. S. Ramakrishnan, and R. S. Tripathi (1983b), *Proc. Indian Acad. Sci. (Plant Sci.)*, *92*, 313–321.

Lamb, D. (1980), *Oecologia, 47*, 257–263.

Legg, C. J., E. Maltby, and M. C. F. Proctor (1992), *J. Ecol.*, *80*, 737–752.

MacArthur, R. H. and E. O. Wilson (1967), *The theory of island biogeography*, Princeton University Press, Princeton, NJ.

Mallik, A. U., R. J. Hobbs, and C. J. Legg (1984). *J. Ecol.*, *72*, 855–871.

Miranda, A. C., H. S. Miranda, O. F. Dias, and B.F.D. Dias (1993), *J. Trop. Ecol.*, *9*, 313–320.

Mishra, B. K. and P. S. Ramakrishnan (1984), *Plant Soil*, *81*, 37–46.

Mooney, H. A. and W. D. Billings (1960), *Am. J. Bot.*, *47*, 594–598.

Mooney, H. A., T. M. Bonnicksen, N. L. Christensen, J. E. Lotan, and W. A. Reiners (Eds.) (1981), *Fire regimes and ecosystem properties*, USDA For. Serv. Gen. Tech. Rep., WO-26, Honolulu, HI, 593 pp.

Newell, S. J. and E. J. Tramer (1978), *Ecology*, *59*, 228–234.

Nye, P. H and D. J. Greenland (1960). *The soil under shifting cultivation*, Tech. Comm. No. 51, Commonwealth Bureau of Soil, Harpenden, England, 156 pp.

Odum, E. P. (1969), *Science*, *164*, 262–270.

Ogden, J. (1974), *J. Ecol.*, *62*, 291–324.

Ohlson, M. and N. Malmer (1990), *Oikos*, *58*, 100–108.

Olson, M. S. and M. S. Platt (1995), *Vegetatio*, *119*, 101–118.

Pivello, V. R. and L. M. Coutinoh (1992), *J. Trop. Ecol.*, *47*, 487–497.

Purdie, R. O. and E. O. Slatyer (1976), *Austr. J. Ecol.*, *1*, 223–236.

Pyne, S. J. (1982), *Fire in America*, Princeton University Press, Princeton, NJ, 654 pp.

Raison, R. J. (1979), *Plant Soil*, *51*, 73–108.

Ram, S. C. and P. S. Ramakrishnan (1988), *Environ. Conserv.*, *15*, 29–35.

Ram, S. C. and P. S. Ramakrishnan (1992), *Int. J. Wildland Fire*, *2*, 131–138.

Ramakrishnan, P. S. (1992), *Shifting agriculture and sustainable development*, Parthenon Publishing Group Limited, Carnforth, Lancs, UK, 424 pp.

Ramakrishnan, P. S. and B. K. Mishra (1981), *Weed Res.*, *22*, 77–84.

Ramakrishnan, P. S. and S. C. Ram (1988), *Vegetatio*, *74*, 47–53.

Ramakrishnan, P. S. and K. G. Saxena (1984), *Curr. Sci.*, *53*, 107–109.

Ramakrishnan, P. S. and O. P. Toky (1981). *Plant Soil*, *60*, 41–64.

Rao, K. S. and P. S. Ramakrishnan (1989), *J. Appl. Ecol.*, *26*, 625–633.

Saxena, K. G. and P. S. Ramakrishnan (1982), *Proc. Indian. Natl. Sci. Acad.*, B *48*, 807–818.

Saxena, K. G. and P. S. Ramakrishnan (1983), *Can. J. Bot.*, *61*, 1300–1306.

Saxena, K. G. and P. S. Ramakrishnan (1984a), *Weed Res.*, *24*, 135–142.

Saxena, K. G. and P. S. Ramakrishnan (1984b), *Weed Res.*, *24*, 127–136.

Saxena, K. G. and P. S. Ramakrishnan (1984c), *Acta Oecol./Oecol. Plant.*, *5*, 335–346.

Saxena, K. G. and P. S. Ramakrishnan (1986), *Acta Oecol./Oecol. Plant.*, *7*, 319–331.

Saxena, K. G. and P. S. Ramakrishnan (1988), *Int. J. Ecol. Env. Sci.*, *14*, 1–19.

Simmon-Sylvestre, G. (1976), *Ann. Agron.*, *18*, 253–266.

Smith, W., F. H. Bormann, and G. E. Likens (1968), *Soil Sci.*, *106*, 471–473.

Spencer, J. E. (1966), *Shifting cultivation in southeastern Asia*, Publ. Geogr. No. 199, University of California, Berkeley, 247 pp.

Stebbins, G. L. (1974), *Flowering plants—Evolution above the species level*, Harvard University Press, Cambridge, MA.

Swamy, P. S. and P. S. Ramakrishnan (1987a), *Agric. Ecosyst. Environ.*, *18*, 195–204.

Swamy, P. S. and P. S. Ramakrishnan (1987b), *For. Ecol. Manage.*, *22*, 229–237.

Swamy, P. S. and P. S. Ramakrishnan (1988), *J. Appl. Ecol.*, *25*, 653–658.

Thanos, C. A. and P. W. Rundel (1995), *J. Ecol.*, *83*, 207–216.

Tilman, D. (1985), *Am. Nat.*, *125*, 827–852.

Toky, O. P. and P. S. Ramakrishnan (1981), *Environ. Conserv.*, *8*, 313–321.

Toky, O. P. and P. S. Ramakrishnan (1983), *J. Ecol.*, *71*, 747–757.

Uhl, C. and J. B. Kauffman (1990), *Ecology, 71*, 437–449.

Uhl, C. and P. Murphy (1981), *Agro-Ecosystems, 7*, 63–83.

Valbuena, I. and L. Trabaud (1995), *Vegetatio, 119*, 81–90.

Vitousek, P. M. (1982), *Am. Nat.*, *119*, 553–572.

Waldrop, T. A., D. L. White, and S. M. Jones (1992), *For. Ecol. Manage.*, *47*, 195–210.

Watters, R. F. (1971), *Shifting cultivation in Latin America*, FAO Forestry Dev. Paper No. 17, FAO, Rome, 305 pp.

Wong, R. (1964), *J. Rubber Res. Inst.*, *Malaya, 18*, 231–242.

Woodmansee, R. W. and L. S. Wallach (1981), *in Fire regimes and ecosystem properties* (H. A. Mooney, T. M. Bonnicksen, N. L. Christensen, J. E. Lotan, and W. A. Reiners, Eds.), USDA For. Serv. Gen. Tech. Rep., WO-26, pp. 379–400.

Yadav, A. S. and R. S. Tripathi (1981), *Oikos, 23*, 355–361.

Zammit, C. A. and M. Westoby (1988), *Ecology, 68*, 1984–1992.

Zammit, C. A. and P. H. Zedler (1988), *Vegetatio, 75*, 175–187.

Index of Plant Names*

*Includes algae, fungi, bryophytes, pteridophytes, gymnosperms, and bacteria.

Subject Index